Advanced

ENGINEERING PHYSICS

Advanced
ENGINEERING PHYSICS

[For CS / BTech [EE (New) / EEE (New) / EIE (New) / ICE (New)] / Semester-4 / PH(EE)–401, also useful for BSc (Hons), NET, SET, GATE, JEST, IAS, IFS and WBCS examinees]

Prof. (Dr) S P Kuila MSc, PhD (IIT, Kharagpur)

Professor and Head, Department of Applied Physical Sciences
Bengal Institute of Technology and Management, Santiniketan, WB
Former Professor of Physics, Abhedananda Mahavidyalaya
Sainthia, Birbhum, WB

Author of Perspective of Quantum Mechanics, Vector Analysis and its Applications, Principles of Engineering Physics (Vols. I, II), Fundamentals of Quantum Mechanics, Statistical Mechanics and Solid State Physics, Essentials of Solid State Physics etc.

New Central Book Agency (P) Ltd

LONDON
HYDERABAD ERNAKULAM BHUBANESWAR
NEW DELHI KOLKATA PUNE GUWAHATI

NCBA

REGD OFFICE
8/1 Chintamoni Das Lane, Kolkata 700 009, India
email: ncbakolkata@gmail.com

OVERSEAS
NCBA (UK) Ltd, 149 Park Avenue North, Northampton, NN3 2HY, UK
email: ncbauk@yahoo.com

EXPORT
NCBA Exports Pvt. Ltd, 212 Shahpur Jat, New Delhi 110 049
email: ncbaexports@ncbapvtltd.com

BRANCHES

New Central Book Agency (P) Ltd
212 Shahpur Jat, **New Delhi** 110 049
email: ncbadel@ncbapvtltd.com

Shop Nos. 3 & 4, Vinayak Towers
681/B Budhwar Peth, Appa Balwant Chowk
Pune 411 002
email: ncbapune@ncbapvtltd.com

Shop No. 15, Bharati Towers, Ground Floor
Forest Park
Bhubaneswar 751 001
email: ncbabub@gmail.com

House No. 3-1-315, 1st Floor
Nimboliadda, Kachiguda, **Hyderabad** 500 027
email: ncbahydb@ncbapvtltd.com

GSS Shopping Complex, 1st Floor
Opposite College Post Office, Convent Road
Ernakulam 682 035
0484-4021054 / 53

Radhamayee Çomplex [Second Floor]
Opp. Axis Bank, Panbazar, Jaswanta Road
Guwahati 781 001
email: ncbaguwahati@ncbapvtltd.com

ADVANCED ENGINEERING PHYSICS • Prof (Dr) S P Kuila

First Published: May 2014
Reprinted: March 2016, **September 2023**

PUBLISHER AND TYPESETTER
New Central Book Agency (P) Ltd
8/1 Chintamoni Das Lane, Kolkata 700 009

PRINTER
New Central Book Agency (P) Ltd
Web-Offset Division, Dhulagarh, Sankrail, Howrah

TECHNICAL EDITOR
Dipan Roy

PROJECT TEAM
Abhijit Sen, Pradip Biswas, Prodip Baidya and Prabhat Jas

ISBN: 978 81 7381 898 1

Price ₹ 495.00

Dedicated to
JYOTSNA

Contents

Preface

The book *Advanced Engineering Physics* has been written in accordance with the revised syllabus prescribed by the West Bengal University of Technology to be followed by the students of fourth semester pursuing BTech in EE/EEE/EIE/ICE with effect from 2011. Attempts have been made to present a thorough and comprehensive understanding in depth of the subject matter through large number of examples based on mathematical derivation. Some topics not directly included in the syllabus have also been incorporated in the book for the sake of completeness of the subject.

The content of five modules of the syllabus has been divided into ten chapters each of which has been enriched with 'Answers of Some Important Questions', 'Large Number of Solved Problems', etc. The exercise part of each chapter has been divided into 'Objective Type Questions', 'Short Answer Type Questions', 'Numerical Problems' and 'Broad Answer Type Questions'. The questions are directly linked with physical principles and theories, many of which have been taken from recent examination papers of different universities.

I would like to thank my numerous students and colleagues working in different colleges/institutions and universities for their constructive and valuable suggestions. I would also like to extend my sincere thanks to Mr Amitabha Sen (Director), Mr R M Nath, Mr Soumen Pal, Mr Dipan Roy, Mr Abhijit Sen and Mr Prabhat Jas of New Central Book Agency (P) Ltd, Kolkata for bringing out the book in time with precision.

I shall deem my labour amply rewarded if the book proves useful to those for whom it is intended. Any suggestion for the improvement of the book will be thankfully acknowledged.

14.05.2014 Author

Syllabus for B.Tech (Electrical Engineering) Second Year

Revised Syllabus of B.Tech in EE (To be followed from the Academic Session, July 2011, i.e., for the students who were admitted in Academic Session 2010–2011)

PH (EE)-401: 4 Physics

Contacts : 3L+1T
Credits : 4

Topics	No. of Periods
Module I	
Quantum Mechanics:	
— Generalised coordinates, Lagrange's equation of motion and Lagrangian, generalised force potential, moment and energy. Hamilton's Equation of motion and Hamiltonian. Properties of Hamilton and Hamilton's equation of motion.	6
— Concept of probability and probability density, Operator, Commutator, Formulation of quantum mechanics and Basic postulates, Operator correspondence, Time dependent Schrödinger's equation, formulation of time independent Schrödinger's equation by method of separation of variables, Physical interpretation of wave function Ψ (normalisation and probability interpretation), Expectation values, Application of Schrödinger equation — Particle in an infinite square well potential (1D and 3D potential well), Discussion on degenerate levels.	10
Module II	
Statistical Mechanics:	
— Concept of energy levels and energy states. Micro-states, Macro-states and thermodynamic probability, equilibrium macro-state. MD, FD, statistics (no deduction necessary), fermions, bosons (definitions in terms of spin, examples), physical significance and application, classical limits of quantum statistics, Fermi distribution at zero and non-zero temperature.	4
Module III	
Dielectric Properties:	
— Dielectric Material: Concept of Polarisation, the relation between D, E and P, Polarisability, Electronic, Ionic, Orientation and Space charge polarisation, behaviour of Dielectric under alternating field, Dielectric losses.	3
The Magnetic Properties:	
— Magnetisation M, relation between B, H and M. Bohr megneton, Diamagnetism-Larmour frequency and susceptibility. Curie law, Weiss molecular field theory and Curie-Weiss law, Hysteresis loss, Antiferromagnetism, Ferromagnetism and Ferrites (analitative).	4
Module IV	
Crystal Structure:	
— Crystal structure — Bravais lattice, Miller indices.	1
— Crystal diffraction (qualitative), Bragg's law and reciprocal lattice, Brillouin zones. (Qualitative description)	2
— Free electron theory of metal — Calculation of Fermi energy, density of states.	2
— Band theory of solids — Bloch's theorem, Kronig-Penney model.	3
— Electronic conduction in solids — Drude's theory, Boltzmann equation, Wiedemann Frantz law.	3
— Semiconductor — Band structure, concept of electron and holes, Fermi level, density of states.	3

MODULE 1

Chapter 1 : Classical Mechanics

Chapter 2 : Quantum Mechanics

Chapter 1

Classical Mechanics [Lagrangian Formulation]

1.1 Constraints

The limitations or geometrical restrictions on the motion of a particle or system of particles are generally known as the constraints. The forces responsible for the constraints are called the constraint forces.

Some of the examples of constraint motion are as follows:

1. The motion of a rigid body such that the distance between any two particles remains unchanged.
2. The motion of a train confined along the rails.
3. The motion of the bob of a simple pendulum restricted to oscillate under Gravity in the vertical plane passing through the string.
4. The motion of the gas molecules within a container restricted by the walls of the same.
5. The beads of an abacus constrained to move along the supporting wires.
6. A particle placed on the surface of a solid sphere restricted, so that it can move only either on the surface or outside the sphere.

1.2 Role of Constraints in Dynamics

The constraints play two important roles in dynamics as follows:

1. Every constraint limits the number of degrees of freedom of the system and
2. The constraint forces lead to certain relations which may be either in the form of equality or inequality.

As the constraints limit the degrees of freedom which in turn determine the number of independent coordinates which are to be known to determine the dynamical system.

1.2.1 Classification of constraints

Following is the different types of constraints:

1. *Holonomic and non-holonomic constraints.* If the condition of constraints be expressed in terms of equality, then it is *holonomic*, otherwise *non-holonomic*.
   Suppose, $\vec{r}_1, \vec{r}_2, \ldots, \vec{r}_n$ be the position coordinates of particles and t denotes the time. If the condition of all the constraints be expressed in terms of equations like,
   $$f(\vec{r}_1, \vec{r}_2, \ldots, \vec{r}_n, t) = 0, \tag{1.1}$$
   then the constraints are *holonomic.* If the constraints are not expressible in the above form, then the constraints are called *non-holonomic.*

(a) *Examples of holonomic constraints:*

(i) The constraints involved in the motion of rigid bodies are *holonomic* because the equation of constraints are $(\vec{r}_i - \vec{r}_j)^2 - c_{ij}^2 = 0$, where $\vec{r}_i$ and $\vec{r}_j$ are respectively the position vectors of the ith and jth particles and c_{ij} is the distance between them.

(ii) The point mass of a simple pendulum moves such that, $(\vec{r} - \vec{a})^2 = l^2$, where $\vec{r}$ is the position vector of the point mass, $\vec{a}$ that of the point of suspension and l is the effective length. Hence the constraint involved in the present case is *holonomic.*

(b) *Examples of non-holonomic constraints:*

(i) The motion of a particle on the surface of a sphere of radius a is restricted to the condition, $r^2 - a^2 \geq 0$, where r is the distance of the particle from the centre of the sphere. The condition $r^2 - a^2 = 0$ holds so long as the particle moves on the surface of the sphere and the condition $r^2 - a^2 > 0$ is true when it leaves the surface of the sphere. Hence it is clearly a case of *non-holonomic constraint.*

(ii) If a gas container is spherical in shape of radius a and $\vec{r}$ is the position vector of a molecule, then the condition of constraint for the motion of the molecules is $|r| \leq a$ or, $r - a \leq 0$.

It is thus another example of *non-holonomic* constraint. Rolling of an object on a rough surface without slipping involves *non-holonomic constraints.*

Constraints are also restricted on the velocities rather than on the coordinates. For example, let a cylinder of radius a be rolling and sliding down an inclined plane with its axis always horizontal. If s and θ be respectively the distance of the axis of the cylinder from the edge A of the inclined plane and the angle that a fixed radius has rotated from the radius to the point of contact with the inclined plane, then for rolling without slipping [Fig. 1.1],

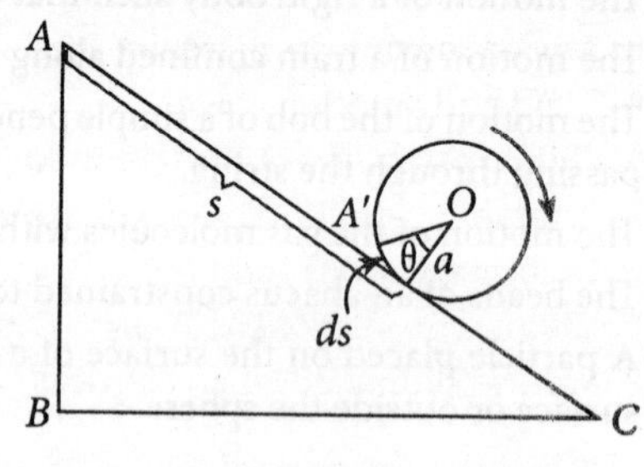

Fig. 1.1: A cylinder rolling on an inclined plane AC.

$$\dot{s} = a\dot{\theta} \text{ or } \frac{ds}{dt} = a\frac{d\theta}{dt}.$$

Integrating w.r.t. time,

$$(s - a\theta) = \text{a constant.} \tag{1.2}$$

Thus the constraint involved in the description of the rolling without slipping is *holonomic.*

There are, however, cases in which equations involving velocities and are non-integrable. Such systems involve *non-holonomic* constraints.

Note. The term *holonomic* is derived from the word *holos* (= integer) in Greek and *whole* or *integrable* in Latin.

2. *Scleronomic and Rheonomic constraints.* If the equations of constraint do not contain time explicitly, then they are called *scleronomic.* On the other hand, if the equations of constraint contain time explicitly, then they are called *rheonomic.*

(a) *Examples of scleronomic constraints:*

(i) The motion of a point mass in a simple pendulum with rigid support is an example of *scleronomic* constraint because the equation of constraint, viz., $(\vec{r} - \vec{a})^2 = l^2$ does not contain time.

(ii) A bead moving in a circular wire is expressed by the equation, $r^2 = a^2$, which does not contain time explicitly. Hence, the constraint involved is *scleronomic*.

(b) *Examples of rheonomic constraints*:

(i) A bead sliding on a moving wire is an example of *rheonomic* constraint.

(ii) If the effective length l of a simple pendulum changes with temperature at different times t of a day, then

$$l = f(t) \neq \text{constant}.$$

It is thus also an example of *rheonomic* constraint.

3. *Bilateral and unilateral constraints*. If the constraint relations are expressed in the form of *equations*, then they are *bilateral constraints*. On the other hand, if the constraint relations are in the form of *inequalities*, then the constraints are *unilateral*.
4. *Conservative and dissipative constraints*. During constraint motion if the constraint forces do not do any work so that the total mechanical energy remains constant, then the constraint is *conservative*. If, on the other hand, the constraint forces do work and the total mechanical energy is not conserved, then the constraint is said to be *dissipative*.

Table 1.1: Some systems associated with the type of constraint

Systems	*Types of constraints*
1. A particle sliding down a sphere under gravity	non-holonomic
2. A bead moving on a circular wire	holonomic, scleronomic, bilateral
3. Rigid body motion	holonomic, scleronomic, bilateral, conservative.
4. Simple pendulum with rigid support	holonomic, scleronomic, bilateral, conservative
5. Pendulum with time-dependent length	holonomic, rheonomic, bilateral, dissipative.
6. Deformable body motion	holonomic, rheonomic, bilateral, dissipative
7. Hollow sphere filled with gas	holonomic, scleronomic, unilateral, conservative
8. Particle sliding down an ellipsoid	non-holonomic, rheonomic

1.2.2 Difficulties introduced by constraints

Constraints introduce *two* types of difficulties in solving problems in dynamics. These are as follows:

1. We consider a system of interacting particles in which the force on the ith particle is given by

$$\vec{F}_i = \vec{F}_i^e + \sum_{j=1}^{n} \vec{F}_{ij},$$

where $\vec{F}_i^e$ is the *external force* and $\vec{F}_{ij}$ is the *internal force* on the ith particle due to the jth particle. According to *Newton's second law of motion*, the equation of motion for the ith particle is

$$m_i \frac{d^2\vec{r}_i}{dt^2} = \vec{F}_i^e + \sum_{i=1}^{N} \vec{F}_{ij}, \tag{1.3}$$

where $i = 1, 2, \ldots, N$. (1.3) thus represents N equations. There are N values of the position vector $\vec{r}_i$, which are connected by the equation of constraint as $f(\vec{r}_1, \vec{r}_2, \ldots, \vec{r}_N, t) = 0$.

Thus the N number of coordinates $\vec{r}_1, \vec{r}_2, \ldots, \vec{r}_N$ are not independent but are connected by the above equation. Hence the N-equations represented by (1.3) are not all independent. So to have a solution of the problem these equations are to be rewritten by considering the effect of the constraints. Herein lies the first difficulty.

2. In many situations initially the nature of the constraint forces remained unknown and they belong to the unknowns of the problem. These unknown constraint forces, however, may be obtained from the solution of the problem which are to be found out. Thus these unknown constraint forces introduce complications in deriving the solution of the problem. As for example, the force of constraint which the wire exerts on a moving bead not known initially and thus is a difficulty towards the solution of the problem.

   The above mentioned difficulties may, however, be solved in the following way:

   The first difficulty in case of holonomic constraint can be solved by introducing the concept of *generalised coordinates*. If K be the number of equation of constraints in an N-particle system, then the number of generalised coordinates necessary to describe the motion of the system is $f = 3N - K$, where f is the *degree of freedom* of the system. Thus the interconnected N-equations of motion may be represented by f-equations which are independent.

   For the removal of the second difficulty, i.e., the difficulty arising due to the initially unknown forces of constraint, we need to recast the mechanics in such a way that the forces of constraint disappear. *d'Alembert* provided such an approach by introducing the concept of *virtual displacement* and *virtual work*, which shall be discussed later in the present chapter of the text.

1.3 Degrees of Freedom

A mechanical system can execute motion in a number of independent ways without violating any constraint which are supposed to be imposed. *These number of independent ways of executing motion without violating the effect of the constraints is called the degrees of freedom of the dynamical system.*

If a particle moves in a straight line along X-direction, the number of coordinates necessary to specify its position is one. The system is said to have *one* degree of freedom. Two coordinates are necessary if the particle moves in the XY-plane and the system is said to have *two* degrees of freedom. Similarly, for a particle moving in X-Y-Z space the degrees of freedom is *three* as three coordinates (x, y, z) are necessary to specify the position of the particle.

Hence, *degree of freedom of a dynamical system is defined as the number of coordinates necessary to specify the position (or, configuration) of a dynamical system.*

Again, the kinetic energies (KE) of the particle when it moves along X-axis, in XY-plane and X-Y-Z space are given by $\frac{1}{2}m\dot{x}^2$, $\frac{1}{2}m(\dot{x}^2+\dot{y}^2)$ and $\frac{1}{2}m(\dot{x}^2+\dot{y}^2+\dot{z}^2)$ respectively. So, *the degrees of freedom may also be defined as the number of squared terms occurring in the expression for the KE of the system.*

Generalised coordinates. A set of independent coordinates sufficient in number to specify unambiguously the system configuration is called *generalised coordinates*. The generalised coordinates are denoted by $q_1, q_2, \ldots, q_k, \ldots, q_f$, where f is the degrees of freedom of the system. These f-number of coordinates is the minimum number of coordinates necessary to describe the motion of the system.

Example of generalised coordinates. The number of generalised coordinates necessary for the motion of a particle along the circumference of a circle of radius r is only one, i.e., θ.

For, if a particle P moves along the circumference of a circle of radius r in the XY-plane [Fig. 1.2], then at any instant t the position of the particle is denoted by $x = r\cos\theta$ and $y = r\sin\theta$.

Since r is constant, the generalised coordinates in this case is only one, viz., $q_1 = \theta$.

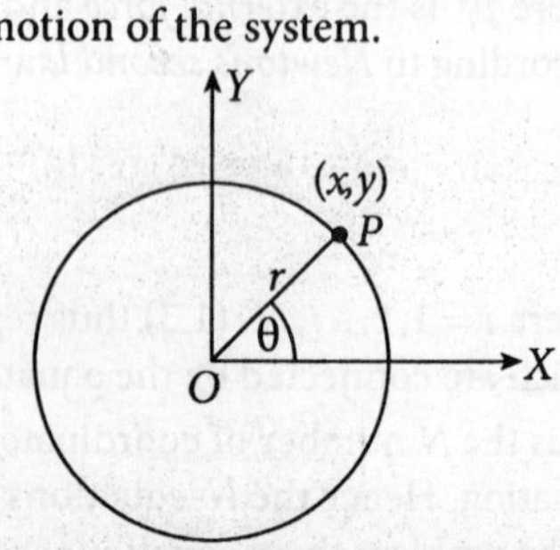

Fig. 1.2: Particle P moves in a circular path of radius r.

If a particle moves on the surface of a sphere, then the number generalised coordinates will be two, viz., $q_1 = \theta$ and $q_2 = \phi$.

Similarly, if the system of N-particles moves under K-number of constraints, then the number of generalised coordinates necessary to describe the motion is

$$f = 3N - K. \tag{1.4}$$

1.3.1 Characteristics of generalised coordinates

Following are the features of the generalised coordinates:

1. The generalised coordinates may not essentially be rectangular, spherical or cylindrical.
2. The quantities like length, (length)^2, angle, energy or any dimensionless quantity may be used as generalised coordinates provided these completely describe the state of the system.
3. The constraints have no effect on the f generalised coordinates.

1.3.2 Transformation equations

If x_i, y_i, z_i be the coordinates of the ith particle of the system w.r.t. an arbitrary origin, then these Cartesian coordinates can be expressed as functions of generalised coordinates $q_1, q_2, \ldots, q_f$ as follows:

$$\left.\begin{aligned} x_i &= x_i\left(q_1, q_2, \ldots, q_f; t\right) \\ y_i &= y_i\left(q_1, q_2, \ldots, q_f; t\right) \\ z_i &= z_i\left(q_1, q_2, \ldots, q_f; t\right) \end{aligned}\right\}, \tag{1.5}$$

where t denotes time.

If $\vec{r}_i = \hat{\imath}x + \hat{\jmath}y + \hat{k}z$, where $\hat{\imath}$, $\hat{\jmath}$ and $\hat{k}$ are the unit vectors along X-, Y- and Z-directions respectively, then

$$\vec{r}_i = \vec{r}_i\left(q_1, q_2, \ldots, q_f; t\right). \tag{1.6}$$

Equations (1.5) and (1.6) are called *transformation equations* which contain the constraints explicitly.

(1.5) can also be expressed as below:

$$\left.\begin{aligned} q_1 &= q_1\left(x_1, y_1, z_1; x_2, y_2, z_2; \ldots; x_f, y_f, z_f; t\right) \\ q_2 &= q_2\left(x_1, y_1, z_1; x_2, y_2, z_2; \ldots; x_f, y_f, z_f; t\right) \\ &\cdots\cdots\cdots\cdots\cdots\cdots\cdots\cdots\cdots\cdots\cdots \\ q_f &= q_f\left(x_1, y_1, z_1; x_2, y_2, z_2; \ldots; x_f, y_f, z_f; t\right) \end{aligned}\right\} \tag{1.7}$$

The necessary and sufficient condition that (1.6) holds good is

$$J\left(\frac{q_1, q_2, \ldots, q_f}{x_1, y_1, \ldots, z_f}\right) = \frac{\partial\left(q_1, q_2, \ldots, q_f\right)}{\partial\left(x_1, y_1, \ldots, z_f\right)} = \begin{vmatrix} \dfrac{\partial q_1}{\partial x_1} & \dfrac{\partial q_2}{\partial x_1} & \cdots & \dfrac{\partial q_f}{\partial x_1} \\ \dfrac{\partial q_1}{\partial x_2} & \dfrac{\partial q_2}{\partial x_2} & \cdots & \dfrac{\partial q_f}{\partial x_2} \\ \cdots & \cdots & \cdots & \cdots \\ \dfrac{\partial q_1}{\partial z_f} & \dfrac{\partial q_2}{\partial z_f} & \cdots & \dfrac{\partial q_f}{\partial z_f} \end{vmatrix} \neq 0.$$

1.3.3 Configuration space

If a single particle moves in space, then a set of three coordinates is required to describe its motion, i.e., the space for the description of motion for a single particle must be *three-dimensional.* For the motion of two particles a set of $2 \times 3 = 6$ coordinates is needed. So the problem of two particles can be reduced to that of a single particle in *six-dimensional space.* In this way the motion of a system of N-particles requires $3N$ independent coordinates for its description. In other words, the motion of $3N$ particles in three-dimensional space may be considered as the motion of a single particle in $3N$-*dimensional space.*

If the condition of constraints of a system of N-particles be represented by K equations, then the degrees of freedom $f = 3N - K$, i.e., f-coordinates will be required to describe the motion of the system. Hence the problem of N-particles can be considered as that of a single particle in f-*dimensional space.*

In case of a system of N-particles, the $3N$-dimensional space, in absence of constraints, and the f-dimensional space, taking into account the presence of constraints, is called the *configuration space.*

1.3.4 Generalised notations

1. *Generalised displacement.* Let $\vec{r}_i$ $(i = 1,2,\ldots,N)$ be the position vector of the ith particle of a system of N-particles w.r.t. an arbitrary origin. If $\delta\vec{r}_i$ be the small displacement of the ith particle with time t fixed, then since $\vec{r}_i = \vec{r}_i\left(q_1,q_2,\ldots,q_f,t\right)$ $[\because f = 3N - K]$

$$\delta\vec{r}_i = \sum_{j=1}^{f} \frac{\partial \vec{r}_i}{\partial q_j}\delta q_j \; [\because \delta t = 0], \tag{1.8}$$

where $\vec{r}_i$ represents $3N$ coordinates $\left(x_1,y_1,z_1;\ldots,x_N,y_N,z_N\right)$, δq_j in (1.8) are the *generalised displacement.*

2. *Generalised velocity.* The position vector of the ith particle w.r.t. an arbitrary origin for a system of N-particles is $\vec{r}_i = \vec{r}_i\left(q_1,q_2,\ldots,q_f;t\right)$ (in absence of constraints).

$$\therefore \dot{\vec{r}}_i = \sum_{j=1}^{f} \frac{\partial \vec{r}_i}{\partial q_j}\dot{q}_j + \frac{\partial \vec{r}_i}{\partial t}. \tag{1.9}$$

$\dot{q}_j$ is the *generalised velocity.* (1.9) shows that *the Cartesian components of velocity $\dot{\vec{r}}_i$ are linear functions of the generalised velocity $\left(\dot{q}_j\right)$ components*, or, $\dot{\vec{r}}_i = \sum_{j=1}^{f} a_{ij}\dot{q}_j + \vec{b}_i$, where $\vec{a}_{ij} = \frac{\partial \vec{r}_i}{\partial q_l}$ and $\vec{b}_i = \frac{\partial \vec{r}_i}{\partial t}$.

If $\vec{r}_i$ does not depend explicitly on time, then $\vec{b}_i = \frac{\partial \vec{r}_i}{\partial t} = 0$. Hence,

$$\dot{\vec{r}}_i = \sum_{j=1}^{f} a_{ij}\dot{q}_j, \tag{1.10}$$

$\dot{\vec{r}}_i$ represents the velocity of ith particle in terms of generalised velocity $\dot{q}_j$ $(j = 1,2,3,\ldots,f)$.

If generalised coordinates are expressed in Cartesian coordinates and time, then

$$q_j = q_j(\vec{r}_1,\vec{r}_2,\ldots,\vec{r}_f,t) \tag{1.11}$$

$$\text{and } dq_j = \sum_{i=1}^{f} \frac{\partial q_j}{\partial \vec{r}_i} d\vec{r}_i + \frac{\partial q_j}{\partial t} dt. \tag{1.12}$$

Then generalised velocity is given by

$$\dot{q}_j = \frac{dq_j}{dt} = \sum_{i=1}^{f} c_{ij}\dot{\vec{r}}_i + d_j, \tag{1.13}$$

where $c_{ij} = \dfrac{\partial q_j}{\partial \vec{r}_i}$ and $d_j = \dfrac{\partial q_j}{\partial t}$. If q_j does not depend explicitly on time, then $d_j = \dfrac{\partial q_j}{\partial t} = 0$.

$$\therefore \dot{q}_j = \sum_{i=1}^{f} c_{ij}\dot{\vec{r}}_i. \tag{1.14}$$

As q_j *need not have the dimension of length,* $\dot{q}_j$ *need not have the dimension of velocity.*

3. *Generalised acceleration.* We have $\dot{\vec{r}}_i = \sum_{j=1}^{f} \dfrac{\partial \vec{r}_i}{\partial q_j}\dot{q}_j + \dfrac{\partial \vec{r}_i}{\partial t}$.

Differentiating again w.r.t. time,

$$\ddot{\vec{r}}_i = \frac{d}{dt}\left[\sum_{j=1}^{f}\frac{\partial \vec{r}_i}{\partial q_j}\dot{q}_j + \frac{\partial \vec{r}_i}{\partial t}\right] = \sum_{j=1}^{f}\frac{d}{dt}\left(\frac{\partial \vec{r}_i}{\partial q_j}\right)\dot{q}_j + \sum_{j=1}^{f}\frac{\partial \vec{r}_i}{\partial q_j}\ddot{q}_j + \frac{d}{dt}\left(\frac{\partial \vec{r}_i}{\partial t}\right)$$

$$= \sum_{j=1}^{f}\frac{\partial \dot{\vec{r}}_i}{\partial q_j}\dot{q}_j + \sum_{j=1}^{f}\frac{\partial \vec{r}_i}{\partial q_j}\ddot{q}_j + \frac{\partial \dot{\vec{r}}_i}{\partial t}$$

$$= \sum_{j=1}^{f}\frac{\partial}{\partial q_j}\left(\sum_{k=1}^{f}\frac{\partial \vec{r}_i}{\partial q_k}\dot{q}_k + \frac{\partial \vec{r}_i}{\partial t}\right)\dot{q}_j + \sum_{j=1}^{f}\frac{\partial \vec{r}_i}{\partial q_j}\ddot{q}_j + \frac{\partial}{\partial t}\left(\sum_{k=1}^{f}\frac{\partial \vec{r}_i}{\partial q_k}\dot{q}_k + \frac{\partial \vec{r}_i}{\partial t}\right)$$

$$\left[\because \dot{\vec{r}}_i = \sum_{j=1}^{f}\frac{\partial \vec{r}_i}{\partial q_j}\dot{q}_j + \frac{\partial \vec{r}_i}{\partial t}\right]$$

$$= \sum_{j=1}^{f}\sum_{k=1}^{f}\frac{\partial^2 \vec{r}_i}{\partial q_j \partial q_k}\dot{q}_k\dot{q}_j + \sum_{j=1}^{f}\frac{\partial^2 \vec{r}_i}{\partial q_j \partial t}\dot{q}_j + \sum_{j=1}^{f}\frac{\partial \vec{r}_i}{\partial q_j}\ddot{q}_j + \sum_{k=1}^{f}\frac{\partial^2 \vec{r}_i}{\partial q_k \partial t}\dot{q}_k + \frac{\partial^2 \vec{r}_i}{\partial t^2}.$$

$$\therefore \ddot{\vec{r}}_i = \sum_{j=1}^{f}\frac{\partial \vec{r}_i}{\partial q_j}\ddot{q}_j + \sum_{j=1}^{f}\sum_{k=1}^{f}\frac{\partial^2 \vec{r}_i}{\partial q_j \partial q_k}\dot{q}_j\dot{q}_k + 2\sum_{j=1}^{f}\frac{\partial^2 \vec{r}_i}{\partial q_j \partial t}\dot{q}_j + \frac{\partial^2 \vec{r}_i}{\partial t^2}. \tag{1.15}$$

(1.15) shows that *the Cartesian components of acceleration* $\ddot{\vec{r}}_i$ *are not linear functions of the components of generalised acceleration* $\ddot{q}_j$ *alone.* $\ddot{\vec{r}}_i$ *depends quadratically and linearly on* $\dot{q}_j$ *components.*

4. *Generalised KE and momentum.* The KE of the system of N-particles is

$$T=\sum_{i=1}^{N}\frac{1}{2}m_i\dot{r}_i^2=\sum_{i=1}^{N}\frac{1}{2}m_i\left(\dot{\vec{r}}_i\cdot\dot{\vec{r}}_i\right)=\sum_{i=1}^{N}\frac{1}{2}m_i\left[\sum_{j=1}^{f}\frac{\partial\vec{r}_i}{\partial q_j}\dot{q}_j+\frac{\partial\vec{r}_i}{\partial t}\right]\cdot\left[\sum_{k=1}^{f}\frac{\partial\vec{r}_i}{\partial q_k}\dot{q}_k+\frac{\partial\vec{r}_i}{\partial t}\right]$$

$$\left[\because\ \dot{\vec{r}}_i=\sum_{j=1}^{f}\frac{\partial\vec{r}_i}{\partial q_j}\dot{q}_j+\frac{\partial\vec{r}_i}{\partial t}\right]$$

$$=\frac{1}{2}\sum_{i=1}^{N}\sum_{j=1}^{f}\sum_{k=1}^{f}m_i\frac{\partial\vec{r}_i}{\partial q_j}\cdot\frac{\partial\vec{r}_i}{\partial q_k}\dot{q}_j\cdot\dot{q}_k+\frac{1}{2}\sum_{i=1}^{N}m_i\left[\sum_{j=1}^{f}\frac{\partial\vec{r}_i}{\partial q_j}\dot{q}_j+\sum_{k=1}^{f}\frac{\partial\vec{r}_i}{\partial q_k}\dot{q}_k\right]\cdot\frac{\partial\vec{r}_i}{\partial t}+\frac{1}{2}\sum_{i=1}^{N}m_i\left(\frac{\partial\vec{r}_i}{\partial t}\right)$$

$$\text{or } T=\frac{1}{2}\sum_{i=1}^{N}\sum_{j=1}^{f}\sum_{k=1}^{f}m_i\frac{\partial\vec{r}_i}{\partial q_j}\cdot\frac{\partial\vec{r}_i}{\partial q_k}\dot{q}_j\dot{q}_k+\sum_{i=1}^{N}\sum_{k=1}^{N}m_i\frac{\partial\vec{r}_i}{\partial q_k}\dot{q}_k\cdot\frac{\partial\vec{r}_i}{\partial t}+\frac{1}{2}\sum_{i=1}^{N}m_i\left(\frac{\partial\vec{r}_i}{\partial t}\right)^2$$

$$\left[\because\ \sum_{j=1}^{f}\frac{\partial\vec{r}_i}{\partial q_j}\dot{q}_j=\sum_{k=1}^{f}\frac{\partial\vec{r}_i}{\partial q_k}\dot{q}_k\right] \tag{1.16}$$

(1.16) gives the *generalised KE of the system.* Now the momentum p_k associated with q_k is

$$p_k=\frac{\partial T}{\partial\dot{q}_k}=\sum_{i=1}^{N}\sum_{j=1}^{f}\frac{1}{2}m_i\frac{\partial\vec{r}_i}{\partial q_j}\cdot\frac{\partial\vec{r}_i}{\partial q_k}\dot{q}_j+\sum_{i=1}^{N}m_i\frac{\partial\vec{r}_i}{\partial q_k}\cdot\frac{\partial\vec{r}_i}{\partial t}. \tag{1.17}$$

Thus *the generalised momentum p_k is a linear function of $\dot{q}_j$.*

5. *Generalised force.* If $\sum_i\delta\vec{r}_i$ be the displacement of a system of N-particles due to the application of total force $\sum_i\vec{F}_i$, then the amount of works done,

$$\delta W=\sum_{i=1}^{N}\vec{F}_i\cdot\delta\vec{r}_i=\sum_{i=1}^{N}\vec{F}_i\cdot\sum_{j=1}^{f}\frac{\partial\vec{r}_i}{\partial q_j}\delta q_j=\sum_{j=1}^{f}\left(\sum_{i=1}^{N}\vec{F}_i\cdot\frac{\partial\vec{r}_i}{\partial q_j}\right)\delta q_j \text{ or } \delta W=\sum_{j=1}^{f}Q_j\delta q_j, \tag{1.18}$$

where

$$Q_j=\sum_{i=1}^{N}\vec{F}_i\cdot\frac{\partial\vec{r}_i}{\partial q_j}. \tag{1.19}$$

Q_j is the *generalised force* associated with generalised coordinate q_j. It is clear from (1.19) that Q_j *depends on q_j and possibly on time t*. It is to be noted from (1.18) that the product $Q_j\delta q_j$ must have the dimensions of work, whatever be the dimension of Q_j. Thus *the generalised force Q_j need not always have the dimension of force.*

6. *Generalised potential (velocity dependent potential).* If the force $\vec{F}$ is derivable from scalar potential V, where V is the PE of the system, then the work done by the force $\vec{F}$ for an infinitesimal displacement $\delta\vec{r}_i$ is

$$\delta W=-\delta T \text{ [where } T \text{ is the KE of the system]}$$

$$=-\sum_{i=1}^{N}\left[\frac{\partial V}{\partial x_i}\delta x_i+\frac{\partial V}{\partial y_i}\delta y_i+\frac{\partial V}{\partial z_i}\delta z_i\right].$$

Also,

$$\delta W = -\delta V = -\sum_{k=1}^{f} \frac{\partial V}{\partial q_k} \delta q_k = \sum_{k=1}^{f} Q_k \delta q_k,$$

where

$$Q_k = -\frac{\partial V}{\partial q_k}. \tag{1.20}$$

Q_k is the generalised force component.

If, however, the system is not *conservative and depends on generalised velocity $\dot{q}$*, then the *generalised force* is given by

$$Q_k = -\frac{\partial V}{\partial q_k} + \frac{d}{dt}\left(\frac{\partial U}{\partial \dot{q}_k}\right), \tag{1.21}$$

where U is usually called the *generalised potential* or *velocity dependent potential.*

1.4 Limitations of Newton's Laws of Motion

Newton's laws, in solving mechanical problems, suffer from the following limitations:

1. Newton's laws are valid only in inertial frame. In non-inertial frame (e.g., in the rotating frame of reference of the earth) *pseudo-forces like Coriolis force, centrifugal force* arise and as such the form of the Newton's second law of motion undergoes a change.
2. A complete knowledge of the forces acting on a system is needed to solve the mechanical problems using Newton's laws. But in many situations the nature of the hidden forces (constraint forces) are not known.
3. Newton's laws of motions are valid in Cartesian frame but it becomes more inconvenient to use in spherical or other coordinate systems.
4. The equation of motion involves vector quantities like force, momentum, etc., which introduce complexity in solving the problems.
5. The greatest limitation is faced that the mechanical problems are always tried to resolve geometrically rather than analytically. In case of constraint motion, the determination of all the reaction forces is difficult in Newtonian method.

1.5 Virtual Displacement

A system of particles may be imagined to have two possible configurations at any instant of time which are *consistent with the forces and constraints.* The system can transfer itself from one configuration to the other if each particle of the system is imagined to be displaced by an infinitesimal amount $\delta \vec{r}_i$ from previous to new position. *The displacement $\delta \vec{r}_i$, which does not involve time interval dt, is called the virtual displacement.*

The virtual displacement $\delta \vec{r}_i$ differs from the real displacement $d\vec{r}_i$ in the sense that real displacement involves time interval dt. During this time interval the forces and constraints may change.

1.6 Principle of Virtual Work

Statement. A system of particles will be in equilibrium if and only if the total virtual work done by the actual or applied forces in any arbitrary infinitesimal virtual displacement is zero.

If a system of particles is in equilibrium, then the resultant force acting on each particle of the system must be zero, i.e., $\vec{F}_i = 0$, where $\vec{F}_i$ represents the force on the ith particle of the system. Then the virtual work, $\delta W = \vec{F}_i \cdot \delta \vec{r}_i = 0$.

For the whole system of N-particles,

$$W = \sum_{i=1}^{N} \vec{F}_i \cdot \delta \vec{r}_i = 0. \tag{1.22}$$

If $\vec{f}_i$ and $\vec{F}_i^a$ be respectively the *force of constraint* and the *actual force* on the ith particle, then

$$\vec{F}_i = \vec{F}_i^a + \vec{f}_i. \tag{1.23}$$

Using (1.23) in (1.22), we get

$$\sum_{i=1}^{N} \left(\vec{F}_i^a + \vec{f}_i\right) \cdot \delta \vec{r}_i = 0 \text{ or } \sum_{i=1}^{N} \vec{F}_i^a \cdot \delta \vec{r}_i + \sum_{i=1}^{N} \vec{f}_i \cdot \delta \vec{r}_i = 0. \tag{1.24}$$

Now, if a body moves such that the displacement $\delta \vec{r}_i$ is *perpendicular to the force of constraint* $\vec{f}_i$, then

$$\sum_{i=1}^{N} \vec{f}_i \cdot \delta \vec{r}_i = 0.$$

This happens if a body moves over a spherical surface, where the constraint force is perpendicular to the surface, i.e., $\vec{f}_i$ is perpendicular to $\delta \vec{r}_i$. This does not hold for the system consisting of *frictional forces* and as such we should not consider the system consisting of frictional forces. (1.24) is applicable to systems which are in *statical equilibrium*.

So (1.24) reduces to

$$\sum_{i}^{N} \vec{F}_i^a \cdot \delta \vec{r}_i = 0. \tag{1.25}$$

(1.25) is the mathematical form of the *principle of virtual work* expressing condition of equilibrium. The advantage of applying this principle while considering the equilibrium of a system is two-fold:

(i) *the nature of the constraint force need not be known.*

(ii) *to determine the condition of equilibrium, a single equation is sufficient which can be solved easily.*

1.7 D'Alembert's Principle

Statement. A system of particles moves in such a way that the total virtual work done by the effective forces is zero.

If $\vec{p}_i$ be the momentum of the ith particle of the system, then according to *Newton's second law of motion* the force $\vec{F}_i$ acting on it is

$$\vec{F}_i = \frac{d\vec{p}_i}{dt} = \dot{\vec{p}}_i \text{ or } \vec{F}_i - \dot{\vec{p}}_i = 0. \tag{1.26}$$

(1.26) represents that a moving system of particles can be considered to be in *equilibrium* under the force $\left(\vec{F}_i - \dot{\vec{p}}_i\right) = \vec{F}_i + \left(-\dot{\vec{p}}_i\right)$, i.e., the actual force $\vec{F}_i$ plus an additional force $-\dot{\vec{p}}_i$. Here $-\dot{\vec{p}}_i$ is called the *reversed effective force* on the ith particle.

Now from the principle of virtual work,

$$\sum_{i} \left(\vec{F}_i - \dot{\vec{p}}_i\right) \cdot \delta \vec{r}_i = 0. \tag{1.27}$$

If the forces of constraint are present, then

$$\vec{F}_i = \vec{F}_i^a + \vec{f}_i. \tag{1.28}$$

Using (1.26) in (1.27), we get

$$\sum_{i=1}^{N} \left(\vec{F}_i^a - \dot{\vec{p}}_i\right) \cdot \delta \vec{r}_i + \sum_{i=1}^{N} \vec{f}_i \cdot \delta \vec{r}_i. \tag{1.29}$$

If the virtual work done by the constraint forces is zero, i.e.,

$$\sum_{i=1}^{N} \vec{f}_i \cdot \delta \vec{r}_i = 0,$$

then

$$\sum_{i=1}^{N} \left(\vec{F}_i^a - \dot{\vec{p}}_i\right) \cdot \delta \vec{r}_i = 0. \tag{1.30}$$

This is *D'Alembert's principle.*

Dropping the superscript a, we have

$$\sum_{i=1}^{N} \left(\vec{F}_i - \dot{\vec{p}}_i\right) \cdot \delta \vec{r}_i = 0. \tag{1.31}$$

This principle is valid for all *rheonomic* and *scleronomic systems* that are either *holonomic* or *non-holonomic.*

LAGRANGIAN FORMULATION OF CLASSICAL MECHANICS

1.8 Derivation of Lagrange's Equations of Motion

Lagrange's equation of motion may be derived in various methods. Here we propose to derive it using *D'Alembert's principle.*

According to the coordinate transformation equation, the position vectors of the particles $\vec{r}_1, \vec{r}_2, \ldots, \vec{r}_i, \ldots$ are expressed as the functions of generalised coordinates $q_1, q_2, \ldots, q_j, \ldots, q_f$ and time t in the form:

$$\vec{r}_i = \vec{r}_i \left(q_1, q_2, \ldots, q_j, \ldots, q_f; t\right). \tag{1.32}$$

If $\vec{F}_i$ be the *external force* acting on the ith particle of mass m_i, then according to *D'Alembert's principle,*

$$\sum_{i=1}^{N} \left(\vec{F}_i - \dot{\vec{p}}_i\right) \cdot \delta \vec{r}_i = 0, \tag{1.33}$$

where $\dot{\vec{p}}_i$ is the *inertial force* for the ith particle and $\delta \vec{r}_i$ is the *virtual displacement* of the ith particle due to $\vec{F}_i$.

$$\therefore \sum_{i=1}^{N} \left(\vec{F}_i - m_i \ddot{\vec{r}}_i\right) \cdot \delta \vec{r}_i = 0. \tag{1.34}$$

Differentiating (1.32) partially w.r.t. t, we get

$$\dot{\vec{r}}_i = \frac{d\vec{r}_i}{dt} = \frac{\partial \vec{r}_i}{\partial q_1}\dot{q}_1 + \frac{\partial \vec{r}_i}{\partial q_2}\dot{q}_2 + \cdots + \frac{\partial \vec{r}_i}{\partial q_j}\dot{q}_j + \cdots + \frac{\partial \vec{r}_i}{\partial q_f}\dot{q}_f + \frac{\partial \vec{r}_i}{\partial t} = \sum_j \frac{\partial \vec{r}_i}{\partial q_j}\dot{q}_j + \frac{\partial r_i}{\partial t}. \tag{1.35}$$

Again, the virtual displacement $\partial \vec{r}_i$ in terms of *generalised coordinates* is given by

$$\delta \vec{r}_i = \sum_j \frac{\partial \vec{r}_i}{\partial q_j}\delta q_j. \tag{1.36}$$

This equation does not involve the variation of time. Now, (1.34) becomes

$$\sum_i \left(\vec{F}_i - m_i\ddot{\vec{r}}_i\right)\cdot\sum_j \frac{\partial \vec{r}_i}{\partial q_j}\delta q_j = 0 \text{ or } \sum_i\sum_j \vec{F}_i\cdot\frac{\partial \vec{r}_i}{\partial q_j}\delta q_j = \sum_i\sum_j m_i\ddot{\vec{r}}_i\cdot\frac{\partial \vec{r}_i}{\partial q_j}\delta q_j. \tag{1.37}$$

Again,

$$\because \frac{d}{dt}\left(\dot{\vec{r}}_i\cdot\frac{\partial \vec{r}_i}{\partial q_j}\right) = \ddot{\vec{r}}_i\cdot\frac{\partial \vec{r}_i}{\partial q_j} + \dot{\vec{r}}_i\cdot\frac{d}{dt}\left(\frac{\partial \vec{r}_i}{\partial q_j}\right), \quad \therefore \ddot{\vec{r}}_i\cdot\frac{\partial \vec{r}_i}{\partial q_j} = \frac{d}{dt}\left(\dot{\vec{r}}_i\cdot\frac{\partial \vec{r}_i}{\partial q_j}\right) - \dot{\vec{r}}_i\cdot\frac{d}{dt}\left(\frac{\partial \vec{r}_i}{\partial q_j}\right). \tag{1.38}$$

Using (1.38) in (1.37), we get

$$\sum_i\sum_j\left(\vec{F}\cdot\frac{\partial \vec{r}_i}{\partial q_j}\right)\delta q_j = \sum_i\sum_j m_i\left[\frac{d}{dt}\left(\dot{\vec{r}}_i\cdot\frac{\partial \vec{r}_i}{\partial q_j}\right) - \dot{\vec{r}}_i\cdot\frac{d}{dt}\left(\frac{\partial \vec{r}_i}{\partial q_j}\right)\right]\delta q_j. \tag{1.39}$$

Also, since

$$\dot{\vec{r}}_i = \sum_j \frac{\partial \vec{r}_i}{\partial q_j}\dot{q}_j + \frac{\partial \vec{r}_i}{\partial t}, \left[\therefore \frac{\partial \dot{\vec{r}}_i}{\partial \dot{q}_j} = \frac{\partial \vec{r}_i}{\partial q_j}\right] \tag{1.40}$$

and

$$\frac{d}{dt}\left(\frac{\partial \vec{r}_i}{\partial q_j}\right) = \sum_k \frac{\partial}{\partial q_k}\left(\frac{\partial \vec{r}_i}{\partial q_j}\right)\frac{dq_k}{dt} + \frac{\partial}{\partial t}\left(\frac{\partial \vec{r}_i}{\partial q_j}\right)\frac{dt}{dt} = \sum_k \frac{\partial^2 \vec{r}_i}{\partial q_k \partial q_j}\dot{q}_k + \frac{\partial^2 \vec{r}_i}{\partial t \partial q_j}$$

$$= \sum_k \frac{\partial^2 \vec{r}_i}{\partial q_j \partial q_k}\dot{q}_k + \frac{\partial^2 \vec{r}_i}{\partial q_j \partial t} = \frac{\partial}{\partial q_j}\left[\sum_k \frac{\partial \vec{r}_i}{\partial q_k}\dot{q}_k + \frac{\partial \vec{r}_i}{\partial t}\right] = \frac{\partial}{\partial q_j}\left(\frac{d\vec{r}_i}{dt}\right) = \frac{\partial \vec{v}_i}{\partial q_j}\left[\because \vec{v}_i = \frac{d\vec{r}_i}{dt}\right]. \tag{1.41}$$

Substituting (1.40) and (1.41) in (1.39), we get

$$\sum_i\sum_j\left(\vec{F}_i\cdot\frac{\partial \vec{r}_i}{\partial q_j}\right)\delta q_j = \sum_i\sum_j\left[\frac{d}{dt}\left(m_i\vec{v}_i\cdot\frac{\partial \vec{v}_i}{\partial \dot{q}_j}\right) - m_i\vec{v}_i\cdot\frac{\partial \vec{v}_i}{\partial q_j}\right]\delta q_j\left[\because \frac{\partial \vec{r}_i}{\partial q_j} = \frac{\partial \dot{\vec{r}}_i}{\partial \dot{q}_j} = \frac{\partial \vec{v}_i}{\partial \dot{q}_j}\right]$$

$$= \sum_j\left[\frac{d}{dt}\left\{\frac{\partial}{\partial \dot{q}_j}\left(\sum_i \frac{1}{2}m_i v_i^2\right)\right\} - \frac{\partial}{\partial q_j}\left(\sum_i \frac{1}{2}m_i v_i^2\right)\right]\delta q_j$$

$$\text{or} \sum_j Q_j\delta q_j = \sum_j\left[\frac{d}{dt}\left(\frac{\partial T}{\partial \dot{q}_j}\right) - \frac{\partial T}{\partial q_j}\right]\delta q_j = 0,$$

where $Q_j = \sum_i \vec{F}_i \cdot \frac{\partial \vec{r}_i}{\partial q_j}$ is the *generalised force component* and $\sum_i \frac{1}{2} m_i v_i^2 = T$ is the *total KE of the system*

$$\text{or } \sum_j \left[\frac{d}{dt}\left(\frac{\partial T}{\partial \dot{q}_j}\right) - \frac{\partial T}{\partial q_j} - Q_j \right] \delta q_j = 0. \tag{1.42}$$

The generalised displacements δq_j are independent due to linear independence of generalised coordinates q_j. Then the (1.42) holds only if the coefficients of δq_j vanish separately, i.e.,

$$\frac{d}{dt}\left(\frac{\partial T}{\partial \dot{q}_j}\right) - \frac{\partial T}{\partial q_j} = Q_j, \tag{1.43}$$

where $j = 1, 2, 3, \ldots, N$.

These equations are *second order* and their number is equal to the degrees of freedom of the system. (1.43) is derived if the system does not involve constraints and is known as the general form of *Lagrange's equation*. (1.43) also holds for systems involving holonomic constraints because such constraints may be used to minimise the degrees of freedom of the system and as such independent of generalised coordinates.

Case I. *When the system is wholly conservative.*

If the system under consideration be *conservative*, then the forces will be derivable from a *scalar potential function V*. For the ith particle,

$$\vec{F}_i = -\Delta V_i = -\frac{\partial V_i}{\partial r_i}. \tag{1.44}$$

$$\therefore Q_k = \sum_i \vec{F}_i \cdot \frac{\partial \vec{r}_i}{\partial q_k} = -\sum_i \frac{\partial V_i}{\partial r_i} \cdot \frac{\partial \vec{r}_i}{\partial q_k} = -\sum_i \frac{\partial V_i}{\partial q_k} = -\frac{\partial}{\partial q_k} \sum_i V_i = -\frac{\partial V}{\partial q_k}, \tag{1.45}$$

where $V = \sum_i V_i$ = the *total PE of the system.*

Now from (1.43), we get

$$\frac{d}{dt}\left(\frac{\partial T}{\partial \dot{q}_j}\right) - \frac{\partial T}{\partial q_j} = -\frac{\partial V}{\partial q_j} \left[\because Q_k = -\frac{\partial V}{\partial q_k}\right]$$

$$\text{or } \frac{d}{dt}\left(\frac{\partial T}{\partial \dot{q}_j}\right) - \frac{\partial}{\partial q_j}(T - V) = 0 \text{ or } \frac{d}{dt}\left[\frac{\partial (T-V)}{\partial \dot{q}_j}\right] - \frac{\partial (T-V)}{\partial q_j} = 0 \because V \neq f(\dot{q}_j), \therefore \frac{\partial V}{\partial \dot{q}_j} = 0;$$

The *Lagrangian of the system* $L = T - V$.

$$\therefore \frac{d}{dt}\left(\frac{\partial L}{\partial \dot{q}_j}\right) - \frac{\partial L}{\partial q_j} = 0 \tag{1.46}$$

$$\text{or } \sum_j \left[\frac{d}{dt}\left(\frac{\partial L}{\partial \dot{q}_j}\right) - \frac{\partial L}{\partial q_j}\right] = 0. \tag{1.47}$$

(1.47) is called the *Lagrange's equation* of motion of a system of particles.

Case II. *When the system is non-conservative.*

If the system is *not conservative*, then also the Lagrangian can exist if Q_k are derivable from a *velocity dependent or generalised potential function* $U\left(q_j, \dot{q}_j\right)$ in the form:

$$Q_j = -\frac{\partial U}{\partial q_j} + \frac{d}{dt}\left(\frac{\partial U}{\partial \dot{q}_j}\right). \tag{1.48}$$

Using this value of Q_j in (1.46), we get

$$\frac{d}{dt}\left(\frac{\partial T}{\partial \dot{q}_j}\right) - \frac{\partial T}{\partial q_j} = -\frac{\partial U}{\partial q_j} + \frac{d}{dt}\left(\frac{\partial U}{\partial \dot{q}_j}\right) \text{ or } \frac{d}{dt}\left[\frac{\partial (T-U)}{\partial \dot{q}_j}\right] - \frac{\partial}{\partial q_j}(T-U) = 0.$$

Putting $L = T - U$, where *U is the generalised potential*, then

$$\sum_j \left[\frac{d}{dt}\left(\frac{\partial L}{\partial \dot{q}_j}\right) - \frac{\partial L}{\partial q_j}\right] = 0. \text{ [Ref. (1.47)]}$$

This type of equations are used in the calculation of Lagrangian for the case of *electromagnetic forces on moving charges.*

Case III. *When the force is non-potential.*

If Q'_j denotes the *non-potential force*, then the Lagrange's equations of motion can be written as

$$\frac{d}{dt}\left(\frac{\partial L}{\partial \dot{q}_j}\right) - \frac{\partial L}{\partial q_j} = Q'_j \; (j = 1, 2, \ldots, f)$$

$Q'_j = Q'_j\left(q_1, q_2, \ldots, q_f, \dot{q}_1, \dot{q}_2, \ldots, \dot{q}_f, t\right)$ and $L = T - V$ or $L = T - U$.

Note. The Coriolis force $\vec{F} = 2m(\vec{v} \times \vec{\omega})$ and the Lorentz force $\vec{F} = q\left(\vec{v} \times \vec{B}\right)$ are the examples of non-potential force and these forces cannot be derived from potential.

Case IV. *When the force is dissipative.*

We know that the dissipative (e.g., frictional) force is proportional to the velocity of the particle. So,

$$\vec{F}_i^d = -\lambda_i \dot{\vec{r}}_i, \tag{1.49}$$

where $\vec{F}_i^d$ = the dissipative force, r_i = the displacement and λ_i = the constant of proportionality for the ith particle of the system.

Now *Rayleigh's dissipation function* is defined as

$$R = \frac{1}{2}\sum_i \lambda_i \dot{r}_i^2. \tag{1.50}$$

$$\therefore \frac{\partial R}{\partial \dot{r}_i} = \lambda_i \dot{r}_i = -\vec{F}_i^d \text{ or } \vec{F}_i^d = -\frac{\partial R}{\partial \dot{r}_i}. \tag{1.51}$$

The Lagrange's equation is then

$$\frac{d}{dt}\left(\frac{\partial L}{\partial \dot{r}_i}\right) - \frac{\partial L}{\partial r_i} + \frac{\partial R}{\partial \dot{r}_i} = 0. \tag{1.52}$$

Again,

$$Q_k^d = \sum_i \vec{F}_i^d \cdot \frac{\partial \vec{r}}{\partial q_k} = \sum_i -\lambda_i \dot{\vec{r}}_i \cdot \frac{\partial \vec{r}_i}{\partial q_k} = -\sum_i \lambda_i \dot{\vec{r}}_i \cdot \frac{\partial \dot{\vec{r}}_i}{\partial \dot{q}_k} \left[\because \frac{\partial \vec{r}}{\partial q_k} = \frac{\partial \dot{\vec{r}}_i}{\partial \dot{q}_k}\right]$$

$$= -\sum_i \lambda_i \frac{\partial}{\partial \dot{q}_k}\left(\frac{1}{2}\dot{r}_i^2\right) = \frac{\partial}{\partial \dot{q}_k}\sum_i -\frac{1}{2}\lambda_i \dot{r}_i^2 \text{ or } Q_k^d = \frac{\partial R}{\partial \dot{q}_k}. \quad (1.53)$$

Thus the *Lagrange's equation of motion* for a system containing dissipative forces is

$$\frac{d}{dt}\left(\frac{\partial L}{\partial \dot{q}_k}\right) - \frac{\partial L}{\partial q_k} + \frac{\partial R}{\partial \dot{q}_k} = 0. \quad (1.54)$$

1.8.1 Characteristics of Lagrange's equation

The Lagrange's equation has the following characteristics:

1. For a system of particles, the Lagrange's equations of motion have been described by the consideration of energy and not force.
2. The effects of constraint forces have been eliminated in the Lagrangian formulation
3. $L = T - V$ is a scalar quantity and the solutions of the Lagrange's equation of motion is easy. But Newton's equation of motion involves force having vectorial sense.
4. Since the Lagrange's equation of motion is second order equation, its solution contains two arbitrary constants and the form of the equation is independent of the coordinates used.

Finally, under certain conditions, Lagrange's equations of motion are basically improved form of Newton's equation of motion.

1.9 Lagrangian Formulation of Conservation Theorems

First we define *generalised momentum* and *cyclic coordinates* in view of Lagrangian of a system.

Generalised momentum. We know that for a *conservative system,* force $\vec{F} = -\Delta V$, where the potential function V is a function of position coordinates only. For such a system,

$$\frac{\partial L}{\partial \dot{q}_j} = \frac{\partial (T-V)}{\partial \dot{q}_j} = \frac{\partial T}{\partial \dot{q}_j} - \frac{\partial V}{\partial \dot{q}_j} = \frac{\partial T}{\partial \dot{q}_j} \left[\because \frac{\partial V}{\partial \dot{q}_j} = 0\right]$$

$$= \frac{\partial}{\partial \dot{q}_j}\sum_j \frac{1}{2} m_j \dot{q}_j^2 = m_j \dot{q}_j = p_j. \quad (1.55)$$

$p_j = \dfrac{\partial L}{\partial \dot{q}_j}$ is called the *generalised momentum associated with the generalised coordinate* q_j.

Cyclic coordinates (or, ignorable coordinates): The Lagrangian (L) of a physical system is in general a function of all the generalised coordinates (q_j) and time t, i.e., $L = L(q_1, q_2, \ldots, q_j, \ldots, q_f, t)$.

If some of the generalised coordinates do not appear explicitly in the Lagrangian function, then those coordinates are called the cyclic or *ignorable coordinates.*

As the Lagrangian is not the function of *cyclic coordinates* q_j,

$$\therefore \frac{\partial L}{\partial q_j} = 0.$$

1. *Conservation theorem of generalised momentum.* We have the Lagrange's equation of motion for a conservative system

$$\frac{d}{dt}\left(\frac{\partial L}{\partial \dot{q}_j}\right)-\frac{\partial L}{\partial q_j}=0.$$

If q_j is cyclic, then $\frac{\partial L}{\partial q_j}=0.$

$$\therefore\ \frac{d}{dt}\left(\frac{\partial L}{\partial \dot{q}_j}\right)=0 \text{ or } \frac{\partial L}{\partial \dot{q}_j}=\text{constant or } p_j=\text{constant.} \tag{1.56}$$

Hence, *the conservation theorem of generalised momentum states that the generalised momentum corresponding to a cyclic or ignorable coordinate is conserved.*

2. *Conservation theorem of energy.* We consider a conservative system, where the potential V is a function of position only and the constraints do not change with time.

$$L=L\left(q_j,\dot{q}_j\right),$$

$$\therefore\ \frac{dL}{dt}=\sum_j\frac{\partial L}{\partial q_j}\frac{dq_j}{dt}+\sum_j\frac{\partial L}{\partial \dot{q}_j}\frac{d\dot{q}_j}{dt}.$$

Again,

$$\because\ \frac{d}{dt}\left(\frac{\partial L}{\partial \dot{q}_j}\right)-\frac{\partial L}{\partial q_j}=0,\quad \therefore\ \frac{\partial L}{\partial q_j}=\frac{d}{dt}\left(\frac{\partial L}{\partial \dot{q}_j}\right).$$

So,

$$\frac{dL}{dt}=\sum_j\left[\frac{d}{dt}\left(\frac{\partial L}{\partial \dot{q}_j}\right)\dot{q}_j+\frac{\partial L}{\partial \dot{q}_j}\frac{d\dot{q}_j}{dt}\right]=\sum_j\frac{d}{dt}\left(\dot{q}_j\frac{\partial L}{\partial \dot{q}_j}\right)=\sum_j\frac{d}{dt}\left(\dot{q}_j\frac{\partial T}{\partial \dot{q}_j}\right).$$

$$\left[\because \text{ for a conservative system, } \frac{\partial V}{\partial \dot{q}_j}=0,\ \therefore\ \frac{\partial L}{\partial \dot{q}_j}=\frac{\partial T}{\partial \dot{q}_j}-\frac{\partial V}{\partial \dot{q}_j}=\frac{\partial T}{\partial \dot{q}_j}.\right]$$

Also $p_j=\frac{\partial T}{\partial \dot{q}_j}$.

$$\therefore\ \frac{dL}{dt}=\sum_j\frac{d}{dt}\left(\dot{q}_jp_j\right) \text{ or } \frac{dL}{dt}-\sum_j\frac{d}{dt}\left(\dot{q}_jp_j\right)=0$$

$$\text{or } \frac{d}{dt}\left(\sum_j\dot{q}_jp_j-L\right)=0 \text{ or } \sum_j\dot{q}_jp_j-L=\text{constant }(J). \tag{1.57}$$

The constant J is called the *Jacobi's integral* for the system and it is one of the first integrals of equations of motion. It can be shown that J is the Hamiltonian (H) of the system.

$$\therefore\ H=\sum_j\dot{q}_jp_j-L=\text{constant.} \tag{1.58}$$

If the system is *holonomic*, then the KE of the system is a *quadratic function of generalised velocities*, i.e.,

$$T=\sum_{jk}a_{jk}\dot{q}_j\dot{q}_k.$$

If f is a homogeneous function of order n of a set of variables q_j, $j = 1,2,\ldots$, then according to *Euler's theorem,*

$$\sum_j q_j \frac{\partial f}{\partial \dot{q}_j} = nf.$$

Here $n = 2$ and $f = T$.

$$\therefore \sum_j \dot{q}_j \frac{\partial T}{\partial \dot{q}_j} = 2T \text{ or } \sum_j \dot{q}_j p_j = 2T.$$

Now from (1.58),

$$H = 2T - L = 2T - (T - V) \text{ or } H = T + V. \tag{1.59}$$

Thus *if the Lagrangian function does not depend explicitly on time, then the total energy of a conservative system is constant.*

3. *Conservation theorem of linear momentum.* We consider a conservative system for which the PE is a function of position only and the KE is independent of position. Then, $\frac{\partial V}{\partial \dot{q}_j} = 0$ and $\frac{\partial T}{\partial q_j} = 0$.

For such a system, the Lagrange's equation of motion is

$$\frac{d}{dt}\left(\frac{\partial L}{\partial \dot{q}_j}\right) - \frac{\partial L}{\partial q_j} = 0.$$

But,

$$\frac{\partial L}{\partial q_j} = \frac{\partial T}{\partial q_j} - \frac{\partial V}{\partial q_j} = -\frac{\partial V}{\partial q_j} \left[\because \frac{\partial T}{\partial q_j} = 0\right].$$

$$\therefore \frac{d}{dt}\left(\frac{\partial L}{\partial \dot{q}_j}\right) + \frac{\partial V}{\partial q_j} = 0.$$

So,

$$\dot{p}_j = \frac{d}{dt}\left(\frac{\partial L}{\partial \dot{q}_j}\right) = -\frac{\partial V}{\partial q_j} = Q_j. \tag{1.60}$$

Now,

$$Q_j = \sum_i \vec{F}_i \cdot \frac{\partial \vec{r}_i}{\partial q_j}.$$

If $\hat{n}$ be the unit vector along the translation of the system, then $\delta \vec{r}_i = \hat{n}\delta q_j$ and $\frac{\partial \vec{r}_i}{\partial q_j} = \hat{n}$.

$$\therefore Q_j = \sum_i \vec{F}_i \cdot \hat{n} = \hat{n} \cdot \vec{F}. \tag{1.61}$$

This shows, *the component of total force* is along the direction of motion $(\hat{n})$.

The KE of the system is

$$T = \frac{1}{2}\sum_i m_i \dot{r}_i^2.$$

Now the component of *generalised momentum*

$$p_j = \frac{\partial L}{\partial \dot{q}_j} = \frac{\partial T}{\partial \dot{q}_j} = \sum_i m_i \dot{\vec{r}}_i \cdot \frac{\partial \dot{\vec{r}}_i}{\partial q_j}, \quad \left[\because T = \frac{1}{2}\sum_i m_i \dot{\vec{r}}_i^2, \therefore \frac{\partial T}{\partial \dot{q}_j} = \frac{1}{2}\sum_i m_i \frac{1}{2}\dot{\vec{r}}_i \cdot \frac{\partial \dot{\vec{r}}_i}{\partial q_j} = \sum_i m_i \dot{\vec{r}}_i \cdot \frac{\partial \dot{\vec{r}}_i}{\partial q_j}\right]$$

$$\text{or } p_j = \sum_i m_i \vec{v}_i \cdot \frac{\partial \vec{v}_i}{\partial q_j} = \sum_i m_i \vec{v}_i \cdot \hat{n} = \hat{n} \cdot \sum \vec{p}_i. \tag{1.62}$$

(1.62) shows that *the component of total linear momentum is along the direction ($\hat{n}$) of translation of the system.*

Hence we can say that (1.60), i.e., $\dot{p}_j = Q_j$ *is the equation of motion for total linear momentum of the system*. If $Q_j = 0$, then $\dot{p}_j = 0$ and p_j = constant.

Hence, *if a given component of the total applied force vanishes, the corresponding component of the linear momentum is conserved.*

4. *Conservation theorem of angular momentum.* The theorem states that, *if the rotation coordinates q_j is cyclic, i.e., if the component of applied torque along the axis of rotation vanishes, then the component of total angular momentum along the axis of rotation is conserved.*

If V does not depend on $\dot{q}_j$ and T does not depend on q_j, then for a conservative system,

$$\frac{\partial V}{\partial \dot{q}_j} = 0 \text{ and } \frac{\partial T}{\partial q_j} = 0.$$

The Lagrangian equation for such a system is

$$\frac{d}{dt}\left(\frac{\partial L}{\partial \dot{q}_j}\right) - \frac{\partial L}{\partial q_j} = 0$$

$$\text{or } \frac{d}{dt}\left(\frac{\partial L}{\partial \dot{q}_j}\right) = \frac{\partial L}{\partial q_j} = -\frac{\partial V}{\partial q_j}$$

or $\dot{p}_j = Q_j$ [cf. (1.60)]. (1.63)

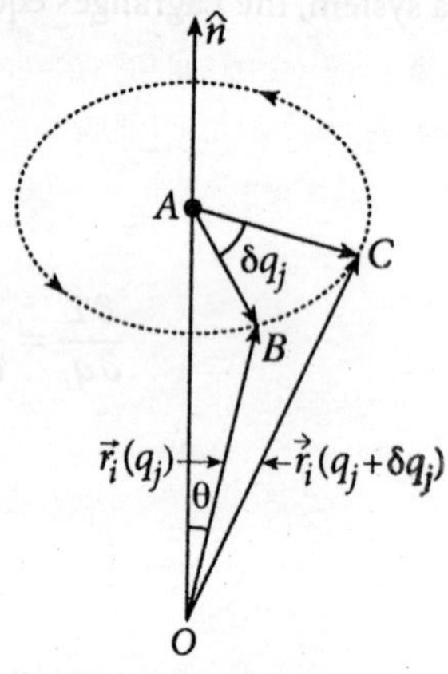

Fig. 1.3: Conservation of angular momentum.

If now we can show that with q_j, the generalised force Q_j is the component of total applied torque about the axis of rotation and the generalised momentum p_j is the component of total angular momentum along the same axis, then (1.63) will represent the *equation of motion for the total angular momentum*.

The generalised force

$$Q_j = \sum_i \vec{F}_i \cdot \frac{\partial \vec{r}_i}{\partial q_j}.$$

The magnitude of change in $\vec{r}_i(q_j)$ due to a change in rotation coordinate q_j is given by

$$|\delta \vec{r}_i| = \left|\overrightarrow{AB}\right| \delta q_j = |\vec{r}_i| \sin\theta\, \delta q_j \text{ or } \left|\frac{\partial \vec{r}_i}{\partial q_j}\right| = r_i \sin\theta. \tag{1.64}$$

If $\hat{n}$ be the unit vector along the axis of rotation, then

$$\frac{\partial \vec{r}_i}{\partial q_j} = \hat{n} \times \vec{r}_i. \tag{1.65}$$

Now the generalised force

$$Q_j = \sum_i \vec{F}_i \cdot \hat{n} \times \vec{r}_i = \sum_i \hat{n} \cdot (\vec{r}_i \times \vec{F}_i) = \sum_i \hat{n} \cdot N_i = \hat{n} \cdot \vec{N}, \tag{1.66}$$

where $\vec{N}$ is the *total torque*. *Thus Q_j is the component of the total torque along the axis of rotation.*

Again,

$$p_j = \frac{\partial T}{\partial \dot{q}_j} = \sum_i m_i \vec{v}_i \cdot \frac{\partial \vec{r}_i}{\partial q_j} = \sum_i m_i \vec{v}_i \cdot \hat{n} \times \vec{r}_i = \sum_i \hat{n} \cdot \vec{r}_i \times m_i \vec{v}_i \text{ or } p_j = \hat{n} \cdot \sum_i \vec{L}_i = \hat{n} \cdot \vec{L}. \tag{1.67}$$

$$\vec{L} = \sum_i \vec{L}_i = \text{total angular momentum.}$$

(1.67) shows that, *p_j is the component of the total angular momentum along the axis of rotation.*

So, $\dot{p}_j = Q_j$ represents the equation of motion of the system.

If q_j is cyclic, then

$$\frac{\partial L}{\partial q_j} = 0 \text{ or } \frac{\partial T}{\partial q_j} - \frac{\partial V}{\partial q_j} = 0 - \frac{\partial V}{\partial q_j} \text{ or } \frac{\partial L}{\partial q_j} = -\frac{\partial V}{\partial q_j}.$$

$$\therefore\ \hat{n} \cdot \vec{N} = Q_j = -\frac{\partial V}{\partial q_j} = \frac{\partial L}{\partial q_j} = 0.$$

$$\therefore\ \dot{p}_j = Q_j = -\frac{\partial V}{\partial \dot{q}_j} = 0 \text{ or } p_j = \text{constant or } p_j = \hat{n} \cdot \vec{L} = \text{constant.} \tag{1.68}$$

(1.68) represents the *conservation law of angular momentum for a conservative system.*

1.9.1 Advantage of Lagrangian approach over Newtonian approach

1. **In *Lagrangian formulation* we have only to deal with two scalar functions T and V which greatly simplifies the problem. But in *Newtonian formulation* we have to work with many forces and accelerations which are vector quantities.**
2. **Lagrangian formulation, unlike Newtonian formulation, is not restricted to Cartesian (or its equivalent) coordinates as well as the forces of constraints.**
3. **T and V are scalar functions and are invariant under coordinate transformations. In Newtonian approach force, momentum, torque, etc. vector quantities are involved which introduce difficulty in solving problems.**
4. **When constraints are involved, it may not be possible to know the forces but KE (T) and PE (V) may be given. Under such situation, Lagrangian formulation is far more convenient and useful.**
5. **Newtonian approach is valid only in *inertial frame.* But Lagrangian approach is valid in both *inertial and non-inertial frames of reference*; only one must calculate T in an inertial frame.**

1.10 The Hamiltonian Formulation

The Lagrangian formulation is described in terms of generalised coordinate (q_j), generalised velocity $(\dot{q}_j)$ and time which are independent variables. Here the generalised velocity $(\dot{q}_j)$ being the time derivative of $q_j, \dot{q}_j$ are treated ultimately as *dependent variables*. In Hamiltonian formulation this dependence is removed by introducing a new independent variable p_j, where p_j is the *generalised momentum*. This means that the basis has been changed from $(q_j, \dot{q}_j, t)$ to (q_j, p_j, t). Like the Lagrangian $L = L(q_j, \dot{q}_j, t)$ a new function H is defined as $H = H(q_j, p_j, t)$, where H is called the *Hamiltonian function*.

1.10.1 The Hamiltonian Function (*H*)

We know that the Lagrangian of a system is $L = L(q_j, \dot{q}_j, t)$.

$$\therefore \frac{dL}{dt} = \sum_j \frac{\partial L}{\partial q_j}\dot{q}_j + \sum_j \frac{\partial L}{\partial \dot{q}_j}\ddot{q}_j + \frac{\partial L}{\partial t}. \tag{1.69}$$

The Lagrangian equation is

$$\frac{d}{dt}\left(\frac{\partial L}{\partial \dot{q}_j}\right) - \frac{\partial L}{\partial q_j} = 0 \text{ or } \frac{\partial L}{\partial q_j} = \frac{d}{dt}\left(\frac{\partial L}{\partial \dot{q}_j}\right) = p_j. \tag{1.70}$$

Using (1.70) in (1.69), we get

$$\frac{dL}{dt} = \sum_j \frac{d}{dt}\left(\frac{\partial L}{\partial \dot{q}_j}\right)\dot{q}_j + \sum_j \frac{\partial L}{\partial \dot{q}_j}\ddot{q}_j + \frac{\partial L}{\partial t} = \sum_j \frac{d}{dt}\left[\frac{\partial L}{\partial \dot{q}_j}\dot{q}_j\right] + \frac{\partial L}{\partial t}$$

$$\text{or } \frac{d}{dt}\left[L - \sum_j \frac{\partial L}{\partial \dot{q}_j}\dot{q}_j\right] = \frac{\partial L}{\partial t} \tag{1.71}$$

$$\text{or } \frac{d}{dt}\left[L - \sum_j p_j\dot{q}_j\right] = \frac{\partial L}{\partial t} \text{ or } \frac{d}{dt}\left[\sum_j p_j\dot{q}_j - L\right] = -\frac{\partial L}{\partial t}. \tag{1.72}$$

We now introduce a new function such that

$$H = \sum_j p_j\dot{q}_j - L(q_j, \dot{q}_j, t), \tag{1.73}$$

where H is called the *Hamiltonian function* of the independent variables q_j, p_j, t. Hence,

$$\frac{dH}{dt} = -\frac{\partial L}{\partial t}. \tag{1.74}$$

1.10.2 Hamilton's canonical equations of motion

We know that the Hamiltonian is a function of q_j, p_j, t, i.e., $H = H(q_j, p_j, t)$.

$$\therefore dH = \sum_j \frac{\partial H}{\partial q_j}dq_j + \sum_j \frac{\partial H}{\partial p_j}dp_j + \frac{\partial H}{\partial t}dt. \tag{1.75}$$

Again, since $H=\sum_j p_j\dot{q}_j - L,$

$$\therefore\ dH=\sum_j \dot{q}_j dp_j+\sum_j p_j d\dot{q}_j - dL. \tag{1.76}$$

Again, $\because\ L=L(q_j,\dot{q}_j,t),$

$$\therefore\ dL=\sum_j \frac{\partial L}{\partial q_j}dq_j+\sum_j \frac{\partial L}{\partial \dot{q}_j}d\dot{q}_j+\frac{\partial L}{\partial t}dt=\sum_j\left[\frac{\partial L}{\partial q_j}dq_j+p_j d\dot{q}_j\right]+\frac{\partial L}{\partial t}dt. \tag{1.77}$$

Using this value of dL in (1.76), we get

$$dH=\sum_j \dot{q}_j dp_j+\sum_j p_j d\dot{q}_j-\sum_j \frac{\partial L}{\partial q_j}d\dot{q}_j-\sum_j p_j d\dot{q}_j-\frac{\partial L}{\partial t}dt$$

$$dH=\sum_j \dot{q}_j dp_j-\sum_j \dot{p}_j d\dot{q}_j-\frac{\partial L}{\partial t}dt\quad\left[\because\ \dot{p}_j=\frac{\partial L}{\partial q_j}\right]. \tag{1.78}$$

Comparing the coefficients of dp_j, dq_j and dt in (1.75) and (1.78), we get

$$\left.\begin{aligned}\dot{q}_j&=\frac{\partial H}{\partial p_j}\\ \dot{p}_j&=-\frac{\partial H}{\partial q_j}\end{aligned}\right\} \tag{1.79}$$

$$-\frac{\partial L}{\partial t}=\frac{\partial H}{\partial t}. \tag{1.80}$$

The equations (1.79) are called *Hamilton's canonical equations of motion.* These are $2N$ first order equations of motion for a system containing N-particles in absence of holonomic constraints in place of $3N$ Lagrangian equations of the 2nd order. Integrating $2N$ first order (partial) differential equations, we get $6N$ constants which can be found out from the knowledge of the initial conditions of the problem.

1.10.3 Physical significance of H

Though both L and H have the dimension of energy but H is not equal to the total energy in all situations. This is due to the following facts:

1. *For conservative system the PE is coordinate dependent and not velocity dependent.*
2. *The coordinate transformation equations are independent of time, so that* $\sum_j p_j\dot{q}_j=2T$.

If the above conditions are fulfilled, then only H can represent the total energy E.

If the coordinates transformation equations involve time explicitly, L may still be independent of time. Then

$$H=\sum_j p_j\dot{q}_j-L\left(q_j,\dot{q}_j\right)$$

is a constant of motion but not equal to total energy.

We have the Hamiltonian function $H=H\left(q_1,q_2,\ldots,q_j,\ldots,\ p_1,p_2,\ldots,p_j,\ldots,t\right)$.

$$\therefore\ \frac{dH}{dt}=\sum_j\frac{\partial H}{\partial q_j}\dot{q}_j+\sum_j\frac{\partial H}{\partial p_j}\dot{p}_j+\frac{\partial H}{\partial t}=-\sum_j\dot{p}_j\dot{q}_j+\sum_j\dot{q}_j\dot{p}_j+\frac{\partial H}{\partial t}=\frac{\partial H}{\partial t}.$$

But

$$\frac{\partial H}{\partial t}=-\frac{\partial L}{\partial t}, \therefore \frac{dH}{dt}=-\frac{\partial L}{\partial t}. \tag{1.81}$$

If $L \neq L(t)$, then $\dfrac{\partial L}{\partial t}=0$ and so

$$\frac{dH}{dt}=0 \text{ or } H=\text{constant}. \tag{1.82}$$

Thus, *if the Lagrangian is not an explicit function of time, then the Hamiltonian is a constant of motion.*

Now *for a conservative system*, V does not depend on $\dot{q}_j$, i.e., $\dfrac{\partial V}{\partial \dot{q}_j}=0$.

Also,

$$H=\sum_j p_j\dot{q}_j-L=\sum_j \frac{\partial L}{\partial \dot{q}_j}\dot{q}_j-L\left[\because p_j=\frac{\partial L}{\partial \dot{q}_j}\right]$$

$$=\sum_j \dot{q}_j\left[\frac{\partial}{\partial \dot{q}_j}(T-V)\right]-L=\sum_j \dot{q}_j\frac{\partial T}{\partial \dot{q}_j}-L\left[\because \frac{\partial V}{\partial \dot{q}_j}=0\right]$$

$$=\sum_j \dot{q}_j\frac{\partial}{\partial \dot{q}_j}\left(\frac{1}{2}m\dot{q}_j^2\right)-L\left[\because T=\frac{1}{2}m\dot{q}_j^2\right]$$

$$=\sum_j m\dot{q}_j^2-L=2T-L=2T-(T-V)=T+V=E,\text{ the total energy.} \tag{1.83}$$

Hence, *if the coordinate transformation is independent of time for a conservative system, then the Hamiltonian represents the total energy of the system.*

1.10.4 If a given coordinate is cyclic in Lagrangian, it will also be cyclic in Hamiltonian

If q_j is cyclic in L, then $\dot{p}_j=\dfrac{\partial L}{\partial q_j}=0$; so p_j is constant of time.

Again, if p_j is constant, then $\dot{p}_j=-\dfrac{\partial H}{\partial q_j}=0$.

This shows that q_j does not appear in H, i.e., q_j is also cyclic in H. **A cyclic coordinate reduces the number of variables by two in Hamiltonian formulation.**

1.10.5 Characteristics of Hamiltonian function

1. The Hamiltonian function is defined as $H\left(p_j,q_j,t\right)=\sum_j \dot{q}_j p_j - L\left(q_j,\dot{q}_j,t\right)$.

   Thus H is a function of p_j, q_j and t. But Lagrangian is a function of $q_j, \dot{q}_j, t$.
2. If the KE T of the system is a homogeneous quadratic function of velocities and PE V is independent of velocity and time, then only the Hamiltonian (H) is equal to the total energy of the system, i.e., $H=T+V$.
3. If the constraints are time-dependent (*rheonomic*) or the transformation equations $r_i=r_i(q_i,t)$ contains time explicitly, then $H \neq E$ but still E is constant of time.

1.10.6 Advantage of Hamiltonian formulation over Lagrangian formulation

1. The two variables q_j and $\dot{q}_j$ are not given equal status in Lagrangian formulation since $\dot{q}_j$ is simply the time derivative of q_j. But in Hamiltonian formulation both coordinate $\left(q_j\right)$ and momentum $\left(p_j\right)$ are given equal status. This provides a frequent freedom of choosing coordinates and momenta.

2. The fact that the numerical value of H is equal to the total energy of the conservative system is very important in the cases of energy changes in atoms and molecules.
3. The *equality status* of p_j and q_j in Hamiltonian formulation provides a convenient basis for the development of *quantum mechanics* and *statistical mechanics.*
4. In quantising a dynamical system, the knowledge of the Hamiltonian is extremely important. In setting up Schrödinger equation in *wave mechanics* we replace generalised momenta by corresponding differential operators.

1.11 Answers of Some Important Questions

1.1 What do you mean by constraints? What role do constraints play in dynamics?

Ans. The limitations or geometrical restrictions on the motion of a particle or a system of particles are generally known as *constraints.*

The forces responsible for the constraints are called the *constraint forces.*

The motion of a train along the rails is an example of constraint motion. The force of reaction due to the rails is the force of constraints.

Constraints play two important roles in dynamics, viz.,

(a) *Every constraint limits the number of degrees of freedom of the system and*

(b) *The constraint forces lead to certain relations which may be either in the form of equality or inequality.*

1.2 Identify the nature of constraint and the origin of constraint force in case of a rigid body motion.

Ans. The nature of constraint in case of a rigid body motion is *conservative, scleronomic, holonomic* and *bilateral.*

During motion, the distance between any two particles remains constant by the action of the *interatomic forces.* The effect of such forces is to maintain the separation distance constant between any two particles but the magnitudes of these forces are quite unknown.

These internal (interatomic) forces which maintain the separation distance between any two particles in a rigid body are called *forces of constraint.*

1.3 How would you differentiate holonomic and non-holonomic constraints in view of velocity?

Ans. We know that the holonomic constraints are *velocity independent* but *non-holonomic are not.*

In many situations the equation of constraint may contain velocity terms and it may appear that the related constraint is non-holonomic. But if the velocity term can be eliminated then the corresponding constraint is holonomic. If, however, it cannot be eliminated at all, then the constraint is definitely non-holonomic.

For example, let a cylinder of radius a be rolling and sliding down over an inclined plane with its axis always horizontal. If s and θ be the distance of the axis of the cylinder from the inclined plane and the angle that the fixed radius has rotated from the radius to the point of contact respectively, then $\dot{s} = a\dot{\theta}$ or, $\frac{ds}{dt} = a\frac{d\theta}{dt}$.

Integrating w.r.t. time t, we get $(s - a\theta) =$ a·constant.

This equation is *independent of velocity* and hence the constraint is *holonomic.*

However, there are cases in which the equations involving velocities are *non-integrable.* The related constraints are, therefore, *non-holonomic.*

1.4 What are the difficulties introduced by constraints in solving mechanical problems?

Ans. Constraints introduce two types of difficulties in solving mechanical problems, viz.,—

(a) The equation of constraint for a system having N number of particles with $\vec{r}_1, \vec{r}_2, \ldots, \vec{r}_N$ position coordinates is $f(\vec{r}_1, \vec{r}_2, \ldots, \vec{r}_N, t) = 0$.

Now, these N number of coordinates are not all independent but are connected by the above equation. Hence, the N-equations connected by the above relation are not all independent. So, to have a solution of the problem, these equations are to be rewritten by considering the effect of the constraints. Herein lies the first difficulty.

(b) In many situations the nature of constraint forces remained unknown and they belong to the unknown of the problems. These unknown forces of constraint introduce complications in the solution of the problem. This is the second difficulty.

1.5 What do you mean by degrees of freedom of a dynamical system?

Ans. By the term *degrees of freedom* of a dynamical system we mean *the number of independent ways in which a dynamical system can execute motion consistent with the constraints imposed on it.*

It is defined as *the minimum number of independent coordinates, which can specify the position of the system completely. It is also defined as the number of squared terms in the expression for the KE of the system.*

If a system of N-particles has K number of constraints, then the minimum number of coordinates to specify the position is $(3N - K)$. So, the degrees of freedom of the system is $f = 3N - K$.

1.6 Show that the work done by constraint force is zero.

Ans. Let the position vectors of the mass points m_i and m_j be $\vec{r}_i$ and $\vec{r}_j$ respectively w.r.t. the origin O. Then $\vec{r}_i - \vec{r}_j$ = the distance between m_i and m_j.

If $\vec{F}_{ij}$ be the force on the ith particle due to the jth particle, then the total work done on the ith particle is $W_i = \sum_j \vec{F}_{ij} \cdot \Delta \vec{r}_i$.

The total work done due to all the particles of the rigid body is

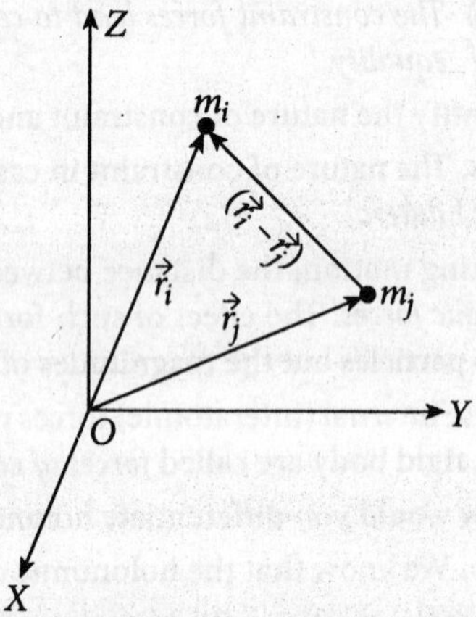

Fig. 1.4: Position vectors of m_i and m_j w.r.t. the origin O.

$$W = \sum_i \sum_j \vec{F}_{ij} \cdot \Delta \vec{r}_i = \sum_j \sum_i \vec{F}_{ji} \cdot \Delta r_j$$

$$\text{or } W = \frac{1}{2} \sum_i \sum_j \left(\vec{F}_{ij} \cdot \Delta \vec{r}_i + \vec{F}_{ji} \cdot \Delta \vec{r}_j \right)$$

$$= \frac{1}{2} \sum_i \sum_j \vec{F}_{ij} \cdot \Delta \left(\vec{r}_i - \vec{r}_j \right) \left[\because \vec{F}_{ij} = -\vec{F}_{ji} \right].$$

Again, the equation of constraint of a rigid body is

$$\left| \vec{r}_i - \vec{r}_j \right|^2 = \text{constant or } 2 \left(\vec{r}_i - \vec{r}_j \right) \cdot \Delta \left(\vec{r}_i - \vec{r}_j \right) = 0.$$

Also, $\vec{F}_{ij} = c_{ij} \left(\vec{r}_i - \vec{r}_j \right)$ as $\vec{F}_{ij}$ is along $\left(\vec{r}_i - \vec{r}_j \right)$ and c_{ij} is a real constant.

$$\therefore W = \frac{1}{2} \sum_i \sum_j c_{ij} \left(\vec{r}_i - \vec{r}_j \right) \cdot \Delta \left(\vec{r}_i - \vec{r}_j \right) = 0.$$

So, the work done by the forces of constraint is zero.

1.7 What are generalised coordinates? Indicate their advantage.

Ans. If N be the number of particles of a rigid body and K, the number of constraints, then the degrees of freedom is $f = (3N - K)$.

This suggests that the system can be described by a minimum of f number of independent coordinates, where the effects of constraints are included. These f number of independent coordinates $q_1, q_2, \ldots, q_f$ are called the *generalised coordinates.*

Definition. *The generalised coordinates are a set of independent coordinates which completely specify the position of the system.*

Advantages of generalised coordinates

Following are the advantages of introduction of generalised coordinates:

(a) The generalised coordinates may not essentially be rectangular, spherical or, polar, etc. That is the dependence of the description of the configuration of the dynamical system is eliminated.

(b) The individual variation (Δq_j) may be considered as the generalised coordinates q_j are independent of each other.

(c) A dynamical system described in terms of generalised coordinates may be treated as an otherwise free system.

1.8 What is configuration space?

Ans. A system of N-particles having K number of constraints has degrees of freedom $f = 3N - K$ or, $f = 3N$ (if $K = 0$).

Hence, the configuration space of a system having N particles is defined as *the $3N$-dimensional space (in absence of constraints, i.e., $K = 0$) and f-dimensional space (taking into account the presence of constraints).*

An f-dimensional configuration space is represented by f number of generalised coordinates viz., $q_1, q_2, \ldots, q_f$.

1.9 Explain what do you mean by virtual displacement.

Ans. See Art. 1.5 (P. 11).

1.10 What do you mean by cyclic coordinate?

Ans. The Lagrangian (L) of a physical system is in general a function of all the generalised coordinates (q_j) and time t, i.e., $L = L(q_1, q_2, \ldots, q_j, \ldots, q_f; t)$.

If some of the generalised coordinates do not appear explicitly in the Lagrangian function, then those coordinates are called *cyclic* or *ignorable coordinates.*

Hence, if q_j is cyclic, then $\frac{\partial L}{\partial q_j} = 0$.

1.11 What are the limitations of Newtonian mechanics?

Ans. In solving dynamical problems we must first construct the differential equation of motion of the system considering the forces acting on it. To investigate the nature of motion we must solve the differential equation. Newtonian mechanics suffers from the following limitation in doing so:

(a) Newton's laws of motion are valid only in inertial frames moving with constant velocities.

(b) To write the differential equation of motion, only a particular coordinate system is to be selected from the symmetry of the problem.

(c) The form of the differential equation of motion is changed due to transformation of coordinates, i.e., *Newtonian mechanics is not invariant under coordinate transformation.*

(d) Newton's laws of motion demand complete specification of the forces acting on the system at all time which is impossible.

(e) Newtonian mechanics does not involve constraint forces and hence dynamical problems involving constraints cannot be solved.

1.12 Discuss the advantages of Lagrangian formulation over Newtonian formulation of mechanics.
Ans. See Art. 1.9.1 (P. 21)

1.13 Discuss the physical significance of the Hamiltonian of a system.
Ans. See Art. 1.10 (P. 22)

1.14 What are the advantages of the Hamiltonian formulation over Lagrangian formulation mechanics.
Ans. See Art. 1.10.6 (P. 24)

1.15 State the conservation theorem of generalised momentum.
Ans. The conservation theorem of generalised momentum states that, *the generalised momentum corresponding to a cyclic (or, ignorable) coordinate is conserved.*

If q_j be a cyclic coordinate, then $\frac{\partial L}{\partial q_j} = 0$ and the Lagrange's equation of motion for a conservative system,

$$\frac{d}{dt}\left(\frac{\partial L}{\partial \dot{q}_j}\right) - \frac{\partial L}{\partial q_j} = 0$$

reduces to

$$\frac{d}{dt}\left(\frac{\partial L}{\partial \dot{q}_j}\right) = 0 \text{ or } \frac{\partial L}{\partial \dot{q}_j} = \text{constant, i.e., } p_j = \text{constant.}$$

The pair (q_j, p_j) is called the *canonical* or *conjugate variables.*

1.12 Solved Problems

1.1 Determine the degrees of freedom of the following:

(a) The bob of a simple pendulum oscillating in a plane.
(b) Dumb-bell moving in space.
(c) Two particles connected by a rigid rod moving freely in plane.
(d) A rigid body rotating in space with one point fixed.
(e) Three particles connected by rigid rods moving in space. [WBUT 2006]
(f) A particle moving on the circumference of a circle.
(g) NH_3-molecule.

Solution. (a) The motion of the bob of a simple pendulum can be described by one coordinate θ. The equation of constraint of the bob is $|r| = l$, where l is the effective length of the simple pendulum. Hence the bob of an oscillating simple pendulum has one degree of freedom.

(b) A dumb-bell constitutes two heavy particles connected by a rigid massless rod. So the equation of constraint is $(x_1 - x_2)^2 + (y_1 - y_2)^2 + (z_1 - z_2)^2 = \text{constant}$.

Here $N = 2$, $K = 1$.

$\therefore$ degrees of freedom $f = 3 \times 2 - 1 = 5$.

(c) Since the motion is in plane, two particles need (x_1, y_1) and (x_2, y_2), i.e., 4 coordinates. Here the equation of constraint $K = 1$, viz., $(x_2 - x_1)^2 + (y_2 - y_1)^2 = l^2$.

So the number of degrees of freedom $f = 2 \times 2 - 1 = 3$.

(d) When a rigid body rotates about a fixed axis passing through a fixed point, the equation of constraint is $x^2 + y^2 = r^2$, if r be the radius of the circle of motion about the fixed z-axis. So the degree of freedom $f = 2 \times 1 - 1 = 1$, since the particles move in a plane in circular path.

1.2 (e) Three particles may be connected in two ways:

(i) *Linear arrangement*. Here the distance between particles 1 and 2, and 2 and 3 remains constant. Hence the number of constraints $K = 2$.

The degrees of freedom $f = 3 \times 3 - 2 = 7$.

(ii) *Triangular arrangement*. In this case the number of constraint is 3 as there are three equations of the sides of the triangle to be satisfied. So, $K = 3$. The degrees of freedom is $f = 3 \times 3 - 3 = 6$.

(f) The equation of constraint relating to the motion of a particle on the circumference of a circle is $x^2 + y^2 = a^2$ or, $r = a$ (constant). In Cartesian coordinates one variable either x or y and in polar coordinates one variable θ is sufficient to locate the particle at any instant.

Hence, the degrees of freedom is one.

(g) In NH_3 molecule, three H atoms are linked with central N atom. So the number of atoms is 4. There are six restrictions, three for fixed bond length and three for bond angles. Thus the degrees of freedom $f = 4 \times 3 - 6 = 6$.

1.3 Discuss the nature of constraints and specify the constraint force in the following:

(a) A pendulum with variable length.
(b) A particle moving on an ellipsoid under gravity.
(c) Simple pendulum with rigid support.
(d) Gas-filled hollow sphere.
(e) An expanding or contracting spherical container of gas.
(f) The motion of a body in an inclined plane under gravity.
(g) A particle sliding down a sphere from a point near the top under gravity.
(h) A sphere rolling down an inclined plane under gravity.
(i) A bead moving on a circular wire.

Solution. (a) If the length of the pendulum is variable, then $l = l(t)$ and the equation of constraint is $|\vec{r}_1 - \vec{r}_2|^2 = l^2(t)$, which is time-dependent. So, the constraint is *rheonomic and holonomic*.

The constraint force is the *tension* in the string.

(b) The particle moving on an ellipsoid under gravity will leave the ellipsoid after reaching a certain point on the ellipsoid, so the constraint is *non-holonomic*.

The force of reaction of the ellipsoid surface on the particle is the constraint force.

(c) The motion of the simple pendulum has the equation of constraint $|\vec{r}| = l$. The constraint is *scleronomic, holonomic, bilateral and conservative*.

The force of constraint is the *tension* in the string.

(d) If a be the radius of the sphere and $\vec{r}$ the position vector of a gas molecule in the sphere, then the condition of motion is $r^2 \le a^2$. So the constraint is *scleronomic, holonomic, conservative and unilateral.*

(e) The condition of motion of a gas molecule in an expanding or contracting spherical container of gas is $r^2 \le a^2(t)$, where $a(t)$ is the radius of the sphere at any time t and r is the position vector of the gas molecule at t.

The constraint is *rheonomic, holonomic, unilateral and dissipative.*

(f) The motion of a body on an inclined plane under gravity is such that it is confined to move on the inclined plane surface. So the constraint is *holonomic, scleronomic, bilateral and conservative.* The equation of constraint is $y - x\tan\theta = 0$, where slope $\tan\theta = y/x$. The force of constraint is the *reaction of the plane* in a direction normal to the inclined plane.

(g) If a be the radius of the sphere and $\vec{r}$, the position vector of the particle with reference to the centre of the sphere, then the condition of motion is $r^2 \geq a^2$. So the constraint is *non-holonomic, unilateral and scleronomic.*

The reaction of the surface of the sphere on the particle is the force of constraint.

(h) Two sets of coordinates are necessary to describe the motion of the body, viz., (i) angular coordinates to specify the orientation of the sphere and (ii) a set of coordinates to locate the position of the sphere. These coordinates are not independent. A change in the orientation of the sphere means a change in the location. The constraint is *non-holonomic as we cannot minimise* the number of coordinates.

(i) The bead always remains at a constant distance a from the centre of the circular wire, where a is the radius of the circular wire. The condition of constraint is $r = a$, r is the position vector of the bead. So, the constraint is *holonomic, scleronomic, bilateral.*

The force of constraint is the *reaction of the wire* acting on the bead moving on the circular wire.

1.4 Show that the constraints relating to a rigid body are conservative.

Solution. If the force on the ith particle due to the jth particle of a rigid body be $\vec{F}_{ij}$, then according to *Newton's third law of motion*, $\vec{F}_{ij} = -\vec{F}_{ji}$, where $\vec{F}_{ji}$ is the force on the jth particle due to the ith particle. If $d\vec{r}_i, d\vec{r}_j$ be the displacement of the ith, jth particles respectively due to the action of $\vec{F}_{ij}, \vec{F}_{ji}$, then the work done is

$$W = \sum \int \vec{F} \cdot d\vec{r} = \sum_{i,j} \int \left(\vec{F}_{ij} \cdot d\vec{r}_i + \vec{F}_{ji} \cdot d\vec{r}_j \right)$$

$$\text{or } W = \sum_{i,j} \int \vec{F}_{ij} \cdot d\left(\vec{r}_i - \vec{r}_j\right) \quad \left[\because \vec{F}_{ij} = -\vec{F}_{ji}\right]$$

$$= \sum_{i,j} \int c_{ij}\left(\vec{r}_i - \vec{r}_j\right) \cdot d\left(\vec{r}_i - \vec{r}_j\right), \tag{1}$$

since the internal force $\vec{F}_{ij} = c_{ij}\left(\vec{r}_i - \vec{r}_j\right)$ is conservative and c_{ij} are constants for the rigid body.

But $r_{ij}^2 = \left|\vec{r}_i - \vec{r}_j\right|^2 = \text{constant}$.

$$\therefore d\left(r_{ij}^2\right) = d\left[\left(\vec{r}_i - \vec{r}_j\right) \cdot \left(\vec{r}_i - \vec{r}_j\right)\right] = \left(\vec{r}_i - \vec{r}_j\right) \cdot d\left(\vec{r}_i - \vec{r}_j\right) + d\left(\vec{r}_i - \vec{r}_j\right) \cdot \left(\vec{r}_i - \vec{r}_j\right)$$

$$\text{or } 0 = 2\left(\vec{r}_i - \vec{r}_j\right) \cdot d\left(\vec{r}_i - \vec{r}_j\right), \therefore \left(\vec{r}_i - \vec{r}_j\right) \cdot d\left(\vec{r}_i - \vec{r}_j\right) = 0. \tag{2}$$

Now using (2) in (1), we get $W = 0$.

Hence the constraint forces in case of motion of a rigid body are *conservative.*

1.5 What are the generalised coordinates needed to specify the motion of the following:

(a) A particle constraint to move on the circle in the XY-plane.

(b) A particle constraint to move on an ellipse in XY-plane.

(c) A particle moving on the surface of a sphere.

(d) Beads on an abacus.
(e) A particle moving on inside the surface of a cone.
(f) Two masses of a double pendulum constraint to move in a plane.
(g) A circular cylinder rolling down an inclined plane.

Solution. (a) For a particle moving in a circular path of radius r in the XY-plane, the position coordinates are

$$x = r\cos\theta$$
$$y = r\sin\theta.$$

Since r is constant, the motion of the particle can be specified by the generalised coordinate θ alone.
(b) For the particle moving on the ellipse in XY-plane with a and b as the semi-major and semi-minor axes, the position coordinates are

$$x = a\cos\theta$$
$$y = b\sin\theta.$$

Since a and b are constants, the generalised coordinate is θ alone.
(c) For a particle moving on the surface of a sphere the Cartesian coordinates are

$$x = a\cos\theta\sin\phi$$
$$y = a\sin\theta\sin\phi$$
$$z = a\cos\theta,$$

where a is the radius of the sphere and is constant. Hence, the generalised coordinates are θ and ϕ.
(d) The generalised coordinate for the motion of a bead on an abacus along the horizontal direction along the X-axis is x in Cartesian coordinates.
(e) The generalised coordinates for the motion of a particle on the inside surface of a cone are r, θ, where r = radius vector drawn from the vertex as origin to the position of the particle, θ = the angle made by the radius vector with a fixed slant edge of the cone.
(f) Here if θ_1 and θ_2 be the angles made by the length l_1 and l_2 of the first and the second pendulum respectively with the vertical line, then these two coordinates θ_1 and θ_2 are sufficient to specify the motion of the double pendulum.
So the generalised coordinates are θ_1, θ_2.
(g) On the inclined plane, the position of the cylinder can be completely specified by the distance x travelled by the centre of mass and the angle θ of rotation turned through by the cylinder about its axis.
If the cylinder rolls without slipping, then x and θ are related. So the generalised coordinate in this case is one, either x or θ.

1.6 Discuss the constraint motion of the following and determine the equation and nature of the constraints:

(a) A particle is moving in a cubical box.
(b) A disc (or a sphere) is rolling down without slipping in an inclined plane.
(c) A disc is rolling without slipping in a horizontal plane.

Solution. (a) Here the particle can move only within the box and can reach to the position of the walls of the box. So the equations of constraint are

$$0 \le x \le a$$
$$0 \le y \le a$$
$$0 \le z \le a,$$

where a is the length of the side of the cube along X-, Y- and Z-axes.

Here the constraint is *non-holonomic, unilateral* and *conservative*.

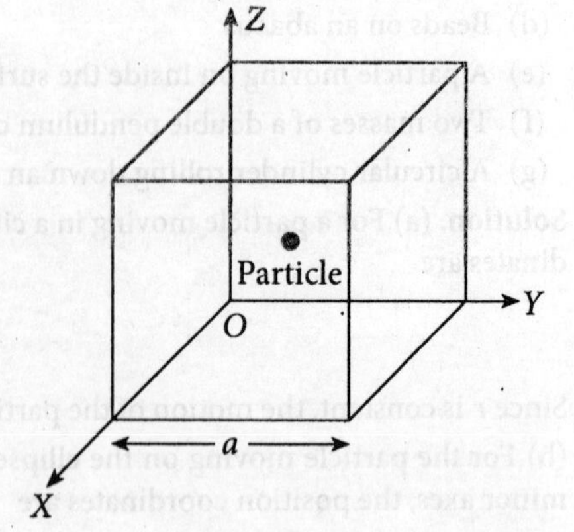

Fig. 1.5: Particle moving in a cubical box.

(b) Let a disc of radius a is rolling down an inclined plane as shown in Fig. 1.6. If $d\theta$ be the angle rotated by the radius OA in time dt so that

$$v = \frac{ds}{dt} = a\frac{d\theta}{dt}.$$

Integrating, $s = a\theta$ = constant or, $s - a\theta$ = constant.

This equation is obtained from the relationship of two velocities (i.e., $\frac{ds}{dt}$ and $\frac{d\theta}{dt}$) and involve position coordinates only after integration. The associated constraint is, therefore, *holonomic.*

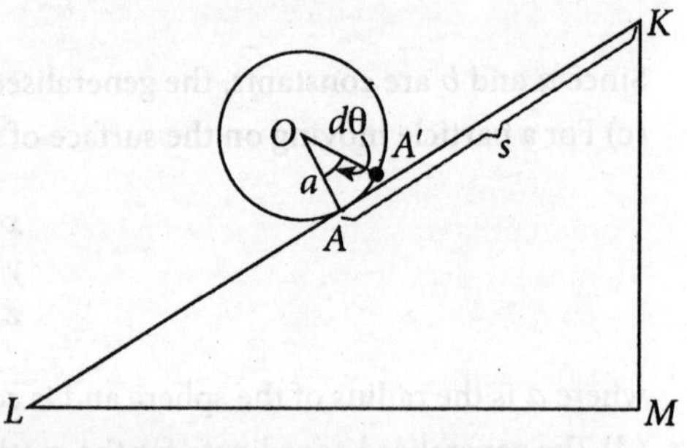

Fig. 1.6: A disc rolling down an inclined plane.

(c) Let a disc be rolling on the XY-plane with its plane always vertical. The axis of the disc makes an angle θ with the X-axis and ϕ is the angle of rotation of a fixed radius from CA to CA'. The velocity v of the disc is perpendicular to the axis of the disc. Here v makes an angle $(90° - \theta)$ with the X-axis [Fig. 1.7].

$$\therefore v = a\frac{d\phi}{dt}$$

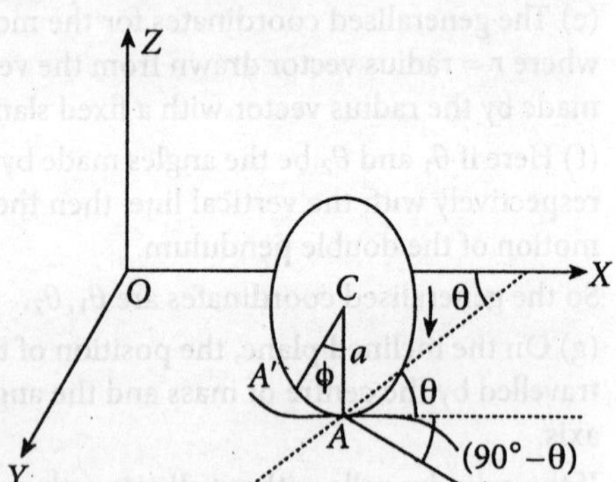

Fig. 1.7: Vertical disc rolling on a horizontal plane.

Now, the components of v along X- and Y-axes are

$$v_x = \frac{dx}{dt} = v\cos(90° - \theta) = v\sin\theta = a\frac{d\phi}{dt}\sin\theta$$

$$\text{and } v_y = \frac{dy}{dt} = -v\cos\theta = -a\frac{d\phi}{dt}\cos\theta.$$

Hence, $dx - a\sin\theta\, d\phi = 0$ and $dy + a\cos\theta\, d\phi = 0$.

These two expressions cannot be integrated without solving the problem and as such the associated constraints are *non-holonomic.*

1.7 What are superfluous coordinates?

Solution. While dealing motion in rectangular Cartesian coordinates we have noted that several coordinates are redundant (or superfluous) if there are holonomic constraints. Complicity arises due to this redundancy and non-independence of the coordinates to solve problems by using generalised coordinates.

We make the role of superfluous coordinates by taking an example as below:

When a ball is thrown vertically upward under gravity, we require the Cartesian coordinates to specify it as

$$x=0$$
$$y=0$$
$$z=v_0t-\tfrac{1}{2}gt^2,$$

where x and y are in the horizontal plane and z is in the vertical direction. v_0 is the velocity of projection. With the increase of t, only z changes and x and y remain constant. Here one coordinate $q=z$ is necessary to specify the motion and x, y are *redundant* or *superfluous coordinates.*

1.8 Two particles of masses m_1 and m_2 are located on a frictionless double inclined plane and connected by an inextensible massless string passing over a smooth peg. Using the principle of virtual work show that at equilibrium, $\frac{\sin\theta_1}{\sin\theta_2}=\frac{m_2}{m_1}$, where θ_1 and θ_2 are the angles of the incline.

Solution. Let $\vec{r}_1$ and $\vec{r}_2$ be the position vector of masses m_1 and m_2 respectively relative to O. If $\vec{F}_i^a$ be the externally applied force on the ith particle and $\delta\vec{r}_i$ the virtual displacement, then according to the *principle of virtual work*,

$$\sum_i \vec{F}_i^a\cdot\delta\vec{r}_i=0. \tag{1}$$

Since there are two masses m_1 and m_2,

$$\therefore\ \vec{F}_1^a\cdot\delta\vec{r}_1+\vec{F}_2^a\cdot\delta\vec{r}_2=0, \tag{2}$$

where $\vec{F}_1^a$ and $\vec{F}_2^a$ are the actual forces on m_1 and m_2 respectively due to gravity.

$$\vec{F}_1^a=m_1\vec{g}\sin\theta_1,\ \vec{F}_2^a=m_2\vec{g}\sin\theta_2. \tag{3}$$

$\therefore$ from (2),

$$m_1g\sin\theta_1\delta r_1+m_2g\sin\theta_2\delta r_2=0. \tag{4}$$

Since the string is inextensible, $r_1+r_2=$ constant.

$$\therefore\ \delta r_1+\delta r_2=0 \text{ or } \delta r_2=-\delta r_1.$$

Then from (4),

$$m_1g\sin\theta_1\delta r_1=m_2g\sin\theta_2\delta r_1$$
$$\text{or } m_1g\sin\theta_1=m_2g\sin\theta_2\ [\delta r_1\neq 0]$$
$$\text{or } \frac{\sin\theta_1}{\sin\theta_2}=\frac{m_1}{m_2} \text{ (Proved).}$$

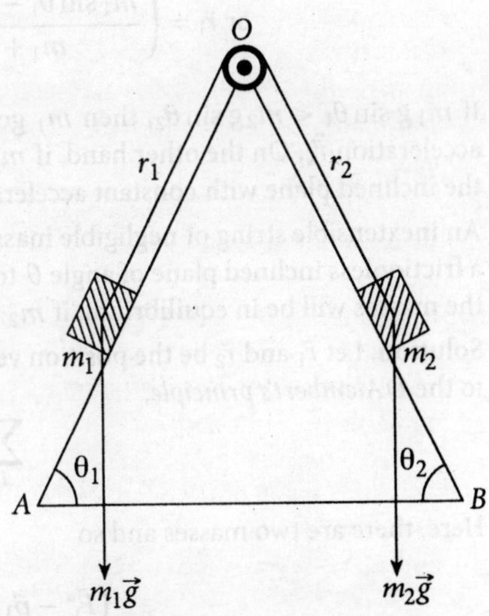

Fig. 1.8: Motion of masses m_1 and m_2 on a double inclined plane.

1.9 In Problem 1.7 use D'Alembert's principle to describe the motion of the masses.

Solution. According to D'Alembert's principle

$$\sum_i \left(\vec{F}_i^a - \dot{\vec{p}}_i\right)\cdot\delta\vec{r}_i = 0.$$

Here,

$$\left(\vec{F}_1^a - \dot{\vec{p}}_1\right)\cdot\delta\vec{r}_1 + \left(\vec{F}_2^a - \dot{\vec{p}}_2\right)\cdot\delta\vec{r}_2 = 0, \tag{1}$$

where $\dot{\vec{p}}_1$ and $\dot{\vec{p}}_2$ are the *reversed effective forces* acting on m_1 and m_2 respectively.

$$\vec{F}_1^a = m_1\vec{g}\sin\theta_1 \text{ and } \vec{F}_2^a = m_2\vec{g}\sin\theta_2 \text{ and } \dot{\vec{p}}_1 = m_1\ddot{\vec{r}}_1 \text{ and } \dot{\vec{p}}_2 = m_2\ddot{\vec{r}}_2.$$

Now from (1),

$$\left(m_1\vec{g}\sin\theta_1 - m_1\ddot{\vec{r}}_1\right)\cdot\delta\vec{r}_1 + \left(m_2\vec{g}\sin\theta_2 - m_2\ddot{\vec{r}}_2\right)\cdot\delta\vec{r}_2 = 0. \tag{2}$$

$\because r_1 + r_2 = \text{constant}$,

$$\therefore \delta r_1 + \delta r_2 = 0 \text{ or } \delta r_2 = -\delta r_1,$$

also $\ddot{r}_1 + \ddot{r}_2 = 0$,

$$\therefore \ddot{r}_2 = -\ddot{r}_1.$$

Putting these values in the above equation, we get

$$(m_1 g\sin\theta_1 - m_1\ddot{r}_1)\,\delta r_1 + (m_2 g\sin\theta_2 + m_2\ddot{r}_1)(-\delta r_1) = 0$$

$$\text{or } m_1 g\sin\theta_1 - m_1\ddot{r}_1 - m_2 g\sin\theta_2 - m_2\ddot{r}_1 = 0 \ (\because \delta r_1 \neq 0)$$

$$\text{or } (m_1 + m_2)\ddot{r}_1 = (m_1\sin\theta_1 - m_2\sin\theta_2)\,g$$

$$\text{or } \ddot{r}_1 = \left(\frac{m_1\sin\theta_1 - m_2\sin\theta_2}{m_1 + m_2}\right) g.$$

If $m_1 g\sin\theta_1 < m_2 g\sin\theta_2$, then m_1 goes up and m_2 goes down the inclined plane with constant acceleration $\ddot{r}_1$. On the other hand, if $m_1 g\sin\theta_1 > m_2 g\sin\theta_2$, then m_1 goes down and m_2 goes up the inclined plane with constant acceleration $\ddot{r}_1$.

1.10 An inextensible string of negligible mass hanging over a smooth peg at A connects one mass m_1 on a frictionless inclined plane of angle θ to another mass m_2. Use D'Alembert's principle to prove that the masses will be in equilibrium, if $m_2 = m_1\sin\theta$.

Solution. Let $\vec{r}_1$ and $\vec{r}_2$ be the position vectors of masses m_1 and m_2 relative to the peg A. According to the *D'Alembert's principle*,

$$\sum_i \left(\vec{F}_i^a - \dot{\vec{p}}_i\right)\cdot\delta r_i = 0. \tag{1}$$

Here, there are two masses and so

$$\left(\vec{F}_1^a - \dot{\vec{p}}_1\right)\cdot\delta\vec{r}_1 + \left(\vec{F}_2^a - \dot{\vec{p}}_2\right)\cdot\delta\vec{r}_2 = 0. \tag{2}$$

Here $\vec{F}_1^a, \vec{F}_2^a$ are the actual forces acting on m_1 and m_2 respectively, $-\dot{\vec{p}}_1, -\dot{\vec{p}}_2$ are the corresponding reversed effective forces and $\delta\vec{r}_1, \delta\vec{r}_2$ are the virtual displacements of m_1 and m_2.

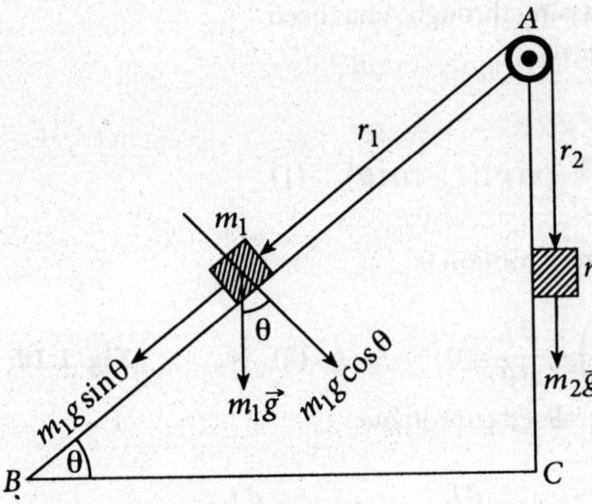

Fig. 1.9: Motion of mass m_1 on an inclined plane.

In the present case,

$$\vec{F}_1^a = m_1 \vec{g} \sin\theta \text{ and } \vec{F}_2^a = m_2 \vec{g}\, \dot{\vec{p}}_1 = m_1 \ddot{\vec{r}}_1 \text{ and } \dot{\vec{p}}_2 = m_2 \ddot{\vec{r}}_2.$$

Now from (2),

$$\left(m_1 \vec{g} \sin\theta - m_1 \ddot{\vec{r}}_1\right) \cdot \delta \vec{r}_1 + \left(m_2 \vec{g} - m_2 \ddot{\vec{r}}_2\right) \cdot \delta \vec{r}_2 = 0. \tag{3}$$

Since the string is inextensible, $r_1 + r_2 = $ constant or, $\delta r_1 + \delta r_2 = 0$ or, $\delta r_1 = -\delta r_2$.
Also, $\ddot{\vec{r}}_1 + \ddot{\vec{r}}_2 = 0$ or, $\ddot{\vec{r}}_2 = -\ddot{\vec{r}}_1$.
Hence (3) becomes

$$-\left(m_1 g \sin\theta + m_1 \ddot{r}_2\right) \delta r_2 + \left(m_2 g - m_2 \ddot{r}_2\right) \delta r_2 = 0$$
$$\text{or } \left(m_2 g - m_1 g \sin\theta\right) - \left(m_1 + m_2\right) \ddot{r}_2 = 0 \; [\because \delta r_2 \neq 0].$$

For the masses to be in equilibrium, $\ddot{r}_2 = 0$.

$$\therefore m_2 g - m_1 g \sin\theta = 0 \text{ or } g\left(m_2 - m_1 \sin\theta\right) = 0$$
$$\text{or } m_2 - m_1 \sin\theta = 0 \; [\because g \neq 0].$$
$$\therefore m_2 = m_1 \sin\theta \; \textbf{(Proved)}$$

1.11 (a) Obtain the equation of motion of a simple pendulum from its Lagrangian representation. [WBUT 2006]

(b) Deduce the formula for its time period for small amplitude oscillations.

Solution. (a) Let at any instant of time during oscillation the string OB make an angle θ with its rest position of equilibrium OA.

The KE of the bob is $T = \frac{1}{2} m v^2 = \frac{1}{2} m \left(l\dot{\theta}\right)^2 = \frac{1}{2} m l^2 \dot{\theta}^2$, where l is the length of the string and v, the velocity of the bob being $v = l\dot{\theta}$.

In passing from B to A the bob falls freely through a vertical distance CA. The PE of the bob is thus

$$V = mg(OA - OC)$$
$$\text{or } V = mg(l - l\cos\theta) = mgl(1 - \cos\theta).$$

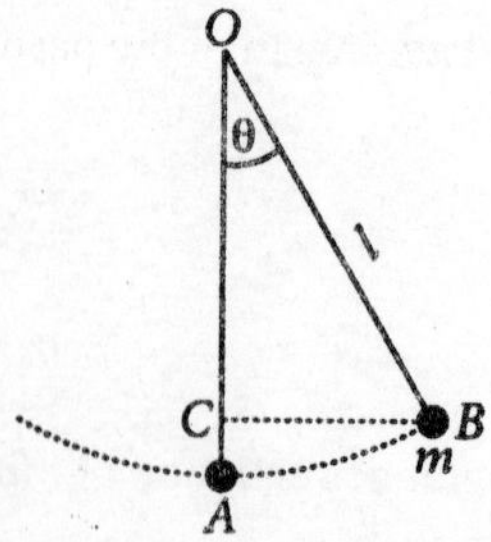

Fig. 1.10: A simple pendulum.

The horizontal plane passing through A has been taken as the reference level.

The Lagrangian,

$$L = T - V = \tfrac{1}{2} m l^2 \dot{\theta}^2 - m g l (1 - \cos\theta). \quad (1)$$

The Lagrange's equation of motion is

$$\frac{d}{dt}\left(\frac{\partial L}{\partial \dot{\theta}}\right) - \frac{\partial L}{\partial \theta} = 0. \quad (2)$$

Here θ is taken as generalised coordinate.

From (1),

$$\frac{\partial L}{\partial \dot{\theta}} = m l^2 \dot{\theta} \text{ and } \frac{\partial L}{\partial \theta} = -m g l \sin\theta.$$

Using these values of $\dfrac{\partial L}{\partial \dot{\theta}}$ and $\dfrac{\partial L}{\partial \theta}$ in (2), we get

$$\frac{d}{dt}\left(m l^2 \dot{\theta}\right) + m g l \sin\theta = 0 \text{ or } \ddot{\theta} + \frac{g}{l}\sin\theta = 0. \quad (3)$$

This is the required differential equation of motion of the simple pendulum.

(b) For small oscillation of the bob, $\sin\theta \simeq \theta$.

$\therefore$ (3) becomes

$$\ddot{\theta} + \frac{g}{l}\theta = 0 \text{ or } \ddot{\theta} + \omega^2\theta = 0, \quad (4)$$

where $\omega = \sqrt{g/l}$.

(4) represents a simple harmonic motion whose time period,

$$T = \frac{2\pi}{\omega} = 2\pi\sqrt{\frac{l}{g}}.$$

1.12 A particle of mass m is moving in a plane under the inverse square law of attractive force. Set up the Lagrangian and hence obtain the equation describing its motion. [cf. BUH 2000]

Solution. If (r, θ) be the plane polar coordinates of the particle of mass m, then the KE T of the particle is $T = \frac{1}{2} m \left(\dot{r}^2 + r^2\dot{\theta}^2\right)$.

If the force acting on the particle be F, then

$$F \propto \frac{1}{r^2} \text{ or } F = \frac{k}{r^2},\ k \text{ is a constant of proportionality.}$$

Again,

$$F = -\frac{dV}{dr},\ V \text{ is the PE}$$

$$\text{or } dV = -F\,dr.$$

$$\therefore V = -\int_\infty^r F\,dr = \int_\infty^r \frac{k}{r^2}\,dr = -\frac{k}{r}.$$

Now the Lagrangian of the particle is

$$L = T - V = \frac{1}{2}m\left(\dot{r}^2 + r^2\dot{\theta}^2\right) + \frac{k}{r}.$$

$$\therefore \quad \frac{\partial L}{\partial \dot{r}} = m\dot{r}, \quad \frac{\partial L}{\partial r} = mr\dot{\theta}^2 - \frac{k}{r^2}, \quad \frac{\partial L}{\partial \theta} = 0 \text{ and } \frac{\partial L}{\partial \dot{\theta}} = mr^2\dot{\theta}.$$

$\therefore$ the Lagrange's equation of motion in r is

$$\frac{d}{dt}\left(\frac{\partial L}{\partial \dot{r}}\right) - \frac{\partial L}{\partial r} = 0 \text{ or } \frac{d}{dt}(m\dot{r}) - \left(mr\dot{\theta}^2 - \frac{k}{r^2}\right) = 0 \text{ or } m\ddot{r} - \left(mr\dot{\theta}^2 - \frac{k}{r^2}\right) = 0$$

$$\text{or } \boldsymbol{m\ddot{r} - mr\dot{\theta}^2 + \frac{k}{r^2} = 0.} \tag{1}$$

The Lagrange's equation of motion in θ is

$$\frac{d}{dt}\left(\frac{\partial L}{\partial \dot{\theta}}\right) - \frac{\partial L}{\partial \theta} = 0 \text{ or } \frac{d}{dt}\left(mr^2\dot{\theta}\right) = 0 \text{ or } \boldsymbol{mr^2\ddot{\theta} + 2mr\dot{r}\dot{\theta} = 0.} \tag{2}$$

1.13 Write down the Lagrange's equations when the Lagrangian function has the form $L = \dot{q}_k q_k - \sqrt{1-\dot{q}_k^2}$.

Solution.

$$L = \dot{q}_k q_k - \sqrt{1-\dot{q}_k^2}. \tag{1}$$

$$\therefore \quad \frac{\partial L}{\partial \dot{q}_k} = q_k - \frac{1}{2}\left(1-\dot{q}_k^2\right)^{-1/2}(-2\dot{q}_k) = q_k + \frac{\dot{q}_k}{\sqrt{1-\dot{q}_k^2}}.$$

Now the Lagrange's equation of motion is

$$\frac{d}{dt}\left(\frac{\partial L}{\partial \dot{q}_k}\right) - \frac{\partial L}{\partial q_k} = 0 \text{ or } \frac{d}{dt}\left[q_k + \frac{\dot{q}_k}{\sqrt{1-\dot{q}_k^2}}\right] - \dot{q}_k = 0 \left[\because \frac{\partial L}{\partial q_k} = \dot{q}_k\right]$$

$$\text{or } \dot{q}_k + \frac{\ddot{q}_k\sqrt{1-\dot{q}_k^2} - \frac{1}{2}\left(1-\dot{q}_k^2\right)^{1/2}(-2\dot{q}_k)\ddot{q}_k\dot{q}_k}{\left(1-\dot{q}_k^2\right)} - \dot{q}_k = 0$$

$$\text{or } \ddot{q}_k\sqrt{1-\dot{q}_k^2} + \frac{\dot{q}_k^2\ddot{q}_k}{\sqrt{1-\dot{q}_k^2}} = 0 \text{ or } \frac{\ddot{q}_k\left(1-\dot{q}_k^2\right) + \dot{q}_k^2\ddot{q}_k}{\sqrt{1-\dot{q}_k^2}} = 0$$

$$\text{or } \ddot{q}_k\left(1-\dot{q}_k^2 + \dot{q}_k\right) = 0 \text{ or } \ddot{q}_k = 0.$$

Thus the particle moves in such a way that the generalised acceleration $\ddot{q}_k = 0$.

1.14 Find the equation of motion of a simple pendulum whose point of suspension is moving along X-axis with acceleration f.

Solution. The point of suspension O' moves along X'-axis with acceleration f. Hence O', $X'Y'$ is non-inertial frame and O, X, Y is the inertial frame of reference.

The constants are expressed by

$$x' = l\sin\theta$$
$$y' = l\cos\theta.$$

Here the motion is described by the generalised coordinates θ.

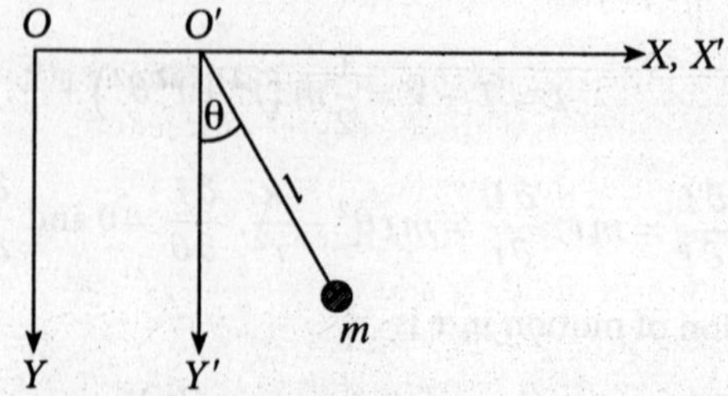

Fig. 1.11: A simple pendulum whose point of suspension is moving.

These constrained relations can be transferred from the non-inertial frame to inertial frame of reference in the following way:

$$x = x' + \tfrac{1}{2} f t^2 \text{ or } x = l\sin\theta + \tfrac{1}{2} f t^2 \text{ and } y = y' = l\cos\theta.$$

Again, $\dot{x} = l\cos\theta\,\dot{\theta} + ft$ and $\dot{y} = -l\sin\theta\,\dot{\theta}$.
KE of the bob is

$$\begin{aligned} T &= \tfrac{1}{2} m\left(\dot{x}^2 + \dot{y}^2\right) = \tfrac{1}{2} m\left[\left(l\cos\theta\dot{\theta} + ft\right)^2 + \left(-l\sin\theta\dot{\theta}\right)^2\right] \\ &= \tfrac{1}{2} m\left[l^2\cos^2\theta\dot{\theta}^2 + f^2t^2 + 2lft\cos\theta\dot{\theta} + l^2\sin^2\theta\dot{\theta}^2\right] \\ &= \tfrac{1}{2} m\left[l^2\dot{\theta}^2 + f^2t^2 + 2lft\cos\theta\dot{\theta}\right]. \end{aligned}$$

PE of the bob is $V = -mgy = -mgl\cos\theta$, assuming V to be zero at $y = 0$ plane.
∴ Lagrangian of the system

$$L = T - V = \tfrac{1}{2} m\left[l^2\dot{\theta}^2 + f^2t^2 + 2lft\cos\theta\dot{\theta}\right] + mgl\cos\theta.$$

Now the Lagrange's equation of motion is

$$\frac{d}{dt}\left(\frac{\partial L}{\partial \dot{\theta}}\right) - \frac{\partial L}{\partial \theta} = 0$$

$$\text{or } \frac{d}{dt}\left(ml^2\dot{\theta} + 2\times\frac{1}{2} lftm\cos\theta\right) - \left(-\frac{1}{2} m2lft\sin\theta\dot{\theta} - mgl\sin\theta\right) = 0,$$

$$\because \frac{\partial L}{\partial \dot{\theta}} = ml^2\dot{\theta} + 2\times\frac{1}{2} lftm\cos\theta \text{ and } \frac{\partial L}{\partial \theta} = -\frac{1}{2} m2lft\sin\theta\dot{\theta} - mgl\sin\theta.$$

$$\text{or } ml^2\ddot{\theta} + lfm\cos\theta - lftm\sin\theta\dot{\theta} + mlft\sin\theta\dot{\theta} + mgl\sin\theta = 0$$

$$\text{or } ml^2\ddot{\theta} + lfm\cos\theta + mgl\sin\theta = 0$$

$$\text{or } \boldsymbol{\ddot{\theta} + \frac{f}{l}\cos\theta + \frac{g}{l}\sin\theta = 0}.$$

This represents the equation of motion of a simple pendulum whose point of suspension is moving along the X-axis with acceleration f.

1.15 Obtain the equation of motion of a system of two masses, connected by an inextensible string passing over a small smooth pulley.

Solution. The problem of *Atwood's machine* is a good example of conservative system involving *holonomic and scleronomic constraints* with one degree of freedom.

Fig. 1.12 shows the principle of the Atwood's machine comprising masses m_1 and m_2 suspended over a frictionless pulley A of radius a and connected by a flexible string of constant length l. Then, if $PA = x$, $QA = l - x$. Here x is the only independent coordinate.

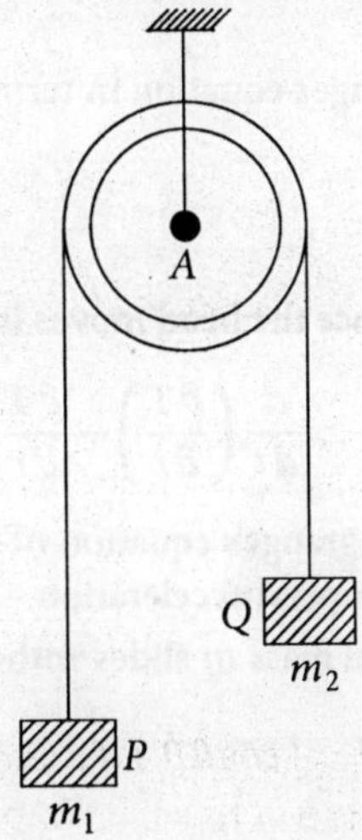

Fig. 1.12: Atwood's machine.

Now

$$\text{KE of mass } m_1 = \tfrac{1}{2} m_1 \dot{x}^2$$

$$\text{and KE of mass } m_2 = \tfrac{1}{2} m_2 \dot{x}^2.$$

Total KE of the system is

$$T = \tfrac{1}{2} m_1 \dot{x}^2 + \tfrac{1}{2} m_2 \dot{x}^2 = \tfrac{1}{2}(m_1 + m_2)\dot{x}^2.$$

With reference to the horizontal plane passing through A, the PE of $m_1 = -m_1 g x$ and that of $m_2 = -m_2 g(l - x)$.

$\therefore$ total PE of the system is

$$V = -m_1 g x - m_2 g(l - x).$$

The Lagrangian of the system is

$$L = T - V = \tfrac{1}{2}(m_1 + m_2)\dot{x}^2 + m_1 g x + m_2 g(l - x). \qquad (1)$$

$$\therefore \quad \frac{\partial L}{\partial \dot{x}} = (m_1 + m_2)\dot{x} \qquad (2)$$

$$\frac{\partial L}{\partial x} = (m_1 - m_2)\, g. \qquad (3)$$

Then the Lagrange's equation of motion is

$$\frac{d}{dt}\left(\frac{\partial L}{\partial \dot{x}}\right) - \frac{\partial L}{\partial x} = 0 \text{ or } \frac{d}{dt}[(m_1 + m_2)\dot{x}] - (m_1 - m_2)\, g = 0$$

$$\text{or } (m_1 + m_2)\ddot{x} = (m_1 - m_2)\, g \text{ or } \ddot{x} = \left(\frac{\mathbf{m_1 - m_2}}{\mathbf{m_1 + m_2}}\right) \mathbf{g}. \qquad (4)$$

(4) is the desired equation of motion of the system.

1.16 A bead is sliding on a rotating wire. Determine the Lagrange's equation of motion.

Solution. We assume that the bead is sliding on a uniformly rotating wire in a force-free space. The involved constraint is time-dependent.

Let (x, y) and (r, θ) be the Cartesian and plane polar coordinates of the bead at any time t.

Then, $x = r\cos\theta = r\cos\omega t$ and $y = r\sin\theta = r\sin\omega t$, where θ is the angle rotated and ω is the angular velocity of rotation of the wire.

$\therefore \ \theta = \omega t.$

The KE of bead is $T=\frac{1}{2}m\left(\dot{x}^2+\dot{y}^2\right)$.

But $\dot{x}=\dot{r}\cos\omega t-r\omega\sin\omega t,\ \dot{y}=\dot{r}\sin\omega t+r\omega\cos\omega t$.

$$\therefore\ T=\frac{1}{2}m\left(\dot{r}^2+r^2\omega^2\right) \text{ and } \frac{\partial T}{\partial r}=mr\omega^2 \text{ and } \frac{\partial T}{\partial \dot{r}}=m\dot{r}.$$

$\therefore$ Lagrange's equation in terms of r is

$$\frac{d}{dt}\left(\frac{\partial T}{\partial \dot{r}}\right)-\frac{\partial T}{\partial r}=Q_r.$$

Here, since the bead moves in a force-free space, $Q_r=0$.

$$\therefore\ \frac{d}{dt}\left(\frac{\partial T}{\partial \dot{r}}\right)-\frac{\partial T}{\partial r}=0 \text{ or } \frac{d}{dt}(m\dot{r})-mr\omega^2=0 \text{ or } m\ddot{r}-mr\omega^2=0 \text{ or } \ddot{r}=r\omega^2$$

is the Lagrange's equation of motion of the bead which shows that the bead moves outward due to the centripetal acceleration.

1.17 A bead of mass m slides without friction on a frictionless wire in the shape of cycloid with equations

$$x=a(\theta-\sin\theta)$$
$$y=a(1+\cos\theta),$$

where $0\le\theta\le 2\pi$. Find the Lagrangian function and the Lagrange's equation of motion.

Solution. Let m = mass of the bead sliding on the frictionless wire in the shape of a cycloid as shown in Fig. 1.13.

The KE of the bead is

$$T=\frac{1}{2}m\left(\dot{x}^2+\dot{y}^2\right)$$
$$=\frac{1}{2}ma^2\left[\left\{(1-\cos\theta)\dot{\theta}\right\}^2+\left\{(-\sin\theta)\dot{\theta}\right\}^2\right]$$
$$=ma^2(1-\cos\theta)\dot{\theta}^2.$$

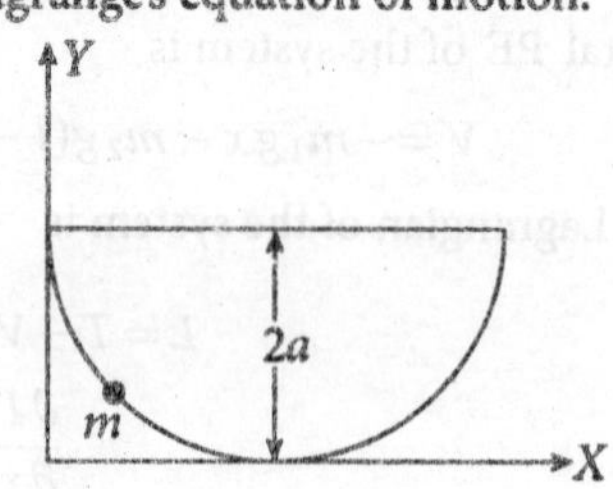

Fig. 1.13: Motion of bead on a cycloid-shaped wire.

PE, $V=mgy=mga(1+\cos\theta)$.

The Lagrangian,

$$L=T-V=ma^2(1-\cos\theta)\dot{\theta}^2-mga(1+\cos\theta)^2.$$

This is the Lagrangian function for the sliding mass m.

The Lagrangian equation of motion of the bead

$$\frac{d}{dt}\left(\frac{\partial L}{\partial \dot{\theta}}\right)-\frac{\partial L}{\partial \theta}=0$$

$$\text{or } \frac{d}{dt}\left[2ma^2(1-\cos\theta)\dot{\theta}\right]-\left[ma^2\sin\theta\,\dot{\theta}^2+mga\sin\theta\right]=0$$

$$\text{or } \frac{d}{dt}\left[(1-\cos\theta)\dot{\theta}\right]-\frac{1}{2}\sin\theta\,\dot{\theta}^2-\frac{g}{2a}\sin\theta=0$$

$$\text{or } (1-\cos\theta)\ddot{\theta}+\frac{1}{2}\sin\theta\,\dot{\theta}^2-\frac{g}{2a}\sin\theta=0.$$

This is the desired Lagrangian equation of motion.

1.18 Find the Lagrange's equation of motion of a double pendulum.

Solution. A double pendulum consists of two pendulums having bobs of masses m_1 and m_2 and lengths of the pendulum l_1 and l_2 respectively. The motion of the system is defined by generalised coordinates θ_1 and θ_2. It is thus a problem of two degrees of freedom.

Let (x_1, y_1) and (x_2, y_2) be the coordinates of m_1 and m_2 as shown in Fig. 1.14.

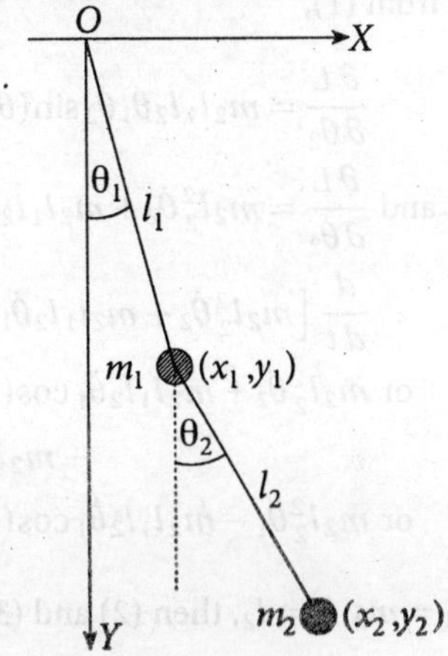

Fig. 1.14: Motion of a double pendulum.

Here

$$y_1 = l_1 \cos\theta_1$$
$$y_2 = l_1 \cos\theta_1 + l_2 \cos\theta_2$$
$$x_1 = l_1 \sin\theta_1$$
and $$x_2 = l_1 \sin\theta_1 + l_2 \sin\theta_2.$$
$$\therefore\ \dot{y}_1 = -l_1 \dot{\theta}_1 \sin\theta_1$$
$$\dot{y}_2 = -l_1 \dot{\theta}_1 \sin\theta_1 - l_2 \dot{\theta}_2 \sin\theta_2$$
and $$\dot{x}_1 = l_1 \dot{\theta}_1 \cos\theta_1$$
$$\dot{x}_2 = l_1 \dot{\theta}_1 \cos\theta + l_2 \dot{\theta}_2 \cos\theta_2.$$

The KE of the system is

$$T = \tfrac{1}{2} m_1 \left(\dot{x}_1^2 + \dot{y}_1^2\right) + \tfrac{1}{2} m_2 \left(\dot{x}_2^2 + \dot{y}_2^2\right)$$
$$= \tfrac{1}{2} m_1 l_1^2 \dot{\theta}_1^2 + \tfrac{1}{2} m_2 \left[l_1^2 \dot{\theta}_1^2 + l_2^2 \dot{\theta}_2^2 + 2 l_1 l_2 \dot{\theta}_1 \dot{\theta}_2 \cos(\theta_1 - \theta_2)\right].$$

Taking the plane passing through a point at a distance $(l_1 + l_2)$ from O, the PE is

$$V = m_1 g (l_1 + l_2 - l_1 \cos\theta_1) + m_2 g \left[l_1 + l_2 - (l_1 \cos\theta_1 + l_2 \cos\theta)\right].$$

The Lagrangian of the system is

$$L = T - V$$
$$= \tfrac{1}{2} m_1 l_1^2 \dot{\theta}_1^2 + \tfrac{1}{2} m_2 \left[l_1^2 \dot{\theta}_1^2 + l_2^2 \dot{\theta}_2^2 + 2 l_1 l_2 \dot{\theta}_1 \dot{\theta}_2 \cos(\theta_1 - \theta_2)\right] - m_1 g \left[l_1 + l_2 - l_1 \cos\theta_1\right]$$
$$- m_2 g \left[l_1 + l_2 - (l_1 \cos\theta_1 + l_2 \cos\theta_2)\right]. \qquad (1)$$

$$\therefore\ \frac{\partial L}{\partial \theta_1} = -m_2 l_1 l_2 \dot{\theta}_1 \dot{\theta}_2 \sin(\theta_1 - \theta_2) - m_1 g l_1 \sin\theta_1 - m_2 g l_1 \sin\theta_1$$
$$= -m_2 l_1 l_2 \dot{\theta}_1 \dot{\theta}_2 \sin(\theta_1 - \theta_2) - (m_1 + m_2) g l_1 \sin\theta_1.$$

$$\frac{\partial L}{\partial \dot{\theta}_1} = m_1 l_1^2 \dot{\theta}_1^2 + m_2 l_1^2 \dot{\theta}_1 + m_2 l_1 l_2 \dot{\theta}_2 \cos(\theta_1 - \theta_2)$$
$$= (m_1 + m_2) l_1^2 \dot{\theta}_1 + m_2 l_1 l_2 \dot{\theta}_2 \cos(\theta_1 - \theta_2).$$

Now the Lagrange's equation in θ_1 is

$$\frac{d}{dt}\left(\frac{\partial L}{\partial \dot{\theta}_1}\right) - \frac{\partial L}{\partial \theta_1} = 0$$
$$= (m_1 + m_2) l_1^2 \ddot{\theta}_1 + m_2 l_1 l_2 \ddot{\theta}_2 \cos(\theta_1 - \theta_2) - m_2 l_1 l_2 \dot{\theta}_2 \sin(\theta_1 - \theta_2)\left(\dot{\theta}_1 - \dot{\theta}_2\right)$$
$$+ m_2 l_1 l_2 \dot{\theta}_1 \dot{\theta}_2 \sin(\theta_1 - \theta_2) + m_1 g l_1 \sin\theta_1 + m_2 g l_1 \sin\theta_1 = 0$$

or $$(m_1 + m_2) l_1^2 \dot{\theta}_1 + m_2 l_1 l_2 \ddot{\theta}_2 \cos(\theta_1 - \theta_2) + m_2 l_1 l_2 \dot{\theta}_2^2 \sin(\theta_1 - \theta_2)$$
$$+ (m_1 + m_2) g l_1 \sin\theta_1 = 0. \qquad (2)$$

The Lagrange's equation in θ_2 is
$$\frac{d}{dt}\left(\frac{\partial L}{\partial \dot{\theta}_2}\right)-\frac{\partial L}{\partial \theta_2}=0.$$

Now from (1),
$$\frac{\partial L}{\partial \theta_2}=m_2 l_1 l_2 \dot{\theta}_1 \dot{\theta}_2 \sin(\theta_1-\theta_2)-m_2 g l_2 \sin\theta_2$$
and
$$\frac{\partial L}{\partial \dot{\theta}_2}=m_2 l_2^2 \dot{\theta}_2+m_2 l_1 l_2 \dot{\theta}_1 \cos(\theta_1-\theta_2).$$
$$\therefore \frac{d}{dt}\left[m_2 l_2^2 \dot{\theta}_2+m_2 l_1 l_2 \dot{\theta}_1 \cos(\theta_1-\theta_2)\right]-m_2 l_1 l_2 \dot{\theta}_1 \dot{\theta}_2 \sin(\theta_1-\theta_2)+m_2 g l_2 \sin\theta_2=0$$
$$\text{or } m_2 l_2^2 \ddot{\theta}_2+m_2 l_1 l_2 \ddot{\theta}_1 \cos(\theta_1-\theta_2)-m_2 l_1 \dot{\theta}_1 \sin(\theta_1-\theta_2)\left(\dot{\theta}_1-\dot{\theta}_2\right)$$
$$-m_2 l_1 l_2 \dot{\theta}_1 \dot{\theta}_2 \sin(\theta_1-\theta_2)+m_1 g l_2 \sin\theta_2=0$$
$$\text{or } m_2 l_2^2 \ddot{\theta}_1-m_2 l_1 l_2 \ddot{\theta}_1 \cos(\theta_1-\theta_2)-m_2 l_1 l_2 \dot{\theta}_1^2 \sin(\theta_1-\theta_2)+m_2 g l_2 \sin\theta_2=0. \quad (3)$$

If $m_1=m_2, l_1=l_2$, then (2) and (3) take the form,
$$2l\ddot{\theta}_1+l\ddot{\theta}_2\cos(\theta_1-\theta_2)+l\dot{\theta}_2^2\sin(\theta_1-\theta_2)=-2g\sin\theta_1 \quad (4)$$
$$l\ddot{\theta}_1\cos(\theta_1-\theta_2)+l\ddot{\theta}_2-l\dot{\theta}_1^2\sin(\theta_1-\theta_2)=-g\sin\theta_2. \quad (5)$$

If now $\sin\theta\simeq\theta$ and $\cos\theta\simeq 1$, when θ is small, and neglecting terms containing $\dot{\theta}^2$ in (4) and (5), we get
$$\mathbf{2l\ddot{\theta}_1+l\ddot{\theta}_2=-2g\theta_1 \text{ and } l\ddot{\theta}_1+l\ddot{\theta}_2=-g\theta_2.}$$

1.19 Construct the Hamiltonian and use Hamilton's canonical equations to obtain the equation of motion of a simple pendulum.

[BUH 1996; WBUT 2006]

Solution. Referring to Fig. 1.15, if m be the mass of the bob and l, the effective length and θ, the angle of small oscillation, then the KE is $T=\frac{1}{2}ml^2\dot{\theta}^2$ and the PE, $V=mgl(1-\cos\theta)$.

$\therefore$ the Lagrangian
$$L=T-V=\tfrac{1}{2}ml^2\dot{\theta}^2-mgl(1-\cos\theta). \quad (1)$$

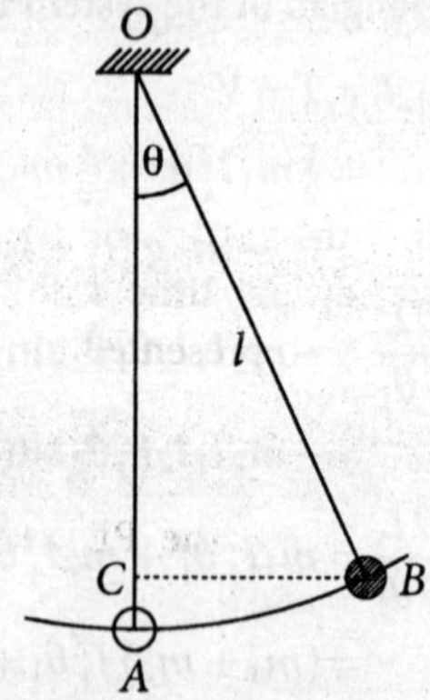

Fig. 1.15: Motion of a simple pendulum.

Now,
$$p_\theta=\frac{\partial L}{\partial\dot{\theta}}=ml^2\dot{\theta}. \quad (2)$$

The Hamiltonian function
$$H=\sum_k p_k\dot{q}_k-L=p_\theta\dot{\theta}-L$$
$$=ml^2\dot{\theta}^2-\left[\tfrac{1}{2}ml^2\dot{\theta}^2-mgl(1-\cos\theta)\right]=\tfrac{1}{2}ml^2\dot{\theta}^2+mgl(1-\cos\theta) \quad (3)$$
$$=T+V=\text{total energy.}$$

From (2),

$$\dot{\theta} = \frac{p_\theta}{ml^2}.$$

Then from (3), we get

$$H = \frac{1}{2}ml^2\left(\frac{p_\theta}{ml^2}\right)^2 + mgl(1-\cos\theta) = \frac{p_\theta^2}{2ml^2} + mgl(1-\cos\theta) \tag{4}$$

$$\therefore \frac{\partial H}{\partial p_\theta} = \frac{p_\theta}{ml^2}, \frac{\partial H}{\partial \theta} = mgl\sin\theta. \tag{5}$$

Now, the Hamilton's equations for $\dot{\theta}$ and $\dot{p}_\theta$ are

$$\dot{\theta} = \frac{\partial H}{\partial p_\theta},\ \dot{p}_\theta = -\frac{\partial H}{\partial \theta}. \tag{6}$$

Here,

$$\dot{\theta} = \frac{p_\theta}{ml^2} \text{ and } \dot{p}_\theta = -mgl\sin\theta. \tag{7}$$

(6) and (7) are the Hamilton's equations for a simple pendulum.
Again, $\because\ p_\theta = ml^2\dot{\theta}$ [from (6)] and $\dot{p}_\theta = ml^2\ddot{\theta}$,
$\therefore$ from (7), we get

$$ml^2\ddot{\theta} = -mgl\sin\theta \text{ or } \ddot{\boldsymbol{\theta}} + \frac{\boldsymbol{g}}{\boldsymbol{l}}\sin\boldsymbol{\theta} = \mathbf{0}. \tag{8}$$

(8) is the equation of motion of a simple pendulum representing a SHM with time period

$$T = 2\pi\sqrt{l/g}.$$

1.20 A particle of mass m is projected with initial velocity u at an angle θ with the horizontal. Use Lagrange's equations to describe the motion of the particle [Neglect air resistance to the motion].

Solution. If the particle be projected at time $t = 0$, then let at any time $t > 0$ the position of the particle be represented by $P(x, y)$ in the XY-plane.

The KE of the particle at any time t is $T = \frac{1}{2}m(\dot{x}^2 + \dot{y}^2)$ and the PE with reference to the X-axis is $V = mgy$.

Then the Lagrangian L is given by

$$L = \frac{1}{2}m(\dot{x}^2 + \dot{y}^2) - mgy. \tag{1}$$

$$\therefore \frac{\partial L}{\partial \dot{x}} = m\dot{x}, \frac{\partial L}{\partial x} = 0, \frac{\partial L}{\partial \dot{y}} = m\dot{y}, \frac{\partial L}{\partial y} = -mg.$$

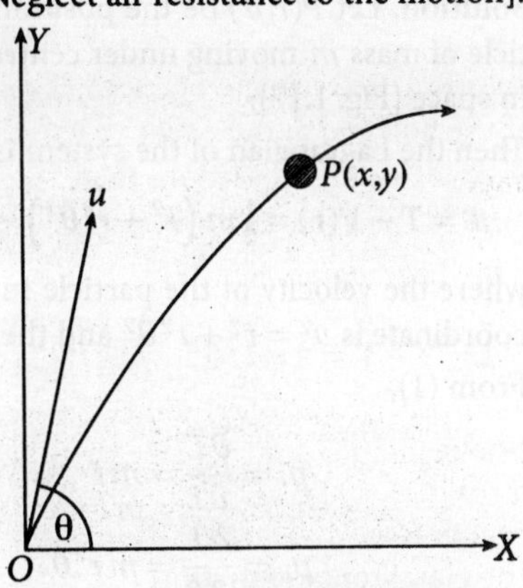

Fig. 1.16: Motion of a projectile.

Then the Lagrange's equations become

$$\frac{d}{dt}\left(\frac{\partial L}{\partial \dot{x}}\right) - \frac{\partial L}{\partial x} = 0,\ \frac{d}{dt}\left(\frac{\partial L}{\partial \dot{y}}\right) - \frac{\partial L}{\partial y} = 0 \text{ or } \ddot{x} = 0, \ddot{y} + g = 0.$$

Integrating w.r.t. t, we get

$$x = At + B \tag{2}$$

$$\text{and } y = -\tfrac{1}{2}gt^2 + ct + D, \tag{3}$$

where A, B, C and D are the constants of integration whose values are to be found out from the condition of the problem.

At $t = 0$,

$$x = 0,\ \dot{x} = u\cos\theta$$

$$y = 0,\ \dot{y} = u\sin\theta.$$

$\therefore$ from (2),

$$x = (u\cos\theta)t \tag{4}$$

$$y = (u\sin\theta)t - \tfrac{1}{2}gt^2. \tag{5}$$

From (4),

$$t = \frac{x}{u\cos\theta}.$$

$\therefore$ from (5),

$$y = u\sin\theta\frac{x}{u\cos\theta} - \frac{1}{2}g\left(\frac{x}{u\cos\theta}\right)^2 = x\tan\theta - \frac{1}{2}g\frac{x^2}{u^2\cos^2\theta}. \tag{6}$$

(6) is the equation of a parabola in the XY-plane. So, the path followed by the projectile is a *parabola*.

1.21 Write down the Hamiltonian and Hamilton's equation of motion for a particle in central force field in space.

Solution. Let $P(r, \theta)$ be the position of the particle of mass m moving under central force field in space [Fig. 1.17].

Then the Lagrangian of the system is

$$L = T - V(r) = \tfrac{1}{2}m\left(\dot{r}^2 + r^2\dot{\theta}^2\right) - V(r), \tag{1}$$

where the velocity of the particle in plane polar coordinate is $v^2 = \dot{r}^2 + r^2\dot{\theta}^2$ and the PE is $V(r)$.

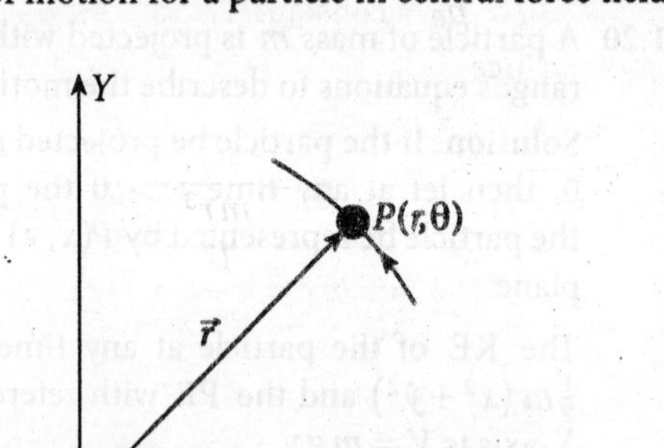

Fig. 1.17: Motion under central force.

From (1),

$$p_r = \frac{\partial L}{\partial \dot{r}} = m\dot{r} \tag{2}$$

$$p_\theta = \frac{\partial L}{\partial \dot{\theta}} = mr^2\dot{\theta}. \tag{3}$$

$\therefore$ the Hamiltonian function

$$H = \sum_k p_k\dot{q}_k - L = p_r\dot{r} + p_\theta\dot{\theta} - L = (m\dot{r})\dot{r} + \left(mr^2\dot{\theta}\right)\dot{\theta} - L$$

$$= m\dot{r}^2 + mr^2\dot{\theta}^2 - \tfrac{1}{2}m\left(\dot{r}^2 + r^2\dot{\theta}^2\right) + V(r) = \tfrac{1}{2}m\left(\dot{r}^2 + r^2\dot{\theta}^2\right) + V(r) = T + V(r). \tag{4}$$

From (2) and (3), we have

$$\dot{r}=\frac{p_r}{m} \text{ and } \dot{\theta}=\frac{p_\theta}{mr^2}.$$

$$\therefore\ H=\frac{1}{2}m\left[\left(\frac{p_r}{m}\right)^2+r^2\left(\frac{p_\theta}{mr^2}\right)^2\right]+V(r)=\frac{1}{2m}\left(p_r^2+\frac{p_\theta^2}{r^2}\right)+V(r). \quad (5)$$

(5) represents the total energy of the system in terms of p_r and p_θ.

The Hamiltonian equations for generalised coordinates q_k and p_k are

$$\dot{q}_k=\frac{\partial H}{\partial p_k} \text{ and } -\dot{q}_k=\frac{\partial H}{\partial q_k}.$$

Here the generalised coordinates are r and θ.

$\therefore$ the Hamilton's equations for r and θ are

$$\dot{r}=\frac{\partial H}{\partial p_r}=\frac{p_r}{m} \quad (6)$$

$$\text{and } \dot{\theta}=\frac{\partial H}{\partial p_\theta}=\frac{p_\theta}{mr^2}. \quad (7)$$

The two equations for p_r and p_θ are

$$\dot{p}_r=-\frac{\partial H}{\partial r}=\frac{p_\theta^2}{mr^2}-\frac{\partial V}{\partial r} \quad (8)$$

$$\text{and } \dot{p}_\theta=-\frac{\partial H}{\partial \theta}=0. \quad (9)$$

$\because \dot{p}_\theta=0, \therefore\ p_\theta$ = constant. Hence, *the angular momentum in case of planetary motion is conserved.*

Again, since

$$\dot{p}_r=\frac{p_\theta^2}{mr^3}-\frac{\partial V}{\partial r}=\frac{l^2}{mr^3}-\frac{k}{r^2}\left[\text{where } p_\theta=l \text{ and } \frac{\partial V}{\partial r}=\frac{k}{r^2}\right]$$

$$\text{or } m\ddot{r}=\frac{l^2}{mr^3}+f(r)\left[\because f(r)=\frac{k}{r^2}\right]. \quad (10)$$

We can also show that,

$$\ddot{r}=-\frac{l^2u^2}{m^2}\frac{d^2u}{d\theta^2}\left[\text{where } \frac{1}{r}=u\right]. \quad (11)$$

Substituting (10) in (11), we get

$$m\left(-\frac{l^2u^2}{m^2}\frac{d^2u}{d\theta^2}\right)=\frac{l^2}{mr^3}+f(r)$$

$$\text{or } -\frac{l^2u^2}{m}\cdot\frac{d^2u}{d\theta^2}=\frac{l^2}{mr^3}+f(r) \text{ or } \frac{d^2u}{d\theta^2}+u=-\frac{m}{l^2u^2}f\left(\frac{1}{u}\right). \quad (12)$$

(12), obtained from Lagrangian formulation, is the equation of motion of a particle under central force.

1.22 Set-up the Hamiltonian of a linear harmonic oscillator and find out the equation of motion.

Solution. The system is conservative and the involved constraint is time independent. In such a case the Hamiltonian H represents the total energy.

The Lagrangian

$$L = T - V = \tfrac{1}{2}m\dot{x}^2 - \tfrac{1}{2}kx^2. \tag{1}$$

$$\therefore \frac{\partial L}{\partial \dot{x}} = \frac{\partial T}{\partial \dot{x}} = p = m\dot{x}.$$

So,

$$\dot{x} = p/m. \tag{2}$$

Again, the Hamiltonian

$$H = T + V = \frac{1}{2}m\left(\frac{p}{m}\right)^2 + \frac{1}{2}kx^2 = \frac{p^2}{2m} + \frac{1}{2}kx^2. \tag{3}$$

The Hamilton's equations of motion are

$$\dot{p} = -\frac{\partial H}{\partial x} = -kx \tag{4}$$

$$\dot{x} = \frac{\partial H}{\partial p} = \frac{p}{m}. \tag{5}$$

From (5),

$$\ddot{x} = \frac{\dot{p}}{m} = -\frac{k}{m}x,$$

$$\therefore m\ddot{x} + kx = 0. \tag{6}$$

This shows that the motion is SHM and the time period $T = 2\pi\sqrt{m/k}$.

1.23 A particle of mass m is constrained to move on the surface of a cylinder. The particle is subjected to a force directed towards the origin (mid-point of the cylinder) and is proportional to the distance of the particle from the origin. Construct the Lagrangian, the Hamiltonian and Hamilton's equations of motion. **[CUH 2002]**

Solution. If r be the distance of the particle from the mid-point of the cylinder, then the force acting on it is $F = -kr$, k is a constant.

$$\text{PE } V = -\int \vec{F}\cdot d\vec{r} = \tfrac{1}{2}kr^2 = \tfrac{1}{2}k\left(x^2 + y^2 + z^2\right) = \tfrac{1}{2}k\left(R^2 + z^2\right),$$

where $R = \sqrt{x^2 + y^2}$ is the radius of the cylinder.

$$\text{KE } T = \tfrac{1}{2}m\left(\dot{x}^2 + \dot{y}^2 + \dot{z}^2\right) = \tfrac{1}{2}m\left(\dot{R}^2 + R^2\dot{\phi}^2 + \dot{Z}^2\right) = \tfrac{1}{2}m\left(R^2\dot{\phi}^2 + \dot{z}^2\right)\ \left[\because \dot{R}^2 = 0\right].$$

Now the Lagrangian of the system is

$$L = T - V = \tfrac{1}{2}m\left(R^2\dot{\phi}^2 + \dot{z}^2\right) - \tfrac{1}{2}k\left(R^2 + z^2\right).$$

Again,

$$p_\phi = \frac{\partial L}{\partial \dot{\phi}} = mR^2\dot{\phi} \text{ and } p_z = \frac{\partial L}{\partial \dot{z}} = m\dot{z}.$$

The Hamiltonian function of the system is

$$H=\sum_K p_k\dot{q}_k-L=p_\phi\dot{\phi}+p_z\dot{z}-L=\frac{1}{2}\frac{p_\phi^2}{mR^2}+\frac{p_z^2}{2m}+\frac{1}{2}k\left(R^2+z^2\right).$$

The Hamilton's equations of motion are

$$\dot{p}_\phi=-\frac{\partial H}{\partial \phi}=0\ [\because\ H\neq f(\phi)]$$

$$\dot{\phi}=\frac{\partial H}{\partial p_\phi}=-\frac{p_\phi}{mk^2}$$

$$\dot{p}_z=-\frac{\partial H}{\partial z}=-kz$$

$$\dot{z}=\frac{\partial H}{\partial p_z}=\frac{p_z}{m}.$$

1.24 Obtain the Hamilton's equations for the spring-mass arrangement as shown in Fig. 1.18. [Assume that the spring is unstretched at $x=0$.]

Solution. The KE of the mass m is $T=\frac{1}{2}m\dot{x}^2$, where x is the displacement of m at any time t.

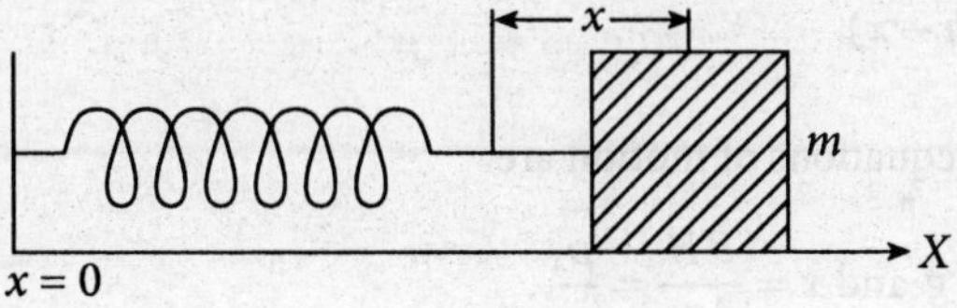

Fig. 1.18: Motion of a spring-mass arrangement.

The PE is $V=\frac{1}{2}kx^2$, where k is the spring constant.

$\therefore$ the Lagrangian of the system is

$$L=T-V=\frac{1}{2}m\dot{x}^2-\frac{1}{2}kx^2. \quad (1)$$

$$\therefore\ p_x=\frac{\partial L}{\partial \dot{x}}=m\dot{x};\ \text{so, } \dot{x}=\frac{p_x}{m}. \quad (2)$$

The Hamiltonian function is

$$H=\sum_k p_k\dot{q}_k-L=p_x\dot{x}-L=p_x\left(\frac{p_x}{m}\right)-\frac{1}{2}m\dot{x}^2+\frac{1}{2}kx^2=\frac{p_x^2}{m}-\frac{1}{2}m\left(\frac{p_x}{m}\right)^2+\frac{1}{2}kx^2. \quad (3)$$

$$\therefore\ H=\frac{p_x^2}{2m}+\frac{1}{2}kx^2. \quad (4)$$

$$\therefore\ \frac{\partial H}{\partial p_x}=\frac{p_x}{m},\ \frac{\partial H}{\partial x}=kx. \quad (5)$$

Now the Hamilton's equations are

$$\dot{x}=\frac{\partial H}{\partial p_x}\ \text{or}\ \dot{x}=\frac{p_x}{m}. \quad (6)$$

$$\dot{p}_x = -\frac{\partial H}{\partial x} \text{ or } \dot{p}_x = -kx. \tag{7}$$

1.25 Derive the Lagrangian and Hamilton's equation of motion for a particle falling freely under the influence of gravity. **[WBUT 2005]**

Solution. We consider that a body of mass m be dropped from the point P at a height h from the ground R. Let it fall to Q from P so that $PQ = x$ under the action of gravity.

Now the PE V at Q is $V = mg(h-x)$ and the KE,

$$\therefore T = \tfrac{1}{2}m\left(\frac{dx}{dt}\right)^2 = \tfrac{1}{2}m\dot{x}^2.$$

$\therefore$ the Lagrangian of the system is

$$L = T - V = \tfrac{1}{2}m\dot{x}^2 - mg(h-x) \text{ and } p_x = \frac{\partial L}{\partial \dot{x}} = m\dot{x} \text{ or } \dot{x} = \frac{p_x}{m}.$$

Hamiltonian,

$$H = p_x\dot{x} - L = \frac{p_x^2}{m} - \left[\frac{1}{2}m\frac{p_x^2}{m^2} - mg(h-x)\right]$$

$$= \frac{p_x^2}{2m} + mg(h-x).$$

So, the Hamilton's equations of motion are

$$\dot{p}_x = -\frac{\partial H}{\partial x} = mg \text{ and } \dot{x} = \frac{\partial H}{\partial p_x} = \frac{p_x}{m}.$$

$$\because p_x = m\dot{x}, \therefore \dot{p}_x = m\ddot{x} \text{ or } mg = m\ddot{x} \; [\because \dot{p}_x = mg]$$

or $\ddot{x} = g$ or $\ddot{x} - g = 0$

is the required equation of motion of the particle falling freely under the influence of gravity.

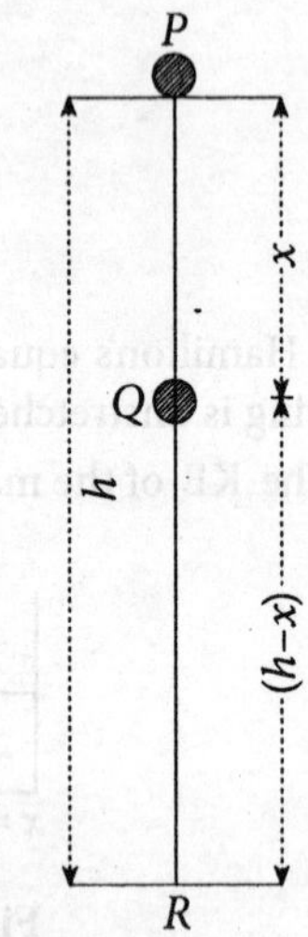

Fig. 1.19: Free fall under gravity.

1.26 Find (a) Lagrangian equation of motion and the (b) Hamilton's equation of motion of a compound pendulum.

Solution. (a) *Lagrangian equation of motion.* We assume that the rigid body oscillates about the horizontal axis through O. When the centre of mass C is displaced to C', then the body rotates through an angle θ [Fig. 1.20].

$$\therefore \text{ KE, } T = \tfrac{1}{2}I\dot{\theta}^2$$

$$\text{PE, } V = -mgl\cos\theta \text{ (w.r.t. 0)},$$

where $OC = l$ and I = moment of inertia of the body about the axis of rotation.

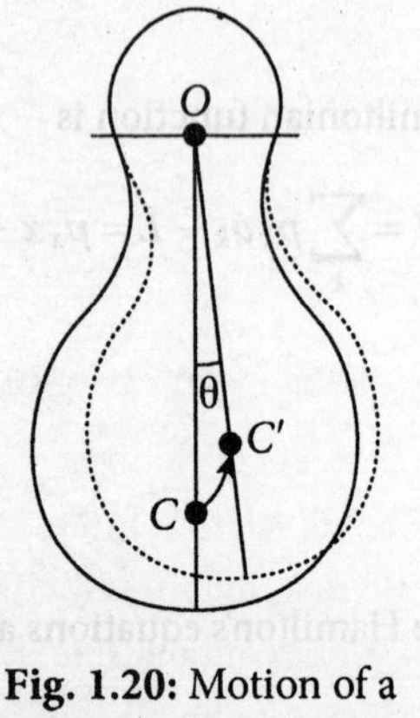

Fig. 1.20: Motion of a compound pendulum.

$$\therefore L = T - V = \tfrac{1}{2}I\dot{\theta}^2 + mgl\cos\theta.$$

Lagrange's equation of motion is

$$\frac{d}{dt}\left(\frac{\partial L}{\partial \dot{\theta}}\right)-\frac{\partial L}{\partial \theta}=0 \text{ or } \frac{d}{dt}\left(I\dot{\theta}\right)+mgl\sin\theta=0 \text{ or } \ddot{\theta}+\frac{mgl}{I}\theta=0. \tag{1}$$

(1) is the Lagrange's equation of motion, which is SHM with time period

$$T=2\pi\sqrt{\frac{I}{mgl}}.$$

(b) ***Hamilton's equations of motion.*** **Here,**

$$p_\theta=\frac{\partial L}{\partial \dot{\theta}}=I\dot{\theta} \text{ or } \dot{\theta}=\frac{p_\theta}{I}.$$

∴ the Hamiltonian

$$H=\sum_k p_k\dot{q}_k-L=p_\theta\dot{\theta}-L=p_\theta\frac{p_\theta}{I}-\frac{1}{2}I\frac{p_\theta^2}{I^2}-mgl\cos\theta=\frac{p_\theta^2}{2I}-mgl\cos\theta.$$

So, the Hamilton's equations of motion are

$$\dot{\theta}=\frac{\partial H}{\partial p_\theta}=\frac{p_\theta}{I} \text{ or } p_\theta=I\dot{\theta} \text{ and } \dot{p}_\theta=-\frac{\partial H}{\partial \theta}=-mgl\sin\theta.$$

Now, $\dot{p}_\theta=-mgl\cos\theta$,

$$\therefore \frac{d}{dt}(p_\theta)=-mgl\sin\theta \text{ or } \frac{d}{dt}\left(I\dot{\theta}\right)+mgl\sin\theta=0 \quad [\because p_\theta=I\dot{\theta}]$$

$$\text{or } \ddot{\theta}+\frac{mgl}{I}\sin\theta=0.$$

Putting $\sin\theta\simeq\theta$,

$$\ddot{\theta}+\frac{mgl}{I}\theta=0. \tag{2}$$

(2) is the Hamilton's equation of motion of the compound pendulum.

Note that, (1) and (2) are the same equation of motion.

EXERCISES

1(A) Objective Type Questions

Choose the correct alternatives.

1.1 A system is scleronomous when constraint equations

(a) do not contain velocity

(b) do not contain time

(c) do not contain velocity and time. **[WBUT 2004]**

1.2 In the rigid body motion the involved constraint is
(a) holonomic
(b) non-holonomic
(c) scleronomic
(d) rheonomic.

1.3 If the constraint relation are or can be made independent of velocity, then the constraint is called
(a) scleronomic
(b) bilateral
(c) conservative
(d) holonomic.

1.4 The constraint involved in the motion of a point mass of a simple pendulum constraint to move at a fixed distance from the point of suspension is
(a) non-holonomic
(b) holonomic
(c) rheonomic
(d) scleronomic.

1.5 When a particle moves along a curve, the involved constraint is
(a) scleronomic
(b) dissipative
(c) bilateral
(d) holonomic.

1.6 When a particle placed on the surface of a sphere moves on the surface, the constraint is
(a) holonomic
(b) non-holonomic
(c) scleronomic
(d) rheonomic.

1.7 The constraint associated with the motion of a pendulum is
(a) holonomic
(b) bilateral
(c) dissipative
(d) rheonomic.

1.8 Constraint involving rolling without sliding
(a) holonomic
(b) non-holonomic
(c) rheonomic
(d) conservative.

1.9 The characteristics of generalised coordinates are that
(a) they are independent of each other
(b) they are necessarily spherical coordinates
(c) they may be Cartesian coordinates
(d) none of these.

1.10 The generalised coordinates in a conservative system may have the dimension of
(a) force
(b) torque
(c) energy
(d) angular momentum.

1.11 The dimensions of generalised momentum
(a) are those of linear momentum
(b) are those of angular momentum
(c) may be those of linear momentum
(d) may be those of angular momentum.

1.12 If the Lagrangian is free of a particular generalised coordinate q_j, then q_j is
(a) superfluous
(b) cyclic
(c) Cartesian
(d) spherical.

1.13 The number of generalised coordinate necessary to describe the motion of a particle in a circular wire of radius a is

(a) one (c) three
(b) two (d) none of these.

1.14 The number of generalised coordinates necessary to describe the motion of a double pendulum is
(a) one (c) three
(b) two (d) none of these.

1.15 If a particular generalised coordinate q_j be absent in the Lagrangian, then the corresponding conserved quantity is
(a) energy (c) velocity
(b) momentum (d) force.

1.16 Virtual displacement $\delta \vec{r}_i$
(a) does not involve time (c) may or may not involve time
(b) may involve time (d) none of these.

1.17 A system consists of three point masses. If the mutual distances between the masses do not vary with time, then the degree of freedom of such system is
(a) zero (c) six
(b) three (d) nine. [WBUT 2006]

1.18 If a given component of the total applied force vanishes, then the corresponding component of linear momentum is
(a) not conserved (c) conserved
(b) zero (d) none of these.

1.19 If the generalised coordinate q_j is cyclic, then the corresponding generalised momentum p_j is
(a) $\dfrac{\partial L}{\partial \dot{q}_j}$ (c) $\dfrac{d}{dt}\left(\dfrac{\partial L}{\partial \dot{q}_j}\right)$
(b) $\dfrac{\partial L}{\partial q_j}$ (d) $\dfrac{d}{dt}\left(\dfrac{\partial L}{\partial q_j}\right)$.

1.20 If a system is conservative, then the Hamiltonian is
(a) $T-V$ (c) $V-T$
(b) $T+V$ (d) $T \div V$.

1.21 The Lagrangian of a system is a function of
(a) $q_j, \dot{q}_j, t$ (c) $\dot{p}_j, q_j, t$
(b) p_j, q_j, t (d) $p_j, \dot{q}_j, t$.

1.22 The Hamiltonian of a system is a function of
(a) q_j, p_j, t (c) $q_j, \dot{q}_j, t$
(b) $p_j, \dot{q}_j, t$ (d) $p_j, \dot{p}_j, t$.

1.23 If q_j is cyclic in the Lagrangian, then in Hamiltonian it is
(a) not cyclic (c) cyclic
(b) may or may not be cyclic (d) unpredictable.

1.24 The Lagrange's equation $\dfrac{d}{dt}\left(\dfrac{\partial L}{\partial \dot{q}_j}\right) - \dfrac{\partial L}{\partial q_j} = 0$ for a conservative system is a set of

(a) $2N$ 2nd order differential equations
(b) N 2nd order differential equations
(c) N first order differential equations
(d) $2N$ first order differential equations.

1.25 The Hamilton's equations of motion of a system consists of
(a) N second order differential equations
(b) N first order differential equations
(c) $2N$ first order differential equations
(d) $2N$ second order differential equations.

1.26 The variables q_i and $\dot{q}_i$ are given equal status in
(a) Lagrangian approach
(b) Hamiltonian approach
(c) Newtonian approach
(d) none of these.

1.27 If the constraints are time-dependent, then the relation between the Hamiltonian H and the total energy E is given by
(a) $H=E$
(b) $H\neq E$
(c) $H=E-V$
(d) $H=E-T$.

1.28 In which of the following groups a rigid body has constraints classified?
(a) Rheonomic and Holonomic
(b) Rheonomic and non-holonomic
(c) Scleronomic and Holonomic
(d) Scleronomic and non-holonomic. [WBUT 2007]

1.29 Generalised coordinate can be of any dimension. However,
(a) the product of generalised coordinate and generalised momentum is always ml^2/t
(b) dimension of the work done depends on the dimension of the generalised coordinate
(c) the generalised momentum is always of the dimension ml/t^2
(d) the generalised force is always of the dimension ml/t^2. [WBUT 2007]

ANSWERS

1.1 (b)	1.6 (b)	1.11 (c),(d)	1.16 (a)	1.21 (a)	1.26 (b)
1.2 (a),(c)	1.7 (a),(b),(c),(d)	1.12 (b)	1.17 (c)	1.22 (a)	1.27 (b)
1.3 (d)	1.8 (b)	1.13 (a)	1.18 (d)	1.23 (c)	1.28 (c)
1.4 (b)	1.9 (a),(c)	1.14 (b)	1.19 (a)	1.24 (b)	1.29 (a).
1.5 (d)	1.10 (b)	1.15 (b)	1.20 (b)	1.25 (c)	

1(B) Short Answer Type Questions

1.1 What are constraints. Mention its different types with examples. [BUH 1999; WBUT 2004]

1.2 Distinguish between holonomic and non-holonomic constraints. [WBUT 2006; BUH 2002; CUH 1997]

1.3 Distinguish scleronomous and rheonomous constraints with example. [BUH 2002]

1.4 Explain the following terms with examples: Virtual work, cyclic coordinates. [BUH 2000]

1.5 What is the advantage of using *generalised coordinates*? Obtain the expression of *generalised force*. [WBUT 2004; CUH 2003]

1.6 Write down the equation of constraint in each of the following cases:
(a) a double pendulum
(b) a particle constraint to move on the surface of the sphere
(c) a disc rolling down an inclined plane

(d) a simple pendulum with rigid support.

Also specify the nature of constraint and enumerate the degrees of freedom in each of the above cases. [BUH 1997]

1.7 State the Principle of virtual work.

1.8 State the principle of D'Alembert. [CUH 2002]

1.9 What are the advantages of Lagrange's equation over Newton's equation of motion? [BUH 2003; CUH 2002]

1.10 Prove that a coordinate cyclic in L is also cyclic in H, where L and H have their usual meanings. [BUH 1998]

1.11 Show that the constraints in a rigid body motion are conservative.

1.12 What are the advantages of Hamiltonian approach over Lagrangian approach?

1.13 What are superfluous coordinates?

1.14 Discuss the limitations of Newtonian approach in solving dynamical problems.

1.15 What are the basic characteristics of Lagrangian formulation?

1.16 Write down the Lagrangian for a cylinder rolling down an inclined plane.

1.17 Write down the Hamiltonian for a particle moving in a central force field.

1(C) Numerical Problems

1.1 Use Lagrange's equations of motion to determine the motion of a mass m sliding without friction down an inclined plane.

[Hints. The vertical height described by the mass m is $h = x\sin\alpha$.

KE, $T = \frac{1}{2}m\dot{x}^2$ and PE, $V = -mgh$ or $V = -mgx\sin\alpha$.

[taking $V = 0$ at A]

$\therefore\ L = T - V = \frac{1}{2}m\dot{x}^2 + mgx\sin\alpha$.

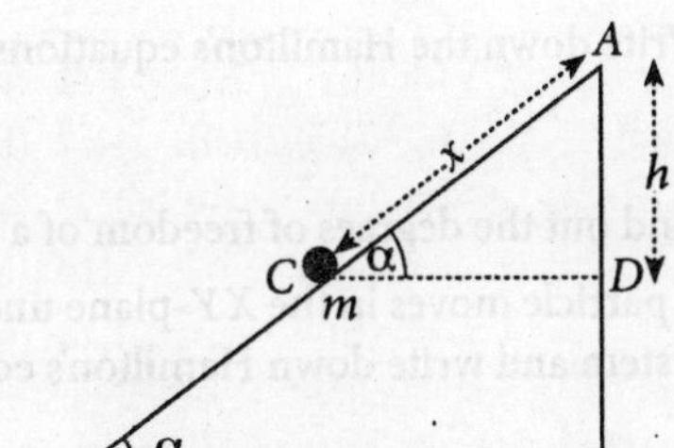

Fig. 1.21: Sliding of mass m on an inclined plane without friction.

$\therefore$ the Lagrange's equation of motion is

$$\frac{d}{dt}\left(\frac{\partial L}{\partial \dot{q}_k}\right) - \frac{\partial L}{\partial q_k} = 0 \text{ or } \frac{d}{dt}\left(\frac{\partial L}{\partial \dot{x}}\right) - \frac{\partial L}{\partial x} = 0$$

or $\frac{d}{dt}(m\dot{x}) - mg\sin\alpha = 0$

or $m\ddot{x} - mg\sin\alpha$ or $\ddot{x} - g\sin\alpha = 0$].

1.2 An inverted pendulum consists of a mass m supported by a rigid massless rod of length l. The pivot O has a vertical motion given by $Z = A\sin\omega t$. Obtain the Lagrange's equation of motion.

(*Ans.* $ml^2\ddot{\theta} + mlA\omega^2\sin\omega t\sin\theta - mgl\sin\theta = 0$.)

1.3 Use Lagrange's equations to set up the equations of motion of a dumb-bell in a vertical plane.

[Hints.

$$(m_1+m_2)\ddot{x}_1 - m_2 l\ddot{\theta}\sin\theta - m_2 l\dot{\theta}^2\cos\theta = 0$$

$$(m_1+m_2)\ddot{y}_1 + m_2 l\ddot{\theta}\cos\theta - m_2 l\dot{\theta}^2\sin\theta + (m_1+m_2)\,g = 0$$

$$m_2 l^2\ddot{\theta} - m_2 l\ddot{x}_1\sin\theta - m_2 l\dot{x}_1\dot{\theta}\cos\theta + m_2 l\ddot{y}_1\cos\theta - m_2 l\dot{y}_1\dot{\theta}\sin\theta + m_2 g l\cos\theta = 0.]$$

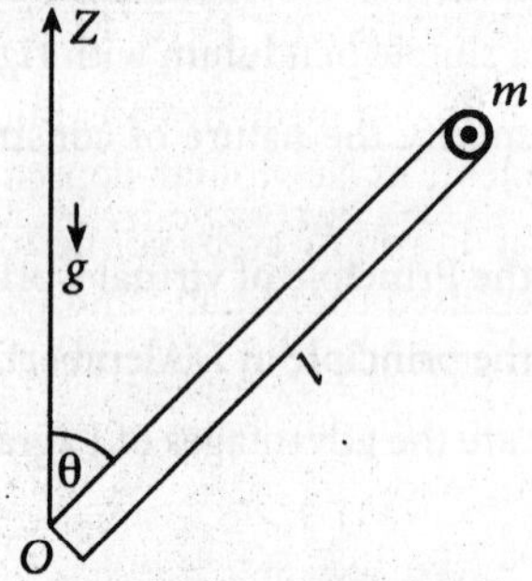

Fig. 1.22: Motion of an inverted pendulum.

1.4 Use Lagrange's equations to find the equations of motion of a compound pendulum which oscillates in a vertical plane about a fixed horizontal axis. Hence find the period of oscillation of the compound pendulum. (*Ans.* $\ddot{\theta}+(gh/k^2)\,\theta=0$; $T=2\pi\sqrt{k^2/gh}$.)

1.5 Use Lagrange's equation of motion to show that the time period of oscillation of a compound pendulum is $T=2\pi\sqrt{\frac{k^2+h^2}{gh}}$, where k is the radius of gyration, h is the distance of the centre of gravity of the body from the point of suspension and g is the acceleration due to gravity.

1.6 A particle of mass m moves in a two-dimensional potential $V(x,y)=-(k_1x^2+k_2y^2)$. Write down the Lagrange's equation of motion.

1.7 The bob of a simple pendulum moves in a horizontal circle. Find the angular frequency of the circular motion in terms of angle θ and the length l of the rod. (*Ans.* $\omega=\sqrt{\frac{g}{l\cos\theta}}$)

1.8 Write the Hamiltonian and equation of motion of a simple pendulum.

(*Ans.* $H=\frac{p_\theta^2}{2ml^2}+mgl(1-\cos\theta)$, $\ddot{\theta}+\frac{g}{l}\sin\theta=0$)

1.9 Write down the Hamilton's equations of motion of a compound pendulum.

(*Ans.* $H=\frac{p_\theta^2}{2I}-mgh\cos\theta$, $\ddot{\theta}+\frac{gh}{k^2}\sin\theta=0$.)

1.10 Find out the degrees of freedom of a CH_4-molecule.

1.11 A particle moves in the XY-plane under the action of a central force. Set up the Hamiltonian for the system and write down Hamilton's equations of motion.

(*Ans.* $H=\frac{1}{2m}\left(p_r^2+\frac{p_\theta^2}{r^2}\right)+V(r)$; $\dot{p}_r=\frac{p_\theta^2}{mr^3}-\frac{\partial V}{\partial r}$, $\dot{p}_\theta=0$.)

1.12 Find out the Hamiltonian if the Lagrangian of a system is given by

$$L(\dot{q},q,t)=\tfrac{1}{2}G(q,t)\dot{q}^2+F(q,t)\dot{q}^2+F(q,t)\dot{q}-V(q,t).$$ [BUH 1998]

1.13 The Lagrangian of a system is given by $L=\frac{1}{2}\dot{x}^2-\frac{1}{2}\omega^2x^2-\alpha x^3+\beta x\dot{x}^2$. Obtain Hamiltonian and hence the canonical equation. [BUH 2003; CUH 1995]

1.14 The Lagrangian of a particle of mass m in one dimension is given by $L=\frac{1}{2}m\,(\dot{x}^2-\omega^2x^2)\,e^{bt}$.

(a) Obtain the canonical momentum and the equation of motion. Is the Hamiltonian a constant of motion?

(b) What is the relation between Lagrangian and Hamiltonian of a system? [BUH 2003]

1(D) Broad Answer Type Questions

1.1 What do you mean by constraint and constrained force? How do they introduce difficulty in solving problems in Newtonian approach? How are these difficulties removed in Lagrangian formulation?

1.2 What do you mean by configuration space and degrees of freedom of a dynamical system? What are generalised coordinates and what are their advantages? Write down the equation of transformation from generalised coordinates to Cartesian coordinates.

1.3 Explain different types of constraints with suitable examples. State the nature of constraint in each of the following cases:

(a) A particle constraint to move along a curve.

(b) The motion of a particle inside the sphere.

(c) An expanding or contracting spherical container of gas.

(d) A particle sliding down an inclined plane under the influence of gravity.

(e) A bead moving on a circular wire.

1.4 (a) What is virtual displacement? How does it differ from real displacement?

(b) Derive the principle of (i) Virtual work and (ii) D'Alembert's principle.

1.5 Obtain Lagrange's equation of motion from D'Alembert's principle. Discuss the effect of dissipative forces.

1.6 (a) What do you mean by (i) cyclic coordinates and (ii) superfluous coordinates?

(b) From the Lagrangian obtain the Hamiltonian. Is the Hamiltonian equal to total energy? Is this equality of the Hamiltonian and total energy valid in general? [CUH 1996]

1.7 The KE of an electron of charge e moving in the field of the nucleus of charge $+Ze$ in spherical polar coordinates is given KE by $T=\frac{1}{2}m\left(r^2+r^2\dot{\theta}^2+r^2\sin^2\theta\dot{\phi}^2\right)$ and PE V by $V=\frac{Ze^2}{r}$. Obtain Lagrange's equation of motion. What is the Hamiltonian of the system? [BUH 2002]

1.8 Establish Lagrange's equation of motion for conservative holonomic system using D'Alembert's principle. [BUH 2001; CUH 2002]

1.9 Obtain Hamilton's canonical equations using the relations between the Hamiltonian and the Lagrangian of a system. [BUH 2000]

1.10 What is holonomic system? Show that if the Lagrangian function L does not contain time explicitly, then the energy is conserved. What are cyclic coordinates.

1.11 The force on a particle of mass m and charge e, moving with a velocity $\vec{v}$ in an electric field $\vec{E}$ and magnetic field $\vec{B}$ is given by $\vec{F}=e\left[\vec{E}+\frac{\vec{v}\times\vec{B}}{c}\right]$, where c is the velocity of light.

The fields are expressed by the relations $\vec{E}=-\Delta\phi-\frac{\partial\vec{A}}{\partial t}$, $\vec{B}=\Delta\times\vec{A}$, ϕ and $\vec{A}$ being scalar and vector potentials respectively. Prove that the Lagrangian defining the motion of such a system is $L=\frac{1}{2}mv^2+e\left(\vec{A}\cdot\vec{v}\right)-e\phi$.

1.12 (a) If a proton is 1800 times heavier than an electron, then show that the centre of mass of hydrogen atom is located practically at proton. Find the reduced mass of this system and write the Lagrangian of the reduced mass.

(b) Show that if a coordinate is cyclic in Lagrangian it will be cyclic in Hamiltonian also.

(c) Two objects of masses 5 kg and 10 kg moving at an angle of 60° w.r.t. each other, symmetric to an axis on a frictionless surface with velocities 20 m/s and 10 m/s respectively, collide and diverge out at angles symmetric to the above axis. The velocity of the first body is 8 m/s after the collision, find the velocity of the second body after collision in the lab frame as well as centre mass frame. Is the collision elastic?

(d) Give example of system with non-holonomic and rheonomic (one each) constraints. **[WBUT 2007]**

1.13 (a) Find out Hamilton's equations of mass for a system comprising of masses m_1 and m_2 connector by a massless spring of length L through a frictionless pulley such that $m_1 > m_2$.

(b) Find out the degrees of freedom of a rigid body constrained to move along the equator and remain on the surface of the earth. **[WBUT 2007]**

Chapter 2

Quantum Mechanics

2.1 Introduction

Classically, a wave equation is a *second order differential equation in space and time coordinates*, the solution of which represents a wave in a medium. For example, the equation

$$\frac{\partial^2 E_y}{\partial x^2} = \frac{1}{c^2}\frac{\partial^2 E_y}{\partial t^2} \tag{2.1}$$

represents an *electromagnetic wave* travelling along the x-direction. E_y is the y-component of the electric intensity.

According to *de Broglie*, a material particle of mass m moving with velocity $\vec{v}$ is associated with a wave whose wavelength is $\lambda = \frac{h}{mv}$. Since the particle has wave aspect, there should be some sort of wave equation whose solution describes the *space-time behaviour* of the particle. *There is, however, no way of derivation of the wave equation from the first principle.* It can, however, be developed on the basis of knowledge of *classical physics. Erwin Schrödinger,* an Austrian theoretical physicist, formulated the wave equation using *de Broglie's concept* of wave properties of matter. This equation is known as *Schrödinger Wave equation*, which plays the same role in *quantum mechanics* as *Newton's second law* does in *classical mechanics*.

2.2 Basic Postulates of Quantum Mechanics

Following are the basic postulates of *Quantum mechanics:*

1. For a moving particle the same *space-time behaviour* can be completely described by a wave function $\psi(\vec{r}, t)$, consistent with the *Heisenberg's uncertainty principle.*
2. Dynamical variables, viz., position vector ($\vec{r}$), linear moments ($\vec{p}$) angular moment $(\vec{L})$, etc. describing the motion of the particle can be represented by mathematical *operator.* These operators are *linear* and *Hermitian.*
3. The results of measurement of a *dynamical variable a*, are the *eigenvalues of the operator* $\hat{A}$ of the eigenvalue equation
$$\hat{A}\psi_n = a_n\psi_n, \tag{2.2}$$
where ψ_n is the *eigenfunction of the operator* $\hat{A}$ *belonging to eigenvalue* a_n.
4. The probability of finding the particle is
$$\rho = \psi^*\psi = |\psi|^2, \tag{2.3}$$
where ψ^* is the *complex conjugate of* ψ.

The probability of finding the particle in a given volume of space is

$$\int_V \rho \, dV = \int_V |\psi|^2 dV. \tag{2.4}$$

Since the particle must be found somewhere in space, the total probability must be unity, i.e.,

$$\int_{-\infty}^{+\infty} |\psi|^2 dV = 1, \tag{2.5}$$

where the integration is taken all over the space. The wave function satisfying (2.5) is said to be *normalised.*

5. The *expectation value* (or expected average) of the results of a large number of measurements of a physical property α of the particle is given by

$$\langle \alpha \rangle = \int_{-\infty}^{+\infty} \psi^* \hat{\alpha} \psi \, dV. \tag{2.6}$$

$\hat{\alpha}$ is the operator relating to the dynamical variable α, and ψ is normalised.
If ψ is not normalised, then

$$\langle \alpha \rangle = \frac{\int_{-\infty}^{+\infty} \psi^* \hat{\alpha} \psi \, dV}{\int_{-\infty}^{+\infty} \psi^* \psi \, dV}. \tag{2.7}$$

2.3 Schrödinger Equation

1. *One-dimensional time-dependent Schrödinger equation for a free particle.* A subatomic particle (viz., *electron, proton, neutron,* etc.) can be represented by a *wave packet* formed due to the superposition of number of harmonic waves having varied wavelength (λ).
A wave packet having wave number $k = \frac{2\pi}{\lambda}$ of angular frequency ω moving along the positive X-direction can be represented by

$$\psi(x,t) = Ae^{i(kx-\omega t)}, \tag{2.8}$$

where A is the amplitude.
According to *de Broglie's hypothesis* the momentum (p) of the particle of wavelength λ is

$$p = \frac{h}{\lambda} = \frac{h}{2\pi}\frac{2\pi}{\lambda} = \hbar k. \tag{2.9}$$

The KE (KE) of the free particle of mass m is

$$E = \frac{p^2}{2m} = \frac{\hbar^2 k^2}{2m} = h\nu = \hbar\omega. \tag{2.10}$$

Differentiating (2.8) w.r.t. x, we get

$$\frac{\partial \psi}{\partial x} = A(ik)e^{i(kx-\omega t)} \text{ and } \frac{\partial^2 \psi}{\partial x^2} = -Ak^2 e^{i(kx-\omega t)} \text{ or } \frac{\partial^2 \psi}{\partial x^2} = -k^2\psi. \tag{2.11}$$

Again differentiating (2.8) w.r.t. time, we get

$$\frac{\partial \psi}{\partial t} = -i\omega\psi. \tag{2.12}$$

Now multiplying both sides of (2.11) by $\left(-\frac{\hbar^2}{2m}\right)$, we get

$$\left(-\frac{\hbar^2}{2m}\right)\frac{\partial^2\psi}{\partial x^2} = -\frac{\hbar^2}{2m}(-k^2\psi) = \frac{\hbar^2 k^2}{2m}\psi = \hbar\omega\psi. \tag{2.13}$$

Multiplying both sides of (2.12) by $(i\hbar)$, we get

$$i\hbar\frac{\partial \psi}{\partial t} = i\hbar(-i\omega)\psi = \hbar\omega\psi. \tag{2.14}$$

From (2.13) and (2.14), we get

$$-\frac{\hbar^2}{2m}\frac{\partial^2\psi}{\partial x^2} = i\hbar\frac{\partial\psi}{\partial t}. \tag{2.15}$$

(2.15) represents a *one-dimensional time-dependent Schrödinger equation for a free particle.*

2. *One-dimensional time-dependent Schrödinger equation for a particle acted upon by a force field.* Let a particle of mass m move under the action of a force $F(x)$ along the X-direction. The PE $V(x)$ is a function of x only and is independent of time, i.e., the force field is a *conservative one.*
If p be the momentum of the particle then its total energy

$$E = \frac{p^2}{2m} + V(x) \text{ or } \hbar\omega = \frac{\hbar^2 k^2}{2m} + V(x), \tag{2.16}$$

where $E = \hbar\omega$ and $p = \hbar k$.
Let the wave function representing the particle be $\psi(x,t) = Ae^{i(kx-\omega t)}$ (2.8).

$$\therefore \frac{\partial\psi}{\partial t} = A(-i\omega)e^{i(kx-\omega t)} = (-i\omega)\psi$$

$$\text{or } i\hbar\frac{\partial\psi}{\partial t} = i\hbar(-i\omega)\psi = \hbar\omega\psi \quad [(2.14)]. \tag{2.17}$$

Differentiating (2.8) twice w.r.t. x, we get

$$\frac{\partial\psi}{\partial x} = A(ik)e^{i(kx-\omega t)} \text{ and } \frac{\partial^2\psi}{\partial x^2} = -k^2\psi = (ik)\psi.$$

Multiplying both sides by $\left(-\frac{\hbar^2}{2m}\right)$, we get

$$\left(-\frac{\hbar^2}{2m}\right)\frac{\partial^2\psi}{\partial x^2} = \left(-\frac{\hbar^2}{2m}\right)(-k^2\psi) = \frac{\hbar^2k^2}{2m}\psi$$

$$\text{or } -\left(\frac{\hbar^2}{2m}\right)\frac{\partial^2\psi}{\partial x^2} + V(x)\psi = \left[\frac{\hbar^2k^2}{2m} + V(x)\right]\psi$$

$$\text{or } -\frac{\hbar^2}{2m}\left(\frac{\partial^2\psi}{\partial x^2}\right) + V(x)\psi = \hbar\omega\psi \quad \left[\because \frac{\hbar^2k^2}{2m} + V(x) = \hbar\omega\right]. \tag{2.18}$$

Now equating the left-hand sides of (2.14) and (2.18), we get

$$i\hbar\frac{\partial\psi}{\partial t}=\left[-\frac{\hbar^2}{2m}\frac{\partial^2}{\partial x^2}+V(x)\right]\psi. \tag{2.19}$$

(2.19) represents a *one-dimensional time-dependent Schrödinger equation for a particle moving under a force field* $F(x)$.

Note. If the particle is free, then $V(x)=0$. (2.19) then reduces to the one-dimensional time-dependent Schrödinger equation for a free particle.

3. *Three-dimensional time-dependent Schrödinger equation for a free particle.* Let a particle move in three-dimension whose wave equation is represented as

$$\psi(\vec{r},t)=Ae^{i(\vec{k}\cdot\vec{r}-\omega t)}, \tag{2.20}$$

where

$$\vec{k}\cdot\vec{r}=k_x x+k_y y+k_z z. \tag{2.21}$$

Let us operate ψ by *Laplacian operator* ∇^2, where

$$\nabla^2=\frac{\partial^2}{\partial x^2}+\frac{\partial^2}{\partial y^2}+\frac{\partial^2}{\partial z^2}.$$

Then from (2.20) and (2.21), we get

$$\frac{\partial\psi}{\partial x}=ik_x\psi \text{ and } \frac{\partial^2\psi}{\partial x^2}=(ik_x)^2\psi=-k_x^2\psi.$$

Similarly,

$$\frac{\partial^2\psi}{\partial y^2}=-k_y^2\psi \text{ and } \frac{\partial^2\psi}{\partial z^2}=-k_z^2\psi.$$

Hence

$$\nabla^2\psi=\frac{\partial^2\psi}{\partial x^2}+\frac{\partial^2\psi}{\partial y^2}+\frac{\partial^2\psi}{\partial z^2}=\left(k_x^2+k_y^2+k_z^2\right)\psi=-k^2\psi.$$

$$\therefore -\frac{\hbar^2}{2m}\nabla^2\psi=\left(-\frac{\hbar^2}{2m}\right)\times\left(-k^2\psi\right) \text{ or } -\frac{\hbar^2}{2m}\nabla^2\psi=\frac{\hbar^2k^2}{2m}\psi=\frac{p^2}{2m}\psi. \tag{2.22}$$

Again, differentiating (2.20) w.r.t. time,

$$\frac{\partial\psi}{\partial t}=A(i\omega)e^{i(\vec{k}\cdot\vec{r}-\omega t)}=(-i\omega)\psi \text{ or } i\hbar\frac{\partial\psi}{\partial t}=i\hbar(-i\omega)\psi=\hbar\omega\psi.$$

But

$$\hbar\omega=\frac{\hbar^2k^2}{2m}=\frac{p^2}{2m}, \quad \therefore i\hbar\frac{\partial\psi}{\partial t}=\frac{p^2}{2m}\psi. \tag{2.23}$$

Equating left-hand sides of (2.22) and (2.23), we get

$$i\hbar\frac{\partial\psi}{\partial t}=-\frac{\hbar^2}{2m}\nabla^2\psi. \tag{2.24}$$

(2.24) represents the *three-dimensional time-dependent Schrödinger equation for a free particle.*

4. *Three-dimensional time-dependent Schrödinger equation for a particle acted upon by a force field.* We consider a particle of mass m acted upon by a force field derivable from a potential $V(\vec{r})$. Let the wave equation of the particle be represented by

$$\psi(\vec{r},t)=Ae^{i(\vec{k}\cdot\vec{r}-\omega t)}, \quad [(2.20)]$$

where

$$\vec{k}\cdot\vec{r}=k_x x+k_y y+k_z z. \quad [(2.21)]$$

Total energy of the particle is $E=\frac{p^2}{2m}+V(r)$.

Now operating (2.20) by

$$\nabla^2=\frac{\partial^2}{\partial x^2}+\frac{\partial^2}{\partial y^2}+\frac{\partial^2}{\partial z^2},$$

we have

$$\frac{\partial^2\psi}{\partial x^2}=-k_x^2\psi,\ \frac{\partial^2\psi}{\partial y^2}=-k_y^2\psi \text{ and } \frac{\partial^2\psi}{\partial z^2}=-k_z^2\psi.$$

Hence,

$$\nabla^2\psi=\frac{\partial^2\psi}{\partial x^2}+\frac{\partial^2\psi}{\partial y^2}+\frac{\partial^2\psi}{\partial z^2}=\left(k_x^2+k_y^2+k_z^2\right)\psi=-k^2\psi.$$

$$\therefore -\frac{\hbar^2}{2m}\nabla^2\psi=\left(-\frac{\hbar^2}{2m}\right)\left(-k^2\psi\right)=\frac{\hbar^2k^2}{2m}\psi \text{ or } -\frac{\hbar^2}{2m}\nabla^2\psi=\frac{p^2}{2m}\psi$$

$$\text{or } -\frac{\hbar^2}{2m}\nabla^2\psi+V(r)\psi=\frac{p^2}{2m}\psi+V(r)\psi \text{ or } -\frac{\hbar^2}{2m}\nabla^2\psi+V(r)\psi=\left[\frac{p^2}{2m}+V(r)\right]\psi \quad (2.25)$$

$$=\hbar\omega\psi\ \left[\because\left(\frac{p^2}{2m}+V(r)\right)=\left(\frac{\hbar^2k^2}{2m}+V(r)\right)=\hbar\omega\right]. \quad (2.26)$$

Again, differentiating (2.20) w.r.t. t, we get

$$\frac{\partial\psi}{\partial t}=A(-i\omega)e^{i(\vec{k}\cdot\vec{r}-\omega t)}=(-i\omega)\psi \text{ and } i\hbar\frac{\partial\psi}{\partial t}=\hbar\omega\psi. \quad (2.27)$$

Now, equating the left-hand sides of (2.26) and (2.27), we get

$$\hbar\frac{\partial\psi}{\partial t}=\left[-\frac{\hbar^2}{2m}\nabla^2+V(r)\right]\psi. \quad (2.28)$$

(2.28) represents a *three-dimensional time-dependent Schrödinger equation for a particle acted upon by a force field derivable from potential* $V(\vec{r})$.

If the particle is free, then $V(r)=0$. (2.28) then reduces to

$$i\hbar\frac{\partial\psi}{\partial t}=-\frac{\hbar^2}{2m}\nabla^2\psi, \text{ same as (2.24).}$$

2.3.1 One-dimensional time-independent Schrödinger equation for a free particle ($V=0$)

We have the *one-dimensional time-dependent Schrödinger equation for a free particle* (i.e., $V=0$),

$$-\frac{\hbar^2}{2m}\frac{\partial^2\psi}{\partial x^2}=i\hbar\frac{\partial\psi}{\partial t} \quad [(2.15)].$$

The above equation is dependent on both space (x) and time (t). Now $\psi(x,t)$ can be expressed as the product of two separate functions, $u(x)$ which is a function of x only and $f(t)$ is a function of t only as below.

$$\psi(x,t)=u(x)f(t) \tag{2.29}$$

$$\therefore \frac{\partial\psi}{\partial x}=f(t)\frac{du}{dx}, \frac{\partial^2\psi}{\partial x^2}=f(t)\frac{d^2u}{dx^2} \text{ and } \frac{\partial\psi}{\partial t}=u(x)\frac{df}{dt}. \tag{2.30}$$

Now according to quantum theory the energy E is related to its frequency v by $E=hv$ or, $v=E/h$.

$$\therefore \omega t=2\pi v t=2\pi\frac{E}{h}t=\frac{Et}{\hbar}$$

and momentum $p=\frac{h}{\lambda}$ or $\frac{1}{\lambda}=\frac{p}{h}$.

$$\therefore kx=\frac{2\pi}{\lambda}x=\frac{2\pi p}{h}x=\frac{px}{\hbar}\left(\because \hbar=\frac{h}{2\pi}\right).$$

Using these values of ωt and kx in the wave function $\psi(x,t)$, we get

$$\psi(x,t)=Ae^{i(kx-\omega t)}=Ae^{i\left(\frac{px}{\hbar}-\frac{Et}{\hbar}\right)}=Ae^{\frac{i}{\hbar}(px-Et)}$$

$$\text{and } \frac{\partial\psi}{\partial t}=-\frac{iE}{\hbar}Ae^{\frac{i}{\hbar}(px-Et)}=-\frac{i}{\hbar}E\psi(x,t)=-\frac{i}{\hbar}Eu(x)f(t). \tag{2.31}$$

Substituting (2.30) and (2.31) in (2.15), we get

$$-\frac{\hbar^2}{2m}f(t)\frac{d^2u}{dx^2}=i\hbar\left(-\frac{i}{\hbar}\right)Eu(x)f(t)$$

$$\text{or } \frac{\hbar^2}{2m}\frac{d^2u}{dx^2}=Eu(x) \text{ or } \frac{d^2u(x)}{dx^2}+\frac{2m}{\hbar^2}Eu(x)=0. \tag{2.32}$$

(2.32) represents a *one-dimensional time-independent Schrödinger equation for a free particle.*

2.3.2 One-dimensional time-independent Schrödinger equation for a particle acted upon by a force field ($V\neq 0$)

We have the one-dimensional *time-dependent* Schrödinger equation for a particle moving under a force field derivable from a potential V is

$$-\frac{\hbar^2}{2m}\frac{\partial^2\psi}{\partial x^2}+V(x)\psi=i\hbar\frac{\partial\psi}{\partial t} \quad [(2.19)].$$

If V is not a function of time, then the above equation can be separated into *time-independent part and time-dependent part.* If $u(x)$ and $f(t)$ be the time-independent and time-dependent part of the wave function $\psi(x,t)$ respectively, then we can write:

$$\psi(x,t)=u(x)f(t) \quad [(2.19)].$$

Now

$$\frac{\partial \psi}{\partial x} = f(t)\frac{du}{dx} \text{ and } \frac{\partial^2 \psi}{\partial x^2} = f(t)\frac{d^2u}{dx^2} \text{ and } \frac{\partial \psi}{\partial t} = u(x)\frac{df}{dt} \quad [(2.30)].$$

Now, according to *quantum theory* the energy E is related to frequency ν as

$$E = h\nu, \therefore \nu = \frac{E}{h} \text{ and } \omega t = 2\pi\frac{E}{\hbar}t = \frac{Et}{\hbar}$$

and

$$\because p = \frac{h}{\lambda}, \therefore \frac{1}{\lambda} = \frac{p}{h} \text{ and } kx = \frac{2\pi}{\lambda}x = \frac{px}{\hbar}.$$

Using these values of ωt and kx in the expression of $\psi(x,t)$

$$\psi(x,t) = Ae^{i(kx-\omega t)} = Ae^{i\left(\frac{px}{\hbar}-\frac{Et}{\hbar}\right)} = Ae^{\frac{i}{\hbar}(px-Et)}.$$

$$\therefore \frac{\partial \psi}{\partial t} = -\frac{iE}{\hbar}Ae^{\frac{i}{\hbar}(px-Et)} = -\frac{i}{\hbar}Eu(x)f(t) \quad [(2.31)].$$

Substituting the value of $\frac{\partial^2 \psi}{\partial x^2}$ and $\frac{\partial \psi}{\partial t}$ from (2.30) and (2.31) in (2.19), we get

$$-\frac{\hbar^2}{2m}f(t)\frac{d^2u}{dx^2} + V(x)u(x)f(t) = (i\hbar)\times -\frac{iE}{\hbar}u(x)f(t) \text{ or } \frac{d^2u}{dx^2} + \frac{2m}{\hbar^2}(E-V)u(x) = 0. \quad (2.33)$$

(2.33) represents *one-dimensional time-independent Schrödinger equation for a particle being acted upon a force field derivable from potential $V(x)$.*

2.3.3 Three-dimensional time-independent Schrödinger equation for a free particle ($V = 0$)

The three-dimensional time-dependent Schrödinger equation for a free particle is given by

$$-\frac{\hbar^2}{2m}\nabla^2\psi = i\hbar\frac{\partial \psi}{\partial t} \quad [(2.24)].$$

Let $u(\vec{r})$ and $f(t)$ be respectively the *time-independent* and *time-dependent parts* of the progressive wave equation $\psi(\vec{r},t)$, then

$$\psi(\vec{r},t) = u(\vec{r})\, f(t). \quad (2.34)$$

$$\therefore \frac{\partial \psi}{\partial x} = f(t)\frac{du}{dx} \text{ and } \frac{\partial^2 \psi}{\partial x^2} = f(t)\frac{d^2u}{dx^2}.$$

Similarly,

$$\frac{\partial^2 \psi}{\partial y^2} = f(t)\frac{d^2u}{dy^2} \text{ and } \frac{\partial^2 \psi}{\partial z^2} = f(t)\frac{d^2u}{dz^2}.$$

Now

$$\nabla^2\psi = \frac{\partial^2 \psi}{\partial x^2} + \frac{\partial^2 \psi}{\partial y^2} + \frac{\partial^2 \psi}{\partial z^2} = \left(\frac{d^2u}{dx^2} + \frac{d^2u}{dy^2} + \frac{d^2u}{dz^2}\right) f(t) = f(t)\nabla^2 u. \quad (2.35)$$

Again, $\because$ energy $E = h\nu \therefore \nu = \frac{E}{\lambda}$ and $\omega t = 2\pi\nu t = \frac{Et}{\hbar}$ and momentum $p = \frac{h}{\lambda}$,

$$\therefore \frac{1}{\lambda} = \frac{p}{\lambda} \text{ and } k_x x = \frac{2\pi x}{\lambda} = \frac{2\pi p_x x}{h} = \frac{p_x x}{\hbar}.$$

Similarly,

$$k_y y = \frac{p_y y}{\hbar} \text{ and } k_z z = \frac{p_z z}{\hbar}.$$

Now the wave equation

$$\psi(\vec{r},t) = Ae^{i(\vec{k}\cdot\vec{r}-\omega t)} = Ae^{i\left(\frac{p_x x + p_y y + p_z z}{\hbar} - \frac{Et}{\hbar}\right)} \text{ or } \psi(\vec{r},t) = Ae^{\frac{i}{\hbar}(\vec{p}\cdot\vec{r}-Et)}.$$

$$\therefore \frac{\partial\psi}{\partial t} = A\left(-\frac{iE}{\hbar}\right)e^{\frac{i}{\hbar}(\vec{p}\cdot\vec{r}-Et)} = -\frac{iE}{\hbar}\psi(\vec{r},t) = -\frac{iE}{\hbar}u(\vec{r})\,f(t). \quad (2.36)$$

Now, using the values of $\nabla^2\psi$ and $\frac{\partial\psi}{\partial t}$ from (2.35) and (2.36), we get from (2.24),

$$-\frac{\hbar^2}{2m}f(t)\nabla^2 u = i\hbar\left(-\frac{iE}{\hbar}\right)u(\vec{r})\,f(t) \text{ or } \nabla^2 u(\vec{r}) + \frac{2mE}{\hbar^2}u(\vec{r}) = 0. \quad (2.37)$$

(2.37) represents the *three-dimensional time-indepednent Schrödinger equation for a free particle.*

2.3.4 Three-dimensional time-independent Schrödinger equation for a particle acted upon by a force field derivable from a potential ($V \neq 0$)

The *three-dimensional time-dependent Schrödinger equation for a particle moving under a potential V* is

$$\left[-\frac{\hbar^2}{2m}\nabla^2 + V(r)\right]\psi = i\hbar\frac{\partial\psi}{\partial t} \quad [(2.28)].$$

If the potential V does not explicitly depend on time, then the total energy E is constant. The wave function $\psi(\vec{r},t)$ can be expressed as the product of *time-independent part* $u(x)$ and *time-dependent part* $f(t)$, i.e., $\psi(\vec{r},t) = u(\vec{r})\,f(t)$. Then

$$\frac{\partial\psi}{\partial x} = f(t)\frac{\partial u}{\partial x} \text{ and } \frac{\partial^2\psi}{\partial x^2} = f(t)\frac{\partial^2 u}{\partial x^2}.$$

Similarly,

$$\frac{\partial^2\psi}{\partial y^2} = f(t)\frac{\partial^2 u}{\partial y^2} \text{ and } \frac{\partial^2\psi}{\partial z^2} = f(t)\frac{\partial^2 u}{\partial z^2}.$$

$$\therefore \nabla^2\psi = \frac{\partial^2\psi}{\partial x^2} + \frac{\partial^2\psi}{\partial y^2} + \frac{\partial^2\psi}{\partial z^2} = \left(\frac{\partial^2 u}{\partial x^2} + \frac{\partial^2 u}{\partial y^2} + \frac{\partial^2 u}{\partial z^2}\right)f(t) = \nabla^2 u\,f(t) \quad [(2.35)].$$

Again, the energy of the particle is $E = h\nu$, $\therefore \nu = \frac{E}{h}$.

So, $\omega t = 2\pi\nu t = 2\pi\left(\frac{E}{h}\right)t = \frac{Et}{\hbar}$ and momentum $p = \frac{h}{\lambda}$. $\therefore \frac{1}{\lambda} = \frac{p}{h}$.

So,

$$\cdot k_x x = \frac{2\pi}{\lambda}x = \frac{p_x x}{\hbar},\; k_y y = \frac{p_y y}{\hbar} \text{ and } k_z z = \frac{p_z z}{\hbar}.$$

Now the wave equation representing the particle is

$$\psi(\vec{r},t) = Ae^{i(\vec{k}\cdot\vec{r}-\omega t)} = A^{i\left(\frac{p_x x + p_y y + p_z z}{\hbar} - \frac{Et}{\hbar}\right)} = Ae^{\frac{i}{\hbar}(\vec{p}\cdot\vec{r}-Et)}.$$

$$\therefore \frac{\partial\psi}{\partial t} = -\frac{iE}{\hbar}Ae^{\frac{i}{\hbar}(\vec{p}\cdot\vec{r}-Et)} = -\frac{-iE}{\hbar}\psi(\vec{r},t) = \frac{iE}{\hbar}u(\vec{r})\,f(t) \quad [(2.36)].$$

Now substituting the values of $\nabla^2\psi$ and $\frac{\partial\psi}{\partial t}$ from (2.35) and (2.36) in (2.28), we get

$$-\frac{\hbar^2}{2m}\nabla^2 u f(t)+V(\vec{r})\,u(\vec{r})\,f(t)=i\hbar\left(-\frac{iE}{\hbar}\right)u(\vec{r})\,f(t)$$

$$\text{or } -\frac{\hbar^2}{2m}\nabla^2 u(r)=(E-V)u(r)$$

$$\text{or } \nabla^2 u(\vec{r})+\frac{2m}{\hbar^2}(E-V)\,u(\vec{r})=0. \qquad (2.38)$$

(2.38) represents *the three-dimensional time-independent Schrödinger equation for a particle moving under a potential field which is independent of time.*

2.4 Characteristics of Schrödinger Equation

1. **Schrödinger equation**

$$-\frac{\hbar^2}{2m}\frac{\partial^2\psi}{\partial x^2}=i\hbar\frac{\partial\psi}{\partial t}, \text{ (for } V=0)$$

contains ψ and its derivatives only linear, i.e., second or higher powers of these quantities do not appear. Consequently not only harmonic waves but a linear combination of these are the solutions of Schrödinger equation.

If $\psi_1(x,t)$ and $\psi_2(x,t)$ are two harmonic functions, then $\psi(x,t)=\psi_1(x,t)+\psi_2(x,t)$ is also a solution of the Schrödinger equation. So the superposition principle of the wave function, holds good.

The equation contains *first order time derivative* and *second order space derivative* terms linearly.

Classical wave equation

$$\frac{\partial^2 y(x,t)}{\partial t^2}=c^2\frac{\partial^2 y(x,t)}{\partial x^2}$$

contains second derivatives of $y(x,t)$ w.r.t. x and t only linearly and the solution $y(x,t)$ obeys superposition principle.

2. **Schrödinger equation contains imaginary unit i but the classical wave equation does not contain i. This differentiates Schrödinger equation from classical wave equation.**
3. **The wave function ψ must be consistent with de Broglie's hypothesis and correspondence principle.**
4. **The solution of Schrödinger equation is a complex function representing matter wave but the solution $y(x,t)$ of classical equation is always real.**

The above points shows the distinctions between *Schrödinger wave equation* and *classical wave equation*.

Table 2.1: Schrödinger Equations

Dimension	*Time-Dependent*	*Time-Independent*
One-Dimensional	$\left(\frac{\hbar^2}{2m}\right)\frac{\partial^2\psi}{\partial x^2}=i\hbar\frac{\partial\psi}{\partial t}$ (when $V=0$)	$\frac{d^2u(x)}{dx^2}+\frac{2mE}{\hbar^2}u(x)=0$ (when $V=0$)
	$-\left[\frac{\hbar^2k^2}{2m}+V(x)\right]\psi=i\hbar\frac{\partial\psi}{\partial t}$ (when $V\neq 0$)	$\frac{d^2u}{dx^2}u(x)+\frac{2m}{\hbar^2}(E-V)u(x)=0$ (when $V\neq 0$)

Three-Dimensional	$-\frac{\hbar^2}{2m}\nabla^2\psi = i\hbar\frac{\partial\psi}{\partial t}$ (when $V=0$)	$\nabla^2 u(x)+\frac{2mE}{\hbar^2}u(x)=0$ (when $V=0$)
	$\left[-\frac{\hbar^2}{2m}\nabla^2+V(\vec{r})\right]\psi = i\hbar\frac{\partial\psi}{\partial t}$ (when $V\neq 0$)	$\nabla^2 u(x)+\frac{2mE}{\hbar^2}(E-V)u(x)=0$ (when $V\neq 0$)

2.5 Physical Significance of the Wave Function ψ

The wave function $\psi(\vec{r},t)$ representing the behaviour of a moving particle in specified position $\vec{r}$ and at time t must describe it completely. The wave function is large, where particle is most likely to be and small elsewhere. In case of one-dimensional motion the wave function $\psi(x,t)$ represents the behaviour of the particle at position x and time t. This implies that the magnitude $|\psi(x,t)|$ is large for certain specified values of x and t, where the particle is most likely to be found and $|\psi(x,t)|$ is small at certain specified values of x and t, where the probability of finding the particle is less.

Probability. Since $\psi(x,t)$ is a *complex quantity* it cannot have any *direct physical meaning.* But

$$\psi^*(x,t)\psi(x,t)=|\psi(x,t)|^2$$

is a *real positive quantity.* Since probability of finding a particle at x and t is also *real* and *positive* we can take $|\psi(x,t)|^2$ as a measure of probability density. Thus $|\psi(x,t)|^2dx = \psi^*(x,t)\psi(x,t)dx$ represents the *probability of finding the particle in the interval x and $x+dx$ at time t.*

Thus the wave function $\psi(\vec{r},t)$ which is a solution of Schrödinger equation must be such that

$$\left|\psi^*(\vec{r},t)\psi(\vec{r},t)\right|^2 d^3\vec{r}$$

represents the *probability* of finding the particle in volume element d^3r at time t.

Limitation of the wave function: Well-behaved function. The most important limitations of the wave function $\psi(\vec{r},t)$ are as follows: (i) $\psi(\vec{r}) = \psi(x,y,z)$ be *finite* for all values of x,y,z (ii) $\psi(r)$ must be *square-integrable* so that $\left|\psi^*\psi\right|^2$ is *finite* (iii) $\psi(r)$ must be continuous everywhere except, where $V(r)=V(x,y,z)=\infty$ (iv) The *partial* derivatives $\frac{\partial\psi}{\partial x},\frac{\partial\psi}{\partial y},\frac{\partial\psi}{\partial z}$ must be *continuous* for all values of x,y,z.

If a wave function $\psi(\vec{r})$ satisfies all the above conditions, then it is called a *well-behaved function.*

2.6 Probability Density

If $\psi(x,t)$ represents a wave group corresponding to a moving particle at x and time t, then the probability of finding the particle at the specified position x and time t is proportional to the *modulus square of the wave function.* Mathematically, the probability of finding the particle at x and t is $|\psi(x,t)|^2$. So the total probability of finding the particle in the range x_1 to x_2 is

$$P\propto\int_{x_1}^{x_2}|\psi(x,t)|^2dx.$$

Then the probability of finding the particle anywhere from $x=-\infty$ to $x=+\infty$ is

$$P \propto \int_{-\infty}^{+\infty} |\psi(x,t)|^2 dx.$$

Since the particle must be found somewhere between $x=-\infty$ to $+\infty$, total probability of finding the particle in the region $x=-\infty$ to $+\infty$ is unity. Now

$$P = \int_{-\infty}^{+\infty} \rho(x,t)\, dx, \tag{2.39}$$

where $\rho(x,t)$ is the *probability density* and is expressed as

$$\rho(x,t) = \frac{|\psi(x,t)|^2}{\int_{-\infty}^{+\infty} |\psi(x,t)|^2 dx}. \tag{2.40}$$

Substituting (2.40) in (2.39), we get

$$P = \frac{\int_{-\infty}^{+\infty} |\psi(x,t)|^2 dx}{\int_{-\infty}^{+\infty} |\psi(x,t)|^2 dx} = 1. \tag{2.41}$$

Since $|\psi(x,t)|^2$ is always *positive* and *real*, the probability density defined by (2.47) is always *positive* and *real.*

2.7 Normalisation of Wave Function

If $\psi_N(x,t) = N\psi(x,t)$, where N is a complex constant such that $\psi_N(x,t)$ satisfies the relation

$$\int_{-\infty}^{+\infty} \left|\psi_N(x,t)\right|^2 dx = |N|^2 \int_{-\infty}^{+\infty} |\psi(x,t)|^2 dx = 1, \tag{2.42}$$

then $\psi_N(x,t)$ is said to be *normalised wave function.*

From (2.49),

$$|N|^2 = \frac{1}{\int_{-\infty}^{+\infty} |\psi(x,t)|^2 dx}, \tag{2.43}$$

here N is called the *normalisation factor* or normalisation constant.

Hence a wave function is normalised, if

$$\int_{-\infty}^{+\infty} |\psi(x,t)|^2 dx \text{ or } \int_V |\psi(r,t)|^2 dV$$

all over the space remains *finite*. This is ***square integrability*** of the wave function. The value of N can be determined from (2.43).

2.8 Probability Current Density: Conservation of Probability

If $\rho(r,t)$ be the probability density of a particle at r and time t, then total probability of finding the particle in volume V is

$$P = \int_V \rho(r,t)\,dV = \int_V |\psi(r,t)|^2 dV = 1 = \text{constant.}$$

So, any decrease of probability at any r, t in a given volume element dV must be equal to the corresponding increase of probability in other element. The rate of decrease of probability density in dV is called the *probability current density* $\vec{J}$ so that

$$\frac{\partial P}{\partial t} + \nabla \cdot \vec{J} = 0.$$

Probability current density. The above equation represents the conservation of probability. To find out an expression for *probability current density* $(\vec{J})$ we consider a finite volume V enclosed by the surface S. The rate of change of probability of finding the particle in volume V is

$$\frac{\partial}{\partial t}\int_V \rho(r,t)\,dV = \frac{\partial}{\partial t}\int_V \psi^*\psi\,dV = \int_V \left(\frac{\partial \psi^*}{\partial t}\psi + \psi^*\frac{\partial \psi}{\partial t}\right) dV. \tag{2.44}$$

Since ψ is a solution of *Schrödinger wave equation*, we have

$$\left[-\frac{\hbar^2}{2m}\nabla^2 + V(r,t)\right]\psi(r,t) = i\hbar\frac{\partial}{\partial t}\psi(r,t). \tag{2.45}$$

Assuming the PE $V(r,t)$ to be real, we have

$$\left[-\frac{\hbar^2}{2m}\nabla^2 + V(r,t)\right]\psi^*(r,t) = -i\hbar\frac{\partial}{\partial t}\psi^*(r,t). \tag{2.46}$$

From (2.45) and (2.46), we get

$$\frac{\partial \psi}{\partial t} = \frac{1}{i\hbar}\left(-\frac{\hbar^2}{2m}\nabla^2 + V\right)\psi \tag{2.47}$$

$$\text{and} \quad \frac{\partial \psi^*}{\partial t} = -\frac{1}{i\hbar}\left(-\frac{\hbar^2}{2m}\nabla^2 + V\right)\psi^*. \tag{2.48}$$

Now

$$\frac{\partial \psi^*}{\partial t}\psi + \psi^*\frac{\partial \psi}{\partial t} = \frac{i}{\hbar}\left[\left(-\frac{\hbar^2}{2m}\nabla^2\psi^* + V\psi^*\right)\psi - \psi^*\left(-\frac{\hbar^2}{2m}\nabla^2\psi + V\psi\right)\right]$$

$$= -\frac{i\hbar}{2m}\left(\psi\nabla^2\psi^* - \psi^*\nabla^2\psi\right). \tag{2.49}$$

But

$$\left(\psi\nabla^2\psi^* - \psi^*\nabla^2\psi\right) = \nabla\cdot\left(\psi\nabla\psi^* - \psi^*\nabla\psi\right)\ \left[\because\ u\nabla^2 v - v\nabla^2 u = \nabla\cdot(u\nabla v - v\nabla u)\right]. \tag{2.50}$$

Now from (2.44), (2.49) and (2.50), we get

$$\frac{\partial}{\partial t}\int_V \rho\, dV = -\frac{i\hbar}{2m}\int_V \nabla\cdot(\psi\nabla\psi^* - \psi\nabla\psi)\, dV = -\int_V \left(\nabla\cdot\vec{J}\right) dV, \tag{2.51}$$

where

$$\vec{J} = \frac{i\hbar}{2m}\left(\psi\nabla\psi^* - \psi^*\nabla\psi\right). \tag{2.52}$$

Here, $\vec{J}$ is called the *probability current density*.

Conservation of probability. From (2.51),

$$\frac{\partial}{\partial t}\int_V \rho\, dV = -\int_V \left(\nabla\cdot\vec{J}\right) dV = -\oint_S \vec{J}\cdot\vec{dS} = -\oint_S J_n dS, \tag{2.53}$$

where $J_n = \vec{J}\cdot\hat{n}dS$ and $\hat{n}$ is the outward drawn unit normal to dS. From (2.53), we have

$$\int_V \frac{\partial\rho}{\partial t} dV = -\int_V \left(\nabla\cdot\vec{J}\right) dV.$$

Since volume is arbitrary,

$$\int_V \left(\frac{\partial\rho}{\partial t} + \nabla\cdot\vec{J}\right) = 0 \text{ or } \frac{\partial\rho}{\partial t} + \nabla\cdot\vec{J} = 0. \tag{2.54}$$

(2.54) represents the *conservation of probability density.*

2.9 Stationary State

If the state of a system does not change with time then it is said to be in a *stationary* or *steady state.* Such a system can be represented by a wave function of space coordinates only $[\psi(x)]$. *The probability of finding the system is independent of time.* The function $\psi(x)$ as well as $\frac{d\psi(x)}{dx}$ must be *finite, single-valued* and *continuous for all values of* x.

We know that for various constant E values of a particle, the general solution of *one-dimensional time-dependent Schrödinger equation* is given by

$$\psi(x,t) = \sum_n C_n\psi_n(x)e^{-\frac{iEt}{\hbar}}.$$

For one-dimensional motion of the particle the probability density $\rho(x,t)$ is

$$\begin{aligned}\rho(x,t) &= \psi^*(x,t)\psi(x,t)\\ &= \sum_n C_n^* C_n \psi_n^*(x,t)\psi_n(x,t) + \sum_n{\sum_m}' C_m^* C_n \psi_m^*(x)\psi_n(x)e^{\frac{i(E_m-E_n)t}{\hbar}}.\end{aligned}$$

The equation shows that $\psi(x,t)$ depends on time t. But it may so happen that all C_n are zero except for one value of energy E_k, then

$$\psi(x,t)=C_k\psi_k(x)e^{-\frac{iE_kt}{\hbar}}$$

$$\text{and } \rho(x,t)=\psi^*(x,t)\psi(x,t)=C_k^*C_k\psi_k^*(x)\psi_k(x).$$

Thus ρ is independent of time and the corresponding state is called *stationary state.*

2.10 Expectation Value

The expectation value of a dynamical quantity is defined as the average of the results of a large number of measurements on independent identical systems (i.e., systems represented by identical wave functions), or single measurements on each of a large number of identical systems.

1. *Expectation value of x.* Suppose we are to find the positions of a large number of electrons (N), all electrons are described by the same wave function. At time t we measure the position of each electron between x and $x+dx$. Then the probability that an electron will have position between x and $x+dx$ is

$$P=\frac{\text{Number of electrons having positions between } x \text{ and } x+dx}{N}.$$

But probability P of having position between x and $x+dx$ is

$$P=\psi^*(x,t)\psi(x,t)\,dx.$$

$\psi(x,t)$ is assumed to be normalised.

$\therefore$ number of electrons having positions between x and $x+dx$ is $N\psi^*(x,t)\psi(x,t)\,dx$.

If $x_1,x_2,\ldots,x_N$ be the positions of the 1st, 2nd, ..., Nth electron, then

$$x_1+x_2+\cdots+x_N=\int_{-\infty}^{a} xN\psi^*(x,t)\psi(x,t)\,dx$$

$$\text{or } \frac{x_1+x_2+\cdots+x_N}{N}=\int_{-\infty}^{+\infty} x\psi^*(x,t)\psi(x,t)\,dx.$$

But $\dfrac{x_1+x_2+\cdots+x_N}{N}=\langle x\rangle$ is the expectation value of position of an electron.

$$\therefore \langle x\rangle=\int_{-\infty}^{+\infty} \psi^*(x,t)x\psi(x,t)\,dx. \quad (2.55)$$

2. *Expectation value of momentum.* If p_x be the x-component of momentum of a particle, then the expectation value of p_x is

$$\langle p_x\rangle=\int_{-\infty}^{+\infty} \psi^*(x,t)\hat{p}_x\psi(x,t)\,dx,$$

where $\hat{p}_x$ is the *operator* corresponding to p_x, where

$$\hat{p}_x=-i\hbar\frac{\partial}{\partial x} \text{ (to be discussed in subsequent sections).}$$

$$\therefore \langle p_x\rangle=\int_{-\infty}^{+\infty} \psi^*(x,t)(-i\hbar)\frac{\partial}{\partial x}\psi(x,t)\,dx=-i\hbar\int_{-\infty}^{+\infty} \psi^*(x,t)\frac{\partial}{\partial x}[\psi(x,t)]\,dx. \quad (2.56)$$

3. *Expectation value of PE $V(x)$:*

$$\langle V\rangle = \int_{-\infty}^{+\infty} \psi^*(x,t)V(x)\psi(x,t)\,dx. \tag{2.57}$$

4. *Expectation value of KE (K):*

$$\langle K\rangle = \int_{-\infty}^{+\infty} \psi^*(x,t)\hat{K}\psi(x,t)\,dx.$$

But $\hat{K} = \dfrac{(\hat{p}_x)^2}{2m} = \dfrac{1}{2m}\left(-i\hbar\dfrac{\partial}{\partial x}\right)^2 = -\dfrac{\hbar^2}{2m}\dfrac{\partial^2}{\partial x^2}$.

$$\therefore\ \langle K\rangle = \int_{-\infty}^{+\infty} \psi^*(x,t)\left(-\frac{\hbar^2}{2m}\frac{\partial^2}{dx^2}\right)\psi(x,t)\,dx \tag{2.58}$$

$$\text{or } \langle K\rangle = -\frac{\hbar^2}{2m}\int_{-\infty}^{+\infty} \psi^*(x,t)\frac{\partial}{\partial x}\left(\frac{\partial\psi}{\partial x}\right)dx = -\frac{\hbar^2}{2m}\left[\left\{\psi^*\frac{\partial\psi}{\partial x}\right\}_{-\infty}^{+\infty} - \int_{-\infty}^{+\infty}\frac{\partial\psi^*}{\partial x}\cdot\frac{\partial\psi}{\partial x}dx\right].$$

But since according to the *condition of normalisation*, $\int_{-\infty}^{+\infty}\psi^*\psi\,dx = 1$, ψ^* is zero both at $x = +\infty$ and $-\infty$.

$$\therefore \left\{\psi^*\frac{\partial\psi}{\partial x}\right\}_{-\infty}^{+\infty} = 0.$$

$$\therefore\ \langle K\rangle = \frac{\hbar^2}{2m}\int_{-\infty}^{+\infty}\frac{\partial\psi^*}{\partial x}\cdot\frac{\partial\psi}{\partial x}dx \tag{2.59}$$

$$\text{or } \langle K\rangle = \frac{\hbar^2}{2m}\int_{-\infty}^{+\infty}\left|\frac{\partial\psi}{\partial x}\right|dx. \tag{2.60}$$

As the physical quantities corresponding to expectation values can be measured experimentally, they are *real quantities*. This leads to the conclusion that *the eigenvalues of quantum mechanical operator related to observable are real numbers.*

5. *Expectation value of energy for an eigenstate $\psi_n(\vec{r},t)$*. The energy of the nth eigenstate E_n is real.

$$\therefore\ \langle E\rangle = \int_V \psi_n^*(\vec{r},t)\hat{E}\psi_n(\vec{r},t)\,dV = \int_V \psi_n^*(\vec{r})\,e^{\frac{iE_nt}{\hbar}}\left(i\hbar\frac{\partial}{\partial t}\right)\psi_n(\vec{r})e^{-\frac{iE_nt}{\hbar}}\,dV$$

$$\left[\because\ \hat{E} = i\hbar\frac{\partial}{\partial t} \text{ and } \psi_n(\vec{r},t) = \psi_n(\vec{r})\,e^{-\frac{iE_nt}{\hbar}}\right]$$

$$= i\hbar\int_V \psi_n^*(\vec{r})\,e^{\frac{iE_nt}{\hbar}}(-iE_n/\hbar)\psi_n(\vec{r})\,e^{-\frac{iE_nt}{\hbar}}\,dV$$

$$= E_n\int_V \psi_n^*(\vec{r})\psi_n(\vec{r})\,dV \left[\because \int_V \psi_n^*(\vec{r})\psi_n(\vec{r})\,dV = 1 \text{ as } \psi_n(\vec{r}) \text{ is normalised}\right].$$

$$\therefore\ \langle E\rangle = E_n. \tag{2.61}$$

This shows that $\langle E\rangle$ for the nth state, E_n, is constant in time as in case of a stationary state.

2.11 Operators in Quantum Mechanics

An operator in quantum mechanics is a symbol for certain mathematical tool which transforms one function into another.

If $\hat{\alpha}$ be an operator and $f(x)$ is a function, then $\hat{\alpha}$ operating on $f(x)$ generates a new function $g(x)$.

$$\hat{\alpha}f(x) = g(x) \tag{2.62}$$

For example, if $f(x) = x^n$ and $\hat{\alpha} = \frac{d}{dx}$, then

$$\hat{\alpha}f(x) = \frac{d}{dx}(x^n) = nx^{n-1}.$$

The new function is $g(x) = nx^{n-1}$.

Each dynamical variable (observable) is represented by an operator.

The following table shows a list of operators:

Table 2.2

Name of operator	*Symbol of the operator*	*Result of n operation on x*
1. Multiplication by a constant k	k	kx^n
2. Taking square root	$\sqrt{\ }$	$x^{n/2}$
3. Differentiation w.r.t. x	$\frac{d}{dx}$	nx^{n-1}
4. Integration w.r.t. x	$\int x^n dx$	$\frac{x^{n+1}}{n+1} + C$
5. Taking logarithms	$\log_e(\ )$	$n \log_e x$

Characteristics of operators. Quantum mechanical operators are *linear operators. **A linear operator has the following properties:***

1. If $\hat{\alpha}$ is a linear operator and $f(x)$ is a function, then

$$\hat{\alpha}[cf(x)] = c\hat{\alpha}f(x), \tag{2.63}$$

where c is a constant or, $\hat{\alpha}c = c\hat{\alpha}$ or, $\hat{\alpha}c - c\hat{\alpha} = 0$, i.e., $\hat{\alpha}$ commutes with c.

2. *Linear operators obey distributive law.* If $f(x)$ and $g(x)$ be two functions of x and $\hat{\alpha}$ a linear operator, then

$$\hat{\alpha}[f(x) + g(x)] = \hat{\alpha}f(x) + \hat{\alpha}g(x). \tag{2.64}$$

Linear operators also obey associative law.

The *Hamiltonian operator* $\hat{H}$ operating on a wave function gives by corresponding observable quantity (eigenvalue) which is the total energy of the system

$$\hat{H}\psi(x) = E\psi(x), \tag{2.65}$$

$\psi(x)$ is the wave function corresponding to the system.

2.11.1 Important operators in quantum mechanics

1. *Cartesian coordinate operator.* The operators corresponding to Cartesian coordinates are $\hat{x}, \hat{y}, \hat{z}$ corresponding to x, y, z respectively.
   When these operators operate on a wave function $f(x)$ the function $f(x)$ is simply multiplied by the corresponding coordinate, e.g.,
   $$\hat{x}f(x) = xf(x), \hat{y}f(x) = yf(x) \text{ and } \hat{z}f(x) = zf(x).$$
2. *Linear momentum operator.* The component of linear momentum operator are represented by $\hat{p}_x, \hat{p}_y, \hat{p}_z$ corresponding to x, y, z respectively. The values can be found out as
   $$\hat{p}_x = -i\hbar\frac{\partial}{\partial x},\ \hat{p}_y = -i\hbar\frac{\partial}{\partial y},\ \hat{p}_z = -i\hbar\frac{\partial}{\partial z} \text{ and } \hat{p} = -i\hbar\nabla, \tag{2.66}$$
   $\hat{p}_x$ operating on ψ, gives the function multiplied by p_x, i.e., $\hat{p}_x\psi = p_x\psi$.
3. PE *operator.* The PE operator corresponding to PE V is represented by $\hat{V}$. When $\hat{V}$ operates on a function ψ, the function gets multiplied by V,
   $$\hat{V}\psi = V\psi. \tag{2.67}$$
4. *KE operator.* If the wave function of a system is $\psi(x,t) = Ae^{\frac{i}{\hbar}(p_x x - Et)}$, then
   $$\frac{\partial\psi}{\partial x} = A\left(\frac{i}{\hbar}\right) p_x e^{\frac{i}{\hbar}(p_x x - Et)} = \frac{i}{\hbar}p_x\psi \text{ or } \frac{\hbar}{i}\frac{\partial\psi}{\partial x} = p_x\psi \text{ or } \frac{\partial\psi}{\partial x} = \frac{i}{\hbar}p_x\psi.$$
   Differentiating $\psi(x,t)$ w.r.t. x again, we get
   $$\frac{\hbar}{i}\frac{\partial^2\psi}{\partial x^2} = p_x\left(\frac{i}{\hbar}\right)p_x\psi \text{ or } \frac{\hbar^2}{i^2}\frac{\partial^2\psi}{\partial x^2} = p_x^2\psi \text{ or } -\hbar^2\frac{\partial^2\psi}{\partial x^2} = p_x^2\psi.$$
   Dividing by $2m$,
   $$-\frac{\hbar^2}{2m}\frac{\partial^2\psi}{\partial x^2} = \frac{p_x^2}{2m}\psi = K\psi,$$
   where $K = \frac{p_x^2}{2m}$ is the KE of the system.
   $$\therefore \hat{K} = -\frac{\hbar^2}{2m}\frac{\partial^2}{\partial x^2}. \tag{2.68}$$
5. *Total energy operator.* If p_x and $V(x)$ be the momentum and PE of a particle moving along x-direction, then its total energy
   $$E = \frac{p_x^2}{2m} + V(x). \tag{2.69}$$
   Now
   $$-\frac{\hbar^2}{2m}\frac{\partial^2\psi}{\partial x^2} = \frac{p_x^2}{2m}\psi = (E - V)\psi \text{ [from (2.69)]}$$
   $$\text{or } -\frac{\hbar^2}{2m}\frac{\partial^2\psi}{\partial x^2} + V\psi = E\psi \text{ or } \left[-\frac{\hbar^2}{2m}\frac{\partial^2}{\partial x^2} + V\right]\psi = E\psi. \tag{2.70}$$

Here

$$-\frac{\hbar^2}{2m}\frac{\partial^2}{\partial x^2}+V=\hat{H}_x,$$

where H_x is the Hamiltonian representing the total energy E of the system.

∴ total energy operator is

$$\hat{E}=\hat{H}_x=-\frac{\hbar^2}{2m}\frac{\partial^2}{\partial x^2}+V. \qquad (2.71)$$

6. *Total energy operator in terms of differential w.r.t. time.* The wave function of a system is $\psi(x,t)=Ae^{\frac{i}{\hbar}(p_x x-Et)}$.

$$\therefore\ \frac{\partial\psi}{\partial t}=A\left(-\frac{iE}{\hbar}\right)e^{\frac{i}{\hbar}(p_x x-Et)}=-\frac{i}{\hbar}E\psi=\frac{1}{i\hbar}E\psi \text{ or } i\hbar\frac{\partial\psi}{\partial t}=E\psi. \qquad (2.72)$$

Thus

$$\hat{E}=i\hbar\frac{\partial}{\partial t}. \qquad (2.73)$$

7. *Angular momentum operator.* If $\vec{r}$ be the position vector of a particle relative to an arbitrary origin O, then the angular momentum $\vec{L}$ is given by $\vec{L}=\vec{r}\times\vec{p}$.

Now

$$\vec{r}=\hat{i}x+\hat{j}y+\hat{k}z$$

and $\vec{p}=\hat{i}p_x+\hat{j}p_y+\hat{k}p_z$..

$$\therefore\ \vec{r}\times\vec{p}=\left(\hat{i}x+\hat{j}y+\hat{k}z\right)\times\left(\hat{i}p_x+\hat{j}p_y+\hat{k}p_z\right)$$
$$=\hat{i}\left(yp_z-zp_y\right)+\hat{j}\left(zp_x-xp_z\right)+\hat{k}\left(xp_y-yp_x\right).$$

$$\therefore\ \hat{i}L_x+\hat{j}L_y+\hat{k}L_z=\hat{i}\left(yp_z-zp_y\right)+\hat{j}\left(zp_x-xp_z\right)+\hat{k}\left(xp_y-yp_x\right). \qquad (2.74)$$

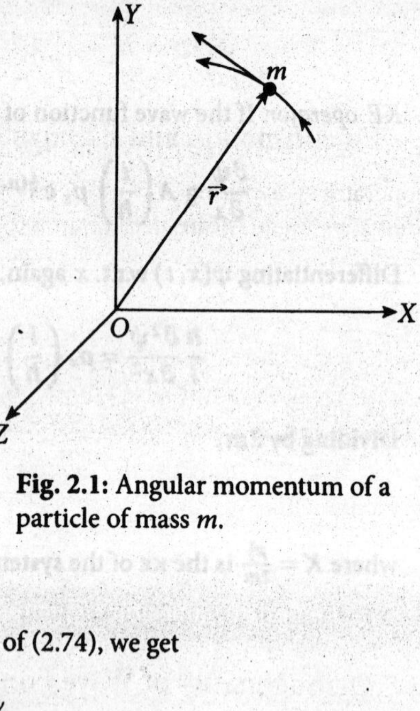

Fig. 2.1: Angular momentum of a particle of mass m.

Comparing the coefficients of $\hat{i},\hat{j}$ and $\hat{k}$ from both sides of (2.74), we get

$$L_x=yp_z-zp_y$$
$$L_y=zp_x-xp_z$$
$$L_z=xp_y-yp_x.$$

∴ the operators corresponding to the x-, y- and z-components of angular momentum are

$$\hat{L}_x=y\left(\hat{p}_z\right)-z\left(\hat{p}_y\right)=y\left(-i\hbar\frac{\partial}{\partial z}\right)-z\left(-i\hbar\frac{\partial}{\partial y}\right),$$

i.e.,

$$\hat{L}_x=-i\hbar\left(y\frac{\partial}{\partial z}-z\frac{\partial}{\partial y}\right). \qquad (2.75)$$

Similarly,

$$\hat{L}_y = -i\hbar\left(z\frac{\partial}{\partial x} - x\frac{\partial}{\partial z}\right) \tag{2.76}$$

$$\hat{L}_z = -i\hbar\left(x\frac{\partial}{\partial y} - y\frac{\partial}{\partial x}\right). \tag{2.77}$$

In spherical polar coordinates (r, θ, ψ),

$$\hat{L}_x = -i\hbar\left[-\sin\phi\frac{\partial}{\partial\theta} - \cot\theta\cos\phi\frac{\partial}{\partial\phi}\right] \tag{2.78}$$

$$\hat{L}_y = -i\hbar\left[\cos\phi\frac{\partial}{\partial\theta} - \cot\theta\sin\phi\frac{\partial}{\partial\phi}\right] \tag{2.79}$$

$$\hat{L}_z = -i\hbar\frac{\partial}{\partial\phi} \tag{2.80}$$

$$\text{and } \hat{L}^2 = \hat{L}_x^2 + \hat{L}_y^2 + \hat{L}_z^2 = -\hbar^2\left[\frac{1}{\sin\theta}\frac{\partial}{\partial\theta}\left(\sin\theta\frac{\partial}{\partial\theta}\right) + \frac{1}{\sin^2\theta}\frac{\partial^2}{\partial\phi^2}\right]. \tag{2.81}$$

2.12 Eigenfunction and Eigenvalue

If an operator $\hat{A}$ operating on a wave function $u(x)$, multiplies $u(x)$ by a constant a, then $u(x)$ is called an *eigenfunction* (eigen means proper) *function of* $\hat{A}$ belonging to the *eigenvalue* a.

For a particular operator $\hat{A}$ there is a set of eigenvalues a_n and a set of eigenfunctions defined by

$$\hat{A}u_n(x) = a_n u_n(x). \tag{2.82}$$

Thus the eigenfunctions $u_n(x)$ form a special set of functions which remains unchanged when operated by $\hat{A}$.

If $u_1(x), u_2(x), \ldots, u_n(x)$ be the *solutions of Schrödinger equation* belonging to the same energy, then their linear combination

$$u(x) = \sum_n u_n(x)$$

is also a solution of *Schrödinger equation* belonging to same energy state.

2.13 Orthogonality of Wave Functions

If a set of functions $u_1(x), u_2(x), \ldots, u_n(x)$ be such that

$$\int_a^b u_m^*(x)\,u_n(x)\,dx = 0 \quad (\text{for } m \neq n), \tag{2.83}$$

then any two functions $u_m(x)$ and $u_n(x)$ are said to be mutually *orthogonal* in the interval a to b.

If, on the other hand,

$$\int_a^b u_m^*(x)\,u_n(x)\,dx = 1 \quad (m = n), \tag{2.84}$$

then $u_m(x)$ and $u_n(x)$ are said to be *normalised* between the interval a to b.

If

$$\int_a^b u_m^*(x)\,u_n(x)\,dx = \delta_{mn}, \tag{2.85}$$

where δ_{mn} is the *Krönecker delta* given by

$$\delta_{mn} = \begin{cases} 1, & \text{for } m = n \\ 0, & \text{for } m \neq n \end{cases}$$

then the wave functions are said to be *ortho-normal* in the interval a to b.

It can be shown that any two eigenfunctions corresponding to different eigenvalues are *orthogonal.*

Proof. Let $u_1(x)$ and $u_2(x)$ be two eigenfunctions of a *Hermitian* operator $\hat{A}$ with eigenvalues a_1 and a_2 respectively, then

$$\hat{A}u_1(x) = a_1 u_1(x)$$
$$\hat{A}u_2(x) = a_2 u_2(x).$$

Taking complex conjugate of the above equations, we get $\hat{A}^* u_1^*(x) = a_1 u_1^*(x)$ [$\because$ a_1 is real, $a_1^* = a_1$].

We now multiply the above equation by $u_2(x)$ and integrate over all space,

$$\int\limits_{\text{all space}} u_2(x)\hat{A}^* u_1^*(x)\,dV = a_1 \int\limits_{\text{all space}} u_1(x)\,u_2^*(x)\,dV$$

Similarly,

$$\int\limits_{\text{all space}} u_1^*(x)\hat{A}u_2(x)\,dV = a_2 \int\limits_{\text{all space}} u_1^*(x)\,u_2(x)\,dV. \tag{2.86}$$

Since $\hat{A}$ is *Hermitian,* the left-hand sides of (2.85) and (2.86) are equal.

$$\therefore a_1 \int\limits_{\text{all space}} u_1^*(x)\,u_2(x)\,dV = a_2 \int\limits_{\text{all space}} u_1^*(x)\,u_2(x)\,dV$$

$$\text{or } (a_1 - a_2) \int\limits_{\text{all space}} u_1^*(x)\,u_2(x)\,dV = 0.$$

$$\because a_1 \neq a_2, \quad \therefore \int\limits_{\text{all space}} u_1^*(x)\,u_2(x)\,dV = 0. \tag{2.87}$$

Hence $u_1(x)$ and $u_2(x)$ are *orthogonal.*

2.14 Hermitian Operator

The Hermitian operators are those which represent such dynamical variables whose expectation values are real.

If $\hat{A}$ represents a dynamical variable A, then the expectation value of A is

$$\langle A\rangle = \int u^*(x)\hat{A}u(x)\,dx, \tag{2.88}$$

where $u(x)$ is normalised.

The complex conjugate of (2.87) is

$$\langle A\rangle^* = \left[\int u^*(x)\hat{A}u(x)\,dx\right]^* = \int u(x)\hat{A}^*u^*(x)\,dx. \tag{2.89}$$

Since the expectation values are real, $\therefore \langle A\rangle = \langle A\rangle^*$.

Hence,

$$\int u^*(x)\hat{A}u(x)\,dx = \int u(x)\hat{A}^*u^*(x)\,dx. \tag{2.90}$$

(2.89) represents the *condition of Hermitian operator.*

2.15 The Eigenvalues of Hermitian Operator are Real

Let $u(x)$ be an eigenfunction of Hermitian operator $\hat{A}$ belonging to eigenvalue a. Then $\hat{A}u(x) = a\,u(x)$.

Taking complex conjugate $\hat{A}^*u^*(x) = a^*u^*(x)$.

$\therefore$ multiplying the above equation by $u^*(x)$ and $u(x)$ respectively, and then integrating from $-\infty$ to ∞, we get

$$\int_{-\infty}^{+\infty} u^*(x)\hat{A}u(x)\,dx = \int_{-\infty}^{+\infty} u^*(x)\,a\,u(x)\,dx = a\int_{-\infty}^{+\infty} u^*(x)\,u(x)\,dx \tag{2.91}$$

and

$$\int_{-\infty}^{+\infty} u(x)\hat{A}^*u^*(x)\,dx = \int_{-\infty}^{+\infty} u(x)\,a^*u^*(x)\,dx = a^*\int_{-\infty}^{+\infty} u(x)u^*(x)\,dx. \tag{2.92}$$

Now since $\hat{A}$ is Hermitian,

$$\therefore \int_{-\infty}^{+\infty} u^*(x)\hat{A}u(x)\,dx = \int_{-\infty}^{+\infty} u(x)\hat{A}^*u^*(x)\,dx.$$

So equating the right sides of (2.91) and (2.92), we get

$$a\int_{-\infty}^{+\infty} u^*(x)\,u(x)\,dx = a^*\int_{-\infty}^{+\infty} u^*(x)\,u(x)\,dx. \tag{2.93}$$

$$\therefore a = a^*. \tag{2.94}$$

Hence *a is real.*

2.16 Hamiltonian Operator

Total energy of a particle of mass m moving with velocity v in a potential field V is

$$H=\frac{1}{2}mv^2+V=\frac{m^2v^2}{2m}+V=\frac{p^2}{2m}+V. \tag{2.95}$$

H is called the *Hamiltonian* of the system.

$$H=\frac{p_x^2+p_y^2+p_z^2}{2m}+V. \tag{2.96}$$

Now

$$p_x^2+p_y^2+p_z^2=\left(-i\hbar\frac{\partial}{\partial x}\right)\left(-i\hbar\frac{\partial}{\partial x}\right)+\left(-i\hbar\frac{\partial}{\partial y}\right)\left(-i\hbar\frac{\partial}{\partial y}\right)+\left(-i\hbar\frac{\partial}{\partial z}\right)\left(-i\hbar\frac{\partial}{\partial z}\right)$$

$$=-\hbar^2\left(\frac{\partial^2}{\partial x^2}+\frac{\partial^2}{\partial y^2}+\frac{\partial^2}{\partial z^2}\right)=-\hbar^2\nabla^2. \tag{2.97}$$

$$\therefore\ \hat{H}=-\frac{\hbar^2}{2m}\nabla^2+V. \tag{2.98}$$

We can prove that $\hat{H}$ *is Hermitian*, i.e.,

$$\int_{-\infty}^{+\infty}u^*(x)\hat{H}u(x)\,dx=\int_{-\infty}^{+\infty}u(x)\hat{H}^*u^*(x)\,dx.$$

2.17 Application of Schrödinger Equation

2.17.1 Particle in a one-dimensional box

Let us consider the motion of a particle in a hollow *rectangular box* along the X-axis in non-relativistic region. The origin is at one corner of the box and X-axis is perpendicular to the parallel opposite walls as shown in Fig. 2.2.

We assume that (i) *the walls of the box are rigid, hard and elastic. Then the particle will rebound in the box without any loss of KE.* (ii) *the walls of the box are non penetrable.* The particle then will remain confined in the box as the energy of the particle is less than the PE outside the box. So

$$V(x)=\begin{cases}0, & \text{in } 0<x<a\\ \infty, & \text{in } x\le 0 \text{ and } x\ge a,\end{cases}$$

where a is the length of the box along X-direction.

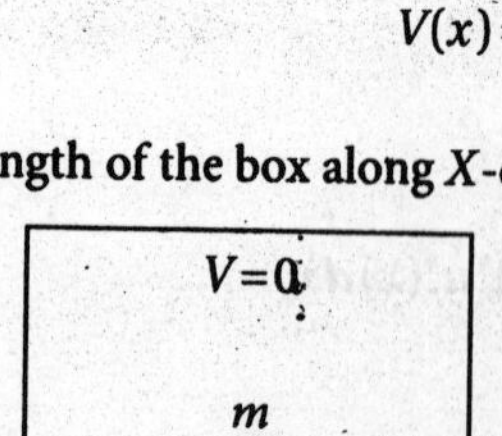

Fig. 2.2: One-dimensional potential box.

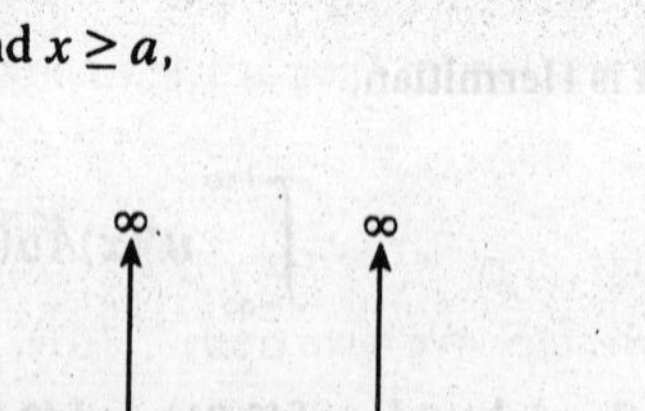

Fig. 2.3: Potential energy curve.

The PE curve [Fig. 2.3] is also called *square well potential of infinite depth.*

1. *Wave equation for the particle.* Since inside the box the PE $V(x)=0$, the particle is free, i.e., it is not acted upon by a force field. The Schrödinger *time-independent equation for the particle is*

$$-\frac{\hbar^2}{2m}\frac{d^2\psi}{dx^2}=E\psi.$$

E is the *total energy* of the particle and $\hbar=\frac{h}{2\pi}$, h being the *Planck's constant.*

$$\text{or } \frac{d^2\psi}{dx^2}+\frac{2mE}{\hbar^2}\psi=0 \text{ or } \frac{d^2\psi}{dx^2}+k^2\psi=0, \tag{2.99}$$

where

$$k=\sqrt{\frac{2mE}{\hbar^2}}. \tag{2.100}$$

This shows that k is constant for a particular value of energy E of the particle.

2. *Solution of* (2.99): The solution of (2.100) is

$$\psi(x)=A\sin kx+B\cos kx, \tag{2.101}$$

where A and B are constants to be found out from the condition of the problem, viz., (a) $\psi(x)=0$, at $x=0$; (b) $\psi(x)=0$ at $x=a$.
From the first condition, $0=A\cdot 0+B\cdot 1$ or, $B=0$.
$\therefore$ (2.101) reduces to

$$\psi(x)=A\sin kx. \tag{2.102}$$

Again at $x=a$ (2.102) gives, $\psi(a)=A\sin ka=0$.
But $A\neq 0$, $\therefore \sin ka=0$ or, $ka=n\pi$.
So

$$k=\frac{n\pi}{a}\quad n=1,2,3,\ldots. \tag{2.103}$$

But $n=0$ is *not possible* because for $n=0$, $k=0$ and $E=0$.
So $\psi(x)=0$ everywhere in the box.
$\therefore$ the wave function for the particle when $0<x<a$ is

$$\psi_n(x)=A\sin\frac{n\pi x}{a}. \tag{2.104}$$

3. *Energy eigenvalues*: We have from (2.101),

$$\sqrt{\frac{2mE_n}{\hbar^2}}=\frac{n\pi}{2} \text{ or } \frac{2mE_n}{\hbar^2}=\frac{n^2\pi^2}{a^2} \text{ or } E_n=\frac{n^2\pi^2\hbar^2}{2ma^2} \tag{2.105}$$

or, $E_n\propto n^2$, that is the energy eigenvalues are discrete.
Since

$$\hbar=\frac{h}{2\pi}, \quad \therefore E_n=\frac{n^2h^2}{8ma^2}. \tag{2.106}$$

Conclusions from (2.105)

(i) Putting $n = 1$, in (2.106), we get the lowest energy of the particle

$$E_1 = \frac{\pi^2\hbar^2}{2ma^2} = \frac{h^2}{8ma^2} \quad \left[\because \hbar = \frac{h}{2\pi}\right]. \qquad (2.107)$$

E_1 is called the *ground state energy* of the particle.

(ii) The energy values for $n = 2, 3, 4, \ldots$ are

$$E_2 = \frac{2\pi^2\hbar^2}{ma^2},\; E_3 = \frac{9\pi^2\hbar^2}{2ma^2},\; E_4 = \frac{8\pi^2\hbar^2}{ma^2}, \text{ etc.,}$$

which are *discrete*.

(iii) The spacing between nth and $(n+1)$th energy level is

$$\Delta E = (n+1)^2 E_1 - n^2 E_1 = (2n+1)E_1.$$

So ΔE increases for higher values of n [Fig. 2.4].

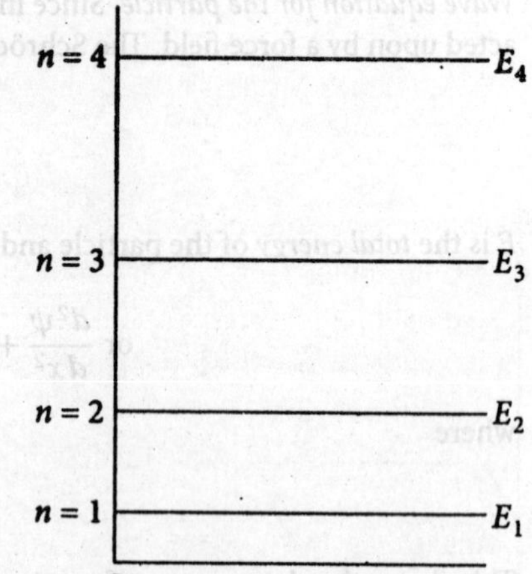

Fig. 2.4: Energy level diagram of a particle in one-dimensional potential box.

4. *Normalisation of* $\psi(x)$: We have for $0 < x < a$

$$\psi_n(x) = A \sin \frac{n\pi x}{a}$$

and for $x < 0$ and $x > a$, $\psi_n(x) = 0$.

Since the total probability of finding the particle in $0 < x < a$ is unity, we have

$$\int_0^a |\psi_n(x)|^2 dx = 1 \text{ or } \int_0^a A^2 \sin^2 \frac{n\pi x}{a} dx = 1$$

$$\text{or } A^2 \int_0^a \frac{1}{2}\left[1 - \cos\frac{2n\pi x}{a}\right] dx = 1$$

$$\text{or } \frac{A^2}{2}\left[x - \frac{a}{2\pi n} \sin\frac{2n\pi x}{a}\right]_0^a = 1$$

$$\text{or } \frac{A^2 a}{2} = 1 \text{ or } A = \sqrt{\frac{2}{a}}.$$

$\therefore$ the normalised wave function for the particle is

$$\psi_n(x) = \sqrt{\frac{2}{a}} \sin \frac{n\pi x}{a}. \qquad (2.108)$$

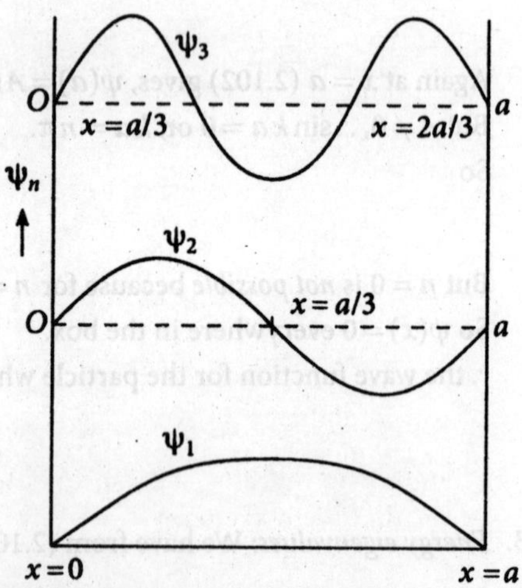

Fig. 2.5: Nature of the wave functions $\psi_1(x)$, $\psi_2(x)$ and $\psi_3(x)$ of a particle in a one-dimensional potential box.

(2.108) shows that, $\psi_1(x) \doteq 0$ for $x = 0$ and $x = a$.

So $\psi_1(x)$ has *two nodes* at $x = 0$ and $x = a$.

$\psi_2(x)$ has *three nodes* at $x = 0$, $x = \frac{a}{2}$, $x = a$.

$\psi_3(x)$ has *four nodes* at $x = 0$, $x = \frac{a}{3}$, $x = \frac{2a}{3}$, $x = a$.

$\psi_n(x)$ has $(n+1)$ *nodes*.

The nature of $\psi_1(x), \psi_2(x)$ and $\psi_3(x)$ are shown in Fig. 2.5.

5. *Eigenvalues of momentum*: For the nth state the momentum p_n is given by

$$p_n^2 = 2mE_n = 2m \times \frac{n^2\pi^2\hbar^2}{2ma^2} \text{ or } p_n^2 = \frac{n^2\pi^2\hbar^2}{a^2}. \quad \therefore p_n = \pm\frac{n\pi\hbar}{a} \tag{2.109}$$

$$\text{or } p_n = \pm\frac{nh}{2a} \quad \left(\because \hbar = \frac{h}{2\pi}\right). \tag{2.110}$$

The $\pm$ sign before $\frac{nh}{2a}$ is due to the fact that the particle moves back and forth in the box.

The momentum is *discrete* but the difference between two consecutive momenta is constant $\left(\frac{h}{2a}\right)$.

6. *Probability of location of the particle*: The probability of finding the particle in the small distance dx is

$$P(x)\,dx = |\psi_n(x)|^2\,dx = \frac{2}{a}\sin^2\frac{n\pi x}{a}\,dx \tag{2.111}$$

or the probability density

$$P(x) = \frac{2}{a}\sin^2\frac{n\pi x}{a}. \tag{2.112}$$

$P(x)$ is maximum when

$$\frac{n\pi x}{a} = (2n+1)\frac{\pi}{2} = \frac{\pi}{2}, \frac{3\pi}{2}, \frac{5\pi}{2}, \ldots \text{ or, at } x = \frac{a}{2n}, \frac{3a}{2n}, \frac{5a}{2n}, \ldots. \tag{2.113}$$

Thus for $n = 1$ state, $P(x)$ is maximum at $x = \frac{a}{2}$

$n = 2$ state, $P(x)$ is maximum at $x = \frac{a}{4}, \frac{3a}{4}$

$n = 3$ state, $P(x)$ is maximum at $x = \frac{a}{6}, \frac{3a}{6}, \frac{5a}{6}$.

Fig. 2.6 shows that the variation of probability density for states $n = 1, 2$, and 3.

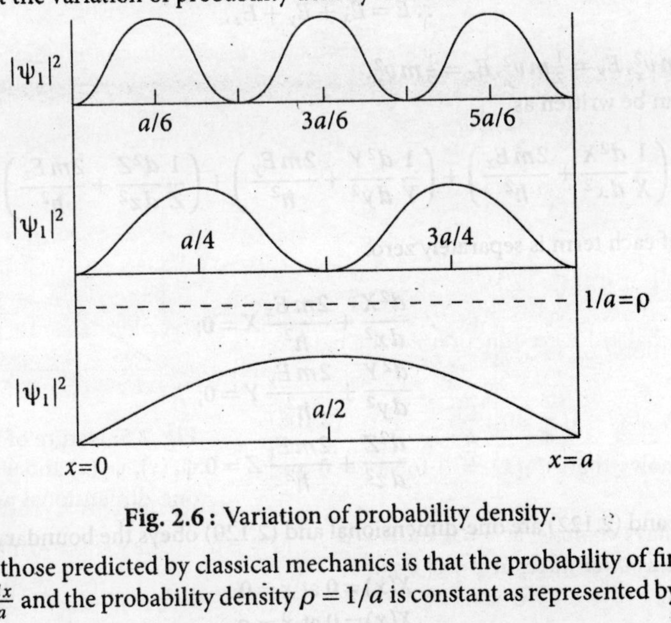

Fig. 2.6 : Variation of probability density.

The results from those predicted by classical mechanics is that the probability of finding the particle in a small dx is $\frac{dx}{a}$ and the probability density $\rho = 1/a$ is constant as represented by a dotted straight line in Fig. 2.6.

2.17.2 Particle in a three-dimensional rectangular box

We consider a particle in motion in a three-dimensional rectangular potential box, the sides of length a, b and c are parallel to X-, Y- and Z-axes respectively. We suppose that motion of the particle is *non-relativistic* and no force acts on it inside the box. We assume that (i) *the walls of the rectangular potential box are rigid, hard and elastic* (ii) *The walls of the box are non-penetrable* (iii) *The potential in the box is given by $V(x)=0$ in $0<x<a, 0<y<b, 0<z<c$ and $V(x)=\infty$ outside the box.*

1. *Wave equation for the particle.* The three-dimensional time-independent Schrödinger equation for the particle in the box is

$$-\frac{\hbar^2}{2m}\nabla^2\psi = E\psi. \tag{2.114}$$

$$\because \nabla^2 = \frac{\delta^2}{\partial x^2}+\frac{\partial^2}{\partial y^2}+\frac{\partial^2}{\partial z^2},$$

$$\therefore -\frac{\hbar^2}{2m}\left(\frac{\partial^2\psi}{\partial x^2}+\frac{\partial^2\psi}{\partial y^2}+\frac{\partial^2\psi}{\partial z^2}\right) = E\psi \text{ or } \frac{\partial^2\psi}{\partial x^2}+\frac{\partial^2\psi}{\partial y^2}+\frac{\partial^2\psi}{\partial z^2}+\frac{2mE}{\hbar^2}\psi = 0. \tag{2.115}$$

Assuming

$$\psi(x,y,z) = X(x)Y(y)Z(z), \tag{2.116}$$

we have from (2.115),

$$\frac{1}{X}\frac{d^2X}{dx^2}+\frac{1}{Y}\frac{d^2Y}{dy^2}+\frac{1}{Z}\frac{d^2Z}{dz^2}+\frac{2mE}{\hbar^2} = 0. \tag{2.117}$$

Since there is no PE (i.e., $V=0$) inside the box of the particle, whole energy E is kinetic, i.e.,

$$E = \tfrac{1}{2}mv^2 = \tfrac{1}{2}mv_x^2+\tfrac{1}{2}mv_y^2+\tfrac{1}{2}mv_z^2.$$

$$\because v^2 = v_x^2+v_y^2+v_z^2,$$

$$\therefore E = E_x+E_y+E_z, \tag{2.118}$$

where $E_x=\frac{1}{2}mv_x^2, E_y=\frac{1}{2}mv_y^2, E_z=\frac{1}{2}mv_z^2$.

Now (2.117) can be written as

$$\left(\frac{1}{X}\frac{d^2X}{dx^2}+\frac{2mE_x}{\hbar^2}\right)+\left(\frac{1}{Y}\frac{d^2Y}{dy^2}+\frac{2mE_y}{\hbar^2}\right)+\left(\frac{1}{Z}\frac{d^2Z}{dz^2}+\frac{2mE_z}{\hbar^2}\right) = 0. \tag{2.119}$$

(2.119) is true if each term is separately zero.

$$\therefore \frac{d^2X}{dx^2}+\frac{2mE_x}{\hbar^2}X = 0; \tag{2.120}$$

$$\frac{d^2Y}{dy^2}+\frac{2mE_y}{\hbar^2}Y = 0; \tag{2.121}$$

$$\frac{d^2Z}{dz^2}+\frac{2mE_z}{\hbar^2}Z = 0. \tag{2.122}$$

(2.120), (2.121) and (2.122) are one-dimensional and (2.120) obeys the boundary conditions

$$X(x)=0 \text{ at } x=0$$
$$X(x)=0 \text{ at } x=a.$$

So, the energy E_x is given by

$$E_x = \frac{n_x^2\pi^2\hbar^2}{2ma^2},. \tag{2.123}$$

where $n_x = 1,2,3,\ldots$.

The *normalised wave function*, as in case of one-dimension box, is

$$X(x) = \sqrt{\frac{2}{a}}\sin\frac{n_x\pi x}{a}. \tag{2.124}$$

Similarly,

$$E_y = \frac{n_y^2\pi^2\hbar^2}{2mb^2} \tag{2.125}$$

$$\text{and } Y(y) = \sqrt{\frac{2}{b}}\sin\frac{n_y\pi y}{b}; \tag{2.126}$$

$$E_z = \frac{n_z^2\pi y}{2mc^2} \tag{2.127}$$

$$\text{and } Z(z) = \sqrt{\frac{2}{c}}\sin\frac{n_z\pi z}{c}. \tag{2.128}$$

2. *Energy eigenvalue*:

$$E_{n_x n_y n_z} = E_x + E_y + E_z = \frac{\pi^2\hbar^2}{2m}\left[\frac{n_x^2}{a^2} + \frac{n_y^2}{b^2} + \frac{n_z^2}{c^2}\right], \tag{2.129}$$

where $n_x, n_y, n_z = 1,2,3,\ldots$.

The energy eigenvalue of the particle represents the energy levels of the same.

3. *Wave functions for the particle.* For the stationary state represented by (n_x, n_y, n_z) the normalised wave function is

$$\psi_{n_x n_y n_z}(x,y,z) = X(x)Y(y)Z(z) = \sqrt{\frac{8}{abc}}\sin\frac{n_x\pi x}{a}\sin\frac{n_y\pi y}{b}\sin\frac{n_z\pi z}{c}. \tag{2.130}$$

It can be proved that

$$\frac{8}{abc}\int_0^a\left(\sin\frac{n_x\pi x}{a}\right)^2 dx \times \int_0^b\left(\sin\frac{n_y\pi y}{b}\right)^2 dy \times \int_0^c\left(\sin\frac{n_z\pi z}{c}\right)^2 dz. \tag{2.131}$$

Conclusions. (i) *A change in sign of the quantum numbers n_x, n_y, n_z does not change the sign of* $E_{n_x n_y n_z}$. So all the stationary states are given by positive values of n_x, n_y, n_z. $n_x, n_y, n_z \neq 0$, because in that case $\psi_{n_x,n_y,n_z}(x,y,z) = 0$, which is not possible. (ii) *The lowest value of energy is*

$$E_{111} = \frac{\pi^2\hbar^2}{2m}\left(\frac{1}{a^2} + \frac{1}{b^2} + \frac{1}{c^2}\right), \tag{2.132}$$

when $n_x = n_y = n_z = 1$.

Cubical box. If the potential box is a cube of side a, then $a = b = c$ and

$$E_{n_x}E_{n_y}E_{n_z} = \frac{\pi^2\hbar^2}{2ma^2}\left(n_x^2 + n_y^2 + n_z^2\right). \tag{2.133}$$

The ground state (or the lowest state) energy is

$$E_{111} = \frac{3\pi^2\hbar^2}{2ma^2}. \tag{2.134}$$

So, there is *only one energy* E_{111} in the ground state and, therefore, the ground state is *non-degenerate*. This state has also one wave function $\psi_{111}(x, y, z)$.

4. *Degeneracy. If two or more quantum states of a particle having different eigenfunctions have the same value of momentum (p) and energy (E), then the states are called degenerate.* For example, for the first excited state the set of values of n_x, n_y and n_z are:

n_x	n_y	n_z
2	1	1
1	2	1
1	1	2

In this case there are *three different eigenfunction*, viz., ψ_{211}, ψ_{121} and ψ_{112} representing three different states but all the states have the *same energy* $E_{211} = E_{121} = E_{112} = \dfrac{6\pi\hbar^2}{2ma^2}$. Thus the *first excited state is triply degenerate*. The following table shows *energy levels* and the *degree of degeneracy* of different stationary states when the particle is confined in a *cubical potential box*.

Table 2.3

States	n_x	n_y	n_z	E	*Degree of Degeneracy*
Ground state		(111)		$\dfrac{3\pi^2\hbar^2}{2ma^2}$	Non-degenerate
1st excited state	(211)	(121)	(112)	$\dfrac{6\pi^2\hbar^2}{2ma^2}$	3-fold degenerate
2nd excited state	(221)	(212)	(122)	$\dfrac{9\pi^2\hbar^2}{2ma^2}$	3-fold degenerate
3rd excited state	(311)	(131)	(113)	$\dfrac{11\pi^2\hbar^2}{2ma^2}$	3-fold degenerate
4th excited state		(222)		$\dfrac{12\pi^2\hbar^2}{2ma^2}$	Non-degenerate
5th excited state	(123) (231)	(132) (312)	(213) (321)	$\dfrac{14\pi^2\hbar^2}{2ma^2}$	6-fold degenerate
6th excited state	(322)	(232)	(223)	$\dfrac{17\pi^2\hbar^2}{2ma^2}$	3-fold degenerate

5. *Momentum*: For the (n_x, n_y, n_z) state, the *eigenvalue of momentum* is given by

$$p_{n_x n_y n_z} = \frac{\pi\hbar}{a}\left(n_x^2 + n_y^2 + n_z^2\right)^{1/2}.$$

(a) For the ground state, $n_x = n_y = n_z = 1$.

$$\therefore p_{111} = \frac{\pi\hbar}{a}\sqrt{3}.$$

(b) For the 1st excited state,

$$p_{211} = \frac{\pi\hbar}{a}\left(2^2+1^2+1^2\right) = \frac{\pi\hbar}{a}\sqrt{6};\ p_{121} = \frac{\pi\hbar}{a}\sqrt{6};\ p_{112} = \frac{\pi\hbar}{a}\sqrt{6}.$$

$$\therefore p_{211} = p_{121} = p_{112} = \frac{\pi\hbar}{a}\sqrt{6}.$$

Similarly,

$$p_{211} = p_{121} = p_{212} = \frac{\pi\hbar}{a}3.$$

2.18 Answers of Some Important Questions

2.1 State the characteristic of Schrödinger wave equation.

Ans. See Art. 2.4. (P. 65)

2.2 Why $|\psi(x,t)|^2$ instead of $\psi(x,t)$ is taken as the probability density?

Ans. Since $\psi(x,t)$ is a complex quantity, it cannot have any direct physical meaning. Again, probability of finding a particle between x and $x+dx$ is a *real positive quantity,* so we take $|\psi(x,t)|^2$ as the measure of probability density as it is a real positive quantity.

2.3 What are the essential conditions to be fulfilled for a function to be well-behaved?

Ans. See Art. 2.5: *Limitations of the wave function.* (P. »)

2.4 Why the Schrödinger wave equation is not valid for relativistic particles? [HPU 1995; GNDU 1997]

Ans. In deriving the Schrödinger wave equation, the non-relativistic (classical) expression for the total energy of the particle, viz., $E = \frac{p^2}{2m} + V$ has been assumed. Here, the momentum $p = mv$ and KE $\frac{1}{2}mv^2$ are the non-relativistic expressions. For the above reasons Schrödinger equation breaks down for relativistic particles.

2.5 Is it possible to derive Schrödinger wave equation?

Ans. No. It is not possible to derive Schrödinger wave equation just as we cannot derive Newton's equation. The validity of these equations lies in the experimental verification of the results obtained from them.

We can at best suggest a way to construct Schrödinger wave equation.

2.6 What is the importance of Schrödinger equation?

Ans. Schrödinger wave equation is the basic equation governing the behaviour of material particles. Schrödinger equation along with the Max Born's probability interpretation forms the foundation of quantum mechanics.

2.7 What do you mean by steady state?

Ans. A steady state (or stationary state) of a system is that when the system does not change with time. The wave function of the steady state is a function of space coordinates, e.g., ψ, $(\vec{r})$ and does not contain time. The wave function $\psi(\vec{r}), \nabla\psi(\vec{r})$ must be *single-valued, finite* and continuous for all values of $\vec{r}$.

2.8 What are observable and operator?

Ans. *Observable: Any physical quantity which can be experimentally measured is called an observable.* For example, position, momentum, energy — all are observables. An observable is, therefore, always a *real quantity* and is the result of actual measurement.

It is to be noted that (i) The observable depends on the *measuring device.* (ii) Simultaneous measurement of conjugate variables like position and momentum, etc. is associated with an *error according to uncertainty principle.* (iii) Certain quantum mechanical observables like spin, symmetry, etc. *do not have equivalent quantities in classical mechanics.*

Operator. An operator is a mathematical rule or prescription. An operator is that which operating on a wave function changes it into another, e.g., $\hat{a}\psi(x) = g(x)$.

Here, $\psi(x)$ being operated by an operator $\hat{a}$ is changed to a new function $g(x)$.

2.9 What do you mean by eigenfunction, eigenvalue and eigenvalue equation?

Ans. The wave function ψ is an eigenfunction of the operator $\hat{\alpha}$, if it operating on ψ gives back the same wave function ψ multiplied by a constant E, i.e., $\hat{\alpha}\psi = E\psi$. *E is the eigenvalue belonging to the operator $\hat{\alpha}$.* The equation $\hat{\alpha}\psi = e\psi$ is called the *eigenvalue equation.*

2.10 Explain the physical significance of $\psi(x, t)$?

Ans. The wave function represents the behaviour of a moving particle in one dimension at a point x and time t. The magnitude of $\psi(x, t)$, i.e., $|\psi(x, t)|$ is large in the region, where the probability of finding the particle is large and $|\psi(x, t)|$ is small in regions, where the probability of finding the particle is small.

Thus $\psi(x, t)$ represents the probability of finding the particle in a certain region.

2.11 What do you mean by degeneracy?

Ans. *The property of two or more quantum states of a particle (having different sets of quantum numbers and different eigenfunctions) having the same value of momentum and energy, is called degeneracy.*

Example. Consider a free particle enclosed in a cubical box. The three sets of quantum numbers for the first excited state is (211), (121) and (112); the corresponding eigenfunctions are ψ_{211}; ψ_{121}; ψ_{112} — all have the same energy

$$E_{211} = E_{121} = E_{112} = \frac{\pi^2\hbar^2}{2ma^2}\cdot 6.$$

Hence, the first excited state is *three-fold degenerate.*

Note. A state is called non-degenerate if there is only one eigenfunction belonging to a given energy eigenvalue.

Example. In the above case, the lowest quantum state (111) is having only one eigenfunction ψ_{111} and only one energy value $E_{111} = \frac{\pi^2\hbar^2}{2ma^r}\cdot 3$.

So, the state (111) is *non-degenerate.*

2.12 Show that a particle cannot exist in a one-dimensional box if its energy is zero.

Ans. The wave function for a particle enclosed in a one-dimensional box is given by

$$\psi_n(x) = \sqrt{\frac{2}{a}}\sin\frac{n\pi x}{a}.$$

The energy of the particle is

$$E_n = \frac{n^2\pi^2\hbar^2}{2ma^2}.$$

Now, if $E_0 = 0$, then $n = 0$ $[\because \pi \neq 0, \hbar \neq 0, m \neq 0, a \neq 0]$.

$$\therefore \psi_0(x) = \sqrt{\frac{2}{a}} \sin \frac{0\pi x}{a} = 0.$$

Hence, the particle cannot be present in the box, if $E = 0$.

2.19 Solved Problems

1.1 Why cannot we represent matter waves associated with a free particle by a wave function $\psi(x,t) = A\sin(\omega t - kx)$? [GNDU 1995]

Solution. The wave, $\psi(x,t) = A\sin(\omega t - kx)$ represents a wave of wavelength $\lambda = \frac{2\pi}{k}$ and frequency $\nu = \frac{\omega}{2\pi}$ associated with a particle travelling along the positive X-direction. If the same particle travels along the negative X-direction, then the wave equation will be $\psi(x,t) = A\sin(\omega t + kx)$.

Now the resultant wave due to their superposition is

$$\psi(x,t) = A\sin(\omega t - kx) + A\sin(\omega t + kx) = 2A\cos kx \sin \omega t. \qquad (1)$$

The above (1) shows that $\psi(x,t) = 0$ at $t = 0$.

As $\psi(x,t)$ represented by (1) might also be a wave function according to principle, but as $\psi(x,t) = 0$ at $t = 0$, the probability of finding the particle is zero everywhere.

So $\psi(x,t) = A\sin(\omega t - kx)$ cannot represent a wave function.

1.2 Out of the given wave functions $\psi(x) = e^x$ and $\psi(x) = e^{-x}$, which one is an acceptable quantum mechanical wave function?

Solution. For the function $\psi(x) = e^x$, we see that $\psi(x) \to \infty$ as $x \to \infty$. For acceptable wave function $\psi(x)$ must have finite value for $x \to \infty$. So, $\psi(x) = e^x$ cannot be an acceptable wave function.

For the function $\psi(x) = e^{-x}$, we see that $\psi(x) \to 0$ as $x \to \pm\infty$. Hence, $\psi(x) = e^{-x}$ is an acceptable wave function.

1.3 Show that the time-dependent part of the wave function $\psi(x,t)$ is given by $f(t) = Ae^{-iEt/\hbar}$.

Solution. $\psi(x,t) = u(x)f(t)$, where $u(x)$ and $f(t)$ represent respectively the time-independent part and time-dependent part of $\psi(x,t)$.

Now

$$\because \psi(x,t) = Ae^{i(kx - \omega t)}$$

$$\therefore \frac{d\psi}{dt} = A(-i\omega)e^{i(kx-\omega t)} = -i\omega\psi \text{ or } u(x)\frac{df}{dt} = -i\omega u(x)f(t)$$

$$\text{or } \frac{df}{dt} = (-i\omega)f(t) \text{ or } \frac{df}{f} = -i\omega dt.$$

Integrating,

$$\ln f = -i\omega t + \ln A \text{ or } \ln\left(\frac{f}{A}\right) = -i\omega t \text{ or } f = Ae^{-i\omega t} = Ae^{-\frac{iEt}{\hbar}} \; [\because E = \hbar\omega] \text{ (Proved)}$$

1.4 Normalise the wave function

$$\psi(x) = \begin{cases} A\sin\left(\frac{\pi x}{a}\right), & 0 < x < a \\ 0, & \text{outside.} \end{cases}$$

[HPU 1995]

Solution. The condition of normalisation is

$$\int_0^a \psi^*\psi\,dx = 1 \text{ or } A^*A\int_0^a \sin^2\left(\frac{\pi x}{a}\right)dx = 1 \text{ or } A^*A\int_0^a \frac{1}{2}\left(1-\cos\frac{2\pi x}{a}\right)dx = 1$$

$$\text{or } \frac{|A|^2}{2}\left[\int_0^a dx - \int_0^a \cos\frac{2\pi x}{a}dx\right] = 1 \text{ or } [x]_0^a - \frac{a}{2\pi}\left[\sin\frac{2\pi x}{a}\right]_0^a = \frac{2}{|A|^2} \text{ or } a = \frac{2}{A^2} \text{ or } A = \sqrt{\frac{2}{a}}.$$

The normalised wave function is $\psi(x) = \sqrt{\frac{2}{a}}\sin\left(\frac{\pi x}{a}\right)$.

1.5 **Normalise the wave function $\psi(x) = Ae^{-\frac{\sigma^2x^2}{2}}e^{ikx}$.**

Solution. From the condition of normalisation

$$\int_{-\infty}^{+\infty}|\psi|^2dx = 1 \text{ or } |A|^2\int_{-\infty}^{+\infty}\left(e^{-\frac{\sigma^2x^2}{2}}\right)^2\left(e^{ikx}\right)\left(e^{-ikx}\right)dx = 1$$

$$\text{or } |A|^2\int_{-\infty}^{+\infty}e^{-\sigma^2x^2}dx = 1 \text{ or } |A|^2\frac{\sqrt{\pi}}{\sigma} = 1\left[\because \int_{-\infty}^{+\infty}e^{-\sigma^2x^2}dx = \frac{\sqrt{\pi}}{\sigma}\right]$$

$$\text{or } A = \left(\frac{\sigma}{\sqrt{\pi}}\right)^{1/2}.$$

The normalised wave function is

$$\psi(x) = \left(\frac{\sigma}{\sqrt{\pi}}\right)^{1/2}e^{-\frac{\sigma^2x^2}{2}}e^{ikx}.$$

1.6 Show that the wave functions given by $\psi_n(x) = \sqrt{\frac{2}{L}}\sin\frac{n\pi x}{L}$ are orthonormal.

Solution. Let $\psi_n(x)$ and $\psi_m(x)$ be the wave functions corresponding to a particle having energies E_n and E_m respectively, n and m being integers.

Then

$$\psi_n(x) = \sqrt{\frac{2}{L}}\sin\frac{n\pi x}{L} \text{ and } \psi_m(x) = \sqrt{\frac{2}{L}}\sin\frac{m\pi x}{L}.$$

$$\therefore \int_0^L \psi_m^*\psi_n\,dx = \frac{2}{L}\int_0^L \sin\frac{n\pi x}{L}\sin\frac{m\pi x}{L}dx = 0 \text{ [if } n \neq m]$$

$$\text{and } \int_0^L \psi_m^*\psi_n\,dx = 1, \text{ when } n = m.$$

Hence *the functions $\psi_n(x)$ and $\psi_m(x)$ are orthonormal.*

1.7 The wave function for a particle is given by $\psi(x) = Ce^{-\alpha^2x^2}$, $-\infty < x < \infty$, where C and α are constants. Calculate the probability of finding the particle in the region $0 < x < \infty$.

Solution. The probability of finding the particle in $0 < x < \infty$ is

$$P = \int_0^\infty \psi^* \psi \, dx = \int_0^\infty C^* C e^{-2\alpha^2 x^2} dx = |C|^2 \int_0^\infty e^{-2\alpha^2 x^2} dx.$$

$$\therefore P = C^2 \frac{\sqrt{\pi}}{2\alpha} \left[\because \int_0^\infty e^{-2\alpha^2 x^2} dx = \frac{\sqrt{\pi}}{2\alpha} \right]$$

Again, to find the value of C we use the normalisation condition

$$\int_{-\infty}^{+\infty} |\psi|^2 d\, dx = \int_{-\infty}^{+\infty} |C|^2 e^{-2\alpha^2 x^2} dx = 1 \text{ or } C^2 = \frac{\sqrt{\pi}}{\alpha}.$$

$$\therefore P = C^2 \frac{\sqrt{\pi}}{2\alpha} = \frac{\alpha}{\sqrt{\pi}} \times \frac{\sqrt{\pi}}{2\alpha} = \frac{1}{2}.$$

$\therefore$ the probability of finding the particle in the region $-\infty < x < \infty$ is 50%.

1.8 The ground state and the first excited state wave functions of an atom are ψ_0 and ψ_1 respectively, the corresponding energies being E_0 and E_1. If the system has a 40% probability of being found in the ground state and 60% probability of being found in the first excited state (a) What is the wave function of the atom? (b) What is the average energy of the atom? [CUH 1995]

Solution. (a) Since the ground state and first excited state wave functions are ψ_0 and ψ_1 respectively, the wave function of the atom is $\psi = C_0\psi_0 + C_1\psi_1$, where C_0 and C_1 are constants.

As the wave functions are normalised,

$$\therefore \int \psi^* \psi \, dV = 1 \text{ or } \int (C_0^* \psi_0^* + C_1^* \psi_1^*)(C_0 \psi_0 + C_1 \psi_1) \, dV = 1$$

$$\text{or } C_0^* C_0 \int \psi_0^* \psi_0 \, dV + C_0 C_1 \int \psi_1^* \psi_1 \, dV = 1 \text{ or } C_0^* C_0 + C_1^* C_1 = 1.$$

But according to the condition of the problem, the probability of finding the particle in the ground state and first excited state are respectively 0.4 and 0.6.

$\therefore C_0^* C_0 = 0.4$ and $C_1^* C_1 = 0.6$.

$\therefore C_0 = \sqrt{0.4}$ and $C_1 = \sqrt{0.6}$.

$\therefore$ the normalised wave function of the atom is $\psi = \sqrt{0.4}\psi_0 + \sqrt{0.6}\psi_1$.

(b) The energy eigenvalues for ψ_0 and ψ_1 are E_0 and E_1.

$\therefore$ the average energy of the atom $\langle E \rangle = 0.4E_0 + 0.6E_1$.

1.9 If $\psi_1(x,t)$ and $\psi_2(x,t)$ are both solutions of Schrödinger time-dependent equation for a given potential $V(x,t)$, show that $\psi = a_1\psi_1 + a_2\psi_2$ in which a_1 and a_2 are arbitrary constants is also a solution. [CUH 1993; PU 1991]

Solution. The time-dependent Schrödinger equation for a particle moving along X-direction in a given potential $V(x,t)$,

$$-\frac{\hbar^2}{2m} \frac{\partial^2}{\partial x^2} \psi(x,t) + V(x,t)\psi(x,t) = i\hbar \frac{\partial}{\partial t} \psi(x,t). \quad (1)$$

Since $\psi_1(x,t)$ and $\psi_2(x,t)$ are solutions of the above equation, we have

$$-\frac{\hbar^2}{2m}\frac{\partial^2}{\partial x^2}\psi_1(x,t)+V(x,t)\psi_1(x,t)=i\hbar\frac{\partial}{\partial t}\psi_1(x,t) \tag{2}$$

$$\text{and } -\frac{\hbar^2}{2m}\frac{\partial^2}{\partial x^2}\psi_2(x,t)+V(x,t)\psi_2(x,t)=i\hbar\frac{\partial}{\partial t}\psi_2(x,t). \tag{3}$$

Multiplying (2) by a_1 and (3) by a_2 and then adding, we get

$$-\frac{\hbar^2}{2m}\left[\frac{\partial^2}{\partial x^2}(a_1\psi_1+a_2\psi_2)+V(a_1\psi_1+a_2\psi_2)\right]=i\hbar\frac{\partial}{\partial t}(a_1\psi_1+a_2\psi_2). \tag{4}$$

Now comparing (1) and (4), we get

$$\psi(x,t)=a_1\psi_1+a_2\psi_2. \tag{5}$$

Thus $\psi=a_1\psi_1+a_2\psi_2$ is also a solution of Schrödinger time-dependent equation.

1.10 If $\psi(x)=\sqrt{a}\,e^{-ax}$ be the one-dimensional wave function of a particle, find the probability of finding it between $x=\frac{1}{a}$ and $x=\frac{2}{a}$.

Solution. The probability of finding the particle between x and $x+dx$ is

$$P(x)\,dx=\psi^*\psi\,dx=\sqrt{a}\,e^{-ax}\times\sqrt{a}\,e^{-ax}\,dx=ae^{-2ax}\,dx.$$

$\therefore$ the probability of finding the particle between $x=1/a$ and $x=2/a$ is

$$P=\int_{1/a}^{2/a}ae^{-2ax}=a\left[\frac{e^{-2ax}}{-2a}\right]_{1/a}^{2/a}=-\frac{1}{2}e^{-2a\times\frac{2}{a}}+\frac{1}{2}e^{-2a\times\frac{1}{a}}=\left(\mathbf{e}^{-2}-\mathbf{e}^{-4}\right)/2.$$

1.11 Using operator representation of the x-component of momentum of a particle, prove that

$$(\hat{x}\hat{p}_x-\hat{p}_x\hat{x})=i\hbar.$$

Solution. If ψ be an arbitrary function, then

$$(\hat{x}\hat{p}_x-\hat{x}\hat{p}_x)\psi=\left[x\left(-i\hbar\frac{\partial\psi}{\partial x}\right)+\left(i\hbar\frac{\partial}{\partial x}(x\psi)\right)\right]=-i\hbar x\frac{\partial\psi}{\partial x}+i\hbar\left(x\frac{\partial\psi}{\partial x}+\psi\right)$$

$$=-i\hbar x\frac{\partial\psi}{\partial x}+i\hbar x\frac{\partial\psi}{\partial x}+i\hbar\psi=i\hbar\psi.$$

$\therefore\ \boldsymbol{\hat{x}\hat{p}_x-\hat{p}_x\hat{x}=i\hbar}.$

1.12 What is commutator? Find the commutator of $\left(\frac{d}{dx}\right)(\hat{x})-(\hat{x})\left(\frac{d}{dx}\right)$.

Solution. If $\hat{\alpha}$ and $\hat{\beta}$ be two operators, then $\hat{\alpha}\hat{\beta}-\hat{\beta}\hat{\alpha}$ is a new operator and is called commutator of the two operators $\hat{\alpha}$ and $\hat{\beta}$. If $\hat{\alpha}$ and $\hat{\beta}$ commute, then $\hat{\alpha}\hat{\beta}-\hat{\beta}\hat{\alpha}=[\hat{\alpha},\hat{\beta}]=0$.

If we write, $\hat{\alpha}=\frac{d}{dx}$ and $\hat{\beta}=\hat{x}$, then we can show that

$$\left(\frac{d}{dx}\right)\hat{x}-\hat{x}\left(\frac{d}{dx}\right)=\left[\frac{d}{dx},\hat{x}\right]=1.$$

Thus $\frac{d}{dx}$ and $\hat{x}$ do not commute.

To prove this let $\psi(x)$ be an arbitrary wave function, then

$$\left(\frac{d}{dx}\right)\hat{x}\psi(x)=\frac{d}{dx}[x\psi(x)]=x\frac{d\psi}{dx}+\psi(x)$$

$$\text{and } \hat{x}\left(\frac{d}{dx}\psi(x)\right)=x\frac{d\psi}{dx}.$$

$$\therefore \left[\left(\frac{d}{dx}\right)\hat{x}-\hat{x}\left(\frac{d}{dx}\right)\right]\psi(x)=x\frac{d\psi}{dx}+\psi(x)-x\frac{d\psi}{dx}$$

$$\text{or } \left[\left(\frac{d}{dx}\right)\hat{x}-\hat{x}\left(\frac{d}{dx}\right)\right]\psi(x)=\psi(x)$$

$$\text{or } \left(\frac{d}{dx}\right)\hat{x}-\hat{x}\left(\frac{d}{dx}\right)=1.$$

Thus the commutator is a unit operator.

1.13 Show that (a) $\left(1+\frac{d}{dx}\right)^2=1+2\frac{d}{dx}+\frac{d^2}{dx^2}$; (b) $\left(\frac{\partial}{\partial x}+\frac{\partial}{\partial y}\right)^2=\frac{\partial^2}{\partial x^2}+2\frac{\partial^2}{\partial x\partial y}+\frac{\partial^2}{\partial y^2}$.

Solution. (a) Let $\psi(x)$ be an arbitrary function, then

$$\left(1+\frac{d}{dx}\right)^2\psi=\left(1+\frac{d}{dx}\right)\left(1+\frac{d}{dx}\right)\psi=\left(1+\frac{d}{dx}\right)\left(\psi+\frac{d\psi}{dx}\right)$$

$$=\psi+\frac{d\psi}{dx}+\frac{d\psi}{dx}+\frac{d^2\psi}{dx^2}=\left(1+2\frac{d}{dx}+\frac{d^2}{dx^2}\right)\psi.$$

$\because \psi$ is arbitrary,

$$\therefore \left(1+\frac{d}{dx}\right)^2=1+2\frac{d}{dx}+\frac{d^2}{dx^2} \text{ (Proved).}$$

(b) $$\left(\frac{\partial}{\partial x}+\frac{\partial}{\partial y}\right)^2\psi=\left(\frac{\partial}{\partial x}+\frac{\partial}{\partial y}\right)\left(\frac{\partial}{\partial x}+\frac{\partial}{\partial y}\right)\psi=\left(\frac{\partial}{\partial x}+\frac{\partial}{\partial y}\right)\left(\frac{\partial\psi}{\partial x}+\frac{\partial\psi}{\partial y}\right)$$

$$=\frac{\partial^2\psi}{\partial x^2}+\frac{\partial^2\psi}{\partial x\partial y}+\frac{\partial^2\psi}{\partial y\partial x}+\frac{\partial^2\psi}{\partial y^2}=\left(\frac{\partial^2}{\partial x^2}+2\frac{\partial^2}{\partial x\partial y}+\frac{\partial^2}{\partial y^2}\right)\psi.$$

$\because \psi$ is arbitrary,

$$\left(\frac{\partial}{\partial x}+\frac{\partial}{\partial y}\right)^2=\frac{\partial^2}{\partial x^2}+2\frac{\partial^2}{\partial x\partial y}+\frac{\partial^2}{\partial y^2} \text{ (Proved).}$$

1.14 Prove that if $[\hat{\alpha},\hat{\beta}]=1$, then $[\hat{\alpha},2\hat{\beta}^2]=2\hat{\beta}$.

Solution.

$$\hat{\alpha}\hat{\beta}-\hat{\beta}\hat{\alpha}=1 \text{ or } \hat{\beta}\hat{\alpha}\hat{\beta}-\hat{\beta}^2\hat{\alpha}=\hat{\beta} \tag{1}$$

$$\text{and } \hat{\alpha}\hat{\beta}^2-\hat{\beta}\hat{\alpha}\hat{\beta}=\hat{\beta}. \tag{2}$$

Adding (1) and (2), we get $\hat{\alpha}\hat{\beta}^2-\hat{\beta}^2\hat{\alpha}=2\hat{\beta}$ or, $[\hat{\alpha},2\hat{\beta}^2]=2\hat{\beta}$ **(Proved).**

1.15 Prove that (a) $[\hat{L}_x,y]=i\hbar z$; (b) $[\hat{L}_x,x]=0$; (c) $[\hat{L}_z,y]=-i\hbar x$.

Solution. (a) We have

$$[\hat{L}_x, \hat{y}] = [(yp_z - zp_y), y] = -i\hbar\left[\left\{y\frac{\partial}{\partial z} - z\frac{\partial}{\partial y}\right\}, y\right]$$

$$= -i\hbar\left[\left(y\frac{\partial}{\partial z} - z\frac{\partial}{\partial y}\right)y - y\left(y\frac{\partial}{\partial z} - z\frac{\partial}{\partial y}\right)\right] = -i\hbar(0 - z.1) = i\hbar z.$$

$\therefore\ [\hat{L}_x, \hat{y}] = i\hbar z$ **(Proved).**

(b) $$[\hat{L}_x, \hat{x}] = [(yp_z - zp_y), x] = -i\hbar\left[\left\{y\frac{\partial}{\partial z} - z\frac{\partial}{\partial y}\right\}x - x\left\{y\frac{\partial}{\partial z} - z\frac{\partial}{\partial y}\right\}\right]$$

$$= -i\hbar[0 - 0] = 0.$$

$\therefore\ [\hat{L}_x, \hat{x}] = 0$ **(Proved).**

(c) $$[\hat{L}_z, \hat{y}] = [L_z y - yL_z] = [(xp_y - yp_x)\,y - y\,(xp_y - yp_x)]$$

$$= -i\hbar\left[\left(x\frac{\partial}{\partial y} - y\frac{\partial}{\partial x}\right)y - y\left(x\frac{\partial}{\partial y} - y\frac{\partial}{\partial x}\right)\right]$$

$$= -i\hbar[x.1 - 0] = -i\hbar x.$$

$\therefore\ [\hat{L}_z, \hat{y}] = -i\hbar x$ **(Proved).**

1.16 Prove that (a) $[\hat{L}_x, \hat{L}_y] = i\hbar L_z$, (b) $[\hat{L}_y, \hat{L}_z] = i\hbar L_x$, (c) $[\hat{L}_z, \hat{L}_x] = i\hbar L_y$. [WBUT 2008]

Solution. (a)

$$[\hat{L}_x, \hat{L}_y] = [L_x L_y - L_y L_z]$$

$$= -\hbar^2\left[\left(y\frac{\partial}{\partial z} - z\frac{\partial}{\partial y}\right)\left(z\frac{\partial}{\partial x} - x\frac{\partial}{\partial z}\right) - \left(z\frac{\partial}{\partial x} - x\frac{\partial}{\partial z}\right)\left(y\frac{\partial}{\partial z} - z\frac{\partial}{\partial y}\right)\right]$$

$$= -\hbar^2\left[y\frac{\partial}{\partial x} - x\frac{\partial}{\partial y}\right] = i\hbar L_z \text{ (Proved).}$$

Similarly, it can be proved that $[\hat{L}_y, \hat{L}_z] = i\hbar L_x$ and $[L_z, L_x] = i\hbar L_y$.

1.17 Prove that $[\hat{L}^2, \hat{L}_z] = 0$.

Solution.

$$[\hat{L}^2, \hat{L}_z] = \left[\left(\hat{L}_x^2 + \hat{L}_y^2 + \hat{L}_z^2\right), \hat{L}_z\right]$$

$$= \left(\hat{L}_x^2\hat{L}_z - \hat{L}_z\hat{L}_x^2\right) + \left(\hat{L}_y^2\hat{L}_z - \hat{L}_z\hat{L}_y^2\right) + \left(\hat{L}_z^2\hat{L}_z - \hat{L}_z\hat{L}_z^2\right)$$

$$= \left(\hat{L}_x\hat{L}_x\hat{L}_z - \hat{L}_z\hat{L}_x\hat{L}_x\right) + \left(\hat{L}_y\hat{L}_y\hat{L}_z - \hat{L}_z\hat{L}_y\hat{L}_y\right),$$

$\because\ \left(\hat{L}_z^2\hat{L}_z - \hat{L}_z\hat{L}_z^2\right) = 0.$

Now adding and subtracting, $\hat{L}_x\hat{L}_z\hat{L}_x$ and $\hat{L}_y\hat{L}_z\hat{L}_y$ to RHS, we get

$$\begin{aligned}
\left[\hat{L}^2,\hat{L}_z\right] &= \left(\hat{L}_x\hat{L}_x\hat{L}_z-\hat{L}_x\hat{L}_z\hat{L}_x+\hat{L}_x\hat{L}_z\hat{L}_x-\hat{L}_z\hat{L}_x\hat{L}_x\right)\\
&\quad+\left(\hat{L}_y\hat{L}_y\hat{L}_z-\hat{L}_y\hat{L}_z\hat{L}_y+\hat{L}_y\hat{L}_z\hat{L}_y-\hat{L}_z\hat{L}_y\hat{L}_y\right)\\
&=\hat{L}_x\left(\hat{L}_x\hat{L}_z-\hat{L}_z\hat{L}_x\right)+\left(\hat{L}_x\hat{L}_z-\hat{L}_z\hat{L}_x\right)\hat{L}_x\\
&\quad+\hat{L}_y\left(\hat{L}_y\hat{L}_z-\hat{L}_z\hat{L}_y\right)+\left(\hat{L}_y\hat{L}_z-\hat{L}_z\hat{L}_y\right)\hat{L}_y\\
&=-i\hbar\hat{L}_x\hat{L}_y-i\hbar\hat{L}_y\hat{L}_x+2i\hbar\hat{L}_x\hat{L}_y=0.
\end{aligned}$$

$\therefore$ $\left[\hat{L}^2,\hat{L}_z\right]=0$ **(Proved).**

1.18 Show that the function $\psi(x)=Ae^{ikx}+Be^{-ikx}$ is an eigenfunction of the KE operator $\left(\hat{E}_k\right)$.

Solution. The KE operator

$$\hat{E}_k=\frac{\hat{p}_x^2}{2m}=\frac{1}{2m}\left(-i\hbar\frac{d}{dx}\right)\left(-i\hbar\frac{d}{dx}\right)=-\frac{\hbar^2}{2m}\frac{d^2}{dx^2}.$$

Now

$$\hat{E}_k\psi=-\frac{\hbar^2}{2m}\frac{d^2\psi}{dx^2}=-\frac{\hbar^2}{2m}\left[(ik)^2Ae^{ikx}+(-ik)^2Be^{-ikx}\right]=\frac{\hbar^2k^2}{2m}\left[Ae^{ikx}+Be^{-ikx}\right].$$

$$\therefore\ \hat{E}_k\psi=\frac{\hbar^2k^2}{2m}\psi.$$

The above is an eigenvalue equation and ψ is an eigenfunction corresponding to $\hat{E}_k$ having eigenvalue $\frac{\hbar^2k^2}{2m}$.

1.19 Show that the expectation value of linear momentum for the wave function

$$\psi(x)=\begin{cases}\left(\frac{2}{L}\right)^{1/2}\sin\frac{\pi x}{L}, & 0<|x|<L\\ 0, & |x|>1.\end{cases}$$

is zero.

Solution.

$$\begin{aligned}
\langle p_x\rangle &=\int_{-\infty}^{+\infty}\psi^*\hat{p}_x\psi\,dx\Big/\int_{-\infty}^{+\infty}\psi^*\psi\,dx\\
&=\int_{-\infty}^{+\infty}\psi^*\left(-i\hbar\frac{\partial}{\partial x}\right)\psi\,dx\Big/\int_{-\infty}^{+\infty}\psi^*(x)\psi(x)\,dx\\
&=\frac{1}{2}\int_{-L}^{L}\left(\frac{2}{L}\right)\sin\frac{\pi x}{L}(-i\hbar)\frac{\pi}{L}\cos\frac{\pi x}{L}\,dx=\frac{\pi\hbar}{2iL^2}\int_{-L}^{+L}2\sin\frac{\pi x}{L}\cos\frac{\pi x}{L}\,dx\\
&=\frac{\pi\hbar}{2iL^2}\int_{-L}^{L}\sin\frac{2\pi x}{L}\,dx=0.
\end{aligned}$$

$\therefore$ $\langle p_x\rangle=0$ **(Proved).**

1.20 The normalised wave function for one-dimensional motion of a particle is given by $\psi = A\sin\alpha x$. What is the expectation value of KE for the particle if it is confined in a region between $x=0$ and $x=5$.

Solution. The operator corresponding to the KE of the particle is

$$\hat{E}_k = \frac{\hat{p}_x^2}{2m} = \frac{1}{2m}\left(-i\hbar\frac{d}{dx}\right)\left(-i\hbar\frac{d}{dx}\right) = -\frac{\hbar^2}{2m}\frac{d^2}{dx^2}.$$

Now, the expectation value of KE is

$$\langle E_k\rangle = \int_0^5 \psi^* \hat{E}_x \psi\, dx = -\frac{A^2\hbar^2}{2m}\int_0^5 (\sin\alpha x)\frac{d^2}{dx^2}(\sin\alpha x)\, dx$$

$$= \frac{\alpha^2 A^2\hbar^2}{2m}\int_0^5 \sin^2\alpha x dx = \frac{\alpha^2 A^2\hbar^2}{2m}\int_0^5 \frac{1-\cos 2\alpha x}{2} dx.$$

$$\therefore \langle E_k\rangle = \frac{\boldsymbol{\alpha^2 A^2\hbar^2}}{\mathbf{4m}}\left[\mathbf{5} - \frac{\sin \mathbf{10\alpha}}{\mathbf{2\alpha}}\right].$$

1.21 Find the expectation value of p_x^2 for the wave function

$$\psi(x) = \begin{cases} \left(\frac{2}{L}\right)^{1/2}\sin\frac{\pi x}{L}, & 0 < x < L \\ 0, & \text{for } |x| > L. \end{cases}$$

Solution. The expectation value of p_x^2 is

$$\langle p_x^2\rangle = \int_{-\infty}^{+\infty} \psi^*\left(-\hbar^2\frac{d^2}{dx^2}\right)\psi(x)\,dx \Big/ \int_{-\infty}^{+\infty} \psi^*(x)\psi(x)\,dx$$

$$= \frac{1}{2}\hbar^2\frac{2}{L}\left(\frac{\pi}{L}\right)^2\int_{-L}^{+L} \sin\frac{\pi x}{L}\sin\frac{\pi x}{L}dx = \frac{\pi^2\hbar^2}{2L^3}\int_{-L}^{L} 2\sin^2\frac{\pi x}{L}dx$$

$$= \frac{\pi^2\hbar^2}{2L^3}\int_{-L}^{+L}\left(1-\cos\frac{2\pi x}{L}\right)dx = \frac{\pi^2\hbar^2}{2L^3}\left[x - \frac{\sin\frac{2\pi x}{L}}{2\pi/L}\right]_{-L}^{+L}$$

$$= \frac{\pi^2\hbar^2}{2L^3}\left[2L - \frac{L}{2\pi}0\right] = \frac{\pi^2\hbar^2}{L^2}.$$

$$\therefore \langle p_x^2\rangle = \frac{\boldsymbol{\pi^2\hbar^2}}{\mathbf{L^2}}.$$

1.22 A particle is located in a one-dimensional potential well of width a and absolutely impenetrable walls at $x=0$ and $x=a$. Find the probability of finding the particle in the region $a/3 < x < 2a/3$.

Solution. The ground state ($n=1$) normalised wave function of the particle in the box is

$$\psi(x) = \sqrt{\frac{2}{a}}\sin\frac{\pi x}{a}.$$

∴ the probability of finding the particle in the interval $a/3 < x < 2a/3$ is

$$P = \int_{a/3}^{2a/3} |\psi(x)|^2 dx = \frac{2}{a}\int_{a/3}^{2a/3} \sin^2 \frac{\pi x}{a} dx$$

$$= \frac{2}{a}\int_{a/3}^{2a/3} \left(1 - \cos\frac{2\pi x}{a}\right) dx = \frac{1}{3} + \frac{\sqrt{3}}{2\pi} = 0.61.$$

1.23 Find the expectation value of x and x^2 of a particle confined in a one-dimensional box of length a if the wave function for the nth state is $\psi_n = \sqrt{\frac{2}{a}} \sin \frac{n\pi x}{a}$.

Solution. The expectation value of x is

$$\langle x \rangle = \int_0^a \psi^* \hat{x} \psi dx = \frac{2}{a}\int_0^a x \sin^2 \frac{n\pi x}{a} dx.$$

$$\therefore \langle x \rangle = \frac{a}{2}.$$

Again,

$$\langle x^2 \rangle = \int_0^a \psi^* \hat{x}^2 \psi dx = \frac{2}{a}\int_0^a x^2 \sin^2 \frac{n\pi x}{a} dx = \left(\frac{a^2}{3} - \frac{a^2}{2n^2\pi^2}\right).$$

1.24 Find the values of momentum and energy for an electron in a box of length 1 Å for $n = 1$ and $n = 2$. [Mass of the electron is 9.1×10^{31} kg and $h = 6.63 \times 10^{-34}$ J-s.] [GNDU 1997]

Solution. For a particle in a one-dimensional potential box of a side a, the momentum in the nth state is $p_n = \frac{n\hbar\pi}{a}$, where $n = 1, 2, 3, \ldots$ and energy $E_n = \frac{p_n^2}{2m} = \frac{n^2\hbar^2\pi^2}{2ma}$.

Momentum: For $n = 1$ and $n = 2$,

$$p_1 = \frac{\hbar\pi}{a} = \frac{h}{2\pi}\frac{\pi}{a} = \frac{h}{2a} = \frac{6.63 \times 10^{-34}}{2 \times 10^{-10}} = \mathbf{3.315 \times 10^{-24}\ kg\text{-}m/s.}$$

For $n = 2$,

$$p_2 = 2\frac{h}{2a} = \frac{h}{a} = \mathbf{6.630 \times 10^{-24}\ kg\text{-}m/s.}$$

Energy: For $n = 1$,

$$E_1 = \frac{\hbar^2\pi^2}{2ma^2} = \frac{h^2\pi^2}{4\pi^2 \cdot 2ma^2} = \frac{h^2}{8ma^2}$$

$$\text{or } E_1 = \frac{(6.63 \times 10^{-34})^2}{8 \times (9.1 \times 10^{-31}) \times (10^{-20})} = 60.4 \times 10^{-19}\ \text{J} = \frac{60.4 \times 10^{-19}}{1.5 \times 10^{-19}} = \mathbf{37.7\ eV.}$$

For $n = 2$,

$$E_2 = 2^2 \frac{h^2}{8ma^2} = 4 \times 37.7 = \mathbf{150.8\ eV.}$$

1.25 Find the values of momentum and energy for the ground state of an electron confined in a cubical box of side 1 Å.

Solution. For the ground state, $n_x = n_y = n_z = 1$.

∴ momentum

$$p_{111}=\frac{\pi\hbar}{a}\sqrt{3}=\frac{\pi h}{2\pi a}\sqrt{3}=\frac{h}{2a}\sqrt{3}=\frac{6.63\times10^{-34}}{2\times10^{-10}}\sqrt{3}=\left(3.315\times10^{-24}\right)\sqrt{3}$$
$$=\mathbf{5.74\times10^{-24}\ kg\text{-}m/s.}$$

Energy

$$E_{111}=\frac{\pi^2\hbar^2}{2ma^2}3=\frac{\pi^2h^2}{8\pi^2ma^3}=\frac{h^2}{8ma^3}\sqrt{3}=\frac{(6.63\times10^{-34})^2\times3}{8\times9.1\times10^{-31}\times10^{-30}}$$
$$=181\times10^{-19}\ \mathrm{J}=\frac{181\times10^{-19}}{1.6\times10^{-19}}\ \mathrm{eV}\doteq\mathbf{113\ eV.}$$

Now, for the 1st excited state one of the quantum numbers is 2.

∴ momentum

$$p_{211}=p_{121}=p_{112}=\frac{\pi\hbar}{a}\sqrt{6}=\frac{h}{2a}\sqrt{6}=3.315\times10^{-34}\sqrt{6}$$
$$=\mathbf{8.13\times10^{-24}\ kg\text{-}m/s.}$$

Energy

$$E_{211}=E_{121}=E_{112}=\frac{\pi^2\hbar^2}{2ma^2}6=\frac{h^2}{8ma^2}6=60.4\times6\times10^{-10}=362\times10^{-19}\ \mathrm{J}$$
$$=\frac{362\times10^{-19}}{1.6\times10^{-19}}\ \mathrm{eV}=\mathbf{226\ eV.}$$

1.26 The wave function of the ground state for H-atom is $\psi(r)=\left(\frac{1}{\pi a^3}\right)^{1/2}e^{-\frac{r}{a}}$, where r is the distance of the electron from the nucleus and a is a constant. Find the expectation value of r.

Solution. The expectation value of r.

$$\langle r\rangle=\int\psi^*r\psi\,dV=\iiint\psi^*r\psi r^2\sin\theta\,d\theta\,d\phi\,dr,$$

where $dV=r^2\sin\theta\,d\theta\,d\phi\,dr$ = the volume element in spherical polar coordinates.

$$\therefore\langle r\rangle=\frac{1}{\pi a^3}\left(\int_0^\infty r^3e^{-2r/a}dr\right)\times\left(\int_0^\pi\sin\theta\,d\theta\right)\times\left(\int_0^{2\pi}d\phi\right)$$
$$=\frac{1}{\pi a^3}\times4\pi\times\int_0^\infty r^3e^{-\frac{r}{2a}}dr=\frac{4}{a^3}\times\frac{3}{8}\times a^4=\frac{3}{2}\boldsymbol{a}.$$

EXERCISES

2(A) Objective Type Questions

Choose the correct alternatives.

2.1 Which of the following wave function is the solution of Schrödinger equation?

(a) $A \sec x$ (c) Ae^{-x^2}
(b) $A \tan x$ (d) Ae^{x^2}. [WBUT 2007]

2.2 What is the value of $\left[\hat{x}, \frac{\partial}{\partial x}\right]$?
(a) −1 (c) −2
(b) 0 (d) 1. [WBUT 2007]

2.3 Which of the wave functions represent(s) a particle moving along the positive X-direction?
(a) $Ae^{-i(kx-\omega t)}$ (c) $Ae^{i(kx-\omega t)}$
(b) $A\cos(kx-\omega t)$ (d) $A\sin(kx-\omega t)$.

2.4 The probability of finding a particle between x and $x+dx$ is
(a) $\psi^*\psi$ (c) $\psi^*\psi dx$
(b) ψ (d) ψ^*/ψ.

2.5 The probability density for a stationary state is
(a) independent of space coordinate (c) depends on time
(b) independent on time (d) none of these.

2.6 Schrödinger wave equation for a moving particle contains
(a) $\frac{\partial \psi}{\partial t}$ (c) ψ
(b) $\frac{\partial^2 \psi}{\partial t^2}$ (d) $\frac{\partial^3 \psi}{\partial t^3}$.

2.7 The normalisation factor of the wave function $\psi_n(x) = A\sin\frac{n\pi x}{a}$ is
(a) $\sqrt{\frac{1}{a}}$ (c) $\sqrt{\frac{a}{2}}$
(b) $\sqrt{\frac{2}{a}}$ (d) $\frac{2}{a}$.

2.8 For a one-dimensional motion of a particle the normalisation condition of the wave function is
(a) $\int_{-\infty}^{+\infty} \psi^*(x)\psi(x)\,dx = 1$ (c) $\int_{-\infty}^{+\infty} \psi^*(x)\psi(x)\,dx = \infty$
(b) $\int_{-\infty}^{+\infty} \psi^*(x)\psi(x)\,dx = 0$ (d) none of these.

2.9 The expectation value of the x-component of momentum of a particle is
(a) 0 (c) $\frac{h}{2a}$
(b) $\frac{\pi h}{a}$ (d) $\frac{2a}{\pi h}$.

2.10 The expectation value of a dynamical variable α is given by
(a) $\int \psi^* \hat{\alpha} \psi\, d^3r$ (c) $\int \hat{\alpha} \psi^* \psi\, d^3r$
(b) $\int \psi^* \psi \hat{\alpha}\, d^3r$ (d) $\alpha \int \psi^* \psi\, d^3r$.

2.11 Energy of a particle confined in a one-dimensional potential box is

(a) $\propto n^2$ (c) $\frac{1}{n}$

(b) $\propto n$ (d) $\propto \frac{1}{n}$, where n is the q. no.

2.12 The energy of a particle in one-dimensional potential box of length a is

(a) an^2 (c) $a\frac{1}{n}$

(b) an (d) $a\frac{1}{n^2}$.

2.13 The number of nodes of the wave function of a particle confined to a one-dimensional box for state 3 is

(a) 3 (c) 1

(b) 4 (d) 2.

2.14 If E_1 and E_n be the energy of a particle in a one-dimensional box, then

(a) $E_n = nE_1$ (c) $E_n = n^2 E_1$

(b) $E_n = (n+1)E_1$ (d) $E_n = \frac{E_1}{n}$.

2.15 The spacing between nth and $(n+1)$th energy levels for a particle in a one-dimensional box is

(a) $(2n+1)E_1$ (c) nE_1

(b) $(n+1)E_1$ (d) $(n-1)E_1$.

2.16 The ground state energy of a particle in a three-dimensional potential box is

(a) $\frac{3\pi^2\hbar^2}{2ma^2}$ (c) $\frac{\pi^2\hbar^2}{6ma^2}$

(b) $\frac{\pi^2\hbar^2}{2ma^2}$ (d) $\frac{6ma^2}{\pi^2\hbar^2}$.

2.17 The degree of degeneracy of energy levels of a particle in a three-dimension rectangular box for n_x, n_y, n_z having either of the values 1, 2, 3 is

(a) 2 (c) 6

(b) 3 (d) 9.

2.18 A system is said to be degenerate, if there is

(a) one energy level with a particular energy value

(b) several energy levels each having different energies

(c) several energy levels each having the same energy

(d) none of these.

2.19 Which one of the following functions is an eigenfunction of the operator $\frac{d^2}{dx^2}$?

(a) x (c) e^{-x^2}

(b) x^2 (d) $\cos x$.

2.20 The operator corresponding to linear momentum (p) for a particle moving along the X-axis is

(a) $\hat{p}_x = i\hbar\dfrac{\partial}{\partial x}$
(c) $\hat{p}_x = \dfrac{1}{i\hbar}\dfrac{\partial}{\partial x}$
(b) $\hat{p}_x = -i\hbar\dfrac{\partial}{\partial x}$
(d) $\hat{p}_x = \hbar\dfrac{\partial}{\partial x}$.

2.21 The operator corresponding to the energy of a moving particle is
(a) $\hat{E} = -i\hbar\dfrac{\partial}{\partial t}$
(c) $\hat{E} = -\dfrac{1}{i\hbar}\dfrac{\partial}{\partial t}$
(b) $\hat{E} = i\hbar\dfrac{\partial}{\partial t}$
(d) $\hat{E} = \hbar\dfrac{\partial}{\partial t}$.

2.22 The eigenvalue of the Hamiltonian operator associated with the eigenfunction $\psi(\vec{r})$ is
(a) total energy (E) of the system
(c) PE (V) of the system
(b) KE (K) of the system
(d) none of these.

2.23 For a stationary state of a system
(a) the wave function is independent of time
(b) the probability density is independent of time
(c) the probability density is a function of time
(d) both the wave function and the probability density are independent of time.

2.24 The wave functions $u_m(x)$ and $u_n(x)$ are said to be orthonormal in the interval a to b if
(a) $\displaystyle\int_a^b u_m^*(x)\,u_n^*(x)\,dx = 0$
(b) $\displaystyle\int_a^b u_m^*(x)\,u_n(x) = 1$
(c) $\displaystyle\int_a^b u_m^*(x)\,u_n(x) = 0$
(d) $\displaystyle\int_a^b u_m^*(x)\,u_n(x) = \delta_{mn}$, where $\delta_{mn} = \begin{cases} 1, & \text{for } m = n \\ 0, & \text{for } m \neq 0. \end{cases}$

2.25 The operator corresponding to the z-component of angular momentum is
(a) $\hat{L}_z = xp_y - yp_z$
(c) $\hat{L}_z = xp_y - yp_x$
(b) $\hat{L}_z = yp_y - zp_z$
(d) $\hat{L}_z = xp_y + yp_z$.

2.26 The expectation value for a particle in one-dimensional potential box of length a is
(a) $\dfrac{2}{3}a^2$
(c) $\dfrac{2}{5}a^2$
(b) $\dfrac{a}{2}$
(d) $\dfrac{a}{3}$.

2.27 The expectation value of the square of linear momentum $\langle p_x^2 \rangle$ of a particle having wave function $\psi(x) = \sqrt{\dfrac{2}{L}}\sin\dfrac{\pi x}{L}$, $0 < |x| < 1$ and $\psi(x) = 0$ for $|x| > L$ is
(a) $\dfrac{\pi^2\hbar^2}{L}$
(c) $\dfrac{\pi^2\hbar^2}{L^2}$
(b) $\dfrac{\pi\hbar^2}{L^2}$
(d) $\dfrac{\pi^2 L^2}{\hbar^2}$.

2.28 The expectation value of the linear momentum of problem (2.27) is

(a) 2 (c) 0

(b) 1 (d) ± 1.

2.29 If $\hat{L}$ and $\hat{L}_x$ be the total angular momentum and the x-component of angular momentum, then

(a) $[\hat{L}^2, \hat{L}_x] = -1$ (c) $[\hat{L}, \hat{L}_x] = \infty$

(b) $[\hat{L}, \hat{L}_x] = 0$ (d) $[\hat{L}, \hat{L}_x] = \pm 1$.

2.30 If $[\hat{\alpha}, \hat{\beta}] = 1$, then

(a) $[\hat{\alpha}, 2\hat{\beta}] = 2\hat{\beta}$ (c) $[\hat{\alpha}, 2\hat{\beta}^2] = 0$

(b) $[\hat{\alpha}, 2\hat{\beta}^2] = \hat{\beta}$ (d) $[\hat{\alpha}, 2\hat{\beta}^2] = 1$.

2.31 For a free particle, the three-dimensional time-dependent Schrödinger equation is

(a) $i\hbar \frac{\partial \psi}{\partial t} = \frac{\hbar^2}{2m} \nabla^2 \psi$ (c) $\frac{\partial \psi}{\partial t} = \frac{\hbar^2}{2m} \nabla^2 \psi$

(b) $-i\hbar \frac{\partial \psi}{\partial t} = \frac{\hbar^2}{2m} \nabla^2 \psi$ (d) $\frac{\partial \psi}{\partial t} = \frac{\hbar^2}{2m} \nabla \psi$.

ANSWERS

2.1 (c)	2.5 (b)	2.9 (a)	2.13 (b)	2.17 (c)	2.21 (b)	2.25 (c)	2.29 (b)
2.2 (a)	2.6 (a)	2.10 (a)	2.14 (c)	2.18 (c)	2.22 (a)	2.26 (b)	2.30 (a)
2.3 (c)	2.7 (b)	2.11 (a)	2.15 (a)	2.19 (d)	2.23 (b)	2.27 (d)	
2.4 (c)	2.8 (a)	2.12 (d)	2.16 (a)	2.20 (b)	2.24 (d)	2.28 (c)	2.31 (b)

2(B) Short Answer Type Questions

2.1 Write down the Schrödinger equation for a particle confined in a one-dimensional box. [WBUT 2006]

2.2 What is the physical significance of probability density? [cf. CUH 1997; WBUT 2003]

2.3 Evaluate the expectation value $\langle x \rangle$ for a one-dimensional potential box of length L in the ground state. [WBUT 2003]

2.4 Write down the time-independent Schrödinger equation for a free particle of mass m and momentum $\vec{p}$. [WBUT 2005]

2.5 If a wave function $\psi(x)$ of a quantum mechanical particle is given by

$$\psi(x) = \begin{cases} a \sin \frac{\pi x}{L}, & \text{for } 0 \le x \le L \\ 0, & \text{otherwise,} \end{cases}$$

then determine the value of a. Also determine the value of x, where the probability of finding the particle is maximum. [WBUT 2007]

2.6 Define eigenvalues and eigenfunctions of an operator. [BUH 2003]

2.7 Calculate the normalisation constant and probability density for a wave function given by $\psi(x) = Ae^{-\frac{\alpha^2 x^2}{2}} e^{ikx}$. [NBUH 2003]

2.8 Which of the following functions do not qualify to be solutions of Schrödinger equation for all values of x? (a) $A \sec x$ (b) $A \tan x$ (c) Ae^{x^2} (d) Ae^{-x^2}. [WBUT 2007; BUH 2006]

2.9 What is the differential operator in quantum mechanics that corresponds p_y, the y-component of linear momentum of a particle? [BUH 2004]

2.10 Evaluate the commutator $[\hat{L}_z, \hat{Z}]$. [BUH 2004]

2.11 Write down the time-dependent Schrödinger equation for a particle moving in three dimensions in a potential $V(r)$.

2.12 What is the probability current density for the wave function $\psi(x) = Ae^{ikx}$? [BUH 2001]

2.13 Calculate the commutator $[x^2, p_x^2]$. [BUH 2001]

2.14 A particle moving in an infinite potential well of width L in one-dimension has the wave function $\psi(x) = a\sin\frac{2\pi x}{L}$ for $-\frac{L}{2} \le x \le \frac{L}{2}$ and $\psi(x) = 0$ outside. What is the average momentum? [BUH 2002]

2.15 What is an observable corresponding to a quantum mechanical system? [BUH 2000]

2.16 Write down the time-dependent Schrödinger equation for a particle of mass m moving in a real potential $V(x, y, z, t)$. [BUH 2000]

2.17 Find out the average value of momentum of a system described by the wave function $\psi(x) = Ae^{ikx} + Be^{-ikx}$ in the region $0 \le x \le \pi/k$ and $\psi(x) = 0$ outside this region. [BUH 1998]

2.18 Explain in words the meaning of (a) normalisation of a wave function (b) the expectation value of an observable. [BUH 1997]

2.19 What does the wave function $\psi(x)$ of a particle in quantum mechanics represent? What are the conditions it must satisfy and why? [BUH 1994]

2.20 State in words the meaning of the expectation value of position variable x in quantum mechanical context. [BUH 1994]

2.21 Evaluate $[x, p_x]$ and $[x, p_x^2]$. [BUH 2004]

2(C) Numerical Problems

2.1 The wave function of a particle is given by

$$\psi(x,y,z) = A\sin\left(\frac{n_x\pi x}{a}\right)\sin\left(\frac{n_y\pi y}{b}\right)\sin\left(\frac{n_z\pi z}{c}\right)$$

where $n_x = n_y = n_z = 1, 2, 3, \ldots$.

Find the value of normalisation constant A. [WBUT 2003]

Solution. The given wave function can be expressed as $\psi(x,y,z) = X(x)Y(y)Z(z)$, where $X(x) = C\sin\left(\frac{n_x\pi x}{a}\right)$.

$$\therefore \int_0^a X^*(x)X(x)\,dx = 1 \text{ or } |C|^2\int_0^a \sin^2\frac{n_x\pi x}{a}\,dx = 1$$

$$\text{or } \frac{|C|^2}{2}\int_0^a\left(1 - \frac{\cos n_x\pi x}{a}\right) = 1 \text{ or } \frac{|C|^2}{2}a = 1.$$

$$\therefore C = \sqrt{\frac{2}{a}}.$$

Similarly, if $Y(y) = D\sin\left(\frac{n_y\pi y}{b}\right)$ and $Z(z) = F\sin\left(\frac{n_z\pi z}{c}\right)$, then we can show that $D = \sqrt{\frac{2}{b}}$ and $F = \sqrt{\frac{2}{c}}$.

$$\therefore \psi(x,y,z) = \frac{2\sqrt{2}}{\sqrt{abc}}\sin\frac{n_x\pi x}{a}\sin\frac{n_y\pi y}{b}\sin\frac{n_z\pi z}{c}$$

So, the normalisation constant is $A = \frac{2\sqrt{2}}{\sqrt{abc}}$.

2.2 A system has two energy eigenvalues ϵ_0 and $3\epsilon_0$, ψ_1 and ψ_2 are the corresponding normalised wave functions. At an instant the system is in a superposed state $\psi = C_1\psi_1 + C_2\psi_2$ and $C_1 = \frac{1}{\sqrt{2}}$.

(a) Find the value of C_2, if ψ is normalised.

(b) What is the probability that an energy measurement would yield a value $3\epsilon_0$?

(c) Find out expectation value of energy. [cf. WBUT 2007]

[Hints. (a) Since ψ_1 and ψ_2 are normalised,

$$\therefore \int \psi^* \psi \, dV = 1 \text{ or } \int \left(C_1^*\psi_1^* + C_2^*\psi_2^*\right)\left(C_1\psi_1 + C_2\psi_2\right) dV = 1$$

$$\text{or } C_1^*C_1 \int \psi_1^*\psi_1 \, dV + C_2^*C_2 \int \psi_2^*\psi_2 \, dV = 1 \text{ or } C_1^2 + C_2^2 = 1.$$

But $C_1 = \frac{1}{\sqrt{2}}$, $\therefore C_2^2 = 1 - \frac{1}{2} = \frac{1}{2}$, $\therefore C_2 = \frac{1}{\sqrt{2}}$.

(b) The probability of both the state 1 and state 2 represented by ψ_1 and ψ_2 respectively is each

$$\left(\frac{1}{\sqrt{2}}\right)^2 = \frac{1}{2} = 0.50 = 50\%.$$

(c) The expectation value of energy $\langle E \rangle = 0.5\epsilon_0 + 0.50 \times 3\epsilon_0 = 2\epsilon_0$.]

2.3 A particle moving along the X-axis has the wave function $\psi = Cx^2$ between $x = 0$ and $x = 2$.

(a) Normalise the wave function over the interval.

(b) Find the probability that the particle can be found between $x = 0.50$ and $x = 0.60$.

(c) Find the expectation value of particle's position x. [BUH 2005]

[Hints. (a) $\int_0^2 \left(Cx^2\right)^2 dx = 1$. $\therefore C = \sqrt{5/32}$.

(b) $P = \int_{0.5}^{0.6} \psi^2 \, dx = 1.5 \times 10^{-3}$.

(c) $\langle x \rangle = \int \psi^* \hat{x} \psi \, dx = \frac{c^2 x^6}{6}$.]

2.4 Compare the lowest three energy states for

(a) an electron in the infinite potential well of width 10 Å.

(b) a grain of dust ($m = 10^{-6}$ gm) moving with a speed of 10^6 m/s in an infinite potential well of width 0.1 mm. What can you conclude from this comparison?

Solution. (a) $E_1 = 0.37$ eV, $E_2 = 1.48$ eV, $E_3 = 3.33$ eV.

(b) $E_1 = 5.49 \times 10^{-51}$ J, $E_2 = 22 \times 10^{-51}$ J, $E_3 = 48.7 \times 10^{-51}$ J, KE = 500 J.

Conclusions.

(a) The energy levels of an electron in an infinite potential well are *quantised*. The difference of energy between successive levels is *large*.

(b) The energy levels of a dust are very close and they constitute a *continuum*.

Thus quantisation of energy is insignificant in case of macroscopic bodies (dust particle).

2.5 Prove that in the case of wave function $\psi = e^{\frac{i\vec{k}\cdot\vec{r}}{r}}$, the probability current density is given by $J = v^2/r$, where v is the particle velocity.

2.6 Normalise the wave function $\psi(x) = e^{-|x|\sin ax}$ in the range $-\infty \leq x < \infty$.

2.7 Calculate the value of lowest energy of an electron in one-dimensional force-free region of length 4 Å. *(Ans. 2.36 eV)*

[HPU 1997]

2.8 Prove quantum mechanically that a particle will not exist in a box, if energy is zero. [PU 1992]

[Hints. $\psi_n(x) = \left(\frac{2}{l}\right)^{1/2} \sin\frac{n\pi x}{l}$; $E_n = \frac{n^2\pi^2\hbar^2}{2ml^2}$.
For $E_0 = 0, n = 0, \therefore \psi(x) = 0$. Since the wave function representing the particle is 0, the particle cannot exist in the box.]

2.9 Think of the nucleus as a box with a size of 10^{-14} m across. Compute the energy of a neutron confined to the nucleus. $[h = 6.62 \times 10^{-34}$ J-s, $m = 1.6 \times 10^{-27}$ kg$]$. *(Ans. 12.88 eV)*

[PU 1995]

2(D) Broad Answer Type Questions

2.1 (a) Obtain time-independent Schrödinger wave equation.

(b) Show that the energy eigenvalues of a particle enclosed in a potential box is quantised. [WBUT 2004]

2.2 (a) Give the physical interpretation of the wave function $\psi(x)$.

(b) Obtain the expression for stationary energy levels for a particle of mass m which is free to move in a region of zero potential between two rigid walls at $x = 0$ and $x = L$. Are the energy levels degenerate?

(c) Find the corresponding wave function pertaining to (b) and normalise. [WBUT 2005]

2.3 (a) What are the basic postulates of Quantum Mechanics? [WBUT 2004(S)]

(b) Deduce time-dependent Schrödinger's equation for a free particle of mass m.

(c) Prove that the coordinate and its conjugate momentum do not commute.

2.4 (a) A system has two energy eigenstates ϵ_0 and $3\epsilon_0$. ψ_1 and ψ_2 are the corresponding normalised wave functions. At an instant the system is in a superposed stated $\psi = C_1\psi_1 + C_2\psi_2$ and $C_1 = \frac{1}{\sqrt{2}}$.

(i) Find the value of C_2, if ψ_2 is normalised.

(ii) What is the probability that an energy measurement would yield a value $3\epsilon_0$?

(iii) Find the expectation value of energy.

(b) Given that for a particle in a rigid box (spanning from $x = 0$ to a) the eigenfunctions are given by $\psi_n = \sqrt{\frac{2}{a}} \sin\frac{n\pi x}{a}$, what are energy eigenvalues in these states? [WBUT 2007]

2.5 (a) Explain the physical significance of ψ-function.

(b) Why normalisation of ψ is required? [NU 1996]

(c) Explain why the wave function ψ must be single-valued and continuous function of position. Why $|\psi|^2$ over all space must be unity? [NU 2000]

2.6 Show that the solution of Schrödinger equation for a particle in an infinite potential well leads to the concept of quantisation of energy. [NU 1999]

2.7 (a) Starting from the time-dependent Schrödinger equation, deduce the time-independent equation from three-dimensional motion. What are stationary states? [NBUH 2003]

(b) What is the physical significance of $\psi(x,t)$?. [CUH 1997]

2.8 (a) Define eigenvalues and eigenfunction of an operator. [BUH 2003]

(b) If $\psi(x,t)$ is a Schrödinger wave function, prove that $\psi^*(x,t)\psi(x,t)$ is necessarily real and either positive or zero. [CUH 1996]

2.9 A particle is free to move one-dimensionally in a region of zero potential between two rigid walls at $x=0$ and $x=a$.

(a) Find the position probability density in the nth state. Does it agree with the result expected classically?

(b) Show that the nth state is an eigenstate of energy as well as momentum.

(c) If E_n is the energy for the nth state and ΔE_n is the energy separation between the $(n+1)$th and the nth states, show that, $\Delta E_n/E_n = (2n+1)/n^2$.

2.10 If $\psi(x)$ be a normalised wave function of a (one-dimensional) system and A be an observable, write down an expression for the expectation value of A in the state represented by ψ. How A is related to the outcome of measurements of A? [BUH 2004]

2.11 (a) Derive the expression for probability current density.

(b) Derive continuity equation from the time-dependent Schrödinger equation. Give physical interpretation of continuity equation. [BUH 2003]

2.12 (a) A particle of mass m is confined within $x=0$ to $x=L$.

(i) Write down Schrödinger equation for such a system.

(ii) Solve the equation to find out the normalised eigenfunctions.

(iii) Show that the eigenfunctions corresponding to two different eigenvalues are orthogonal.

(iv) If p denotes momentum, then find $\langle p\rangle$ as well as $\langle p^2\rangle$ in the ground state.

(b) Using principle of operator correspondence write down the operator $\hat{L}_x, \hat{L}_y, \hat{L}_z$ (components of angular momentum) in terms of position and momentum operators. Then show that, $[\hat{L}_x, \hat{L}_y] = i\hbar\hat{L}_z$. [WBUT 2008]

2.13 (a) A particle is confined in a three-dimensional box with infinitely rigid walls with edges a, b and c. Find the wave equation for the particle.

(b) Write down the energy eigenfunction of a particle in a three-dimensional box of length L (you need not derive this). Hence find out the energy as well as the degeneracies in the ground state and the first excited state. [WBUT 2008]

(c) What are the eigenvalues of momentum for the states (111) and (211)?

MODULE 2

Chapter 3 : Statistical Mechanics

Chapter 3

Statistical Mechanics

3.1 Introduction

Statistical Mechanics deals with systems having a large number of particles and the whole system exists in a state whose *probability of occurring is the maximum*. The above assumptions are true only when the number of particles (n) in the system is very large. All the important physical properties of the system can be deduced just from a knowledge of the *most probable state* (or *equilibrium state*) without considering all other *macro-states whose number is extremely large*.

If, on the other hand, the number of particles in a system is *small*, the statistical method ceases to have any meaning.

3.2 Concept of Energy Levels and Energy States

To study the dynamical behaviour of a *macroscopic* system we apply *Newtonian mechanics*. But for the case of a *microscopic* system, like molecules of a gas, we must apply *quantum mechanics*. According to the *principle of quantum mechanics*, the energy of a dynamical system is *discrete* and the system is said to be only in one of a number of *quantised states* having specified energy value. In tackling statistical problem it is enough to know that *quantised energy levels do exist*.

Let us consider the propagation of transverse waves in a stretched string of length L fixed at both ends. The wavelength λ of the possible stationary waves are $\lambda_1 = 2L$, $\lambda_2 = \frac{1}{2}2L$, $\lambda_3 = \frac{1}{3}2L$, etc.

$$\text{i.e., } \lambda_i = \frac{1}{n_i}2L, \ (n_i = 1,2,3,\ldots). \tag{3.1}$$

The *de Broglie wave* in quantum mechanics due to a moving particle in a box is similar to the stationary waves generated in a *stretched string*. If p be the momentum of the particle and h, the Planck's constant, then *de Broglie wavelength* is

$$\lambda = \frac{h}{p} \text{ or } p = \frac{h}{\lambda} \text{ or } p_i = \frac{h}{\lambda_i}. \tag{3.2}$$

This shows that the momentum of the ith particle of the system is only some one of the set of values. From (3.1) and (3.2),

$$p_i = n_i \frac{h}{2L}. \tag{3.3}$$

If now the particle is *free* and moving in any direction within a *cubical box* of side length L having sides parallel to X-, Y-, and Z-directions of a Cartesian system, then the X-, Y-, and Z-components of linear momentum are

$$p_x = n_x \frac{h}{2L},\ p_y = n_y \frac{h}{2L},\ p_z = n_z \frac{h}{2L}, \tag{3.4}$$

where $n_x, n_y, n_z = 1,2,3,\ldots$ and are called *quantum numbers.*

$$\therefore\ p_i^2 = p_x^2 + p_y^2 + p_z^2 = \left(n_x^2 + n_y^2 + n_z^2\right) \frac{h^2}{4L^2} = n_i^2 \frac{h^2}{4L^2}, \tag{3.5}$$

where $n_i^2 = n_x^2 + n_y^2 + n_z^2$.

The energy associated with the ith particle of momentum p_i is

$$E_i = \frac{p_i^2}{2m} = n_i^2 \frac{h^2}{8mL^2}. \tag{3.6}$$

n_x, n_y, n_z define the *energy state* of the particle and the values of E_i corresponding to different n_i values are called *energy levels.*

Degeneracy. In fact, there may be a number of different states (corresponding to different directions of momentum) which may have the same energy. The energy level is then said to be *degenerate*, the degeneracy of the ith level is denoted by g_i.

Now, the volume of the cubical box of side L is $V = L^3$. $\therefore\ L^2 = V^{2/3}$.

$\therefore$ from (3.6),

$$E_i = n_i^2 \frac{h^2}{8m} V^{-2/3}. \tag{3.7}$$

The *lowest energy level* (for $i = 1$) and $n_x = n_y = n_z = 1$, i.e., $n_i^2 = 1 + 1 + 1 = 3$ is

$$E_1 = \frac{3h^2}{8m} V^{-2/3} \tag{3.8}$$

So, *the lowest state is non-degenerate*, i.e., $g_1 = 1$,

$$p_x = p_y = p_z = \frac{h}{2L}.$$

For level $i = 2$, we only have the following states, where

n_x	n_y	n_z
2	1	1
1	2	1
1	1	2

$\therefore\ p_{2x} = 2\frac{h}{2L}, p_y = \frac{h}{2L}, p_z = \frac{h}{2L}$ for the first of these three states (i.e., $n_x = 2, n_y = 1, n_z = 1$) and

$$n_2^2 = n_x^2 + n_y^2 + n_z^2 = 6 \text{ and } E_2 = \frac{6h^2}{8m} V^{-2/3}. \tag{3.9}$$

Hence, all the three states have the same energy, so the *degeneracy of the level is* $g_i = 3$.

Fig. 3.1 shows the degeneracy (g_i), occupation number (N_i) of a set of energy levels.

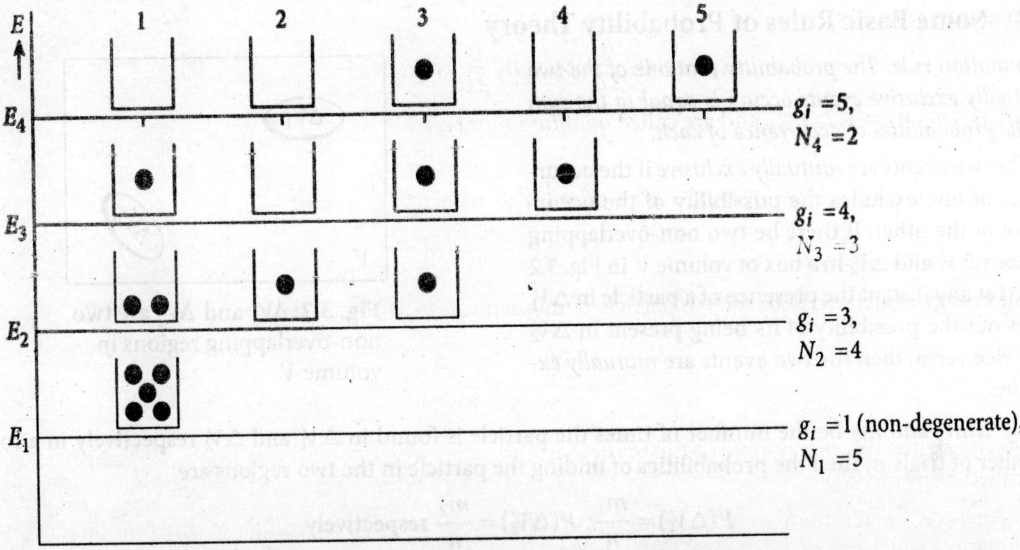

Fig. 3.1: Degeneracy (g_i), occupation number (N_i) of a set of energy levels.

3.3 Basic Concept of Probability

In solving statistical problems the concept of *probability* is used as a tool.

Random events. If we toss a coin it may show *head* or *tail*. It is impossible to predict in advance whether it will show *head* or *tail*. Similarly, if we draw a card at random from a deck, we cannot say in advance which will be the outcome.

These are *random events.*

Probability. The probability of a *sure event* (the event which occurs in an experiment) $P(A)$ is equal to 1; on the other hand, $P(A) = 0$ for an impossible event. Thus the probability of a random event is $0 \le P(A) \le 1$.

Now, the probability $P(A)$ of an event A is defined as

$$P(A) = \frac{\text{Number of events in which } A \text{ occurs}}{\text{Total number of trial}}.$$

This idea may be clarified by taking the case of pulling a card from a thoroughly shuffled pack of 52 cards. What is the probability that a king will occur? Since there are 4 kings in a pack of 52 cards, the probability of occurrence of a king is $\frac{4}{52} = \frac{1}{13}$.

Similarly, as there is only one king of spades, the probability of occurrence of a king of spades is $\frac{1}{52}$.

Hence, probability $P(A)$ and frequency $F(A)$ are not the same but these two quantities are related.

If in tossing a die many times (N), the sit spot appear M times, then the frequency $F(A) = M/N$. If we allow $N \to \infty$, then $F(A)$ approaches to a constant value $P(A)$, the probability of the event.

$$\therefore\ P(A) = \lim_{N\to\infty} F(A), \text{ [where } F(A) = M/N]. \tag{3.10}$$

For this to be true, trials must be conducted until the difference of consecutive M/N ratio differs by a very small value.

3.4 Some Basic Rules of Probability Theory

Summation rule: The probability that one of the two mutually exclusive events occurs is equal to the sum of the probabilities of occurrence of each.

The two events are *mutually exclusive* if the occurrence of one excludes the possibility of the occurrence of the other. If there be two non-overlapping regions ΔV_1 and ΔV_2 in a box of volume V in Fig. 3.2 and if at any instant the presence of a particle in ΔV_1 rules out the possibility of its being present in ΔV_2 and vice versa, then the two events are *mutually exclusive.*

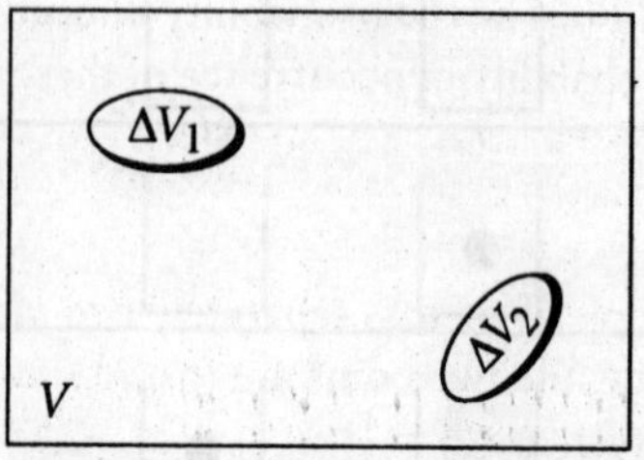

Fig. 3.2: ΔV_1 and ΔV_2 are two non-overlapping regions in volume V.

Proof. If m_1 and m_2 be the number of times the particle is found in ΔV_1 and ΔV_2 respectively in a total number of trials n, then the probabilities of finding the particle in the two regions are

$$P(\Delta V_1) = \frac{m_1}{n};\ P(\Delta V_2) = \frac{m_2}{n} \text{ respectively.}$$

Now, the probability that the particle will be found at least in the two regions is

$$P(\Delta V_1 \text{ or } \Delta V_2) = \frac{m_1+m_2}{n} = \frac{m_1}{n} + \frac{m_2}{n} = P(\Delta V_1) + P(\Delta V_2). \tag{3.11}$$

To generalise the summation rule, let $N_1, N_2, \ldots, N_n$ be the number of times the events $A_1, A_2, \ldots, A_N$ occur in a set of mutually exclusive events in N trials, then

$$N_1 + N_2 + \cdots + N_n = \sum N_n = N.$$

$$\therefore\ P(A_1) + P(A_2) + \cdots + P(A_n) = \frac{N_1}{N} + \frac{N_2}{N} + \cdots + \frac{N_n}{N} = \frac{\Sigma N_n}{N} = \frac{N}{N} = 1.$$

i.e.,

$$P(A_1) + P(A_2) + \cdots + P(A_n) = 1. \tag{3.12}$$

This is known as the *normalisation condition of probability.*

Applying the summation rule, we can show that

1. *The probability that a molecule is in volume ΔV of a box of total volume V is*

$$P(\Delta V) = \frac{\Delta V}{V}. \tag{3.13}$$

2. If out of N trials, N_A is the number of times the event A occurs, N_B is the number of times the event B occurs and N_A and N_B include N_{AB}, where N_{AB} is the number of events in which A and B occur simultaneously, then the total number of events is given by

$$N_{A+B} = N_A + N_B - N_{AB}$$

$$\text{or } \frac{N_{A+B}}{N_A + N_B} = \frac{N_A}{N} + \frac{N_B}{N} - \frac{N_{AB}}{N}$$

$$\text{or } P(A+B) = P(A) + P(B) - P(AB). \tag{3.14}$$

Also if $P(A)$ and $P\left(\bar{A}\right)$ be the probability of occurrence and non-occurrence of an event A, then

$$P(A) + P\left(\bar{A}\right) = 1 \text{ or } P(A) = 1 - P\left(\bar{A}\right). \tag{3.15}$$

Multiplication rule: Joint probability

If $P(A)$ and $P(B)$ be the probability of occurrence of statistically independent events A and B respectively, then the probability of occurrence of these two events simultaneously is

$$P(AB) = P(A) \times P(B). \tag{3.16}$$

For example, if we toss separately two *identical coins*, then the probability of occurrence of *head* of each is 1/2. If now the two coins are tossed *simultaneously*, then the probability of occurrence of *head* of both the coins is $P = \frac{1}{2} \times \frac{1}{2} = \frac{1}{4}$.

Proof. Let A and B are two independent events. Out of total number of trials, let A occur in m_1 times and B in m_2 times, then $P(A) = \dfrac{m_1}{N}$ and $P(B) = \dfrac{m_2}{N}$.

Hence the probability of happening both A and B, i.e., $P(AB) = \dfrac{m_1 m_2}{NN} = \dfrac{m_1}{N} \cdot \dfrac{m_2}{N}$.

$$\therefore\ P(AB) = P(A) \cdot P(B).$$

Conditional probability. The conditional probability of occurrence of event A is defined as the *probability* of occurrence of an event A to occur under the condition that event B has occurred and is denoted by $P(A/B)$.

It can be shown that $P(A/B) = \frac{P(AB)}{P(B)}$ and $P(B/A) = \frac{P(AB)}{P(A)}$.

$$\therefore\ P(AB) = P(A/B)P(B) = P(B/A)P(A). \tag{3.17}$$

3.4.1 Probability of combination

If n *unlike* (or distinguishable) coins are tossed for a large number of times, then the probability of the combination $a^r b^s$ is

$$P(r,s) = \frac{{}^nC_r}{2^n} = \frac{n!}{r!s!}\frac{1}{2^n}\ [s = n - r],$$

where a and b represent respectively heads and tails falling uppermost.

Proof. If n unlike coins are tossed for a large number of times, then the total number of possible events are $(a_1 + b_1)(a_2 + b_2)\ldots(a_n + b_n)$. If all the n coins are *identical* (indistinguishable), then

$$a_1 = a_2 = \cdots = a_n = a \text{ (say)}$$
$$\text{and } b_1 = b_2 = \cdots = b_n = b \text{ (say)}.$$

$\therefore$ total number of possible events = $(a+b)^n$.

Now

$$\begin{aligned}(a+b)^n &= a^n + na^{n-1}b + \frac{n(n-1)}{1\cdot 2}a^{n-2}b^r \\ &\quad + \cdots + \frac{n(n-1)(n-2)\ldots(n-r+1)}{1\cdot 2\cdot 3\ldots r}a^r b^{n-r} + \cdots + b^n \\ &= a^n + {}^nC_1 a^{n-1}b + {}^nC_2 a^{n-2}b^2 + \cdots + {}^nC_{n-r}a^r b^{n-r} + \cdots + {}^nC_n b^n \\ &= \sum_{r=0}^{n} {}^nC_{n-r}a^r b^{n-r} = \sum_{r=0}^{n} {}^nC_r a^r b^{n-r} = \sum_{r=0}^{n} {}^nC_r a^r b^s\ [s = n - r].\end{aligned}$$

The total number of terms in the above expression representing all possible combinations is $\Sigma^n C_r = 2^n$.

$\therefore$ the probability of each equally likely possible complexion $= 1/2^n$.

Hence, the probability of the combination $a^r b^s$ is

$$P(r,s) = \frac{^nC_r}{2^n} = \frac{n!}{r!s!}\frac{1}{2^n}. \quad (3.18)$$

3.4.2 Combinations having maximum probability

If the n coins are similar and tossed for a large number of times, then the number of combination of $a^r b^s$ is $(n+1)$. We are to find which of the $(n+1)$ combinations of $a^r b^s$ will have the *maximum probability*.

Now, if n is even, then the value of ^nC_r is largest, when $r = n/2$.

Hence, the *maximum probability of occurrence* of $a^r b^s$ is given by

$$[P(r,s)]_{\max} = \frac{n!}{\left(\frac{n}{2}\right)!\left(\frac{n}{2}\right)!} \times \frac{1}{2^n} = \frac{n!}{\left[\left(\frac{n}{2}\right)!\right]^2} \times \frac{1}{2^n}. \quad (3.19)$$

3.4.3 Combinations having minimum probability

If n similar coins are tossed for a large number of times, then out of $(n+1)$ combinations of $a^r b^s$, the minimum probability of the said combinations will be $[P(r,s)]_{\min} = \frac{1}{2^n}$.

We know that ^nC_r is minimum, if $r = 0$ or n.

If $r = 0$, then

$$[P(r,s)]_{\min} = \frac{n!}{r!(n-r)!}\frac{1}{2^n} = \frac{n!}{0!(n-0)!}\frac{1}{2^n} = \frac{1}{2^n}. \quad (3.20)$$

If $r = n$, then

$$[P(r,s)]_{\min} = \frac{n!}{r!(n-r)!}\frac{1}{2^n} = \frac{n!}{n!(0!)}\frac{1}{2^n} = \frac{1}{2^n}. \quad (3.21)$$

Hence,

$$[P(r,s)]_{\min} = \frac{1}{2^n}$$

3.4.4 Distribution of molecules in two compartments of a box

Let there be n molecules to be distributed between two compartments of a box with equal probability. A molecule may either go to the first compartment or the second compartment.

Let a and b represent respectively the event of the molecule of being in the first and the second compartments respectively. Then the possible combinations in the distribution of n molecules are

$$(a+b)^n = a^n + {^nC_1}a^{n-1}b + {^nC_2}a^{n-2}b^2 + \cdots + {^nC_r}a^r b^{n-r} + \cdots + {^nC_n}b^n.$$

Here, the total number of terms representing all possible combinations of a and b are

$$\sum {^nC_r} = 2^n.$$

$\therefore$ the probability of distribution $(r, n-r)$ is

$$P(r, n-r) = \frac{n!}{n!(n-r)!}\frac{1}{2^n}. \quad (3.22)$$

$\therefore$ for the most probable distribution, $r = n/2$.

$$\therefore [P(r, n-r)]_{max} = \frac{n!}{\left(\frac{n}{2}!\right)\left(\frac{n}{2}!\right)} \frac{1}{2^n} = \frac{n!}{\left(\frac{n}{2}!\right)^2} \frac{1}{2^n}. \tag{3.23}$$

For the least probable distribution, $r = n$, $n - r = 0$.

$$\therefore [P(r, n-r)]_{min} = \frac{n!}{n!(n-n)!} \frac{1}{2^n} = \frac{1}{2^n}. \tag{3.24}$$

3.4.5 Cases with weightage (General case)

Let a and b represent respectively the probability of *success* and *failure* of an event in a single trial such that $b = 1 - a$. Then we have to find out the probability of r successes of the events n trials if the trials are *independent* of each other.

By the *theorem of probability of composite events,* the probability of r *successes* and $(n-r)$ *failures* in a particular order is

$$[a \times a \times \cdots \times a \text{ (}r\text{ times)}] \times [b \times b \times \cdots \times b (n-r) \text{ times}].$$

But r successes and $(n-r)$ failures can occur in ${}^nC_r$ ways.

$\therefore$ out of n trials, the probability of r successes is

$$P(r) = {}^nC_r a^r b^{n-r} = \frac{n!}{r!(n-r)!} a^r b^{n-r}.$$

Thus the probabilities of $0, 1, 2, \ldots, n$ successes are respectively, b^n, ${}^nC_1 ab^{n-1}$, ${}^nC_2 a^2 b^{n-2}, \ldots, {}^nC_r a^r b^{n-r}$, $\ldots$, ${}^nC_n a^n$, which are the successive terms of the *binomial expansion* of

$$(b+a)^n = b^n + {}^nC_1 ab^{n-1} + {}^nC_2 a^2 b^{n-2} + \cdots + {}^nC_r a^r b^{n-r} + \cdots + a^n.$$

This shows that, ${}^nC_r a^r b^{n-1}$ is the $(r+1)$th term in the expansion of $(b+a)^n$.

Hence, the probability of r successes in n trials is given by the $(r+1)$th term of the binomial expansion of $(b+a)^n$. This is called the *binomial distribution.*

Finally, the probability of r successes in n trials simultaneously is

$$P(r) = \frac{n!}{r!(n-r)!} \frac{1}{2^n} a^r b^{n-r},$$

where a and b have the same significances as discussed earlier.

3.5 Stirling's Approximation

For large values of N, the calculation of $N!$ becomes very laborious. *Stirling's formula* gives an approximate value of $\ln N!$, when $N \to \infty$.

Stirling's formula states that

$$\ln N! = N \ln N - N, \text{ if } N \to \infty. \tag{3.25}$$

Proof. By definition of $N!$, we have $N! = 1 \times 2 \times 3 \times \cdots \times (N-1) \times N$.

$$\therefore \ln N! = \ln 1 + \ln 2 + \ln 3 + \cdots + \ln N = \sum_{m=1}^{N} \ln m.$$

Plotting $\ln m$ against m, we get a curve as shown in Fig. 3.3. The value of $\Sigma \ln m$ is the area S under the stepped line and the axis of abscissa. We have also plotted $\ln x$ against x (smooth curve) taking x to be a continuous variable.

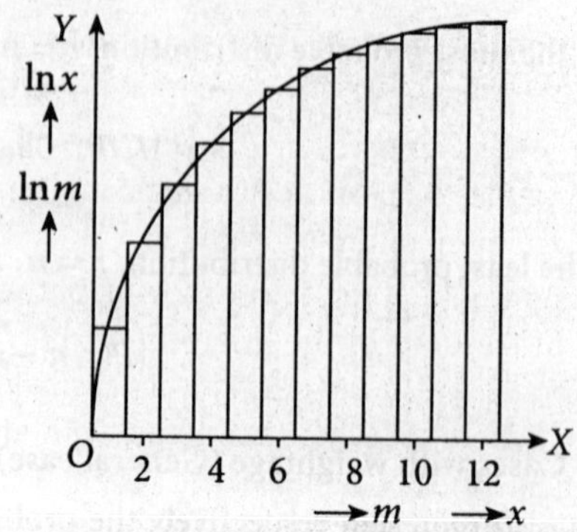

Fig. 3.3: Plot of ln m against m.

The area under the smooth curve is

$$S' = \int_1^N \ln x\, dx = [x \ln x]_1^N - \int_1^N dx = N \ln N - N + 1.$$

$\because N \to \infty, S = S'$,

$$\therefore \ln N! = \sum_{m=1}^{N} \ln m = N \ln N - N + 1.$$

As N is very large, $\ln N! = N \ln N - N$.

According to a more improved treatment,

$$N! = \sqrt{2\pi N}\left(\frac{N}{e}\right)^N \tag{3.26}$$

3.6 Phase Space

Concept of phase space. The total energy of a *one-dimensional simple harmonic oscillator* (SHO) of mass m and spring constant k is given by

$$E = \frac{p^2}{2m} + \frac{1}{2}kq^2, \tag{3.27}$$

where q is the *position coordinate* and p, the *conjugate momentum* [Fig. 3.3]. The total energy of the SHO is constant.

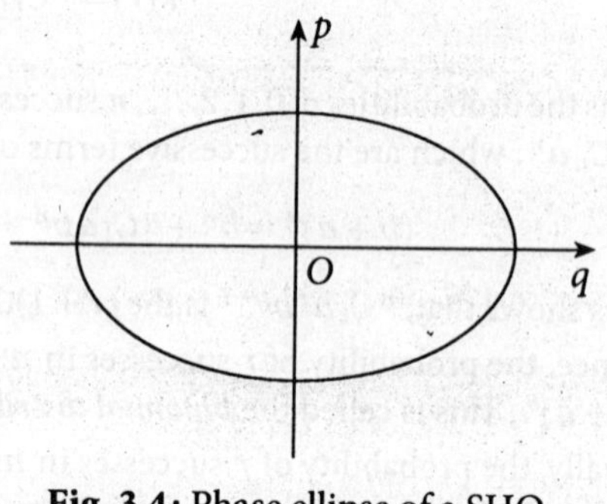

Fig. 3.4: Phase ellipse of a SHO.

Equation can be written as

$$\frac{p^2}{2mE} + \frac{q^2}{2E/k} = 1. \tag{3.28}$$

(3.28) represents an ellipse with semi-major axis $\sqrt{2E/k}$ and semi-minor axis $\sqrt{2mE}$, the corresponding *phase points* (p,q) of the SHO lie on the ellipse. As time passes, the oscillator goes through various phases in physical space represented by p and q. This (p,q) space is called the *phase space* of the SHO. In rectangular coordinate system a moving particle can be represented by three position coordinates x, y, z and three moment coordinates p_x, p_y, p_z. This *six-dimensional space* represented by x, y, z, p_x, p_y, p_z is the *phase space* of the moving particle.

An element of volume represented by $\Delta x\, \Delta y\, \Delta z\, \Delta p_x\, \Delta p_y\, \Delta p_z$ in phase space is called *phase cell.* A path traced by a phase point (also called representative point) is called a *phase curve.* A phase curve *cannot intersect* itself as it leads to a violation of the fundamental characteristics of systems for which a knowledge of (p,q) at $t = 0$ represents the phase at any later time.

μ-space and Γ-space. The instantaneous configuration of a system having f degrees of freedom (number of independent coordinates necessary for the description of the system) can be described by a set of f *generalised position coordinates* $q_1, q_2, q_3, \ldots, q_f$, and f *generalised momentum coordinates* $p_1, p_2, p_3, \ldots, p_f$, i.e., by $2f$ parameters. The state of the system thus can be described by imagining a conceptual space of two dimensions. *This $2f$-dimensional coordinate-momentum orthogonal space is called μ-space.*

The set of numbers $(q_1, q_2, \ldots, q_f; p_1, p_2, \ldots, p_f)$ represent a point in μ-space and is called a *phase point.* For example, the μ-space of a single molecule in a gas is 6-dimensional of which 3 are position coordinates and 3 are momentum coordinates.

In a group of N randomly moving particles each having f-degrees of freedom, the total number of degrees of freedom of the group is $n = Nf$. *The phase space comprising of n position and n momentum coordinates, i.e., a $2n(= 2Nf)$-dimensional orthogonal space is called a Γ-space.* In case of a *macroscopic system,* where the particles interact with any degree of strength, a, Γ-space finds important application to investigate the behaviour of the system.

3.7 Micro-state and Macro-state

Micro-state: A micro-state of a system is defined by specifying the six coordinates $x, y, z; p_x, p_y, p_z$ of each molecule of the system within the limits of the dimension of the cell in which the representative point lies.

Thus to define a *micro-state,* the place of each molecule in the system must be specified within the limit of $dx\,dy\,dz$ and $dp_x\,dp_y\,dp_z$, i.e., it is essential to state to which cell each molecule belongs momentarily.

Macro-state: A macro-state of a system may be defined by the specification of phase points in each cell, i.e., by specifying the number only and overlooking the identities of the molecules. Thus for a macro-state it is sufficient to say that there are n_1 molecules in cell 1, n_2 molecules in cell 2, etc. The macro-state is also called the *thermodynamic state.*

Hence many different micro-states may correspond to the same macro-state.

Example. Let us consider two two-phase cells represented by i and j and four-phase points or molecules a, b, c, d which are *distinguishable* from each other. If N_i and N_j be the number of molecules (or phase points) in the cells i and j respectively, then the possible *macro-states* are five as shown in Table 3.1.

Table 3.1

N_i	4	3	2	1	0
N_j	0	1	2	3	4

Thus the five different macro-states are $(4,0), (3,1), (2,2), (1,3), (0,4)$.

If there are n *distinguishable* molecules, then the possible *macro-states* are: $(n-0), (n-1,1), (n-2,2), \ldots, (1, n-1), (0, n)$. So, for n molecules total number of macro-state is $(n+1)$.

Now, as the molecules a, b, c, d are *distinguishable,* we can have a number of different arrangements of them in each *macro-state* as shown in Table 3.2. *Every distinct arrangement under the same macro-state is called a micro-state of the system.*

Table 3.2

Macro-state	*Possible arrangements*		*No. of micro-states*
	Cell i	Cell j	
(4,0)	abcd	0	1
(3,1)	abc	d	4
	bcd	a	
	cda	b	
	dab	c	

Macro-state	Possible arrangements		No. of micro-states
	Cell i	Cell j	
	ab	*cd*	
	ac	*bd*	
	ad	*bc*	
(2,2)	*bc*	*da*	6
	bd	*ac*	
	cd	*ab*	
	a	*bcd*	
(1,3)	*b*	*cda*	4
	c	*dab*	
	d	*abc*	
(0,4)	0	abcd	1

Thus total number of micro-states is $1+4+6+4+1=16$, which belong to 5 different macro-states of four *distinguishable molecules* a, b, c, d.

3.8 Thermodynamical Probability

Thermodynamical probability of a particular macro-state is defined as the number of possible micro-states corresponding to the given macro-state. Thermodynamical probability is in general a very large number (unlike mathematical probability which is less than one) and is represented by Ω.

Thus the thermodynamical probability of the macro-state $N_i = 3$, $N_j = 1$ is 4 as in previous article, i.e., $\Omega = 4$.

To find out an expression for *thermodynamical probability*, suppose we have N *distinguishable* particles (or phase points) to be distributed in n *containers* (or cells) such that there are N_1 particles in container 1, N_2 in container 2 and so on.

Now total number of distribution of N particles is $N!$. But these $N!$ distributions are not all *relevant* distributions. When more than one particle is present in a container, permuting them among themselves does not give different distributions. Thus the self-distribution of N_1 particles $N_1!$ is the same and one distribution, called *irrelevant* distribution. Similarly, the total number of irrelevant distributions in n containers is $N_1!N_2!N_3!\ldots N_n!$.

The thermodynamical probability (Ω) of the distribution can be found by dividing total number of *possible* distribution ($N!$) by the total number of *irrelevant* distributions:

$$N_1!N_2!N_3!\ldots N_n! \text{ or } \Omega = \frac{N!}{N_1!N_2!N_3!\ldots N_n!} = \frac{N!}{\prod_{i=1}^{n} N_i!}. \tag{3.29}$$

3.9 General Statistical Distribution Law

Suppose we have to distribute N *distinguishable* particular or balls (marbles) in a box having k compartments (or cells) such that $n_1, n_2, n_3, \ldots, n_k$ are the number of particles in compartments $1, 2, 3, \ldots, k$ respectively subject to the condition $n_1 + n_2 + \cdots + n_k = N$.

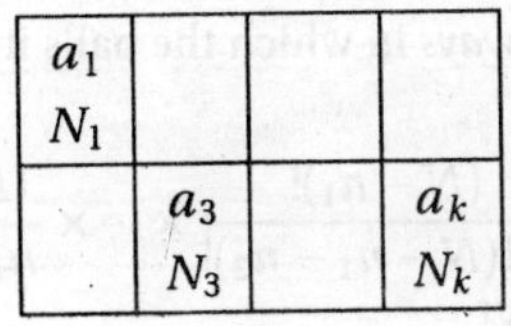

Fig. 3.5: Distribution of N distinguishable particles in k-cells.

Now the question is, what is the *most probable distribution*? The most probable distribution is that for which the *number of balls in each compartment is proportional to the size of the compartment.*

The probability W that the balls be distributed in a particular manner among the compartments depends upon (i) *the a priori probability* (G) *of distribution* and (ii) *the thermodynamical probability* (Ω).

A priori probability. Let the balls be *identical but distinguishable.* Then the *a priori* probability g_i of any ball falling in ith cell is

$$g_i = \frac{\text{area of } i\text{th cell } (a_i)}{\text{total area } (a) \text{ of the box}},$$

i.e., $g_i = \frac{a_i}{a}$, where $a = a_1 + a_2 + \cdots + a_i + \cdots + a_k$.

The sum of *a priori* probabilities must be unity, i.e.,

$$g_1 + g_2 + \cdots + g_i + \cdots + g_k = \sum_{i=1}^{k} g_i = 1, \tag{3.30}$$

as the ball must fall somewhere in the box. The *a priori* probability of two balls falling in the ith cell is $g_i \times g_i = g_i^2$.

In this way, the *a priori* probability of n_i balls falling in the ith cell is $g_i^{n_i}$.

Hence, the *a priori* probability of distribution of N balls among the k cells so that $n_1, n_2, n_3, \ldots, n_i, \ldots, n_k$ balls fall in the cell $1, 2, 3, \ldots, k$ respectively, is

$$G = (g_1)^{n_1} \times (g_2)^{n_2} \times \cdots \times (g_i)^{n_i} \times \cdots \times (g_k)^{n_k} = \prod_{i=1}^{k} (g_i)^{n_i}.$$

If the size of all the cells are the same, i.e., $a_1 = a_2 = \cdots = a_k$, then they all have the same *a priori* probability $G = g^n$.

Thermodynamical probability. First we choose n_1 balls which are to be distributed in the first cell. This can be done in ${}^{N}C_{n_1}$ ways, i.e.,

$${}^{N}C_{n_1} = \frac{N!}{n_1!(N - n_1)!} \text{ ways.}$$

The remaining number of balls is $(N - n_1)$. Now out of $(N - n_1)$ balls we can distribute n_2 balls in cell 2 in ${}^{N-n_1}C_{n_2}$ ways, i.e.,

$${}^{N-n_1}C_{n_2} = \frac{(n - n_1)!}{n_2!(N - n_1 - n_2)!} \text{ ways.}$$

Therefore, *the number of independent ways* in which the balls in different cells may be arranged, keeping the total number in each cell fixed, is

$$\Omega = \frac{N!}{n_1!(N-n_1)!} \times \frac{(N-n_1)!}{n_2!(N-n_1-n_2)!} \times \cdots \times \frac{(N-n_1-n_2\ldots n_{i-1})!}{n_i!(N-n_1-n_2\ldots n_i)!} \times \ldots$$

$$\text{or } \Omega = \frac{N!}{n_1!n_2!\ldots n_k!} = \frac{N!}{\prod_{i=1}^{k} n_i}. \tag{3.31}$$

Hence the total probability of distribution is

$$W = G \times \Omega = \prod_{i=1}^{k} (g_i)^{n_i} \times \frac{N!}{\prod_{i=1}^{k} n_i} \text{ or } W = \frac{N!}{\prod_{i=1}^{k}} \prod_{i=1}^{k} g_i^{n_i}. \tag{3.32}$$

To test the correctness of (3.32) we have the total probability of all possible distributions

$$\sum W = \sum \frac{N!}{n_1!n_2!\ldots n_k!} \times (g_1)^{n_1} \times (g_n)^{n_2} \times \cdots \times (g_k)^{n_k} = N! \prod_{i=1}^{k} \frac{(g_i)^{n_i}}{n_i!}.$$

Taking logarithm on both sides,

$$\ln W = \ln N! + \sum_{i=1}^{k} n_i \ln g_i - \sum_{i=1}^{k} \ln(n_i)!$$

$$= (N \ln N - N) + \sum_{i=1}^{k} n_i \ln g_i - \sum_{i=1}^{k} n_i \ln n_i + \sum_{i=1}^{k} n_i \tag{3.33}$$

[by Stirling's formula, $\ln x! = x \ln x - x$].

Again, $\sum_{i=1}^{k} n_i = n_1 + n_2 + \cdots + n_k = N$.

$$\therefore \ln W = N \ln N + \sum_{i=1}^{k} n_i \ln g_i - \sum_{i=1}^{k} n_i \ln(n_i). \tag{3.34}$$

For the *most probable state, the thermodynamic probability is the maximum*, i.e., $\ln W$ is the maximum, i.e., $d(\ln W) = 0$.

But

$$d(\ln W) = \sum_{i=1}^{k} dn_i \ln g_i - \sum_{i=1}^{k} dn_i \ln(n_i) - \sum n_i \frac{1}{n_i} dn_i$$

$$= \sum_{i=1}^{k} dn_i \ln g_i - \sum_{i=1}^{k} dn_i \ln(n_i), \quad \left[\because \sum_{i=1}^{k} dn_i = dN = 0\right]$$

as N and g_i are constants.

$$\therefore d(\ln W) = \sum_{i=1}^{k} (\ln g_i - \ln n_i)\, dn_i = 0 \quad \left[\because \sum_{i=1}^{k} dn_i = dN = 0\right].$$

Hence, as $d(\ln W) = 0$, the probability given by (3.32) is *the maximum*.

3.10 Density of States

For a single particle in phase space the micro-state is defined by *position coordinates* q_x, q_y, q_z and momentum coordinates p_x, p_y, p_z.

An element of volume in the phase space is

$$\delta q_x \delta q_y \delta q_z \delta p_x \delta p_y \delta p_z = h^3 \quad [\because h = \delta p_i \delta q_i]$$

$\therefore$ the total volume of phase space is

$$= \iiiint\!\!\iint dq_x\, dq_y\, dq_z\, dp_x\, dp_y\, dp_z$$

$$= V \iiint dp_x\, dp_y\, dp_z \quad \left[\because \iiint dq_x\, dq_y\, dq_z = V\right]$$

$$= V \frac{4}{3}\pi p^3 \quad \left[\because \iiint dp_x\, dp_y\, dp_z = \frac{4}{3}\pi p^3\right]$$

$$= \frac{4}{3}\pi V (2mE)^{3/2} \quad \left[\because \text{energy } E = \frac{p^2}{2m}\right].$$

Since the phase volume corresponding to a cell is $\sim h^3$, the number of cells in the above phase volume is

$$g(E) = \frac{\frac{4}{3}\pi V (2mE)^{3/2}}{h^3} = \frac{4\pi V}{3h^3}(2mE)^{3/2}. \tag{3.35}$$

(3.35) represents the *number of cells* in the phase space corresponding to a single particle.

The number of *quantum states* (micro-states) in the energy interval between E and dE is obtained by differentiating (3.35) w.r.t. E, i.e.,

$$g(E)\,dE = \frac{dg(E)}{dE} = \frac{4\pi V}{3h^3}\cdot\frac{3}{2}(2mE)^{1/2} 2m\,dE$$

$$= \frac{4\sqrt{2}\pi V}{h^3} m^{3/2} E^{1/2}\, dE. \tag{3.36}$$

This gives the *density of states* in the energy range E and $E + dE$.

3.11 Ensembles

An ensemble is defined as a collection of a very large number of assemblies which are essentially independent of one another but have been made macroscopically as identical as possible. The term *macroscopically identical* means each assembly is characterised by the same values of set of macroscopic parameters (viz., P, V, T, S) which uniquely determine the *equilibrium state* of the assembly.

In statistical mechanics three types of ensemble are particularly important, viz., (i) *micro-canonical* (ii) *canonical* and (iii) *grand canonical ensembles.* The classification of ensembles is related to the manner in which their systems interact, viz., the exchange of energy or matter or both.

1. ***Micro-canonical ensemble.*** ***A collection of essentially independent assemblies having the same energy E, volume V and number of systems N (assuming only one type of system) is called a micro-canonical ensemble.*** Here the individual assemblies are separated by (a) *rigid* (b) *impermeable* and (c) *well-insulated wall* [Fig. 3.6], but the macroscopic energy of an assembly cannot be actually specified. In this ensemble, exchange of *neither matter nor energy can take place.*

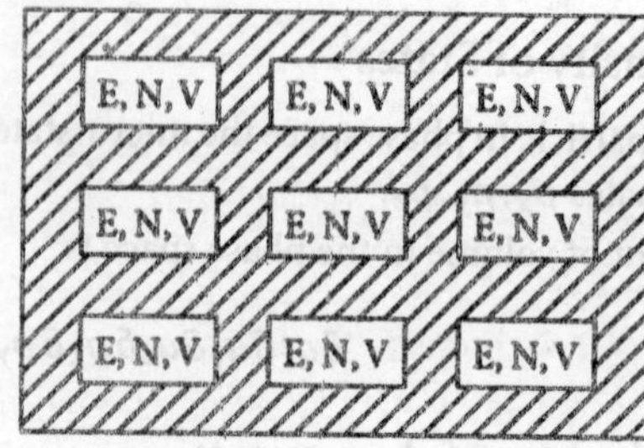

Fig. 3.6 : Micro-canonical ensemble.

The essential properties of the micro-canonical ensemble are:

(a) *The ensemble is in statistical equilibrium.*

(b) *Since it is in statistical equilibrium, the average properties predicted are independent of time.*

(c) *The distribution of phase points representing the system is uniform.*

2. ***Canonical ensemble.*** This is a collection of essentially independent assemblies having the same temperature T, volume V and number of identical systems. The individual assemblies are separated by (a) *rigid* (b) *impermeable* but (c) *diathermic walls.* The different assemblies can assume a common temperature T through exchange energy. Here *the systems can exchange energy but not particles* [Fig. 3.7].

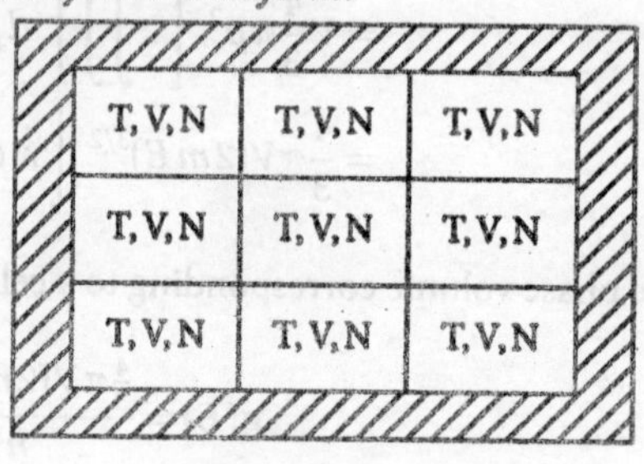

Fig. 3.7: Canonical ensemble.

3. ***Grand canonical ensemble.*** The grand canonical ensemble is a collection of essentially independent assemblies which have the same temperature T, volume V and chemical potential μ. *It is an ensemble in which exchange of energy as well as particles occurs with the heat reservoir* [Fig. 3.8]. Here the individual assemblies are separated by (a) *rigid* (b) *permeable* and (c) *diathermic walls.*

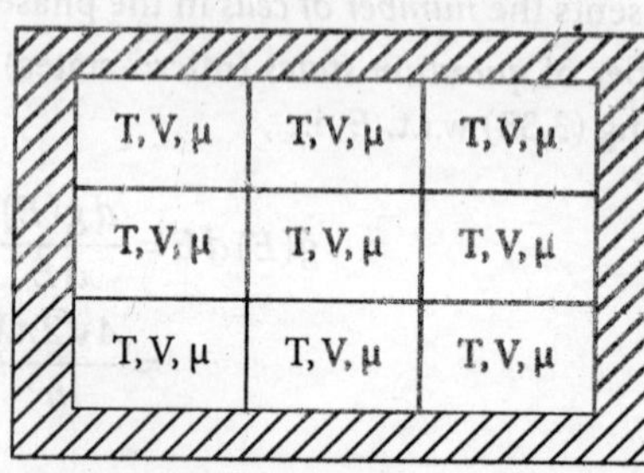

Fig. 3.8: Grand canonical ensemble.

3.12 Ensemble Average and Time Average

Ensemble average. We consider an ensemble consisting of N *identical systems.* Let q_i be the position of coordinate of the particle in the ith system. The average value of the square of q_i is given by

$$\langle q^2 \rangle_e = \frac{1}{N} \sum q_i^2. \tag{3.37}$$

If a particle be found in a particular cell (say, jth cell) in a number of system (N_j) with coordinate q_j, then the contribution by N_j cells to $\Sigma_i q_i^2$ is equal to $N_j q_j^2$.

The total contribution by different N_j cells to $\Sigma_i q_i^2$ is

$$\sum_{i=1}^{N} q_i^2 = \sum_j N_j q_j^2. \tag{3.38}$$

The summation extends to all the cells of each system.

Now from (3.37) and (3.38), we have

$$\langle q^2 \rangle_e = \frac{1}{N}\sum_{i=1}^{N} q_i^2 = \frac{1}{N}\sum_j N_j q_j^2 = \sum_j \left(\frac{N_j}{N}\right) q_j^2.$$

But $N_j/N = P_j$, the probability that the particle may be present in the jth cell.

$$\therefore \langle q^2 \rangle_e = \sum_j P_j q_j^2. \tag{3.39}$$

(3.39) represents *the ensemble average of the square of the position coordinate* (q).

Time average. The position coordinate of a particle changes with time when it jumps from one cell to another of the ensemble. The time average of the coordinate of the particle can be obtained by observing the motion of the particle for a very long time T. The time average of the position coordinate is, then

$$\langle q^2 \rangle_t = \lim_{T\to 0} \frac{1}{T}\int_0^T q^2(t)\,dt. \tag{3.40}$$

The particle, during observation, jumps from cell to cell and let in the ith jump the particle be in a cell with coordinate q_i. If the particle remains in the ith cell for a time Δt_i, then $\Sigma \Delta t_i = T$.

So

$$\sum_{i=1}^{m} q_i^2 \Delta t_i = \int_0^T q^2(t)\,dt, \tag{3.41}$$

where m is the number of jumps of the particle from cell to cell in time T.

Again, if m_j be the number of times the particle arrives in the jth cell with coordinate q_j and remains in that cell for an interval Δt_j, then (3.41) can be written as

$$\int_0^T q^2(t)\,dt = \sum_{i=1}^{m} q_i^2 \Delta t_i = \sum_{j=1}^{M} q_j^2 m_j \Delta t_j = \sum_{j=1}^{M} q_j^2 T_j, \tag{3.42}$$

where $T_j = m_j \Delta t_i$ is the total time of duration of the particle in the jth cell and μ is the total number of cells in a system in the ensemble.

$\therefore$ (3.42) can be written as

$$\langle q^2 \rangle_t = \lim_{T\to\infty} \frac{1}{T}\sum_{j=1}^{\mu} T_j q_j^2 = \lim_{T\to\infty} \sum_{j=1}^{\mu} \frac{T_j}{T} q_j^2.$$

But the particle remains for a time T_j in the jth cell out of total time T. So the probability of its staying in the jth cell is

$$P_j = \lim_{T \to \infty} \frac{T_j}{T}.$$

$$\therefore \left\langle q^2 \right\rangle_t = \sum_{j=1}^{\mu} P_j q_j^2. \tag{3.43}$$

Thus from (3.39) and (3.43), we have

$$\left\langle q^2 \right\rangle_e = \langle q^2 \rangle_t. \tag{3.44}$$

i.e., *the ensemble average is equal to the time average of a physical quantity.* This forms one of the *fundamental postulates of statistical mechanics.*

3.13 Introduction to Statistical Mechanics

When thermodynamical approach came to a standstill, statistical approach proved to be a suitable alternative for the solution of many problems, viz., the *derivation of the equation of state of a macroscopic system,* the problem of *specific heat of solids* and *the problems of thermal radiation,* etc. Statistical mechanics deals with those systems which are *in equilibrium* or *near equilibrium state.*

Statistical mechanics is broadly divided into two categories: (a) *classical statistics* and (b) *quantum statistics.* Quantum statistics is again subdivided into (i) *Bose-Einstein* (BE) *statistics* and (ii) *Fermi Dirac* (FD) *statistics.* The classification of statistical mechanics is shown in the following sketch:

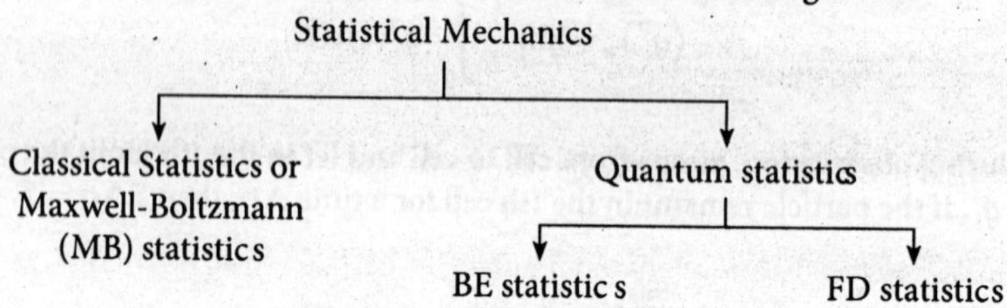

3.14 Fundamental Postulates of Statistical Mechanics

Following are the *fundamental postulates of statistical mechanics*:

1. The molecules of a gas are in *random chaotic motion* and behave like small *elastic spheres.*
2. Total number of molecules in a gas is constant.
3. Total energy of a system is constant in equilibrium condition.
4. All phase cells in the phase space are of equal size.
5. All accessible micro-states corresponding to possible macro-states are equally probable. This is known as the *postulate of equal a priori probability.*
6. The equilibrium state of a system corresponds to states of maximum probability.

3.15 Derivation of Maxwell-Boltzmann (MB) Statistical Distribution Law

We consider a system of N particles (may be ideal gas molecules) which are divided into quantum groups such that $N_1, N_2, \ldots, N_i, \ldots, N_k$ particles be present in groups having approximate energies $E_1, E_2, \ldots, E_i, \ldots, E_k$ respectively. To derive MB *distribution law* we assume that,

1. The particles are *distinguishable* and there is no *symmetry restriction.*
2. Each eigenstate of the ith quantum group may contain $0, 1, 2, \ldots, N_i$ particles.
3. Total number of particles in the entire system is constant, i.e.,

$$N = N_1 + N_2 + \cdots + N_i + \cdots + N_k = \sum_{i=1}^{k} N_i = \text{constant.} \tag{3.45}$$

4. Total energy of the system

$$E = E_1 + E_2 + \cdots + E_i + \cdots + E_k = \sum_{i=1}^{k} E_i = \text{constant.} \tag{3.46}$$

5. When the gas is in equilibrium, *thermodynamical probability is the maximum.*

We have seen earlier that the number of ways in which N particles can be distributed in k cells such that $N_1, N_2, \ldots, N_i, \ldots, N_k$ be the particles in quantum state of energy $E_1, E_2, \ldots, E_i, \ldots, E_k$ is

$$W = \frac{N!}{\prod_{i=1}^{k} N_i!} \prod_{i=1}^{k} g_i^{N_i},$$

where g_i is the number of *quantum states* (statistical weight) of the ith state. W is called the *thermodynamical probability* of the system.

$$\begin{aligned}\therefore \ln W &= \ln N! + \sum_i (N_i \ln g_i - \ln N_i!) = N(\ln N - 1) + \sum_i N_i \ln g_i - \sum_i N_i(\ln N_i - 1)\\ &= N \ln N + \sum_i N_i \ln g_i - \sum_i N_i \ln N_i \; [\because \text{ for } N \to \infty, \ln N - 1 \simeq \ln N].\end{aligned} \tag{3.47}$$

Since the system under consideration is in equilibrium, $N_1, N_2, \ldots, N_k$ are such that W is the maximum subject to the conditions (3.45) and (3.46).

Now taking differential of (3.47),

$$\delta(\ln W) = -\left\{\sum_i \ln \frac{N_i}{g_i} + 1\right\} \delta N_i \; [\because \delta(N \ln N) = 0]$$

For the most *probable distribution,*

$$W = W_{\max}.$$

$$\therefore \delta(\ln W_{\max}) = 0, \text{ i.e., } \sum_i \left\{\ln \frac{N_i}{g_i} + 1\right\} \delta N_i = 0. \tag{3.48}$$

Now multiplying (3.45) by α and (3.46) by β (where α and β are unknown constants to be determined) and then adding the result with (3.48), we get

$$\sum_i \left(\ln \frac{N_i}{g_i} + \alpha + \beta E_i\right) \delta N_i = 0. \left[\because \ln \frac{N_i}{g_i} + 1 \simeq \ln \frac{N_i}{g_i}\right].$$

$\because \delta N_i$ are independent of each other, $\delta N_i \neq 0$.

$$\therefore \ln\frac{N_i}{g_i}+\alpha+\beta E_i=0 \text{ or } \ln\frac{N_i}{g_i}=-(\alpha+\beta E_i) \text{ or } \frac{N_i}{g_i}=e^{-(\alpha+\beta E_i)} \text{ or } N_i=\frac{g_i}{e^{\alpha+\beta E_i}} \tag{3.49}$$

and

$$f(E_i)=\frac{N_i}{g_i}=\frac{1}{e^{\alpha+\beta E_i}}. \tag{3.50}$$

(3.50) is the *MB distribution function* giving the average number of particles per quantum state of the system. (3.49) is called *MB statistical distribution formula.*

Evaluation of α and β

Evaluation of α: From (3.49) we have the number of molecules in the energy range E to $E+dE$ (assuming the energy distribution of molecules to be continuous) is

$$N(E)\,dE=g(E)e^{-\alpha}\,e^{-\beta E}\,dE.$$

Since $E=\frac{p^2}{2m}$ [p is the momentum of a molecule of mass m], we have

$$N(p)\,dp=g(p)e^{-\alpha}\,e^{-\frac{\beta p^2}{2m}}\,dp.$$

Now the number of states in the momentum range p to $p+dp$ is

$$g(p)\,dp=\frac{\iiiint\!\!\iint dx\,dy\,dz\,dp_x\,dp_y\,dp_z}{h^3},$$

where h^3 is the phase volume of each cell and $\iiiint\!\!\iint dx\,dy\,dz\,dp_x\,dp_y\,dp_z$ represents the total phase volume occupied by particles with specified momenta.

But $\iiint dx\,dy\,dz=V$, where V is the volume in the *coordinate space* containing all the particles.

$$\therefore\; g(p)\,dp=\frac{V}{h^3}\iiint dp_x\,dp_y\,dp_z=\frac{V}{h^3}\times 4\pi p^2\,dp,$$

where $\iiint dp_x\,dp_y\,dp_z=4\pi p^2\,dp$ is the volume in the *momentum space* between p and $p+dp$.

$$\therefore\; g(p)\,dp=\frac{4\pi Vp^2\,dp}{h^3} \text{ and } N(p)\,dp=\frac{4\pi Vp^2\,e^{-\alpha}\,e^{-\frac{\beta p^2}{2m}}\,dp}{h^3}. \tag{3.51}$$

Integrating, we get

$$\int_0^\infty N(p)\,dp=N \text{ or } N=\frac{4\pi e^{-\alpha}V}{h^3}\int_0^\infty p^2\,e^{-\frac{\beta p^2}{2m}}\,dp. \tag{3.52}$$

Now let $\frac{\beta p^2}{2m}=Z.$

$$\therefore\ p^2=\frac{2mZ}{\beta} \text{ or } 2p\,dp=\frac{2m}{\beta}dZ \text{ or } dp=\frac{m\,dZ}{\beta^{1/2}\left(\frac{2mZ}{\beta}\right)^{1/2}}=\frac{mZ^{-1/2}\,dZ}{(2m\beta)^{1/2}}.$$

$$\therefore \int_0^\infty p^2\,e^{-\frac{\beta p^2}{2m}}\,dp=\int_0^\infty \frac{2m}{\beta}Ze^{-Z}Z^{-1/2}\frac{m}{(2m\beta)^{1/2}}dZ=\frac{2^{1/2}m^{3/2}}{\beta^{3/2}}\int_0^\infty e^{-Z}Z^{1/2}\,dZ$$

$$=\frac{2^{1/2}m^{3/2}}{\beta^{3/2}}\int_0^\infty e^{-Z}Z^{\frac{3}{2}-1}\,dZ=\frac{2^{1/2}m^{3/2}}{\beta^{3/2}}\Gamma\left(\frac{3}{2}\right)=\frac{2^{1/2}m^{3/2}}{\beta^{3/2}}\frac{\sqrt{\pi}}{2}.$$

$$\left[\because \int_0^\infty e^{-Z}Z^{n-1}\,dZ=\Gamma(n) \text{ and } \Gamma\left(\frac{3}{2}\right)=\frac{1}{2}\Gamma\left(\frac{1}{2}\right)=\frac{1}{2}\sqrt{\pi}\right].$$

Substituting this value in (3.52), we get

$$N=\frac{4\pi e^{-\alpha}V}{h^3}\times\frac{2^{1/2}m^{3/2}}{\beta^{3/2}}\frac{\sqrt{\pi}}{2}=\frac{e^{-\alpha}V}{h^3}\left(\frac{2m\pi}{\beta}\right)^{3/2} \text{ or } e^{-\alpha}=\frac{Nh^3}{V}\left(\frac{\beta}{2\pi m}\right)^{3/2}. \tag{3.53}$$

3.53) gives the value of the constant α.

Evaluation of β : We have $N(E)\,dE=g(E)\,e^{-\alpha}\,e^{-\beta E}\,dE.$

Now since $g(p)\,dp=\frac{4\pi p^2\,dp}{h^3}$,

$$\therefore\ g(E)\,dE=\frac{4\pi V\times 2mE}{h^3}\times\frac{m\,dE}{\sqrt{2mE}}.$$

$$\therefore\ N(E)\,dE=\frac{4\pi V\times 2mE\times m\,dE}{h^3\times\sqrt{2mE}}\times\left[\frac{Nh^3}{V}\left(\frac{\beta}{2\pi m}\right)^{3/2}\right]e^{-\beta E}=\frac{2N\beta^{3/2}}{\sqrt{\pi}}\sqrt{E}\,e^{-\beta E}\,dE. \tag{3.54}$$

Hence the *total energy of the system* is

$$U=\int_0^\infty EN(E)\,dE=\frac{2N\beta^{3/2}}{\sqrt{\pi}}\int_0^\infty E^{3/2}\,e^{-\beta E}\,dE.$$

Now

$$\int_0^\alpha E^{3/2}\,e^{-\beta E}\,dE=\frac{\int_0^\infty e^{-\beta E}(\beta E)^{3/2}\,d(\beta E)}{\beta^{5/2}}=\frac{1}{\beta^{5/2}}\int_0^\infty e^{-\beta E}(\beta E)^{\frac{5}{2}-1}\,d(\beta E)$$

$$=\frac{1}{\beta^{5/2}}\int_0^\infty e^{-x}\,x^{\frac{5}{2}-1}\,dx, \text{ [where } x=\beta E]$$

$$=\frac{1}{\beta^{5/2}}\Gamma\left(\frac{5}{2}\right)=\frac{1}{\beta^{5/2}}\frac{3}{2}\cdot\frac{1}{2}\sqrt{\pi}.$$

$$\therefore\ U=\frac{2N\beta^{3/2}}{\sqrt{\pi}\beta^{5/2}}\times\frac{3}{2}\cdot\frac{1}{2}\sqrt{\pi}=\frac{3}{2}\frac{N}{\beta}. \tag{3.55}$$

Now according to the *law of equi-partition of energy*, the total energy of N particles at T each having *three degrees* of freedom is

$$U = \frac{3}{2} NkT. \tag{3.56}$$

Comparing (3.55) and (3.56), we get $\beta = \frac{1}{kT}$.

1. *Maxwell-Boltzmann law of energy distribution.* Now using the value of $\beta = \frac{1}{kT}$ in (3.54), we get

$$N(E)\,dE = \frac{2\pi N}{(\pi kT)^{3/2}} \sqrt{E}\, e^{-\frac{E}{kT}}\, dE. \tag{3.57}$$

(3.57) represents the *Maxwell-Boltzmann law of energy distribution* among the ideal gas molecules.

2. *Maxwell-Boltzmann law of momentum distribution.* We have the number of molecules in the momentum range p to $p + dp$:

$$N(p)\,dp = \frac{4\pi V p^2 e^{-\alpha} e^{-\frac{\beta p^2}{2m}}\, dp}{h^3} \quad \text{[Ref. (3.51)]}$$

Replacing β by $\frac{1}{kT}$, we get

$$N(p)\,dp = 4\pi N \left(\frac{1}{2\pi mkT}\right)^{3/2} p^2 e^{-\frac{p^2}{2mkT}}\, dp. \tag{3.58}$$

(3.58) is the *Maxwell-Boltzmann law of momentum distribution* among the gas molecules at temperature T.

3. *Maxwell-Boltzmann law of velocity distribution.* The momentum of a molecule of mass m moving with velocity v is $p = mv$.

$$\therefore\ dp = m\,dv. \tag{3.59}$$

Now using these values in (3.48), we get the number of molecules having velocity in the range v to $v + dv$ at temperature T.

$$N(v)\,dv = 4\pi N \left(\frac{1}{2\pi mkT}\right)^{3/2} e^{-\frac{m^2v^2}{2mkT}} m^2 v^2 m\,dv$$

$$= 4\pi N \left(\frac{1}{2m\pi kT}\right)^{3/2} e^{-\frac{mv^2}{2kT}} m^3 v^2\, dv$$

$$= 4\pi N \left(\frac{m}{2\pi kT}\right)^{3/2} e^{-\frac{mv^2}{2kT}} v^2\, dv. \tag{3.60}$$

(3.60) represents *Maxwell-Boltzmann law of velocity distribution.*

Now

$$\frac{N(v)\,dv}{N} = 4\pi \left(\frac{m}{2\pi kT}\right)^{3/2} e^{-\frac{mv^2}{2kT}} v^2\, dv = F(v)\,dv,$$

where

$$F(v) = 4\pi \left(\frac{m}{2\pi kT}\right)^{3/2} e^{-\frac{mv^2}{2kT}} v^2. \tag{3.61}$$

If we plot $F(v)$ against v we get the *velocity distribution curve* as shown in Fig. 3.9.

If we take a strip of thickness dv at a distance v from the origin, then the area of the strip is $F dv$, which represents total number of molecules $\frac{N(v)dv}{N}$ in the velocity range v to $v + dv$. Now from (3.55) we get $N(v)\,dv = 0$ as $v = 0$ or ∞. So in between $v = 0$ and v there is a value of v_m at which $F(v) = \frac{N(v)dv}{N}$ is the maximum, i.e., the maximum number of molecules will have this velocity. This v_m is called the *most probable velocity* of the gas molecules.

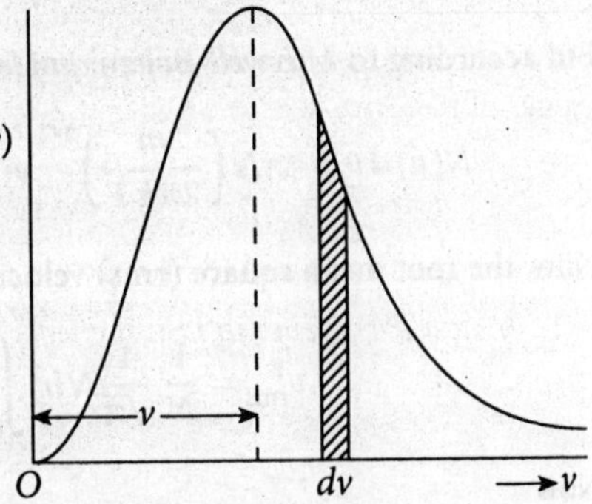

Fig. 3.9: The velocity distribution curve.

3.15.1 Average velocity ($\bar{v}$)

If in the gas, N_1 molecules have velocity v_1, N_2 molecules v_2 and so on, then total velocity is $N_1 v_1 + N_2 v_2 + \ldots$ and total number of molecules is $N_1 + N_2 + \cdots = N$.

$\therefore$ average velocity of a molecule is

$$\bar{v} = \frac{N_1 v_1 + N_2 v_2 + \ldots}{N_1 + N_2 + \ldots} = \frac{\Sigma N_i v_i}{N} = \frac{\int_0^\infty v N(v)\,dv}{N}.$$

But

$$N(v)\,dv = 4\pi N \left(\frac{m}{2\pi kT}\right)^{3/2} e^{-\frac{mv^2}{2kT}} v^2\,dv = \frac{4}{\sqrt{\pi}} N b^3 e^{-b^2 v^2}\,dv \left[\text{where } b = \sqrt{\frac{m}{2kT}}\right].$$

$$\therefore \bar{v} = \frac{1}{N}\frac{4}{\sqrt{\pi}} N b^3 \int_0^\infty e^{-bv^2} v^3\,dv = \frac{4}{\sqrt{\pi}} b^3 \frac{1}{2b^4}.$$

$$\left[\because \text{by the standard integral } \int_0^\infty z^3 e^{-az^2}\,dz = \frac{1}{2a^4} \text{ and so, } \int_0^\infty v^3 e^{-bv^2}\,dv = \frac{1}{2b^4}\right]$$

$$\therefore \bar{v} = \frac{2}{\sqrt{\pi}}\frac{1}{b} = \sqrt{\frac{8kT}{m\pi}}. \tag{3.62}$$

So, $\bar{v} \propto \sqrt{T}$, i.e., *the average velocity of a gas molecule is proportional to its absolute temperature* (T).

3.15.2 Root mean square velocity

It is defined as the *square root of the mean of the squares of the velocities* of all the gas molecules.

The sum of the squares of all the velocities is $N_1 v_1^2 + N_2 v_2^2 + \ldots$, where $N_1, N_2, \ldots$ are the number of molecules having velocities $v_1, v_2, \ldots$.

The total number of molecules is $N_1 + N_2 + \cdots = N$.

$\therefore$ mean square velocity is

$$\bar{v}^2 = \frac{N_1 v_1^2 + N_2 v_2^2 + \ldots}{N_1 + N_2 + \ldots} = \frac{\sum_{i=1}^{\infty} N_i v_i^2}{N} = \frac{1}{N}\int_0^\infty v^2 N(v)\,dv.$$

But according to *Maxwell-Boltzmann law of velocity distribution,*

$$N(v)\,dv = 4\pi N\left(\frac{m}{2\pi kT}\right)^{3/2} e^{-\frac{mv^2}{2kT}} v^2\,dv = \frac{4}{\sqrt{\pi}} N b^3 e^{-b^2v^2} v^2\,dv \quad \left[b = \left(\frac{m}{2kT}\right)^{1/2}\right].$$

Now, the root mean square (rms) velocity of a gas molecule is

$$v_{\text{rms}}^2 = \frac{1}{N}\frac{4}{\sqrt{\pi}} N b^3 \int_0^\infty e^{-b^2v^2} v^4\,dv = \frac{4}{\sqrt{\pi}} b^3 \int_0^\infty e^{-b^2v^2} v^4\,dv. \tag{3.63}$$

Now

$$\int_0^\infty e^{-b^2v^2} v^4\,dv = \int_0^\infty e^{-Z}\left(\frac{Z^2}{b^4}\right)\frac{dZ}{2b\sqrt{Z}} \quad \left[\text{where } Z = b^2v^2,\ \therefore\ dZ = 2b^2 v\,dv \text{ or } dv = \frac{dz}{2b\sqrt{Z}}\right]$$

$$= \int_0^\infty e^{-Z} Z^{3/2}\frac{1}{2b^5}\,dz = \frac{1}{2b^5}\int_0^\infty e^{-Z} Z^{\frac{5}{2}-1}\,dZ = \frac{1}{2b^5}\times\Gamma\left(\frac{5}{2}\right) = \frac{1}{2b^5}\times\frac{3}{2}\times\frac{1}{2}\times\Gamma\left(\frac{1}{2}\right)$$

$$= \frac{1}{2b^5}\times\frac{3}{4}\times\frac{\sqrt{\pi}}{2} \quad \left[\because \Gamma\left(\frac{5}{2}\right) = \frac{3}{2}\Gamma\left(\frac{3}{2}\right) = \frac{3}{2}\cdot\frac{1}{2}\cdot\Gamma\left(\frac{1}{2}\right) \text{ and } \Gamma\left(\frac{1}{2}\right) = \sqrt{\pi}\right]$$

$$= \frac{3}{8b^5}\sqrt{\pi}. \tag{3.64}$$

Now from (3.63),

$$v_{\text{rms}}^2 = \frac{4b^3}{\sqrt{\pi}}\times\frac{3}{8b^5}\sqrt{\pi} = \frac{3}{2b^2} = \frac{3}{2}\times\frac{2kT}{m} \quad \left[\because b = \left(\frac{m}{2kT}\right)^{1/2}\right]$$

$$= \frac{3kT}{m}.$$

$$\therefore\ v_{\text{rms}} = \sqrt{\frac{3kT}{m}}. \tag{3.65}$$

Thus $v_{\text{rms}} \propto \sqrt{T}$, i.e., *the rms velocity of a gas molecule is proportional to its absolute temperature* (T).

3.15.3 Most probable velocity

The most probable velocity is defined as the velocity with which maximum number of gas molecules move.

According to *Maxwell-Boltzmann law of velocity distribution* the number of molecules moving in the velocity range v to $v + dv$ is

$$N(v)\,dv = 4\pi N\left(\frac{m}{2\pi kT}\right)^{3/2} e^{-\frac{mv^2}{2kT}} v^2\,dv.$$

Also, the probability that a molecule will move with velocity between v and $v + dv$ is

$$\frac{N(v)\,dv}{N} = P(v)\,dv = 4\pi\left(\frac{m}{2\pi kT}\right)^{3/2} e^{-\frac{mv^2}{2kT}} v^2\,dv$$

$$\text{or } P(v)\,dv = \frac{4}{\sqrt{\pi}} b^3 e^{-b^2v^2} v^2\,dv \quad \left[\text{where } b = \sqrt{\frac{m}{2kT}}\right]$$

$$\text{or } P(v) = \frac{4}{\sqrt{\pi}} b^3 e^{-b^2v^2} v^2.$$

For $P(v)$ to be maximum, $\frac{dP(v)}{dv}=0$;

$$\therefore \frac{d}{dv}\left[\frac{4}{\sqrt{\pi}}b^3 e^{-b^2v^2}v^2\right]=0 \text{ or } \frac{4b^3}{\sqrt{\pi}}\left[2v e^{-b^2v^2} - v^2 e^{-b^2v^2}\cdot 2b^2 v\right]=0$$

$$\text{or } \frac{8b^3}{\sqrt{\pi}} v e^{-b^2v^2}\left(1-v^2b^2\right)=0 \text{ or } 1-v^2b^2=0 \text{ or } v=\frac{1}{b}=\sqrt{\frac{2kT}{m}}.$$

$\therefore$ most probable velocity

$$v_m=\sqrt{\frac{2kT}{m}}. \tag{3.66}$$

So $v_m \propto \sqrt{T}$, i.e., *the most probable velocity of a gas molecule is proportional to its absolute temperature T.*

3.16 Law of Equi-partition of Energy

Statement. It states that, total KE *of a dynamical system consisting of a large number of particles in thermal equilibrium is equally divided among its all degrees of freedom and the energy associated with one molecule per degree of freedom is equal to* $\frac{1}{2}kT$.

Alternatively, the degree of freedom of a dynamical system is defined as the total number of independent coordinates required to specify completely its position and configuration. It is also defined as the number of squared terms in its expression of KE.

Proof. According to the *Maxwell-Boltzmann law of energy distribution* among the gas molecules, the number of molecules in the energy interval E to $E+dE$, we have

$$N(E)\,dE=\frac{2\pi N}{(\pi kT)^{3/2}}E^{1/2}e^{-E/kT}\,dE.$$

$\therefore$ the total energy E of the system is

$$E=\int_0^\infty EN(E)\,dE=\frac{2\pi N}{(\pi kT)^{3/2}}\int_0^\infty E^{3/2}e^{-E/kT}\,dE. \tag{3.67}$$

Now

$$\int_0^\infty E^{3/2}e^{-E/kT}\,dE=(kT)^{5/2}\int_0^\infty e^{-z}z^{3/2}\,dz \quad \left[\begin{array}{l}\text{where } z=E/kT;\\ \therefore E=kTz \text{ and } dE=(kT)\,dz\end{array}\right]$$

$$=(kT)^{5/2}\int_0^\infty e^{-z}z^{\frac{5}{2}-1}\,dz=(kT)^{5/2}\Gamma\left(\frac{5}{2}\right)=(kT)^{5/2}\frac{3}{2}\cdot\frac{1}{2}\sqrt{\pi}=\frac{3\sqrt{\pi}}{4}(kT)^{5/2}.$$

$\therefore$ from (3.67),

$$E=\frac{2\pi N}{(\pi kT)^{3/2}}\frac{3\sqrt{\pi}}{4}(kT)^{5/2}=\frac{3}{2}NkT.$$

$\therefore$ the average energy of an ideal gas molecule is

$$\langle E\rangle'=\frac{E}{N}=\frac{3}{2}kT. \tag{3.68}$$

Again, each gas molecule has *three degrees of freedom* corresponding to translational motion along X-, Y- and Z-directions. So, the average energy per degree of freedom of an ideal gas molecule is

$$\langle E\rangle = \frac{\langle E\rangle'}{3} = \frac{1}{2}kT, \tag{3.69}$$

where k is the *Boltzmann constant and T is the absolute temperature of the gas.*

3.17 Limitations of Maxwell-Boltzmann Statistics

Maxwell-Boltzmann statistics assumes that the particles under consideration are *distinguishable* which is true for ideal gas molecules. So this statistics could account the distribution of velocity, energy and momentum of *ideal gas molecules* to a fair degree of accuracy.

But in case of *distinguishable particles* like *electrons, photons,* etc. Maxwell-Boltzmann statistics *failed completely* to explain the distribution of energy. Quantum statistics (*Bose-Einstein* and *Fermi-Dirac Statistics*) could, however, solve the problems with a high degree of satisfaction.

3.18 Occupation Index, Degenerate, Non-degenerate and Strongly Degenerate States

Let there be a system in which there are N_i number of identical particles in the g_i quantum state in the ith state of energy E_i, then

(i) the ratio $\frac{N_i}{g_i}$ is called the *occupation number* of the ith state of the system.

(ii) if $\frac{N_i}{g_i} \ll 1$ or, $N_i \ll g_i$, then the system is *non-degenerate.*

(iii) if $\frac{N_i}{g_i} > 1$ or, $N_i > g_i$, then the system is *degenerate.*

(iv) if $\frac{N_i}{g_i} \gg 1$ or, $N_i \gg g_i$, then the system is called a *strongly degenerate* state.

3.19 Bose-Einstein (BE) Statistics

Bose-Einstein statistics is applicable to those particles which are *identical and indistinguishable having integral spin and do not obey Pauli's exclusion principle.* These particles which obey BE statistics are called *Bosons*, e.g., photon (spin 1), π-meson (spin 0).

To arrive at BE distribution formula, the following points are assumed:

1. *The particles are identical and indistinguishable having integral spin angular momentum in units of* $\frac{h}{2\pi}$.
2. *Each eigenstate (i.e., sublevel) of the ith quantum state may contain $0, 1, 2, \ldots, N_i$ identical particles, where n_i is the particles in the ith state, i.e., Pauli's exclusion principle does not hold good.*
3. *The number of phase cells (g_i) is comparable to the number of particles (N_i), i.e., the occupation index $N_i/g_i \simeq 1$. Also the size of a phase cell cannot be less than h^3, where $h = 6.63\times10^{-34}$ J is the Planck's constant.*
4. *Energy is discrete and the total energy of the system is equal to the sum of energies of all the particles in different groups.*

We consider the following array of Fig. 3.10.

Fig. 3.10

Here, the *dots denote the particles* and the *lines, the quantum states containing the particles in the left of the particles.* The problem here is to distribute N_i particles among the g_i quantum states of energy level E_i, there being *no restriction regarding the number of particles in any quantum state.*

The array contains g_i quantum states and N_i particles in the ith state, i.e., $(g_i + N)$ objects. Assuming the first line fixed, the remaining $(g_i + N_i - 1)$ objects can be arranged in $(g_i + N_i - 1)!$ ways. However, among $(g_i + N_i - 1)!$ arrangements $(g_i - 1)!$ and $N_i!$ are *not independent distributions*. So, the total number of *independent distributions of N_i particles in g_i cell is*

$$\frac{(N_i + g_i - 1)!}{N_i!(g_i - 1)!}.$$

As N_i and g_i are very large numbers we can neglect 1 in comparison to them and the above distribution is then

$$\frac{(N_i + g_i)!}{N_i!\,g_i!}.$$

Similar expressions can be obtained for various other quantum states. Therefore, the *thermodynamical probability* can be written as

$$W = \frac{(N_1 + g_1)!}{N_1!\,g_1!} \times \frac{(N_2 + g_2)!}{N_2!\,g_2!} \times \cdots \times \frac{(N_k + g_k)!}{N_k!\,g_k!} = \prod_i^k \frac{(N_i + g_i)!}{N_i!\,g_i!}.$$

$$\therefore \ln W = \sum_i \ln (N_i + g_i)! - \sum_i \ln N_i! - \sum_i \ln g_i!$$

$$= \sum_i [(N_i + g_i)\ln (N_i + g_i) - N_i \ln N_i - g_i \ln g_i] \tag{3.70}$$

[using Stirling's formula, $\ln x! = x\ln x - x$]

$$\text{or } \delta(\ln W) = \sum_i [\ln (N_i + g_i) - \ln N_i]\,\delta N_i$$

$$= -\sum_i \left[\ln\left(\frac{N_i}{N_i + g_i}\right)\right] \delta N_i \ [\because\ g_i \text{ is constant}].$$

Since the system is in equilibrium, $\ln W$ is *maximum*, i.e., $\delta(\ln W) = 0$ under the condition that

$$N = \sum_i N_i = \text{constant}, \ \therefore \delta N = \sum_i \delta N_i = 0 \tag{3.71}$$

and total energy

$$E = \sum N_i E_i = \text{constant or, } \delta E = \sum E_i \delta N_i = 0. \tag{3.72}$$

Also

$$\ln\left(\frac{N_i}{N_i + g_i}\right)\delta N_i = 0. \tag{3.73}$$

Multiplying (3.71) by α and (3.72) by β and then adding to (3.73), we get

$$\sum_{i=1}^{k} [\ln\{N_i/(N_i + g_i)\} + \alpha + \beta E_i]\,\delta N_i = 0, \tag{3.74}$$

where α and β are *undetermined multipliers*.

Since the variations are arbitrary, we get

$$\ln\{N_i/(N_i+g_i)\}+\alpha+\beta E_i=0$$

$$\text{or } \ln\left(\frac{N_i}{N_i+g_i}\right)=-(\alpha+\beta E_i)$$

$$\text{or } 1+\frac{g_i}{N_i}=e^{(\alpha+\beta E_i)} \text{ or } \frac{N_i}{g_i}=1/(e^{\alpha+\beta E_i}-1)$$

$$\text{or } N_i=\frac{g_i}{e^{\alpha+\beta E_i}-1} \tag{3.75}$$

$$\text{and } f(E_i)=\frac{N_i}{g_i}=\frac{1}{e^{\alpha+\beta E_i}-1}. \tag{3.76}$$

(3.75) and (3.76) are respectively known as the *Bose-Einstein distribution law* and *distribution function.*

3.20 Planck's Law of Radiation (Using BE Statistics)

The *black body radiation* inside a hollow enclosure consists of a very large number of *photons* of different energies at a particular temperature *T*. *Planck's law* giving the *distribution of energy* in the black body radiation can be deduced by applying *BE statistics* with the following assumptions:

1. *Radiant energy occurs in energy packets or quanta or photons each of energy $E = h\nu$; where h is the Planck's constant and ν is the frequency of radiations.*
2. *Photons are of zero rest-mass and spin quantum number 1 and so they obey BE statistics.*
3. *In case of photon gas the constancy of number of particles does not apply, i.e., $\Sigma\delta N_i \neq 0$, i.e., $\alpha=0$.*
4. *The number of states per unit volume in the momentum range p and $p+dp$ is*

$$g(p)\,dp=\frac{4\pi p^2}{h^3}\,dp.$$

Since $p=\frac{h\nu}{c}$, where ν is the frequency and c, the velocity of light, we get

$$g(\nu)\,d\nu=g(E)\,dE=\frac{4\pi V}{h^3}\left(\frac{h}{c}\right)^3\nu^2\,d\nu=\frac{4\pi\nu}{c^3}\nu^2\,d\nu. \tag{3.77}$$

According to *BE statistics,*

$$N_i=\frac{g_i}{e^{\beta E_i}-1}=\frac{g_i}{e^{E_i/kT}-1}\quad[\because \alpha=0]. \tag{3.78}$$

In the limit of very close energy levels,

$$N(E)\,dE=\frac{g(E)\,dE}{e^{E/kT}-1} \text{ or } N(\nu)\,d\nu=\frac{4\pi V}{c^3}\frac{\nu^2\,d\nu}{e^{E/kT}-1}. \tag{3.79}$$

(3.79) gives the *number photons* in the enclosure of volume *V*, in the energy interval *E* to $E+dE$.

Since for a photon, there are *two states of polarisation* and energy of a photon is $E = h\nu$, the energy density of radiation in the frequency range $\nu + d\nu$ is

$$E_\nu\,d\nu=2\times\frac{N(\nu)\,d\nu}{V}\times h\nu \text{ or } E_\nu\,d\nu=\frac{8\pi h\nu^3}{c^3}\frac{d\nu}{e^{h\nu/kT}-1}. \tag{3.80}$$

(3.80) is the *Planck's radiation formula* first deduced by *Professor S N Bose* using statistical method.

3.20.1 Rayleigh-Jeans law

For long wavelength (i.e., low frequency) and high temperature $h\nu/kT \ll 1$.

$$e^{h\nu} - 1 \simeq \frac{h\nu}{kT}.$$

Then from (3.80), we get

$$E_\nu \, d\nu = \frac{8\pi kT}{c^3}\nu^2 \, d\nu. \tag{3.81}$$

This is *Rayleigh-Jeans law.*

3.20.2 Wien's law

For short wavelength (i.e., high frequency) and low temperature, $h\nu/kT \gg 1$.

$$e^{h\nu/kT} - 1 \simeq e^{h\nu/kT}.$$

∴ from (3.80), we get

$$E_\nu \, d\nu = \frac{8\pi h\nu^3}{c^3} e^{-h\nu/kT} \, d\nu. \tag{3.82}$$

This is *Wien's distribution law.*

Again, *Planck's law* in terms of λ is

$$E_\lambda \, d\lambda = \frac{8\pi ch}{\lambda^5}\frac{x}{e^x - 1}\, d\lambda, \text{ where } x = \frac{ch}{\lambda kT} \text{ or } E_\lambda = \frac{8\pi ch}{\lambda^5}\frac{x}{e^x - 1}.$$

E_λ is maximum when $\frac{dE_\lambda}{d\lambda} = 0$.

Let at $\lambda = \lambda_m, E_\lambda = (E_\lambda)_{\max}$.

$$\text{or } -\frac{5}{\lambda_m} + \frac{x}{\lambda_m}\frac{e^x}{e^x - 1} = 0 \text{ or } \frac{xe^x}{e^x - 1} = 5 \text{ or } \frac{x}{1 - e^{-x}} = 5 \text{ or } \frac{1}{5}x + e^{-x} = 1.$$

This is a *transcendental equation* and has solution $x = 4.965$

$$\text{or } \frac{ch}{\lambda_m kT} = 4.965$$

$$\text{or } \lambda_m T = \frac{ch}{k \times 4.965} = \frac{2.997 \times 10^{10} \times (6.56 \times 10^{-27})}{(1.38 \times 10^{-16}) \times 4.965} = 0.289 \text{ cm. deg.}$$

$$\text{or } \lambda_m T = 0.289 \tag{3.83}$$

is the *Wien's displacement law.*

3.20.3 Stefan-Boltzmann law

We have from Planck's law

$$E_\nu \, d\nu = \frac{8\pi h\nu^3}{c^3}\frac{d\nu}{e^{h\nu/kT} - 1} \quad [\because \alpha = 0].$$

∴ total energy

$$E = \int_0^\infty E_\nu \, d\nu = \frac{8\pi h}{c^3}\int_0^\infty \frac{\nu^3 \, d\nu}{e^{h\nu/kT} - 1} = \frac{8\pi(kT)^4}{c^3 h^3}\int_0^\infty \frac{x^3}{e^x - 1}\, dx,$$

where $x = h\nu/kT$.

Now $\int_0^\infty \frac{x^3}{e^x-1} = \frac{\pi^4}{15}$.

$$\therefore\ E = \frac{8\pi k^4 T^4}{c^3 h^3} \times \frac{\pi^4}{15} = \frac{8\pi h}{c^3}\left(\frac{kT}{h}\right)^4 \times \frac{\pi^4}{15}$$

$$\text{or } E = bT^4, \tag{3.84}$$

where $b = \frac{8\pi\pi^4 k^4}{15c^3 h^3}$.

(3.84) represents *Stefan-Boltzmann law of total radiation.*

3.21 Relation between Entropy and Probability

Relation between entropy (S) and thermodynamic probability (W): Boltzmann started from the simple idea that a *state of maximum entropy is also a state of maximum probability*. So, entropy is a function of probability, i.e., $S = f(W)$.

To find out the nature of the function, we consider two completely independent states A and B having entropies S_1 and S_2 respectively. Since entropy is an additive quantity, the entropy of systems A and B taken together will be equal to the sum of their respective entropies, i.e.,

$$S = S_1 + S_2. \tag{3.85}$$

If the probability of A and B be respectively W_1 and W_2, then the probability of finding both the systems at their respective conditions is

$$W = W_1 \times W_2. \tag{3.86}$$

$$\therefore\ S = f(W) = f(W_1 W_2). \tag{3.87}$$

Again,

$$S_1 = f(W_1) \tag{3.88}$$

$$S_2 = f(W_2). \tag{3.89}$$

So from (3.85),

$$f(W_1 W_2) = f(W_1) + f(W_2). \tag{3.90}$$

[Substituting the value of S, S_1 and S_2 in (3.85).]

Differentiating partially (3.90) w.r.t. W_1 and W_2, we get

$$W_2 f'(W_1 W_2) = f'(W_1)$$

$$W_1 f'(W_1 W_2) = f'(W_2).$$

$$\therefore\ \frac{f'(W_1)}{f'(W_2)} = \frac{W_2}{W_1} \text{ or } W_1 f'(W_1) = W_2 f'(W_2) = \ldots = k, \tag{3.91}$$

where k is any constant.

$$\therefore\ f'(W_1) = \frac{k}{W_1},\ f'(W_2) = \frac{k}{W_2}. \tag{3.92}$$

Integrating, we get

$$f(W_1) = k \ln W_1 + C_1 \tag{3.93}$$
$$f(W_2) = k \ln W_2 + C_2. \tag{3.94}$$

In general,

$$f(W) = k \ln W + C, \tag{3.95}$$

k is the universal constant and is known as the *Boltzmann constant* and C is found to be zero.

$$\therefore\ S = k \ln W. \tag{3.96}$$

This gives the desired relation between *entropy* and *thermodynamical probability*.]

3.22 Fermi-Dirac (FD) Statistics

FD statistics is obeyed by *indistinguishable particles of half-integral spin having antisymmetric wave functions and obeying Pauli's exclusion principle.* Particles obeying FD statistics are called *fermions.*

Electron (spin $\frac{1}{2}$), neutron (spin $\frac{1}{2}$), proton (spin $\frac{1}{2}$), muon (spin $\frac{1}{2}$), etc. are the examples of fermions.

We consider a system of N *indistinguishable particles* which are divided into *quantum groups* (or levels) such that $N_1, N_2, \ldots, N_k$ be the number of particles in levels of constant energies $E_1, E_2, \ldots, E_k$ respectively. Let g_i denote the degeneracy (number of sub-levels) in the ith level.

Now, the problem is to find the FD *distribution law* of N_i particles in g_i eigenstates (sub-levels), where $g_i > N_i$. We assume that

1. *The particles are indistinguishable so that there is no distinction between the different ways in which N_i particles are chosen.*
2. *Each cell (or sub-level) may contain 0 or one particle following Pauli's exclusion principle and as such g_i must be greater than or equal to N_i.*
3. *Total energy of the system is the sum of the energies of all the particles in different sub-levels, i.e.,*

$$E = \sum_i N_i E_i \tag{3.97}$$

and total number of particles

$$N = \sum_i N_i. \tag{3.98}$$

We can distribute N_i particles in g_i states as follows:

According to *Pauli's exclusion principle* no cell can contain more than one particle. So, out of g_i cells only N_i cell are occupied and $(g_i - N_i)$ are empty. The possible number of such distributions is $g_i!$. But $N_i!$ distributions of filled cells are similar to each other.

Also $(g_i - N_i)!$ distributions of vacant cells are also *irrelevant.*

$\therefore$ the number of *distinguishable arrangements* of the particles among the cells is $\dfrac{g_i!}{N_i!(g_i - N_i)!}$.

Similar expression can also be found for other quantum states. Then the probability of entire distribution is

$$W = \prod_i \frac{g_i!}{N_i!(g_i - N_i)!}.$$

$$\therefore \ln W = \sum_i [\ln g_i! - \ln N_i! - \ln (g_i - N_i)!].$$

$$\text{or } \ln W = \sum_i (g_i \ln g_i - g_i) - (N_i \ln N_i - N_i) - (g_i - N_i)\ln(g_i - N_i) + (g_i - N_i)$$

[by Stirling's formula $\ln x! = x\ln x - x$]

$$= \sum_i g_i(\ln g_i - 1) - N_i(\ln N_i - 1) - (g_i - N_i)\,[\ln(g_i - N_i) - 1]$$

$$= \sum_i [g_i \ln g_i - N_i \ln N_i - (g_i - N_i)\ln(g_i - N_i)]$$

[$\because$ N_i and g_i are very large numbers compared to 1]

$$\therefore \delta(\ln W) = \sum_i \left[-N_i \frac{1}{N_i}\delta N_i - \ln N_i \delta N_i + (g_i - N_i)\frac{1}{(g_i - N_i)}\delta N_i + \ln(g_i - N_i)\,\delta N_i\right]$$

$$\text{or } \delta(\ln W) = \sum_i [\ln(g_i - N_i) - \ln N_i]\,\delta N_i.$$

Since g_i and $N = \Sigma N_i$ are constants,

$$\delta(\ln W) = -\sum_i \ln\left(\frac{N_i}{g_i - N_i}\right)\delta N_i.$$

For *most probable distribution* $W = W_{\max}$, so $\delta(\ln W_{\max}) = 0$.

$$\therefore \sum_i \ln\left(\frac{N_i}{g_i - N_i}\right)\delta N_i = 0. \tag{3.99}$$

Also, $N = \Sigma N_i =$ constant.

$$\therefore \delta N = \Sigma \delta N_i = 0 \tag{3.100}$$

and total energy $E = \sum_i N_i E_i =$ constant.

$$\therefore \delta E = \Sigma E_i \delta N_i = 0. \tag{3.101}$$

Now multiplying (3.100) by α and (3.101) by β and then adding to (3.99), we get

$$\sum_i \left[\ln\left(\frac{N_i}{g_i - N_i}\right) + \alpha + \beta E_i\right]\delta N_i = 0. \tag{3.102}$$

As the variations δN_i are independent of each other,

$$\therefore \ln\left(\frac{N_i}{g_i - N_i}\right) + \alpha + \beta E_i = 0 \text{ or } \frac{N_i}{g_i - N_i} = e^{-(\alpha+\beta E_i)}$$

$$\text{or } \frac{g_i - N_i}{N_i} = e^{\alpha+\beta E_i} \text{ or } \frac{g_i}{N_i} = e^{\alpha+\beta E_i} + 1 \text{ or } N_i = \frac{g_i}{e^{\alpha+\beta E_i} + 1} \tag{3.103}$$

$$\text{and } f(E_i) = \frac{N_i}{g_i} = \frac{1}{e^{\alpha+\beta E_i} + 1}. \tag{3.104}$$

(3.103) and (3.104) are known *FD distribution formula* and *FD distribution function* respectively.

3.22.1 Energy distribution of free electrons in metal

Fermi-Dirac distribution law finds important application in *free electron theory of metals. Fermi-Dirac* derived *energy distribution formula* for free electron gas on the basis of the following assumptions:

1. *Every metal contains large number of free electrons at ordinary temperature which behaves like electron gas.*
2. *The electrons have half-integral spin angular momentum in units of $h/2\pi$ and obey FD statistics.*
3. *The electrons in the metal have quantised energy values and obey Pauli's exclusion principle.*

Considering continuous distribution of energy we may write E in place of E_i. The number of electrons in unit energy interval $\left(E-\frac{1}{2}\right)$ and $\left(E+\frac{1}{2}\right)$ is $N(E)$. The number of *phase cells* in the same energy interval is $g(E)$.

Then, $N(E)\,dE$ and $g(E)\,dE$ will represent the *number of electrons* and *phase cells* respectively in the energy interval E to $E+dE$. (3.105) can now be written as

$$N(E)\,dE=\frac{g(E)\,dE}{e^{\alpha+\beta E}+1}. \tag{3.105}$$

But

$$g(E)\,dE=\frac{8\pi V}{h^3}(2mE)^{1/2}m\,dE=\frac{8\sqrt{2}\pi V}{h^3}m^{3/2}E^{1/2}\,dE.$$

$$\therefore\ N(E)\,dE=\frac{8\sqrt{2}\pi V}{h^3}\frac{m^{3/2}E^{1/2}\,dE}{e^{\alpha+E/kT}+1}\ \left[\because \beta=\frac{1}{kT}\right]. \tag{3.106}$$

(3.106) represents the *energy distribution of free electrons in metal.*

3.22.2 Effect of temperature

(3.106) gives the energy distribution of free electrons in metal at temperature T. With the decrease of T, (3.106) shows that the energy of the electron goes on decreasing. At $T=0$ K the energy of all the electrons is *not zero*, instead have some *finite value*. Since the electrons are fermions (spin 1/2) only one electron with the same set of quantum number can occupy one energy level. In view of this, there will be one electron in the ground state ($E=0$). Thus, if there are N number of electrons at $T=0$ K, only N number of energy levels will be filled each with one electron and the remaining energy levels will remain vacant. Thus, if the energy value of the highest filled level at 0 K be E_f, then the levels with $E<E_f$ will be *fully occupied* (filled with one electron) and levels with $E>E_f$ will remain *empty*.

Fermi level $\left(E_f\right)$: *The highest occupied energy level at 0 K below which all the levels are occupied and above which all the levels are empty is known as the Fermi-energy level* $\left(E_f\right)$.

Now the occupation index

$$\frac{N(E)}{g(E)}=0,\ \text{if } E>E_f \text{ and } \frac{N(E)}{g(E)}=1,\ \text{if } E\le E_f.$$

The above conditions give $\alpha=-\frac{E_f}{kT}$.

For $E<E_f$ and $T=0$ K, $\alpha=-\frac{E_f}{kT}$.

$$\therefore\ \frac{N(E)}{g(E)}=\frac{1}{e^{(E-E_f)/kT}+1}=\frac{1}{e^{-\infty}+1}=1.$$

This shows that at $T=0$ K, all the energy levels with $E<E_f$ are *fully occupied each with one electron.*

For $E > E_f$ and $T = 0$ K, $\alpha = -\frac{E_f}{kT}$.

$$\therefore \frac{N(E)}{g(E)} = \frac{1}{e^{(E-E_f)/kT}+1} = \frac{1}{e^{\infty}+1} = 0.$$

Thus the energy levels which lie above *Fermi level* (i.e., $E > E_f$) are *empty*.

Now substituting $\alpha = -\frac{E_f}{kT}$ in (3.106), we get

$$N(E)\,dE = \frac{8\sqrt{2}\pi V}{h^3}\frac{m^{3/2}E^{1/2}\,dE}{e^{(E-E_f)/kT}+1} \tag{3.107}$$

(3.107) represents the *Fermi-Dirac energy distribution of free electrons*. The variation of occupation index $\frac{N(E)}{g(E)}$ with energy E at $T = 0$ K and $T > 0$ K, are shown in Fig. 3.11.

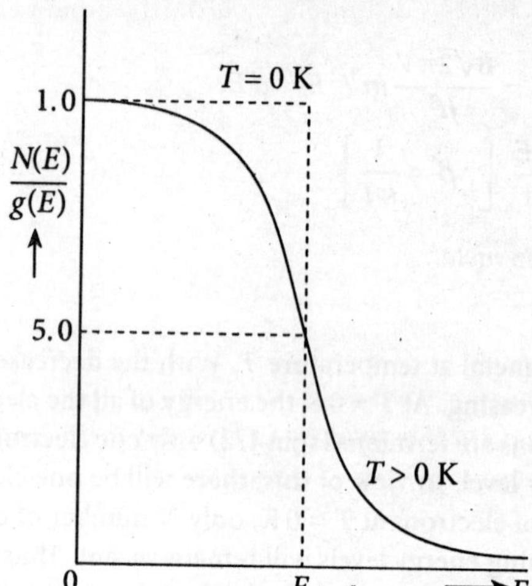

Fig. 3.11: Variation of $\frac{N(E)}{g(E)}$ with E at $T = 0$ K and $T > 0$ K.

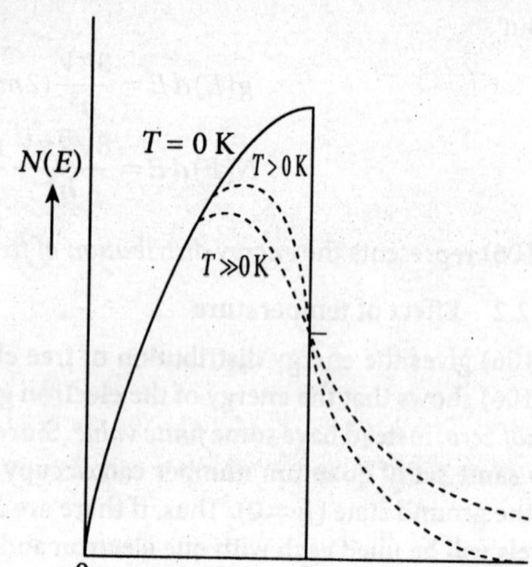

Fig. 3.12: Variation of $N(E)$ with E at $T = 0$ K and $T > 0$ K.

Putting $\alpha = -E_f/kT$ in (3.104), we get

$$f(E_i) = \frac{N_i}{g_i} = \frac{1}{e^{(E_i-E_f)/kT}+1}.$$

The above equation shows that if $T \neq 0$, then $f(E_i) = \frac{1}{2}$ for $E_i = E_f$. *We can thus define Fermi level is that energy level for which the probability of occupation at $T > 0$ is $\frac{1}{2}$.*

At low temperature $N(E)$ undergoes a rapid decrease for $E > E_f$. *At low temperature* when the distribution function $f(E_i) = \frac{N(E)}{g(E)}$ is nearly a step function, $f(E_i)$ is said to be *strongly degenerate* whereas at *high temperature* when the step-like nature is absent, $f(E_i)$ is said to be nearly *non-degenerate*.

3.22.3 FD energy distribution law in terms of Fermi energy (E_f)

Fermi-Dirac law of energy distribution among free electrons in a metal at temperature T is given by (3.107) which can be written after rearrangement as

$$N(E)\,dE = \frac{16\sqrt{2}\pi V}{3Nh^3} m^{3/2} \frac{3N}{2} E^{1/2} \frac{dE}{e^{\alpha}\cdot e^{E/kT}+1}$$

$$= \frac{16\sqrt{2}\pi V}{3Nh^3} m^{3/2} \frac{3N}{2} E^{1/2} \frac{dE}{e^{(E-E_f)/kT}+1} \quad \left[\because \alpha = -\frac{E_f}{kT}\right]$$

$$\text{or } N(E)\,dE = \frac{3}{2} N E_f^{-3/2} E^{1/2} \frac{dE}{e^{(E-E_f)/kT}+1}, \tag{3.108}$$

where $E_f^{3/2} = \frac{3Nh^3}{16\sqrt{2}\pi V m^{3/2}}$.

(3.108) represents *Fermi-Dirac energy distribution law in terms of Fermi energy* (E_f).

3.22.4 Average KE of an electron at $T = 0$ K

We have at $T = 0$ K, energy of an electron E is less than E_f, since by definition, E_f is the maximum energy of an electron at 0 K.

$\therefore$ for $E < E_f$,

$$\frac{E - E_f}{kT} = -\frac{(E_f - E)}{kT} = -\infty \tag{3.109}$$

and $e^{(E-E_f)/kT} = e^{-\infty} = 0$.

Now from (3.108),

$$N(E)\,dE = \frac{3}{2} N E_f^{-3/2} E^{1/2}\,dE \quad \left[\because \frac{1}{e^{(E-E_f)/kT}} = 1\right] \tag{3.110}$$

represents the *energy distribution of free electrons at* $T = 0$ K.

Now the KE of the electrons in the energy interval E to $E + dE$ is given by (3.110) and the *average KE of a single electron at* $T = 0$ K is

$$\langle E\rangle = \frac{\int_0^{E_f} E N(E)\,dE}{\int_0^N N(E)\,dE} = \frac{3}{2}\frac{N}{N} E_f^{-3/2} \int_0^{E_f} E^{3/2}\,dE$$

$$= \frac{3}{2} E_f^{-3/2} \left[\frac{2}{5} E^{5/2}\right]_0^{E_f} = \frac{3}{2} E_f^{-3/2} \frac{2}{5} E_f^{5/2} \text{ or } \langle E\rangle = \frac{3}{5} E_f. \tag{3.111}$$

Thus at $T = 0$ K *the average energy of electron is $\frac{3}{5}$th of Fermi-energy.*

Now the *total energy of all the N electrons at* 0 K is

$$E_t = N\langle E\rangle = \frac{3}{5} N E_f. \tag{3.112}$$

3.22.5 Average speed of electrons at 0 K

If $N(v)\,dv$ be the number of electrons in the velocity range v to $v + dv$, then the *average velocity of an electron* at 0 K is $\langle v_0\rangle = \frac{1}{N}\int_0^{v_f(0)} v N(v)\,dv$, where $v_f(0)$ is called the *Fermi-velocity* at 0 K.

Again,

$$N(E)\,dE = g(E)\,dE = \frac{8\pi V}{h^3}\sqrt{2mE}\,m\,dE = \frac{8\sqrt{2}\pi V}{h^3}m^{3/2}E^{1/2}\,dE.$$

$$\therefore\ N(v)\,dv = \frac{8\sqrt{2}\pi V}{h^3}m^{3/2}\frac{\sqrt{m}}{\sqrt{2}}v\,m\,v\,dv = \frac{8\pi V m^3}{h^3}v^2\,dV.$$

The total number of particles N is given by

$$N = \int_0^{v_f(0)} N(v)\,dv = \int_0^{v_f(0)} \frac{8\pi V m^3}{h^3}v^2\,dv = \frac{8\pi V}{h^3}m^3\int_0^{v_f(0)} v^2\,dv = \frac{8\pi V}{3h^3}m^3\left(v_f(0)\right)^3.$$

$$\therefore\ \left(v_f(0)\right)^3 = \frac{3Nh^3}{8\pi V m^3}\ \text{or}\ \frac{8\pi V m^3}{h^3} = 3N\left(v_f(0)\right)^{-3}.$$

$$\therefore\ N(v)\,dv = \frac{8\pi V m^3}{h^3}v^2\,dv = 3N\left(v_f(0)\right)^{-3}v^2\,dv.$$

And

$$\langle v_0\rangle = \frac{1}{N}\int_0^{v_f(0)} vN(v)\,dv = \frac{1}{N}\int_0^{v_f(0)} 3N\left(v_f(0)\right)^{-3}v^3\,dv = 3\left(v_f(0)\right)^{-3}\frac{v_f(0)^4}{4} = \frac{3}{4}v_f(0).$$

$$\therefore\ \langle v_0\rangle = \frac{3}{4}v_f(0). \tag{3.113}$$

Thus *the average speed of the electron at 0 K is $\frac{3}{4}$th of the Fermi velocity $v_f(0)$.*

3.22.6 Determination of Fermi level

Let in a metal, N be the number of free electrons and E_f, the highest energy at $T = 0$ K. Now, $g(E)$ is the *number of energy states* characterised by the same energy E and is equal to the number of electrons following *Pauli's exclusion principle.*

$\therefore$ at $T = 0$ K,

$$\int_0^{E_f} g(E)\,dE = N.$$

But

$$g(E)\,dE = \frac{8\sqrt{2}\pi V}{h^3}m^{3/2}E^{1/2}\,dE.$$

$$\therefore\ N = \int_0^{E_f}\frac{8\sqrt{2}\pi V}{h^3}m^{3/2}E^{1/2}\,dE = \frac{8\sqrt{2}\pi V m^{3/2}}{h^3}\int_0^{E_f}E^{1/2}\,dE = \frac{16\sqrt{2}\pi V m^{3/2}}{3h^3}E_f^{3/2}$$

$$\text{or}\ E_f = \frac{h^2}{2m}\left(\frac{3N}{8\pi V}\right)^{2/3} = \frac{h^2}{2m}\left(\frac{3n}{8\pi}\right)^{2/3}, \tag{3.114}$$

where $n = \frac{N}{V}$ = the *density of free electrons in the metal.*

Thus, $E_f \propto n^{2/3}$ and is independent of size and volume of the metal. (3.114) *shows that Fermi-energy is independent of temperature of the metal.*

3.23 Classical Limit of Quantum Statistics

In *quantum statistics* the particles obey either BE or FD statistics. We would like to show that *at high temperature and low density limit the two quantum statistics become identical with the classical statistics.*

In quantum statistics, the occupation probability is given by

$$N'(E)=\frac{1}{e^{(\alpha+E/kT)}\pm 1}. \tag{3.115}$$

Here −1 and +1 refer to BE and FD statistics respectively.

In *classical statistics*, the occupation probability is

$$N'(E)=\frac{1}{e^{\alpha+E/kT}}. \tag{3.116}$$

Comparing (3.115) with (3.116), we observe that BE and FD statistics approach to MB statistics, if

$$e^{(\alpha+E/kT)}\gg 1. \tag{3.117}$$

The above relation is true for all values of E. When $E=0$, the above condition reduces to $e^{\alpha}\gg 1$. When $E>0$, the inequality (3.117) is certainly satisfied as $e^{E/kT}>1$.

Again,

$$N(E)\,dE=\frac{CE^{1/2}\,dE}{e^{\alpha+E/kT}+\delta}.$$

$\therefore$ for $\delta=0$, the distribution of particles as a function of E is given by

$$N(E)=cN'(E)E^{1/2}=Ce^{(-\alpha-E/kT)}E^{1/2}, \tag{3.118}$$

where $C=\dfrac{2\pi(2m)^{3/2}V}{h^3}$.

Now, the total number of particles is given by

$$N=\int_0^\infty N(E)\,dE=Ce^{-\alpha}\int_0^\infty e^{-E/kT}E^{1/2}\,dE=Ce^{-\alpha}\cdot(kT)^{3/2}\frac{\sqrt{\pi}}{2}$$

$$\left[\because \int_0^\infty e^{-x}x^{1/2}\,dx=\Gamma\left(\frac{3}{2}\right)=\frac{1}{2}\Gamma\left(\frac{1}{2}\right)=\frac{\sqrt{\pi}}{2}\right]$$

$$\text{or } N=\frac{2\pi(2m)^{3/2}V}{h^3}e^{-\alpha}(kT)^{3/2}\frac{\sqrt{\pi}}{2}$$

$$\text{or } e^{\alpha}=\frac{2\pi^{3/2}(2m)^{3/2}V(kT)^{3/2}}{2h^3N}=\frac{(2m\pi kT)^{3/2}}{h^3}\left(\frac{V}{N}\right)\left[\because e^{\alpha}\gg 1,\ \therefore \frac{(2\pi mkT)^{3/2}}{h^3}\left(\frac{V}{N}\right)\gg 1\right]. \tag{3.119}$$

(3.119) represents *mathematical condition for which BE and FD statistics give the same result as MB statistics.*

Since $kT\simeq$ the average energy of a particle in a gas at temperature T, the term $(2mkT)^{1/2}\simeq$ *the average magnitude of momentum*. The *de Broglie wavelength* associated with this momentum is

$$\lambda=\frac{h}{(2\pi mkT)^{1/2}}. \tag{3.120}$$

The volume occupied by the particle in the *quantum mechanical description* is given by

$$\lambda^3 = \frac{h^3}{(2\pi mkT)^{3/2}}. \tag{3.121}$$

Again, the average volume occupied by a particle in *classical description* is V/N, so classical description will be valid, if

$$\frac{V}{N} \gg \frac{h^3}{(2\pi mkT)^{3/2}} \text{ or } \frac{(2\pi mkT)^{3/2}}{h^3}\frac{V}{N} \gg 1. \tag{3.122}$$

(3.121) and (3.122) are the same. Both the inequalities are valid for $T \to$ large value and V/N is large, i.e., low density (or low pressure). Hence, BE *and* FD *statistics approach to* MB *statistics under high temperature, low particle density* (*i.e., low pressure*).

This is known as the *classical limit of quantum statistics.*

3.24 Comparison of MB, BE and FD Statistics

Table 3.3

MB statistics	*BE statistics*	*FD statistics*
1. Energy is supposed to be continuous.	1. Energy is discrete.	1. Energy is discrete.
2. Particles are identical but distinguishable.	2. Particles are identical and indistinguishable.	2. Particles are identical and indistinguishable.
3. The idea of spin does not apply.	3. Particles have zero or integral spin, e.g., photon, He atoms at low temperature.	3. Particles are of half integral spin, e.g., electrons, protons, neutrons, etc.
4. Distribution formula is $N_i = \frac{g_i}{e^{\alpha+\beta E_i}}$.	4. $N_i = \frac{g_i}{e^{\alpha+\beta E_i} - 1}$.	4. $N_i = \frac{g_i}{e^{\alpha+\beta E_i} + 1}$.
5. No specific wave function for the particle.	5. Particle wave function is symmetric.	5. Particle wave function is antisymmetric.
6. Distribution of two-particles in two cells is as below: [p|q] [q|p] [pq|] [|qp] 4 possible arrangements.	6. Distribution of two-particles in two cells is as under: [•|•] [••|] [|••] 3 arrangements.	6. Distribution of two-particles in two cells is as under: [•|•] only one arrangement.

APPLICATION OF FD STATISTICS

3.25 Derivation of Richardson-Dushman Equation

Richardson-Dushman equation represents a relation between *thermionic current density* with the *absolute temperature of the emitter.*

Let one corner of a metal cube coincides with the origin of a Cartesian coordinate system and the X-axis is perpendicular to the metal surface (S) [Fig. 3.13].

If p_x, p_y, p_z be the momentum components and v_x, v_y, v_z be the velocity components of an electron of energy E along X-, Y-, Z-directions respectively, then according to FD *statistics*, the number of electrons per unit volume in the momentum space between $p_x \to p_x + dp_x$, $p_y \to p_y + dp_y$ and $p_z \to p_z + dp_z$ is

$$dN = \frac{2}{h^3} f(E)\, dp_x\, dp_y\, dp_z, \tag{3.123}$$

where h is the *Planck's constant.*

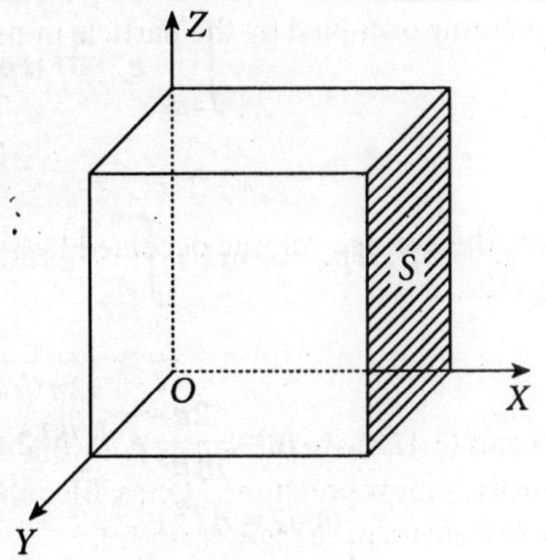

Fig. 3.13

Now, the *FD distribution function* is

$$f(E) = \frac{1}{1 + e^{(E-E_f)/kT}},$$

where E_f *is the Fermi level* and k is *the Boltzmann constant.*

If e be the electronic charge and v_x, the x-component of velocity of an electron, then

$$dJ_x = e v_x\, dN = \frac{2e}{mh^3} f(E)\, p_x\, dp_x\, dp_y\, dp_z, \tag{3.124}$$

where m is the mass of the electron.

If p_{xm} be the *minimum momentum* in the X-direction required to overcome the *surface potential barrier* E_B, then

$$E_B = \frac{p_{xm}^2}{2m}. \tag{3.125}$$

$\therefore$ the total emission current density is

$$J_x = \frac{2e}{mh^3} \int_{p_{xm}}^{\infty} \int_{-\infty}^{\infty} \int_{-\infty}^{\infty} f(E)\, p_x\, dp_x\, dp_y\, dp_z, \tag{3.126}$$

where v_x is replaced by p_x/m.

The emitted electrons from S have energy E much greater than F_f, i.e., $E - E_f \gg kT$.

$$\therefore f(E) = e^{(E_f - E)/kT} \quad \left[\because e^{(E-E_T)/kT} \gg 1\right].$$

$$\therefore J_x = \frac{2e}{mh^3} e^{E_f/kT} \int_{p_{xm}}^{\infty} \int_{-\infty}^{\infty} \int_{-\infty}^{\infty} e^{-\frac{E}{kT}}\, p_x\, dp_x\, dp_y\, dp_z \tag{3.127}$$

$$\text{or } J_x = \frac{2e}{mh^3} e^{E_f/kT} \int_{p_{xm}}^{\infty} e^{-\frac{p_x^2}{2mkT}}\, p_x\, dp_x \times \int_{-\infty}^{\infty} e^{-\frac{p_y^2}{2mkT}}\, dp_y \times \int_{-\infty}^{\infty} e^{-\frac{p_z^2}{2mkT}} \tag{3.128}$$

$$\left[\because E = \frac{p_x^2 + p_y^2 + p_z^2}{2m}\right]. \tag{3.129}$$

Now

$$\int_{-\infty}^{+\infty} e^{-ax^2}\, dx = \sqrt{\frac{\pi}{a}} = \sqrt{2\pi mkT},$$

where $a = \dfrac{1}{2\pi mkT}$.

$$\therefore \int_{-\infty}^{+\infty} e^{-\frac{p_y^2}{2mkT}}\,dp_y = \int_{-\infty}^{+\infty} e^{-\frac{p_z^2}{2mkT}}\,dp_z = \sqrt{2\pi mkT}.$$

And

$$\int_{p_{xm}}^{\infty} e^{-\frac{p_x^2}{2mkT}} p_x\,dp_x = -mkT\int_{\frac{p_{xm}^2}{2mkT}}^{\infty} e^{-\frac{p_x^2}{2mkT}}\,d\left(-\frac{p_x^2}{2mkT}\right) = -mkT\left[e^{-\frac{p_x^2}{2mkT}}\right]_{\frac{p_{xm}^2}{2mkT}}^{\infty} = mkTe^{-\frac{p_{xm}^2}{2mkT}}.$$

$$\therefore J_x = \frac{2e}{mh^3}e^{-E_f/kT}2\pi(mkT)^2 e^{-\frac{p_{xm}^2}{2mkT}} = \frac{4\pi mek^2}{h^3}T^2 e^{-\frac{(E_B-E_f)}{kT}} \tag{3.130}$$

$$\text{or } J_x = AT^2 e^{-\frac{b}{T}}, \tag{3.131}$$

where $b = \dfrac{E_B - E_f}{k}$ and $A = \dfrac{4\pi mek^2}{h^3}$.

(3.131) is known as the *Richardson-Dushman equation.*

3.26 Answers of Some Important Questions

3.1 What is a phase space?

Ans. A static particle can be described by three position coordinates, x, y, z only. But for a particle in motion we require six coordinates, viz., three position coordinates (x, y, z) and three momentum coordinates (p_x, p_y, p_z), since the position of the particle is changing with time. The *combined position and momentum space* of a particle in motion is called the *phase space*. A phase space, therefore, a *six-dimensional space.*

3.2 What do you mean by phase point and phase curve?

Ans. A point in phase space which represents both the position and momentum of a moving particle at any instant of time is called a phase point. A phase point is, therefore, a representative of a dynamical system.

A path traced out by a phase point in phase space is called a phase curve. A phase curve cannot intersect itself as it leads to a violation of the fundamental characteristics of systems for which a knowledge of (p, q) at $t = 0$ represents the phase at any later time. Two phase curves also cannot intersect at a point as the motion of the phase point is uniquely determined by the laws of dynamics.

3.3 What do you mean by cell in a phase space?

Ans. Generally the *a priori* probability of *n distinguishable* particle in *i* number of unequal compartments is not equal because of varying size of the compartments. If a compartment is divided into a very large number of sub-compartments, then size of each *sub-compartment* may be equal and their *a prior* probability will be the same. Such sub-compartments are called *cell.*

A phase cell is, therefore, an element of volume $\Delta x\Delta y\Delta z\Delta p_x\Delta p_y\Delta p_z$ *in phase space. The phase volume* associated a cell is $\sim h^3$, where h is the Planck's constant.

3.4 Define thermodynamical probability.

Ans. See Art. 3.8 (P. 118).

3.5 What is *a priori* probability?

Ans. See Art. 3.9 (P. 119).

3.6 What do you mean by micro-state and macro-state?

Ans. See Art. 3.7 (P. 117).

3.7 Write down the expression for the density of energy state of a system between energy E to $E+dE$.

Ans. The number of energy state in the energy interval of a system is

$$g(E)\,dE = \frac{4\sqrt{2}\pi V}{h^3} m^{3/2} E^{1/2}\,dE,$$

where h = Planck's constant and m = mass of a particle.

3.8 State the fundamental postulates of statistical mechanics.

Ans. See Art. 3.14 (P. 124).

3.9 State what are meant by occupation number, degenerate, non-degenerate and strongly degenerate states.

Ans. See Art. 3.18 (P. 132).

3.10 What is classical limit of quantum statistics?

Ans. Under high temperature ($T \to \infty$) and low particle density ($V/N \to 0$), BE and FD statistics approach MB statistics. This result is known as the *classical limit of quantum statistics*.

3.11 Compare MB, BE and FD statistics.

Ans. See Art. 3.24 (P. 144).

3.12 Define Fermi-level. Give a physical interpretation of Fermi-level.

Ans. *Fermi-level* (E_f) : The highest occupied energy level at 0 K, below which all the levels are occupied, is known as the *Fermi-level*.

***Physical interpretation of E_f*. The *Fermi-Dirac distribution function* is given by**

$$f(E_i) = \frac{N_i}{g_i} = \frac{1}{e^{(E_i-E_f)/kT}+1}.$$

From the above equation we note that, if $T \neq 0$,

$$f(E_i) = \frac{1}{1+1} = \frac{1}{2}\left[\because\ e^{(E_i-E_f)kT} = e^{0/kT} = e^0 = 1 \text{ when } E_i = E_f\right].$$

Thus *Fermi-level is that level for which the probability of occupation at $T > 0$ is $\frac{1}{2}$*, i.e., only 50% of the quantum states are occupied and 50% are empty. This is the physical interpretation of Fermi-level.

3.13 Discuss the temperature variation of Fermi-level (E_f).

Ans. If $E_f(T)$ and $E_f(0)$ be the Fermi-energy at temperature $T = 0$ K, then according to Sommerfeld

$$E_T(T) = E_T(0)\left[1 - \frac{\pi^2 k^2 T^2}{12 E_T^2(0)}\right].$$

The above equation shows that, $E_f(T)$ increases, if T decreases and $E_f(T)$ decreases, if T increases. The variation of $E_f(T)$ is regarded as almost constant at all temperatures and is nearly equal to $E_f(0)$.

3.14 Can an electron have zero energy at 0 K?

Ans. According to Pauli's exclusion principle any level can contain only one electron with the particular set of quantum number. So only one electron can be occupied by the zero energy level and the remaining $(N-1)$ electrons will be occupied by other $(N-1)$ higher energy levels, where N is the total number of electrons. The energy of the electrons in the zero energy level is $E_0 = \frac{3}{5}E_f(0)$, where $E_f(0)$ is the Fermi energy at 0 K.

Hence an electron cannot have zero energy at 0 K.

3.15 **Name the statistics followed by the following particles: electron, ideal gas molecules, proton, photon, helium atom, neutron, π-meson, muon.**

Ans. The statistics followed by the different particles is as below:

MB-statistics. ideal gas

BE-statistics. helium atom, photon, π-meson

FD-statistics. electron, proton, neutron, muon.

3.27 Solved Problems

3.1 **If a die is thrown twice two numbers 4 and 6 are obtained. What is the probability that these numbers 4 and 6 precisely in order?**

Solution. The probability of getting 4 in the 1st throw is 1/6. The probability of getting 6 in the 2nd throw is also 1/6.

These are two independent events.

$\therefore$ the required probability is $\frac{1}{6} \times \frac{1}{6} = \frac{1}{36}$.

3.2 **In a box there are 10 cut up alphabet cards with letters 3*A*'s, 4*C*'s and 3*T*'s. If the cards be drawn one after another and placed on the table in the order they have been drawn, then what is the probability that the word CAT will appear?**

Solution. The probability of the first card being *C* is 4/10.

The probability of the second card being *A* is 3/9. ($\because$ one card is already out).

Likewise, the probability of the third letter being *T* is 3/8. ($\because$ 2 cards are already out).

Hence the probability that the word CAT will appear is

$$P(\text{CAT}) = \frac{4}{10} \times \frac{3}{9} \times \frac{3}{8} = \frac{1}{20}.$$

3.3 **Find the probability that a die will twin to 5 or an even number?**

Solution. The probability of getting 5 is 1/6. The chance of getting an even number (2, 4, 6) is 3/6 = 1/2. Both these events are mutually exclusive as it is 5 which cannot be an even number.

$\therefore$ the required probability $p = 1/6 + 1/2 = 2/3$.

3.4 **A box contains 4 black and 5 white balls. What is the probability that on two successive draws, the balls drawn are both black?**

Solution. Total number of balls in the box is 4 + 5 = 9.

Let *A* be the event, where the first ball drawn is black. This probability is $P(A) = 4/9$. If *B* represents the event, where the second ball drawn is also black. The second ball is selected from $(9 - 1) = 8$ balls after the event *A*. Hence the probability of the event *B* is $P(B/A) = 3/8$.

Here the event *A* affects the probability of *B*. Now, the probability of the event $(A \cap B)$ is

$$P(A \cap B) = P(A) \times P(B/A) = \frac{4}{9} \times \frac{3}{8} = \frac{1}{6}.$$

3.5 **A box contains 5 red balls and 3 white balls. The balls, except their colours, are identical. What is the probability that on two independent draws one ball is red and one ball is white?**

Solution. One red ball and one white ball can be drawn in two ways from the box, viz.,

(a) the draw is such that the first ball is white and the second ball is red.

(b) the first ball is red and the second ball is white.

In the first case, the probability of the first ball being white is $\frac{3}{8}$ and that for the second being red is $\frac{5}{8}$.

Then the combined probability that one ball is white and the other red $P_{12} = \frac{3}{8} \times \frac{5}{8} = \frac{15}{64}$.

As the two cases are independent of each other the total probability is

$$P=\frac{15}{64}+\frac{15}{64}=\mathbf{\frac{15}{32}}.$$

3.6 Six distinguishable particles are distributed over three non-degenerate levels of energies $0, \epsilon$ and 2ϵ. Calculate the total number of micro-states of the system. Find the total energy of the distribution for which the probability is a maximum. [CU 1993]

Solution. Here the energy values $0, \epsilon, 2\epsilon$ are having only one state as they are non-degenerate. Let there be N_1 particles having energy 0, N_2 having ϵ and N_3 having 2ϵ, so that total number of particles is $N_1+N_2+N_3=6$.

The particles being distinguishable, the number of ways of choosing N_1, N_2, N_3 particles out of 6 particles is $W=\frac{6!}{N_1!N_2!N_3!}$.

This is also the number of micro-states.

The thermodynamical probability W is the maximum when $N_1!N_2!N_3!$ is a minimum. It is evident that $N_1!N_2!N_3!$ is minimum if $N_1=N_2=N_3=2$.

Then total energy of distribution is $E=0\times N_1+\epsilon\times N_2+2\epsilon\times N_3=0+2\epsilon+4\epsilon=\mathbf{6\epsilon}$.

3.7 The molecules of a gas have two states of internal energy with statistical weights g_1, g_2 and energies $0, \epsilon$ respectively. Calculate the contribution of these states to the total internal energy.

Solution. The gas molecules obey MB statistics.

$\therefore$ if N_1 and N_2 be the number of molecules in state 1 and state 2, then

$$N_1=g_1 e^{-(\alpha+\beta\epsilon_1)}$$

$$N_2=g_2 e^{-(\alpha+\beta\epsilon_2)}.$$

$$\therefore\ N_2/N_1=\frac{g_2}{g_1}e^{-\beta\epsilon}\ [\because \epsilon_1=0 \text{ and } \epsilon_2=\epsilon].$$

If one mole of the gas be considered, then $N_1+N_2=N$ (Avogadro's number).

$$\therefore\ \frac{N-N_1}{N_1}=\frac{g_2\,e^{-\beta\epsilon}}{g_1}\ \text{or}\ N_1=\frac{N}{1+(g_2/g_1)\,e^{-\beta\epsilon}}\ \text{or}\ N_2=\frac{N\,(g_2/g_1)\,e^{-\beta\epsilon}}{1+(g_2/g_1)\,e^{-\beta\epsilon}}.$$

Then the total energy of the system is

$$E=\sum_i N_i\epsilon_i=0+\frac{N\epsilon\,(g_2/g_1)\,e^{-\beta\epsilon}}{1+(g_2/g_1)\,e^{-\beta\epsilon}}\ [\because \epsilon_1=0, \epsilon_2=\epsilon]$$

$$=\mathbf{\frac{N\epsilon\,(g_2/g_1)\,e^{-\beta\epsilon}}{1+(g_2/g_1)\,e^{-\beta\epsilon}}}.$$

3.8 Consider two identical particles. Each particle can be in one of the three possible quantum states $0, \epsilon$ and 2ϵ. Find the number of micro-states of the system for MB, BE and FD statistics. [CUH 1996]

Solution. Here, the number of particles = 2 and the number of states = 3.

For MB statistics, the number of micro-states is $W=g^N=3^2=\mathbf{9}$.

For BE statistics the number of micro-states is

$$W=\frac{(N+g-1)!}{N!(g-1)!}=\frac{(2+3-1)!}{2!(3-1)!}=\frac{4!}{2!2!}=\frac{4\times3\times2!}{2!2!}=\frac{12}{2}=\mathbf{6}.$$

For FD statistics the number of micro-states is

$$W = \frac{g!}{N!(g-N)!} = \frac{3!}{2!(3-2)!} = \frac{3\times 2!}{2!1!} = 3.$$

3.9 The Fermi energy E for Lithium is 4.72 eV at $T = 0$. Calculate the number of conduction electrons per unit volume in lithium. **[CUH 1997]**

Solution. The Fermi energy at $T = 0\,\text{K}$ is

$$E_f = \frac{h^2}{8m}\left(\frac{3n}{\pi}\right)^{2/3}.$$

∴ conduction electrons per unit volume is

$$n = \frac{\pi}{3}\left[\frac{8m}{h^2}E_f\right]^{3/2} = \frac{\pi}{3}\left[\frac{8\times(9.11\times10^{-31})\times(4.72\times1.6\times10^{-19})}{(6.62\times10^{-34})^2}\right]^{3/2}$$

$$= \frac{\pi}{3}\left[\frac{8\times 9.11\times 4.72\times 1.6}{(6.62)^2}\times 10^{18}\right]^{3/2} = \mathbf{4.66\times10^{28}\ per\ cubic\ metre.}$$

3.10 Three indistinguishable particles (Bosons) are to be distributed in two compartments, the first one having 3 cells and the second one 2 cells. What are the thermodynamic probability for the macro-states (5, 0) and (3, 2).

Solution. The thermodynamic probability

$$W = \prod_{i=1}^{k} \frac{N_i + g_i - 1}{N_i!},$$

where N_i is the number of particles and g_i is the number of cells.

$$W = \prod_{i=1}^{k} \frac{(N_i + g_i - 1)}{N_i!\,(g_i - 1)!} = \frac{(N_1 + g_1 - 1)!}{N_1!\,(g_1 - 1)!} \times \frac{(N_2 + g_2 - 1)!}{N_2!\,(g_2 - 1)!}.$$

Now, for the macro-state (5, 0), $N_1 = 5, N_2 = 0, g_1 = 3, g_2 = 4.$

$$\therefore\ W_{(5,0)} = \frac{(5+3-1)!}{5!(3-1)!} \times \frac{(0+2-1)!}{0!(2-1)!} = \frac{7!}{5!2!} \times \frac{1!}{0!1!} = \frac{7\times6\times5!}{5!2!} \times 1 = \frac{42}{2} = 21.$$

For the macro-state (3, 2), $N_1 = 3, N_2 = 2, g_1 = 3, g_2 = 2$

$$\therefore\ W_{(3,2)} = \frac{(3+3-1)!}{3!(3-1)!} \times \frac{(2+2-1)!}{2!(2-1)!} = \frac{5!}{3!2!} \times \frac{3!}{2!1!} = \frac{5\times4\times3!}{3!2!} \times \frac{3\times2!}{2!1!} = 30.$$

3.11 Find Fermi energy in copper on the assumption that each copper atom contributes one electron to electron gas. **[PU 1991]**

Solution. The density of copper = $8.94\times10^3\ \text{kg/m}^3$.

Atomic mass of copper = 63.5 amu.

$\therefore$ Fermi energy in copper is $E_f = \dfrac{h^2}{2m}\left(\dfrac{3N}{8\pi V}\right)^{2/3}$

Now the number of copper atoms per unit volume

$$n = \frac{N}{V} = \frac{\text{Density of copper}}{\text{Mass of a copper atom}} = \frac{8.94\times 10^3 \text{ kg/m}^3}{63.5\times 1.66\times 10^{-27}}$$

$$= 8.84\times 10^{28} \text{ atoms/m}^3 = 8.84\times 10^{28} \text{ electrons/m}^3.$$

$$\therefore\ E_f = \frac{(6.63\times 10^{-34})^2}{2\times 9.11\times 10^{-31}}\left[\frac{3\times 8.48\times 10^{28}}{8\times 3.14}\right]^{2/3} \text{J}$$

$$= 1.13\times 10^{-18} \text{ J} = 7.04 \text{ eV}.$$

3.12 **A system has 7 particles arranged in two compartments. The first compartment has 8 cells and the second has 10 cells, all cells are of equal size. Calculate the number of micro-states in the macro-state (3,4) when the particles are fermions.** [PU 1993]

Solution. For the fermions obeying FD statistics the thermodynamic probability of the macro-state is given by

$$W(n_1, n_2, \ldots, n_k) = \prod_{i=1}^{k} \frac{g_i!}{N_i!(g_i - N_i)!}.$$

Since in the problem the number of compartments $k = 2$,

$$\therefore\ W(n_1, n_2) = \frac{g_1!}{N_1!(g_1 - N_1)!} \times \frac{g_2!}{N_2!(g_2 - N_2)!}.$$

Here g_1 = number of cells in the first compartment = 8

and g_2 = number of cells in the second compartment = 10.

Now for the macro-state (3,4), $N_1 = 3$, $N_2 = 4$.

$$\therefore\ W(3,4) = \frac{8!}{3!(8-3)!} \times \frac{10!}{4!(10-4)!} = \frac{8!}{3!5!} \times \frac{10!}{4!6!} = \frac{8\times 7\times 6}{3\times 2\times 1} \times \frac{10\times 9\times 8\times 7}{4\times 3\times 2\times 1}$$

$$= 56\times 210 = 11760.$$

3.13 **The number of conduction electrons per cc is 24.2×10^{22} in beryllium and 0.91×10^{27} in Cesium. If the Fermi energy of conduction electrons in beryllium is 14.44 eV, calculate that in cesium.** [HPU 1991]

Solution. Electrons per cc in beryllium and in cesium is

$$\frac{N}{V} = 24.2\times 10^{22} \text{ for beryllium}$$

$$\text{and } \frac{N}{V} = 0.91\times 10^{22} \text{ for cesium.}$$

$$\text{Fermi energy } E_f = \frac{h^2}{2m}\left(\frac{3}{8\pi}\right)^{2/3}\left(\frac{N}{V}\right)^{2/3}.$$

$$\therefore\ \frac{\text{Fermi energy for beryllium}}{\text{Fermi energy for cesium}} = \frac{\frac{h^2}{2m}\left(\frac{3}{8\pi}\right)^{2/3}\times (24.2\times 10^{22})^{2/3}}{\frac{h^2}{2m}\left(\frac{3}{8\pi}\right)^{2/3}\times (0.91\times 10^{22})^{2/3}}.$$

$$\therefore \text{ Fermi energy for cesium} = 14.44\times \left(\frac{0.91}{24.2}\right)^{2/3} = 1.587 \text{ eV}.$$

3.14 Three distinguishable particles have a total energy of 9 units. But the particles are restricted to energy levels from 0 to 4. Calculate the number of macro-states and micro-states.

Solution. Let the total energy of the system be 9ϵ. The particles being distinguishable may be represented by a, b and c. We assume that the energy levels are equispaced. Then the distribution will be as shown below:

Energy	a	b	c
0			
ϵ	1		1
2ϵ			
3ϵ		3	1
4ϵ	2		1

There are five different energy levels.

The number of macro-states = 3:

The number of micro-states in $a = 3$

The number of micro-states in $b = 1$

The number of micro-states in $c = 3! = 6$.

$\therefore$ total number of micro-states $= 3 + 1 + 6 = 10$.

3.15 Find out the possible arrangements of two particles A and B in three cells according to (a) MB, (b) BE and (c) FD statistics.

Solution. (a) We know that in MB statistics the particles are distinguishable and a quantum state (called *cell*) can be accommodated in any number of particles. So the possible arrangements are 9 as shown in table below:

Serial no. of possible arrangements	*Cell 1*	*Cell 2*	*Cell 3*
1	AB	X	X
2	0	AB	0
3	0	0	AB
4	A	B	0
5	B	A	0
6	A	0	B
7	B	0	A
8	0	A	B
9	0	B	A

Hence, the total number of arrangements in the present case is 9.

(b) BE statistics deals with indistinguishable particles ($A = B$) and any number of particles can be accommodated in one cell.

The possible arrangements are shown in table below:

Serial no. of possible arrangements	*Cell 1*	*Cell 2*	*Cell 3*
1	*AA*	*X*	*X*
2	0	*AA*	0
3	0	0	*AA*
4	*A*	*A*	0
5	*A*	0	*A*
6	0	*A*	*A*

Hence, there are 6 possible arrangements.

(c) *FD statistics.* According to FD statistics the particles are indistinguishable and only one particle can be accommodated in each cell.

The possible number of arrangements are shown below:

Serial no. of possible arrangements	*Cell 1*	*Cell 2*	*Cell 3*
1	*A*	*A*	0
2	*A*	0	*A*
3	0	*A*	*A*

There are *three* possible arrangements.

3.16 Find the possible arrangements of 3 particles, A, B, C in 3 cells according to the three statistics.

Solution. (a) *MB statistics.* In this statistics (i) the particles are distinguishable and (ii) any number of particles can be accommodated in one cell.

According to MB statistics, the total number of possible arrangements is

$$W = \frac{N_i! g_i^{N_i}}{N_i!} = \frac{3! 3^3}{3!} = 27.$$

Serial no. of possible arrangements	*Cell 1*	*Cell 2*	*Cell 3*	*No. of possible arrangements*
1	ABC	0	0	
2	0	ABC	0	3
3	0	0	ABC	

4	AB	C	0	
5	AB	0	C	
6	0	AB	C	
7	X	C	AB	
8	C	AB	0	
9	C	0	AB	
10	AC	B	0	
11	AC	0	B	
12	0	AC	B	
13	0	B	AC	18
14	B	0	AC	
15	B	AC	0	
16	BC	A	0	
17	BC	0	A	
18	0	BC	A	
19	0	A	BC	
20	A	0	BC	
21	A	BC	0	
22	A	B	C	
23	A	C	B	
24	B	A	C	
25	B	C	A	6
26	C	A	B	
27	C	B	A	

So there are $3+18+6=27$ possible arrangements.

(b) *BE statistics.* In this statistics (i) the particles are indistinguishable (i.e., $A=B=C$) and (ii) each cell can accommodate any number of particles.

The number of possible arrangements according to BE statistics is

$$W=\frac{(N_i+g_i-1)!}{N_i!\,(g_i-1)!}=\frac{(3+3-1)!}{3!(3-1)!}=10.$$

Serial no. of possible arrangements	*Cell 1*	*Cell 2*	*Cell 3*	*No. of possible arrangements*
1	AAA	0	0	
2	0	AAA	X	3
3	0	0	AAA	
4	AA	A	X	
5	AA	0	A	
6	A	AA	0	6
7	A	0	AA	
8	0	AA	A	
9	0	A	AA	
10	A	A	A	1

Hence, the total number of possible arrangements is 10.

(c) *FD statistics.* In this statistics (i) the particles are indistinguishable and (ii) each cell can accommodate only one particle.

Total number of possible arrangements is $W = \dfrac{g_i!}{N_i!(g_i - N_i)!} = \dfrac{3!}{3!(3-3)!} = 1.$

Serial no. of possible arrangements	*Cell 1*	*Cell 2*	*Cell 3*	*No. of possible arrangements*
1	A	A	A	1

Thus there is only one possible arrangement.

3.17 Consider a two-particle system each of which can exist in three states ϵ_1, ϵ_2 and ϵ_3. What are the possible states if the particles are (a) Bosons (b) Fermions? **[WBUT 2006]**

Solution. (a) *BE statistics.* Here (i) the particles are indistinguishable (say A, A) and (b) each cell can accommodate any number of particles. The possible states are as below:

Micro-states \ *Quantum states*	ϵ_1	ϵ_2	ϵ_3
1	AA	0	0
2	0	AA	0
3	0	0	AA
4	A	A	0
5	A	0	A
6	0	A	A

Here the number of possible distribution = 6.

(b) *FD statistics.* In this statistics (i) the particles are indistinguishable (say A, A) and (ii) each cell can accommodate only one particle. The arrangement of the particles is as below:

Micro-states \ Quantum states	ϵ_1	ϵ_2	ϵ_3
1	A	A	0
2	A	0	A
3	0	A	A

Thus total number of arrangements of the fermions is 3.

3.18 Three distinguishable particles each of which can be in one of the ϵ, 2ϵ, 3ϵ, 4ϵ energy states have total energy 6ϵ. Find all possible number of distributions of all particles in the energy states. Find the number of micro-states in each case. [WBUT 2007]

Solution.

Serial no. of combinations	$E_i \rightarrow$ $N_i \rightarrow$	ϵ N_1	2ϵ N_2	3ϵ N_3	4ϵ N_4	Total energy
1		2	0	0	1	$2\epsilon+4\epsilon=6\epsilon$
2		X	3	0	0	$3\times 2\epsilon=6\epsilon$
3		1	1	1	0	$\epsilon+2\epsilon+3\epsilon=6\epsilon$

So the number of macro-states = 3, viz., $(2,0,0,1), (0,3,0,0)$ and $(1,1,1,0)$.

Let the three particles be represented by A, B and C. The number of possible micro-states of the particle is as follows:

Micro-state	ϵ	2ϵ	3ϵ	4ϵ	No. of micro-states
	AB	0	0	C	
(2,0,0,1)	AC	0	0	B	3
	BC	0	0	A	
(0,3,0,0)	0	ABC	0	0	1
	A	B	C	0	
	B	C	A	0	
	C	A	B	0	
(1,1,1,0)	B	A	C	0	6
	A	C	B	0	
	C	B	A	0	

Thus the total number of macro-states = 3 and the total number of micro-states $=3+1+6=10$.

3.19 If v_x, v_y, v_z represent the three Cartesian components of velocity of a molecule in a gas, then using symmetry considerations and equipartition theorem, deduce expressions for the following mean values in terms of k, T and m: (a) $\langle v_x\rangle$ (b) $\langle v_x^2\rangle$ (c) $\langle v_x v_z\rangle$ (d) $\langle v_x + b v_y\rangle$.

Solution. (a) We have the mean value for the x-component of velocity

$$\langle v_x\rangle = \int_{-\infty}^{+\infty} v_x\left(\frac{m}{2\pi kT}\right)^{1/2} e^{-\frac{mv_x^2}{2kT}}\,dv_x = \left(\frac{m}{2\pi kT}\right)^{1/2}\int_{-\infty}^{+\infty} v_x\,e^{-\frac{mv_x^2}{2kT}}\,dv_x$$

$$= 0 \left[\because \int_{-\infty}^{+\infty} v_x\,e^{-\frac{mv_x^2}{2kT}}\,dv_x = 0\right].$$

(b) $\langle v_x^2\rangle = \int_{-\infty}^{+\infty} v_x^2 P(v_x)\, dv_x = \int_{-\infty}^{+\infty}\left(\frac{m}{2\pi kT}\right)^{1/2} v_x^2\, e^{-\frac{mv_x^2}{2kT}}\, dv_x$

$$= \left(\frac{m}{2\pi kT}\right)^{1/2} \frac{1}{2}\left[\frac{\pi}{\left(\frac{m}{2kT}\right)^3}\right]^{1/2} \quad \left[\because \int_{-\infty}^{+\infty} v_x^2\, e^{-\frac{mv_x^2}{2kT}}\, dv_x = \frac{1}{2}\left[\frac{\pi}{\left(\frac{m}{2kT}\right)^3}\right]^{1/2}\right].$$

$$\therefore \langle v_x^2\rangle = \frac{\boldsymbol{kT}}{\boldsymbol{m}}.$$

(c) $\langle v_x v_z\rangle = \int_{-\infty}^{+\infty}\int_{-\infty}^{+\infty} v_x v_z P(v_x) P(v_z)\, dv_x\, dv_z$

$$= \int_{-\infty}^{+\infty} v_x \left(\frac{m}{2\pi kT}\right)^{1/2} e^{-\frac{mv_x^2}{2kT}}\, dv_x \times \int_{-\infty}^{+\infty} v_z \left(\frac{m}{2\pi kT}\right)^{1/2} e^{-\frac{mv_z^2}{2kT}}\, dv_z$$

$$= \left(\frac{m}{2\pi kT}\right) \int_{-\infty}^{+\infty} v_x\, e^{-\frac{mv_x^2}{2kT}}\, dv_x \times \int_{-\infty}^{+\infty} v_z\, e^{-\frac{mv_z^2}{2kT}}\, dv_z$$

$$= \mathbf{0}.$$

(d) $\left\langle \left(v_x + b v_y\right)^2\right\rangle = \left\langle\left(v_x^2 + 2b v_x v_y + v_z^2\right)\right\rangle = \langle v_x^2\rangle + 2b\langle v_x v_y\rangle + b^2\langle v_z^2\rangle$

$$= \frac{kT}{m} + 0 + b^2\left(\frac{kT}{m}\right) = \left(\mathbf{1} + \boldsymbol{b}^2\right)\frac{\boldsymbol{kT}}{\boldsymbol{m}}.$$

3.20 **If the rms speed of hydrogen molecule at NTP is 1.84 km/s, calculate the rms speed of oxygen molecule at NTP. [Molecular weights of hydrogen and oxygen molecules are 2 and 32 respectively.]**

Solution.

$$(v_{rms})_H = \left(\frac{3kT}{m_H}\right)^{1/2} \text{ and } (v_{rms})_O = \left(\frac{3kT}{m_O}\right)^{1/2}.$$

$$\therefore \frac{(v_{rms})_O}{(v_{rms})_H} = \sqrt{\frac{m_H}{m_O}} = \sqrt{\frac{2}{32}} = \sqrt{\frac{1}{16}}.$$

$$\therefore (v_{rms})_O = \sqrt{\frac{1}{16}} \times (v_{rms})_H = \sqrt{\frac{1}{16}} \times 1.84 = \mathbf{0.46\ km/s}.$$

Here $(v_{rms})_H$ and $(v_{rms})_O$ are the rms velocities of hydrogen and oxygen molecules having masses m_H and m_O respectively.

3.21 **The first excited state of hydrogen atom is 10.2 eV above its ground state. What temperature is needed to excite hydrogen atoms to the first excited state.**

Solution. The average KE per molecule at absolute temperature T is

$$\tfrac{1}{2} m v^2 = 3 \times \tfrac{1}{2} kT \quad \left[\begin{array}{l}\text{according to the theorem}\\ \text{of equi-partition of energy}\end{array}\right]$$

Given, $\frac{1}{2} m v^2 = 10.2\,\text{eV} = 10.2 \times 1.6 \times 10^{-19}$ J.

$$\therefore \frac{3}{2} kT = 10.2 \times 1.6 \times 10^{-19} \text{ or } T = \frac{2}{3} \times \frac{10.2 \times 1.6 \times 10^{-19}}{k}$$

$$= \frac{2}{3} \times \frac{10.2 \times 1.6 \times 10^{-19}}{1.38 \times 10^{-23}} \quad [\because k = 1.38 \times 10^{-23}\ \text{J/K}]$$

$$= \mathbf{7.88 \times 10^4\ K}.$$

3.22 Calculate the fraction of molecules of a gas possessing speed within 1% of the most probable speed. [LU 1992]

Solution. The most probable speed for a molecule in the gas is $v_m = \sqrt{\frac{2kT}{m}}$.
The fraction of the molecules having speeds between v and $v+dv$ is

$$\frac{N(v)\,dv}{N} = 4\pi\left(\frac{m}{2\pi kT}\right)^{3/2} e^{-\frac{mv^2}{2kT}} v^2\,dv.$$

Putting $v = v_m = \left(\frac{2kT}{m}\right)^{1/2}$, we get

$$\frac{N(v)\,dv}{N} = 4\pi\left(\frac{m}{2\pi kT}\right)^{1/2} e^{-\frac{m}{2kT}\cdot\frac{2kT}{m}} \frac{2kT}{m}\,dV.$$

Since v varies within 1% of v_m, it lies between $0.99v_m$ and $1.01v_m$.

$$\therefore\ dv = (1.01-0.99)\,v_m = 0.02v_m = 0.02\left(\frac{2kT}{m}\right)^{1/2}.$$

$$\therefore\ \frac{N(v)\,dv}{N} = \frac{4}{\sqrt{\pi}}\left(\frac{m}{2kT}\right)^{3/2} e^{-1}\frac{2kT}{m}\left(\frac{2kT}{m}\right)^{1/2} \times 0.02 = 0.045e^{-1} = 0.016.$$

EXERCISES

3(A) Objective Type Questions

Choose the correct alternatives.

3.1 A coin and a six-faced die are thrown. The probability that the coin shows tail and the dice shows 5 is

(a) $\frac{7}{12}$ (c) $\frac{1}{12}$

(b) $\frac{1}{8}$ (d) $\frac{1}{6}$. [WBUT 2005]

3.2 Which of the following statements is wrong?

(a) Maxwell-Boltzmann statistics deals with distinguishable particles.

(b) Fermi-Dirac statistics deals with indistinguishable particles.

(c) Bose-Einstein statistics deals with indistinguishable particles.

(d) Bosons are subjected to Pauli's exclusion principle. [WBUT 2006]

3.3 Fermi-Dirac distribution approaches Maxwell-Boltzmann distribution at

(a) low temperature and high pressure

(b) low temperature and low pressure

(c) high temperature and high particle mass

(d) high density and low energy range. [WBUT 2007]

3.4 Average energy of an electron in a metal at $T = 0$ K is

(a) E_F (c) $\frac{3}{5}E_F$

(b) $\frac{1}{2}E_F$ (d) $\frac{5}{3}E_F$. [WBUT 2007]

3.5 The spin angular momentum of a photon is

(a) 0
(b) $\frac{1}{2}\hbar$
(c) $\hbar$
(d) $\frac{3}{2}\hbar$.

3.6 The number of meaningful ways in which 4 Fermions can be arranged in 5 compartments is

(a) 1
(b) 4
(c) 5
(d) 9. [HPU 1995]

3.7 The rest-mass of a photon is

(a) ∞
(b) 0
(c) 9.11×10^{-31} kg
(d) none of these.

3.8 RMS velocity of a gas molecule is

(a) $\sqrt{\frac{2kT}{m}}$
(b) $\sqrt{\frac{8kT}{m\pi}}$
(c) $\sqrt{\frac{3kT}{m}}$
(d) $\sqrt{\frac{m}{8kT}}$.

3.9 $v_{rms} : v_{av} : v_m$ for an ideal gas molecule is

(a) $\sqrt{\frac{8}{\pi}} : \sqrt{2} : \sqrt{3}$
(b) $\sqrt{2} : \sqrt{\frac{8}{\pi}} : \sqrt{3}$
(c) $\sqrt{3} : \sqrt{\frac{8}{\pi}} : \sqrt{2}$
(d) $\sqrt{2} : \sqrt{\frac{\pi}{8}} : \frac{1}{\sqrt{3}}$.

3.10 The thermodynamic probability of distribution of fermions is

(a) $\prod_i \frac{N_i!}{g_i!(g_i - N_i)!}$
(b) $\prod_i \frac{(g_i - N_i)!}{N_i! g_i}$
(c) $\prod_i \frac{N_i! g_i!}{(g_i - N_i)!}$
(d) $\prod_i \frac{g_i!}{N_i!(g_i - N_i)!}$.

3.11 BE statistical distribution formula is given by

(a) $N_i = \frac{g_i}{e^{\alpha + \beta E_i}}$
(b) $N_i = \frac{g_i}{e^{\alpha + \beta E_i} - 1}$
(c) $N_i = \frac{g_i}{e^{\alpha + \beta E_i} + 1}$
(d) $N_i = \frac{g_i}{e^{\alpha + \beta E_i} \pm 1}$.

3.12 The average velocity $\langle v \rangle$ of an electron in terms of velocity v_f at the Fermi level is

(a) $\frac{2}{3} v_f$
(b) $\frac{3}{2} v_f$
(c) $\frac{1}{3} v_f$
(d) $\frac{3}{4} v_f$.

3.13 Total energy of a system of N electrons at 0 K is

(a) $E = \frac{2}{5} N E_f$
(b) $E = \frac{3}{5} N E_f$
(c) $E = \frac{5}{3} N E_f$
(d) $E = \frac{3}{5N} E_f$.

3.14 Energy of electrons at 0 K is not zero because

(a) all the electrons do not have zero velocity at 0 K;

(b) according to Pauli's exclusion principle only one electron can occupy zero energy level and all other electrons occupy higher energy levels;

(c) the energy of the electrons does not depend on temperature;
(d) none of these.

3.15 Particles obeying MB statistics can have
(a) any spin
(b) integral spin
(c) half integral spin
(d) none of these.

3.16 Energy of a system is supposed to be discrete in
(a) MB statistics
(b) BE statistics
(c) FD statistics
(d) Maxwell's law of equi-partion of energy.

3.17 The size of a phase cell in MB statistics can be as small as we like but that for BE or FD statistics is of the order of
(a) $\frac{h^3}{2}$
(b) h^3
(c) $\frac{1}{h^3}$
(d) h^2.

3.18 In which probability the occupation index (N_i/g_i) can tend to zero?
(a) MB
(b) BE
(c) FD
(d) none of these.

3.19 Fermi energy (E_f) depends on
(a) size of the conductor
(b) volume of the conductor
(c) electron concentration of the conductor
(d) none of these.

3.20 MB statistics can be applied to the case of semiconductor because
(a) a semiconductor contains holes
(b) electron and hole recombine to produce photon
(c) the particle number density is small when doping is not large
(d) none of these.

3.21 The density of states in the momentum range p to $p+dp$ is
(a) $4\pi p^2\,dp h^3$
(b) $\frac{p^2\,dp h^3}{4\pi}$
(c) $\frac{4\pi p^2\,dp}{h^3}$
(d) $\frac{h^3}{4\pi p^2\,dp}$.

3.22 The number of meaningful ways in which 4 fermions can be arranged in 5 compartments is
(a) 3
(b) 2
(c) 4
(d) 5.

3.23 The number of micro-states for n particles in two compartments obeying MB statistics is
(a) $n+1$
(b) n
(c) 2^n
(d) n^2.

3.24 The statistics obeyed by photon is

(a) MB statistics (c) FD statistics
(b) BE statistics (d) none of these.

3.25 The statistics obeyed by $_2He^4$ atom is

(a) MB statistics (c) FD statistics
(b) BE statistics (d) none of these.

3.26 For the degeneracy of an energy state, N_i/g_i is

(a) >1 (c) $=0$
(b) <1 (d) $=\infty$.

3.27 If the occupation index $\frac{N_i}{g_i} \gg 1$, then the system obeys

(a) MB statistics (c) FD statistics
(b) BE statistics (d) none of these.

3.28 Planck's law can be obtained by the application of

(a) MB statistics (c) FD statistics
(b) BE statistics (d) none of these.

3.29 Law of equipartition of energy can be derived from

(a) MB statistics (c) FD statistics
(b) BE statistics (d) none of these.

3.30 The probability of occupation of an electron in Fermi-level is $\frac{1}{2}$, if

(a) $T=0\,\text{K}$ (c) $T\to\infty\text{K}$
(b) $T>0\,\text{K}$ (d) none of these.

3.31 If E_f and k represent respectively Fermi-energy and Boltzmann constant, then Fermi-temperature

(a) $\theta_F=kE_f$ (c) $\theta_F=\frac{k}{E_f}$
(b) $\theta_F=E_f/k$ (d) none of these.

3.32 The number of possible arrangements of two fermions in 3 cells are

(a) 9 (c) 3
(b) 6 (d) 1.

3.33 If n_i is the number of identical and indistinguishable particles in the ith energy state with degeneracy g_i, then classical statistics can be applied, if

(a) $\frac{n_i}{g_i}=1$ (c) $\frac{n_i}{g_i}\gg 1$
(b) $\frac{n_i}{g_i}\ll 1$ (d) $g_i=0$.

3.34 The possible arrangements of 4 fermions in 3 cells are

(a) 5 (c) 2
(b) 6 (d) 0.

ANSWERS

3.1 (c)	3.6 (c)	3.11 (b)	3.16 (b),(c)	3.21 (c)	3.26 (a)	3.31 (b)
3.2 (d)	3.7 (b)	3.12 (d)	3.17 (b)	3.22 (d)	3.27 (a)	3.32 (c)
3.3 (c)	3.8 (c)	3.13 (b)	3.18 (d)	3.23 (c)	3.28 (b)	
3.4 (d)	3.9 (c)	3.14 (b)	3.19 (c)	3.24 (b)	3.29 (a)	3.33 (c)
3.5 (c)	3.10 (d)	3.15 (a)	3.20 (c)	3.25 (b)	3.30 (b)	3.34 (d).

3(B) Short Answer Type Questions

3.1 What do you understand by Fermi-energy? [WBUT 2005]

3.2 What do you mean by a phase space? [BUH 2003]

3.3 Define thermodynamic probability. [BUH 2000]

3.4 What are phase point and phase trajectory? [BUH 1999].

3.5 Define (a) Micro-state and (b) Macro-state with suitable example. [CUH 1996]

3.6 Explain the terms ensemble and ensemble average. [BUH 1999]

3.7 Define thermodynamic probability and obtain the relation between entropy and this probability making reasonable assumptions. [BUH 1998]

3.8 Explain what do you mean by micro-canonical ensemble. [BUH 1998]

3.9 Write down Maxwell's law of distribution of velocity and show that it corroborates the equipartition law of energy. [BUH 2003, cf. WBUT 2002]

3.10 What is the significance of rms velocity of gas molecules? [WBUT 2002]

3.11 Name the statistics obeyed by each of the following particles: neutrino, alpha, muon, photon, neutron, π-meson, phonon. [CUH 1992]

3.12 Write down the assumptions in deriving (i) MB-, (ii) BE-, (iii) FD-statistics. [CUH 1997]

3.13 Write down the basic distribution functions for MB-, BE-, and FD-statistics clearly explaining the notations.

3.14 Sketch the Fermi distribution function for $T = 0$ and $T.0$ K. [CUH 1997]

3.15 How does Fermi-energy of free electron gas varies with electron density (n)? [BUH 1997]

3.16 Under what conditions do BE and FD statistics yield classical statistics? [AU 1995]

3.17 Do electrons have zero energy at 0 K? If not, why? [GNDU 1994]

3.18 Give a comparative study of MB, BE and FD statistics.

3.19 Write down Bose-Einstein distribution function and obtain an expression for the total number of particles. [BUH 2003]

3.20 Show that Planck's law reduces to Wien's law for $hv \gg kT$ and the Rayleigh-Jeans law for $hv \gg kT$ (h = Planck's constant, k = Boltzmann constant, T = temperature, v = frequency). [CUH 2005]

3.21 What do you mean by Fermi-temperature (T_f) and velocity of an electron (v_f) in Fermi-level?

3.22 State the fundamental postulates of statistical mechanics.

3(C) Numerical Problems

3.1 In a random distribution of 10 particles between two boxes with equal probability, calculate the following:

(a) Probability of distribution (3, 7).

(b) Ratio of probability of distribution with the distribution of the largest probability.

(c) Total nummber of micro-states and the number of macro-states.

(d) The number of micro-states in macro-state (3,7).

Solution. (a) The probability of distribution of n particles between two boxes with equal probability is $P(r,s) = \frac{n!}{r!s!}\frac{1}{2^n}$, where $s = n - r$, r and s represent the number of particles in box 1 and 2 respectively.

Here $r = 3, s = 7, n = 10$.

$$\therefore\ P(3,7) = \frac{10!}{3!7!}\frac{1}{2^{10}} = 120 \times 10^{-10}.$$

(b) For the distribution, where the probability is the largest, we have $r = \frac{n}{2} = \frac{10}{2} = 5$ and $s = 5$. So, the largest probability distribution is given by

$$P(5,5) = \frac{10!}{5!5!} \times \frac{1}{2^{10}} = 252 \times 10^{-10}.$$

$$\therefore\ \frac{P(3,7)}{P(5,5)} = \frac{120 \times 10^{-10}}{252 \times 10^{-10}} = \frac{10}{21}.$$

(c) Total number of micro-states $= 2^n = 2^{10}$.

Total number of macro-states = number of terms in binomial expansion $(a+b)^n$ = number of binomial terms in $(a+b)^{10} = 11$.

$\therefore$ number of micro-states in macro-state (3,7)

$$= \frac{n!}{r!(n-r)!} = \frac{10!}{3!7!} = 120.$$

3.2 Consider 100 molecules and 10 cells of equal energy. Find $\ln W$ for

(a) the most probability distribution

(b) the least probable distribution.

[**Hints.** (a) Total number of molecules $N = 100$. For the most probable distribution each cell contains equal number of particles, i.e.,

$$n_1 = n_2 = \cdots = n_{10} = \frac{100}{10} = 10.$$

$$\therefore \ln W = \ln\left[\frac{100}{(10!)^{10}}\right] = \ln 100! - 10\ln 10! = 100\ln 100 - 100 - 10(10\ln 10 - 10) \text{ [by Stirling's formula]}$$

$$= 100 \times 2.3 \times 2 - 100 - 10(10 \times 2.3 \times 1 - 10) = 460 - 100 - 230 + 100 = 230.$$

(b) $\ln W = \ln \frac{100!}{100!(0!)}$ [For least possible distribution $N = 100, n_1 = 100, n_2 = 0 \ldots n_{10} = 0$]

$\ln 1 = 0$.]

3.3 Two states with energy difference 4.83×10^{-7} J occur with relative probability e^2. Calculate the temperature. [Given: $k = 1.38 \times 10^{-23}$ J/K]

[**Hint.** Let E_1 and E_2 be the energy of the two states with probabilities P_1 and P_2 respectively. Then $P_1 = Ae^{-E_1/kT}$ and $P_2 = Ae^{-E_2/kT}$.

$$\therefore \frac{P_1}{P_2} = \frac{Ae^{-E_1/kT}}{Ae^{-E_2/kT}} = e^{(E_2-E_1)/kT}.$$

$$\therefore \ln(P_1/P_2) = \frac{E_2 - E_1}{kT} \text{ or } T = \frac{E_2 - E_1}{k\ln(P_2/P_1)}.$$

But, $\frac{P_1}{P_2} = e^2$ and $E_2 - E_1 = 4.83 \times 10^{-7}$ J.

$$\therefore\ T = \frac{4.83 \times 10^7}{k \times 2} = \frac{4.83 \times 10^{-7}}{1.38 \times 10^{-23} \times 2} = 175\,\text{K.}]$$

3.4 A number of identifiable particles are distributed in a two-level system having energies E and $2E$ respectively. Determine the average energy for the most probable distribution, if the degeneracy of each level is the same.

[**Hints.** Here $g_1 = g_2 = g$ (say).

For the most probable distribution, the number of particles in state E and $2E$ are

$$N_1 = \frac{g}{e^{\alpha} e^{E/kT}} = g e^{-\alpha} e^{-E/kT}$$

and $N_2 = g e^{-\alpha} e^{-2E/kT}$ respectively.

$$\therefore \text{ the average energy } \langle E \rangle = \frac{EN_1 + 2EN_2}{N_1 + N_2} = \frac{Ee^{-\frac{E}{kT}}\left(1 + 2e^{-E/kT}\right)}{Ee^{-\frac{E}{kT}}\left(1 + 2e^{-E/kT}\right)} = \frac{E\left(1 + 2e^{-E/kT}\right)}{\left(1 + e^{-E/kT}\right)}.]$$

3.5 A die having six equally like faces marked as 1, 2, 3, 4, 5, 6 is thrown. Calculate the probability of getting number (a) 6 uppermost (b) either 3 or 6 uppermost (c) either 2 or 3 or 4 uppermost.

(*Ans.* (a) $\frac{1}{6}$ (b) $\frac{1}{3}$ (c) $\frac{1}{2}$)

3.6 Consider a two-stage system each of which can exist in states ϵ_1, ϵ_2 and ϵ_3. What are the possible sates if the particles are (a) Bosons (b) Fermions? [WBUT 2006]

3.7 (a) 3 distinguishable particles, each of which can be in one of the $\epsilon, 2\epsilon, 3\epsilon, 4\epsilon$ energy states, have total energy 6ϵ. Find all possible distributions of particles in the energy states.

(b) Find the number of micro-states in each case.

3.8 Assume that in Tungsten (At. wt. 183.8 and density 19.3 gm/cc) there are two free electrons per atom. Calculate the Fermi-energy and electron density. [Mass of electron 9.05×10^{-28} gm; Avogadro's number = 6.02×10^{23} and $h = 6.62 \times 10^{-27}$ erg-sec.] [GNDU 1991]

3.9 (a) Calculate the number of different arrangements of 6 Bosons among 4 cells of equal a priori probability. [GNDU 1984]

(b) Calculate the number of ways of arranging 10 Fermions in 15-phase space cells. [GNDU 2003]

3.10 8 distinguishable particles are distributed in two compartments. The first compartment is divided into 4 cells and the second into 2 cells. Each cell is of equal *a priori* probability and there is no restriction on the number of particles that can be contained in each cell.

Calculate the thermodynamical probability of (a) the most probable state (b) the macro-state (8, 0).

(*Ans.* Two most probable states (6, 2) and (5, 3) (a) $W(6,2) = 7 \times 10^{16}$; $W(5,3) = 7 \times 10^{16}$, (b) $W(8,0) = 2^{16}$.)

3.11 Calculate Fermi-energy at 0 K of sodium containing one free electron per atom. The density and atomic weight of sodium is 9.7×10^2 kg/m^3 and 23 respectively. What is the Fermi-temperature?

(*Ans.* 3.2 eV, 37101 K)

3.12 Calculate the total number of particles in a fermionic gas in terms of the Fermi level at absolute zero temperature. [WBUT 2007]

3.13 A system has non-degenerate single-particle states with 0, 1, 2, 3 energy of the system in 3 units. Find the number of micro-states if the particles obey (a) MB statistics, (b) BE statistics (c) FD statistics.

3(D) Broad Answer Type Questions

3.1 (a) What do you mean by phase space and density of states?

(b) What is thermodynamic probability? How does it differ from mathematical probability?

3.2 (a) What do you mean by *ensemble*. What are micro-canonical, canonical and grand canonical ensembles. Indicate their difference.

(b) What are macro-state and micro-state? Discuss with example.

3.3 (a) Deduce expressions for the thermodynamic probability of a macro-state following BE, FD and MB statistics.

(b) Write down expression for the mean occupation number in an energy state ϵ_i, if the particles obey (a) BE (b) FD statistics. [BUH 2000]

3.4 Write down MB distribution formula. Apply BE statistics to obtain (a) Maxwell's law of velocity distribution among ideal gas molecules (b) the law of equipartition of energy.

3.5 (a) Apply BE statistics to photon gas and deduce Planck's law of spectral energy density of black body radiation.

(b) Hence, deduce the Stefan-Boltzmann law for the total energy density of black body radiation. [Given: $\int_0^\infty \frac{x^3\,dx}{e^x-1} = \frac{\pi^4}{15}$] [BUH 2000]

3.6 Obtain Planck's formula for black body radiation using Bose-Einstein Statistics. Show that Wien's formula and Rayleigh-Jeans formula can be deduced as particular cases of Planck's law.
Deduce the Stefan-Boltzmann law of radiation from Planck's law. [BUH 2006]

3.7 (a) Write down the Fermi-Dirac distribution function and define Fermi-level at absolute zero and at finite temperature.

(b) What is Fermi-temperature? Estimate it for a metal if the Fermi-energy $E_f = 4$ eV.

(c) Obtain an expression for the saturation thermionic current density and discuss the role of work function in this process. [BUH 2003]

3.8 (a) For a system of N-particles, define the following: (a) 6-dimensional μ-space and $6N$-dimensional Γ-space, (b) Ensemble and ensemble average.

(b) Define thermodynamic probability W making reasonable assumptions, obtain a relation between the entropy S and thermodynamic probability W. [BUH 1997]

3.9 (a) Write down distribution functions of Bose-Einstein and Fermi-Dirac states.

(b) Apply appropriate distribution function to derive the Richardson-Dushman formula for the electron current density in thermionic emission.

(c) Explain under what conditions quantum statistics reduces to classical states. [BUH 1997]

3.10 (a) Compare MB, FD, and BE statistics mentioning at least three characteristics.

(b) Draw the Fermi-distribution curve of (a) $T = 0$ K and (b) $T > 0$ K. Explain their significance.

(c) Indicate which statistics will be applicable for (a) H^1 (b) e^- (c) π° (d) photon. [WBUT 2005]

3.11 (a) What are Fermions and Bosons? Give two examples of each.

(b) State the basic postulates of Fermi-Dirac statistics and derive an expression for the probability distribution of particles obeying FD statistics. [WBUT 2004]

3.12 Derive the expression for Bose-Einstein statistics and hence obtain Planck's formula for black body radiation. [WBUT 2002]

3.13 (a) Find how Fermi-energy of a free-electron gas varies with electron density (n). [BUH 2000]

(b) Sketch the Fermi distribution function for $T = 0$ and $T > 0K$. [CUH 1997]

3.14 Write down the basic distribution functions for FD- and BE-statistics clearly explaining the notations. Indicate the constraints imposed upon the system. Show that in the limit of low pressure both the distributions go over to classical Boltzmann distribution, provided the temperature is not too low. What happens at very low temperatures? [CUH 1996]

3.15 (a) Sketch the Fermi distribution for $T = 0$ and $T > 0$ and explain.

(b) What is the occupation probability at $E = E_f$?

(c) Calculate how the degeneracy function $g(E)$ depends on E for a fermionic gas.

(d) Express the Fermi level in a metal in terms of free electron density. [WBUT 2008]

MODULE 3

Chapter 4 : Dielectrical Properties of Solids

Chapter 5 : Magnetic Properties of Solids

Chapter 4

Dielectrical Properties of Solids

4.1 Introduction

Dielectrics are basically insulators. All the electrons in dielectrics are bound to their parent molecules and hence could be affected by externally applied electric fields. There is no free charge and as such electrical conduction could not take place in dielectrics.

Dielectrics are non-metals of high specific resistance and negative temperature coefficients. Glass, mica, ebonite, paraffin wax and dry air are some of the common examples of Dielectric.

4.2 Polar and Non-polar Molecules

The dielectric materials can be classified into the following two categories as regards their behaviour in an electric field:

1. *Non-polar molecules. If the centre of gravity of positive charges exactly coincides with that of the negative charges and the net dipole moment is zero, then the molecule is called a non-polar molecules.* H_2, O_2, N_2, CO_2 and CH_4 are the examples of non-polar molecules.
2. *Polar molecules. If the centre of gravity of positive charges does not coincide with that of the negative charges, then the molecule is called a polar molecules.* Such molecules, therefore, constitute a permanent dipole and have a permanent dipole moment.

Examples of polar molecules are HCl, H_2O, NH_3, etc. These molecules consist of dissimilar atoms and their dipole moments are of the order of 10^{29} C-m.

4.2.1 Polar molecules in absence of electric field

In a polar dielectric the molecular dipoles are oriented at random with their dipole moment ($\vec{p}$) pointing all possible directions [Fig. 4.1]. As a result the dipole moments cancel the effect of each causing the dipole moment per unit volume of the dielectric zero. Hence, though the individual molecules possess dipole moment, the dipole moment of the dielectric as a whole is zero.

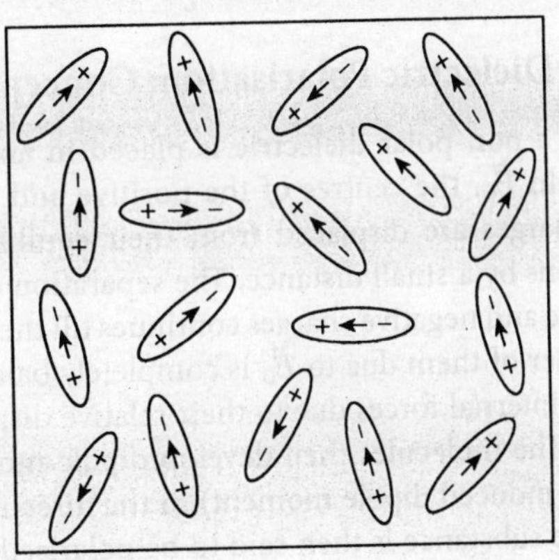

Fig. 4.1: Polar dielectric in absence of $\vec{E}$.

4.2.2 Polar dielectric in presence of electric field

The dipole moments tend to align themselves when an electric field $\vec{E}$ is applied to the dielectric. The alignments take place along the direction of $\vec{E}$ [Fig. 4.2].

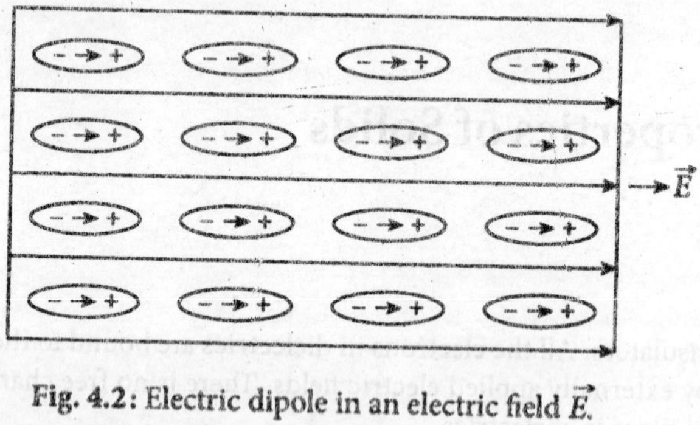

Fig. 4.2: Electric dipole in an electric field $\vec{E}$.

The torque experienced by a dipole of moment $\vec{p}$ is given by $\vec{\tau} = \vec{p} \times \vec{E}$.

This torque tends to align the dipoles along $\vec{E}$ which is opposed by thermal agitation of the molecules. So, the degree of alignment is determined by

(i) the magnitude of $\vec{E}$ and

(ii) the temperature of the dipole causing thermal agitation.

If $\vec{E}$ be increased the separation between the centre of gravity of positive and negative charges will be increased causing an increase in dipole moment.

As a result the magnitude of the torque $\vec{\tau}$ will increase and hence the degree of alignment will be greater.

If the temperature of the dielectric be decreased, then the thermal agitation will be less and the alignment will be greater. When the molecules of a dielectric are completely or partially aligned in the direction of the electric field, the dielectric is said to be polarised [Fig. 4.2].

4.3 Dielectric Polarisation: Concept of Polarisation

When a non-polar dielectric is placed in an electric field $\vec{E}_0$, the centres of the positive and negative charges are displaced from their equilibrium positions by a small distance. The separation of the positive and negative charges continues till the force on either of them due to $\vec{E}_0$ is completely balanced by the internal forces due to their relative displacement. The molecules then develop dipole moment (in the induced dipole moment) in the direction of $\vec{E}_0$. The substance is then said to be polarised with a polarisation $\vec{p}$. *The polarisation is defined as the dipole moment developed per unit volume of the material.*

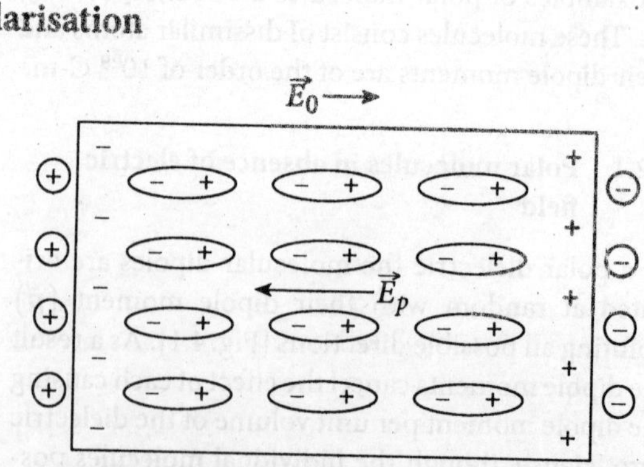

Fig. 4.3: A dielectric slab is placed in $\vec{E}_0$ produced by fixed charges outside the slab and $\vec{E}_p$ is produced by bound charges.

Due to polarisation of the dielectric a polarisation field $\vec{E}_p$ is developed in the opposite direction of the applied field. So, the magnitude of the resultant field is less than the applied field, i.e., $|\vec{E}| < |\vec{E}_0|$.

∴ the resultant field $\vec{E}$ is given by

$$\vec{E} = \vec{E}_0 + \vec{E}_p. \tag{4.1}$$

4 Some Definitions

1. *Dielectric constant.* Dielectric constant of a material is defined as the ratio between the permittivity of the medium (ϵ) and the permittivity of the free space (ϵ_0), i.e.,

$$\epsilon_r = \frac{\epsilon}{\epsilon_0} \tag{4.2}$$

ϵ_r is also called the *relative permittivity* and being the ratio of two same quantities, ϵ_r has no unit. ϵ_r gives a measure of polarisation in the dielectric materials.

2. *Electric polarisation.* If an atom be subjected to an external electric field, the centre of positive charge is displaced along the applied field direction while the centre of the negative charge is displaced in opposite direction. Thus an electric dipole is produced, When a dielectric material is placed in an electric field such dipoles are created in all the atoms of the material.

The process of producing electric dipoles oriented along the applied electric field is called polarisation of dielectric.

3. *Polarisability.* If the strength of applied electric field $\vec{E}$ is increased, then the dipole moment $\vec{\mu}$ also increases, i.e.,

$$\vec{\mu} = \alpha \vec{E}, \tag{4.3}$$

where α is a constant of proportionality and is known as the *polarisability*.

Hence, *polarisability of a dielectric is defined as the dipole moment induced per unit electric field.*

4. *Polarisation vector* $(\vec{P})$. The polarisation vector $(\vec{P})$ is defined as the dipole moment induced per unit volume of the dielectric material.

If N be the number of molecules per unit volume in a dielectric and $\vec{\mu}$, the average dipole moment per molecule, then the polarisation vector is

$$\vec{P} = N\vec{\mu}. \tag{4.4}$$

5. *Electric flux density* $(\vec{D})$. The *electric flux density or electric displacement* is the number of electric lines of force passing normally through unit area. It is also known as the *electric displacement* $(\vec{D})$. Electric flux density at a point in a material is given by

$$\vec{D} = \epsilon_r \epsilon_0 \vec{E}, \tag{4.5}$$

where $\vec{E}$ is the electric field strength, ϵ_0 is the permittivity of free space and ϵ_r is the relative permittivity of the material.

As the polarisation measures the additional flux density arising from the presence of the material as compared to free space, it has the same units as $\vec{D}$ and is related to $\vec{D}$ as

$$\vec{D} = \epsilon_0 \vec{E} + \vec{P}. \tag{4.6}$$

But $\vec{D} = \epsilon_r \epsilon_0 \vec{E}$;

$$\therefore \epsilon_r \epsilon_0 \vec{E} - \epsilon_0 \vec{E} = \vec{P} \text{ or } \vec{P} = \epsilon_0(\epsilon_r - 1)\vec{E}. \tag{4.7}$$

$\vec{D}$ is similar to magnetic induction $\vec{B}$ in magnetostatics.

6. *Electric susceptibility.* The ratio of polarisation per unit volume and the net electric field $\vec{E}$ as modified by the induced charges on the surface of the dielectric is called electric susceptibility and is denoted by χ_e, i.e.,

$$\chi_e = \frac{P}{\epsilon_0 E} \text{ or } \chi_e = \frac{\epsilon_0(\epsilon_r - 1)E}{\epsilon_0 E} \text{ [by (4.7)]} \tag{4.8}$$

$$\therefore \chi_e = (\epsilon_r - 1). \tag{4.9}$$

4.5 Derivation of $\vec{D} = \epsilon_0 \vec{E} + \vec{P}$

We consider a parallel plate condenser in which the space between the plates is filled with an isotropic dielectric. The electric field at any point between the plates is

$$E = \frac{\sigma_f}{\epsilon_0}, \tag{4.10}$$

where $\pm\sigma_f$ are the surface densities of free charges on the plates of the condenser [Fig. 4.4(a)].

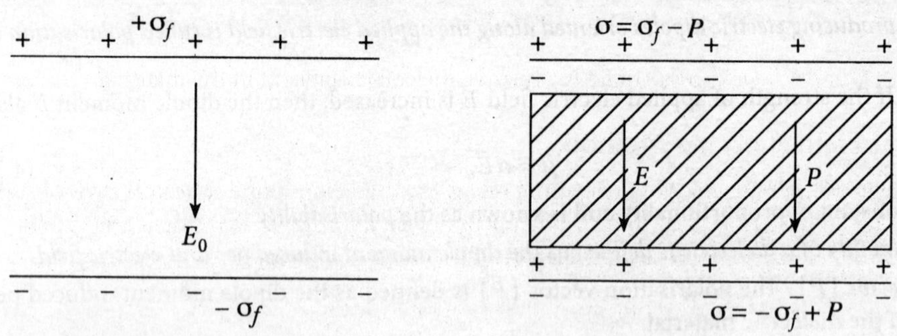

Fig. 4.4 (a) : Capacitor without dielectric. **Fig. 4.4 (b) :** Capacitor with dielectric.

When the dielectric is introduced between the plates of the condenser, it gets polarised and let the density of polarisation (i.e., dipole moment per unit volume) be P. A negative charge density $-P$ is induced on the dielectric surface near the positive charged plate [Fig. 4.4(b)] and a similar positive charge $+P$ on the surface near the positive plate. There will be no resultant charge density at any point within the dielectric. This is because all the individual dipoles are aligned parallel to the applied field with negative charge of one dipole is next to the positive charge of the following dipole.

The densities of charges on the condenser plate becomes, $\sigma = \pm(\sigma_f - P)$.

The new electric field in between the condenser plates is then

$$E = \frac{\sigma}{\epsilon_0} = \frac{\sigma_f - P}{\epsilon_0} \text{ or } \sigma_f = \epsilon_0 E + P. \tag{4.11}$$

The quantity on the right-hand side of (4.11) is called the *electric displacement D*. Thus,

$$D = \epsilon_0 E + P. \tag{4.12}$$

Since D, E and P are in the same direction, the more general form of (4.12) is expressed as

$$\vec{D} = \epsilon_0 \vec{E} + \vec{P}. \tag{4.13}$$

4.6 Measurement of Relative Dielectric Constant

The relative dielectric constant of a dielectric material can be measured by using a parallel plate capacitor.

Let a voltage V be applied between the plates of the capacitor when the space between the plates is vacuum; then the capacity of the condenser is given by

$$C = \frac{\epsilon_0 A}{d}, \tag{4.14}$$

where A is the surface area of the plates and d is the separation of the plates.

If now a dielectric material be inserted between the plates, the charge on the plates increases due to polarisation in the material. Now the capacitance of the condenser is given by

$$C' = \frac{\epsilon_0 \epsilon_r A}{d}. \tag{4.15}$$

From (4.14) and (4.15), the relative dielectric constant is found to be

$$\epsilon_r = \frac{C'}{C}. \tag{4.16}$$

Thus measuring C and C', we can calculate the relative dielectric constant of the material.

4.7 The Local Field (Lorentz Method)

We consider a dielectric slab be placed between the plates of a capacitor and a spherical cavity inside the dielectric around an atom A [Fig. 4.5]. If the diameter of the cavity is larger than the atomic radius, then the local field (also called the *internal field*) at the site of the atom A is given by

$$\vec{E}_{\text{loc}} = \vec{E}_1 + \vec{E}_2 + \vec{E}_3 + \vec{E}_4. \tag{4.17}$$

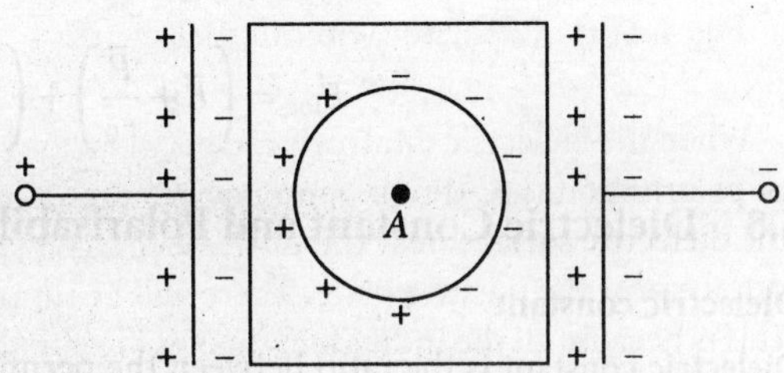

Fig. 4.5: Calculation of $\vec{E}_{\text{loc}}$ for a cubic structure.

Field $\vec{E}_1$. The field $\vec{E}_1$ arises due to the charge density on the capacitor plates. Thus,

$$\vec{E}_1 = \frac{\vec{D}}{\epsilon_0} = \frac{\vec{P} + \epsilon_0 \vec{E}}{\epsilon_0} \quad [\because \vec{D} = \vec{P} + \epsilon_0 \vec{E}]$$

$$\text{or } \vec{E}_1 = \vec{E} + \frac{\vec{P}}{\epsilon_0}. \tag{4.18}$$

Field $\vec{E}_2$. $\vec{E}$ arises due to the induced charge density on the two sides of the dielectric and is given by

$$\vec{E}_2 = -\frac{\vec{P}}{\epsilon_0}. \tag{4.19}$$

Field $\vec{E}_3$. The field $\vec{E}_3$ is the field intensity at A due to the other atoms in the cavity. Since the structure is cubic, $\vec{E}_3 = 0$ because of symmetry.

Field $\vec{E}_4$. $\vec{E}_4$ is the field intensity at A due to the polarisation charges on the surface of the cavity. Fig. 4.6 shows an enlarged view of the cavity inside the dielectric around A.

If dS be the surface area of the sphere of radius r lying between θ and $\theta + d\theta$, where θ is the angle made by AQ with the direction of the applied field $\vec{E}$, then

$$dS = 2\pi (PQ)(QR).$$

Since

$\sin\theta = PQ/r, \quad \therefore PQ = r\sin\theta$

and $d\theta = QR/r, \quad \therefore QR = r\,d\theta.$

$\therefore dS = 2\pi r\sin\theta \times r\,d\theta = 2\pi r^2 \sin\theta\, d\theta.$

Now, the charge on the surface dS is

$dq = \sigma dA = (P\cos\theta)dA$

$[\because \sigma = \text{the surface density of charge} = P\cos\theta]$

$= P(2\pi r^2 \sin\theta\cos\theta\, d\theta).$

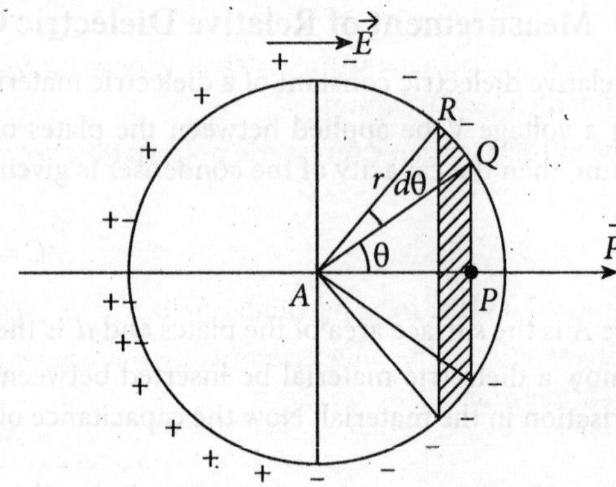

Fig. 4.6: Enlarged cavity around *A*.

$\therefore$ the field at *A* due to dq is in the direction of $\vec{E}$ (i.e., $\theta = 0$),

$$dE_4 = \frac{dq\cos\theta}{4\pi\epsilon_0 r^2} = \frac{P\cos\theta \times 2\pi r^2 \sin\theta\cos\theta\, d\theta}{4\pi\epsilon_0 r^2}$$

$$= \frac{P}{2\epsilon_0}\cos^2\theta\sin\theta\, d\theta.$$

Integrating from $\theta = 0$ to π, we get

$$E_4 = \frac{P}{2\epsilon_0}\int_0^{\pi}\cos^2\theta\sin\theta\, d\theta = \frac{P}{2\epsilon_0}\int_0^{\pi}\cos^2\theta\, d(-\cos\theta)$$

$$= -\frac{P}{2\epsilon_0}\left[\frac{\cos^3\theta}{3}\right]_0^{\pi} = -\frac{P}{6\epsilon_0}[-1-1] = \frac{P}{3\epsilon_0}. \tag{4.20}$$

$\therefore$ the local field (or *Lorentz field*) is

$$\vec{E}_{\text{loc}} = \left(\vec{E} + \frac{\vec{P}}{\epsilon_0}\right) + \left(-\frac{\vec{P}}{\epsilon_0}\right) + 0 + \frac{\vec{P}}{3\epsilon_0} = \vec{E} + \frac{\vec{P}}{\epsilon_0}. \tag{4.21}$$

4.8 Dielectric Constant and Polarisability

Dielectric constant

Dielectric constant is the ratio between the permittivity of the medium (ϵ) to the permittivity of the free space (ϵ_0), i.e., $\epsilon_r = \epsilon/\epsilon_0$.

For an isotropic (or cubic) medium, the electric displacement vector is defined as

$$\vec{D} = \epsilon_0\epsilon_r\vec{E} = \epsilon_0\vec{E} + \vec{P} \text{ [cf. (4.13)]}, \tag{4.22}$$

where ϵ_r is the relative permeability (or, dielectric constant) of the medium. ϵ_r is always dimensionless and a constant quantity.

From (4.9),

$$\epsilon_0\epsilon_r\vec{E} = \epsilon_0\vec{E} + \vec{P} \text{ or } \epsilon_r = \frac{\epsilon_0\vec{E} + \vec{P}}{\epsilon_0\vec{E}} = 1 + \frac{\vec{P}}{\epsilon_0\vec{E}} \tag{4.23}$$

$$\text{or } \epsilon_r = 1 + \chi_e \left[\because \frac{\vec{P}}{\epsilon_0\vec{E}} = \chi_e\right]. \tag{4.24}$$

Thus ϵ_r is also a measure of *degree of polarisation* of the medium. If $\vec{P}/\epsilon_0\vec{E}$ is larger, then ϵ_r will be larger for the dielectric medium.

From (4.9) we can express the polarisation vector as

$$\vec{P} = \epsilon_0(\epsilon_r - 1)\vec{E}. \tag{4.25}$$

This shows that for a particular field $\vec{E}$, $\vec{P}$ is greater, if ϵ_r is greater.

Polarisability. The polarisability (α) *of an atom is defined as the dipole moment developed per unit local field at the atom.*

$$\therefore \vec{p} = \alpha\vec{E}_{\text{loc}}. \tag{4.26}$$

It is to be noted that ϵ_r depends on the arrangements of the atoms within the crystal and hence is a macroscopic property whereas α is an atomic property.

If there are N atoms per unit volume and all atoms have the same value of α, then the polarisation (i.e., dipole moment per unit volume) is given by

$$\vec{P} = N\vec{p} = N\alpha\vec{E}_{\text{loc}}. \tag{4.27}$$

In general,

$$\vec{P} = \sum_j N_j\alpha_j\vec{E}_{\text{loc}}(j), \tag{4.28}$$

where $\sum_j$ represents summation over all the atoms.

From (4.18) and (4.27), we get

$$\vec{P} = N\alpha\left(\vec{E} + \frac{\vec{P}}{3\epsilon_0}\right). \tag{4.29}$$

4.9 Polarisation Process

When a dielectric is placed in a dc electric field $(\vec{E})$, it gets polarised by the following four processes:

(i) Electronic polarisation (ii) Ionic polarisation
(iii) Dipolar or oriental polarisation (iv) Space charge polarisation.

The contribution of a particular process depends on the nature of the dielectric as well as on the frequency of the applied electric field provided the applied field be an ac field.

4.9.1 Electronic polarisability

When an electric field is applied to the dielectric, the positively charged nucleus and the electrons of an atom are displaced in opposite direction and electronic polarisation occurs. The electron cloud shifts towards the positive end of the field [Fig. 4.7 (a), (b)]. Dipole moment is created in each atom due to the relative shift of the centre of the electron cloud and the nucleus. The shift is proportional to the applied electric field $\vec{E}$ and the induced dipole moment of the atom is

$$\vec{p}_e = \alpha_e\vec{E}, \tag{4.30}$$

where α_e is the electronic polarisability which is independent of temperature of the dielectric.

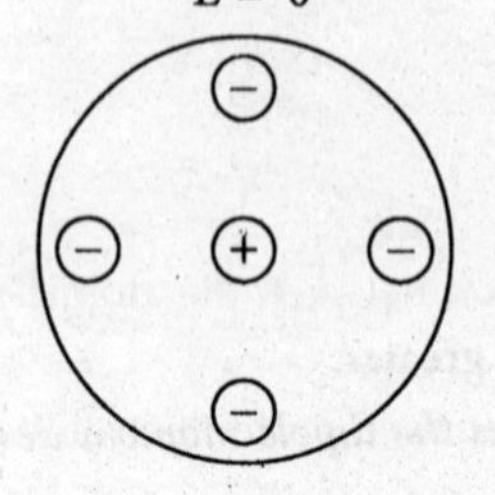

Fig. 4.7(a) : Unpolarised.

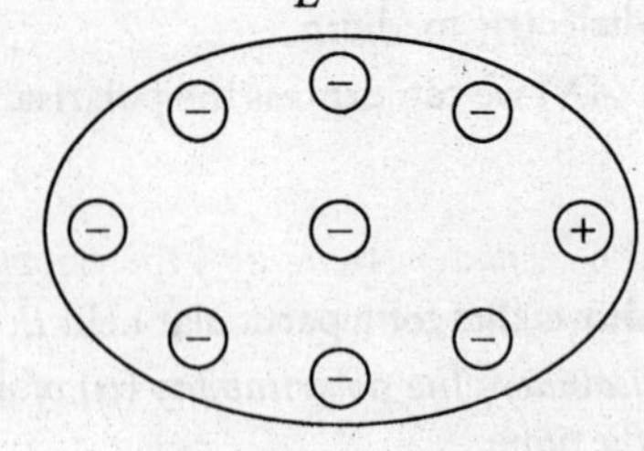

Fig. 4.7(b) : Polarised.

Classical theory of electronic polarisability

If ω_0 be the natural frequency of vibration of an electron bound harmonically to the atom, then it exhibits resonant absorption at frequency ω_0 given by

$$\omega_0 = \sqrt{\beta/m}, \tag{4.31}$$

where β is the force constant and m, the mass of the electron.

Under equilibrium condition, the force on the electron due to local field $\left(-e\vec{E}_{loc}\right)$ is equal to the force of restitution (βx), where x is the displacement of the electron from the mean position, i.e.,

$$-e\vec{E}_{loc} = \beta x = m\omega_0^2 x. \tag{4.32}$$

The *static electron polarisability* is given by

$$\alpha_e = \frac{p}{E_{loc}} = \frac{-ex}{\left(\frac{m\omega_0^2 x}{-e}\right)} = \frac{e^2}{m\omega_0^2}. \tag{4.33}$$

To find the *frequency dependence of* α_e we assume the electron motion to be simple harmonic and the local field $\vec{E} = \vec{E}_{loc} \sin\omega t$, where ω is the frequency of the field $\vec{E}$. Then the equation of motion of the electron is

$$m\frac{d^2x}{dt^2} + m\omega_0^2 x = -eE_{loc}\sin\omega t. \tag{4.34}$$

Substituting $x = x_0 \sin\omega t$, we get from (4.34)

$$m\frac{d^2}{dt^2}(x_0\sin\omega t) + m\omega_0^2 x_0 \sin\omega t = -eE_{loc}\sin\omega t$$

$$\text{or } m\left(\omega_0^2 - \omega^2\right)x_0 = -eE_{loc}. \tag{4.35}$$

The amplitude of the dipole is

$$p_0 = -ex_0 = \frac{e^2 E_{loc}}{m\left(\omega_0^2 - \omega^2\right)}.$$

∴ the frequency dependent electronic polarisability is given by

$$\alpha_e = \frac{p_0}{E_{loc}} = \frac{e^2/m}{\left(\omega_0^2 - \omega^2\right)}. \tag{4.36}$$

The corresponding quantum mechanic expression is

$$\alpha_e = \frac{e^2}{m}\sum_j \frac{f_{ij}}{\left(\omega_{ij}^2 - \omega^2\right)}. \tag{4.37}$$

f_{ij} is the *oscillator strength* of the electronic dipole transition between the atomic states i and j. α_e is independent of temperature.

If N be the number of atoms per unit volume of the dielectric then static electronic polarisation is

$$P_e = N p_e = N\alpha_e E_{\text{loc}}. \tag{4.38}$$

Also from (4.25),

$$P_e = \epsilon_0(\epsilon_r - 1)E_{\text{loc}}.$$

$$\therefore \epsilon_0(\epsilon_r - 1)E = N\alpha_e E_{\text{loc}} \text{ or } \alpha_e = \frac{\epsilon_0(\epsilon_r - 1)}{N}. \tag{4.39}$$

4.9.2 Ionic polarisability

The ionic polarisability arises due to the displacement of cations and anions in opposite directions in an ionic solid.

Let an electric field be applied along the positive directions when the positive ion moves towards right by x_1 and the negative ion by $-x_2$. If each unit cell of the dielectric contains one positive and one negative ion then the dipole moment due to ionic displacement is

$$\vec{p} = e(\vec{x}_1 + \vec{x}_2).$$

If $\vec{F}$ be the force due to applied field $\vec{E}$, and β_1 and β_2, the force constants of cation and anion, then

$$\vec{F} = \beta_1\vec{x}_1 = \beta_2\vec{x}_2, \;\therefore\; \vec{x}_1 = \vec{F}/\beta = \frac{e\vec{E}}{m\omega_0^2},$$

where $\vec{F} = e\vec{E}$ and the restoring force constant $\beta_1 = m\omega_0^2$.

Similarly, $\vec{x}_2 = \dfrac{e\vec{E}}{M\omega_0^2}$, m and M are the masses of the anion and cation respectively.

$$\therefore\; \vec{x}_1 + \vec{x}_2 = \frac{e\vec{E}}{\omega_0^2}\left(\frac{1}{M} + \frac{1}{m}\right) \text{ and } \vec{p} = e(x_1 + x_2) = \frac{e^2\vec{E}}{\omega_0^2}\left(\frac{1}{M} + \frac{1}{m}\right).$$

$\therefore$ ionic polarisability

$$\alpha_i = \frac{\vec{p}}{\vec{E}} = \frac{e^2}{\omega_0^2}\left(\frac{1}{M} + \frac{1}{m}\right) \tag{4.40}$$

$$\text{or } \alpha_i = \frac{e^2}{\omega_0^2\mu} = \frac{e^2}{\beta}, \tag{4.41}$$

where $\beta = \omega_0^2\mu$ and $\mu = \left(\frac{1}{M} + \frac{1}{m}\right)^{-1}$ is the *reduced mass* of the anion and cation.

Thus α_i is inversely proportional to the square of the natural frequency (ω_0) of the molecule and its reduced mass μ.

4.9.3 Dipolar or oriental polarisability

The molecules like H_2O, CH_3Cl, etc. having a permanent dipole moment are called *dipolar* or *polar molecules*. The substance comprising of such molecules is called *dipolar substance.*

In absence of external electric field $(\vec{E})$ the dipoles are randomly oriented and the net dipole moment of the substance is zero. But in presence of an external electric field $\vec{E}$ the dipoles orient along $\vec{E}$ producing oriental polarisability. The thermal agitation of the molecules tends to disorient the dipoles and in equilibrium condition the different dipoles are oriented at various angles $\theta = 0$ to π radiation with the direction of the external field $(\vec{E})$.

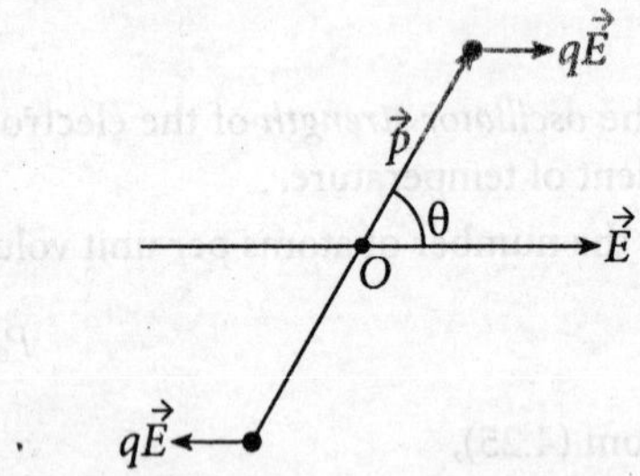

Fig. 4.8: A dipole moment $\vec{p}$ in an electric field $\vec{E}$.

The PE of such a dipole oriented at angle θ with $\vec{E}$ as shown in Fig. 4.8 is given by

$$U = -\vec{p}\cdot\vec{E} = -pE\cos\theta. \qquad (4.42)$$

Now, the number of dipoles oriented between θ and $d\theta$, i.e., within solid angle $d\Omega = 2\pi\sin\theta\, d\theta$ is proportional to

$$e^{-\frac{U}{kT}}\, d\Omega = e^{(pE\cos\theta/kT)} 2\pi \sin\theta\, d\theta,$$

where k is the *Boltzmann constant* and T is the absolute temperature of the material. The contribution of the above number of dipoles to the total polarisation is given by

$$2\pi p e^{(pE\cos\theta/kT)} \cos\theta \sin\theta\, d\theta,$$

where $p\cos\theta$ is the component of the dipole moment along $\vec{E}$.

If there are N number of such dipoles in the material, then the average contribution to polarisation is

$$\langle p\rangle = \frac{\text{Total polarisation due to } N \text{ dipoles}}{\text{Total number of dipoles } (N)} = \frac{p\int_0^\pi e^{(pE\cos\theta/kT)}\cos\theta\sin\theta\, d\theta}{\int_0^\pi e^{(pE\cos\theta/kT)}\sin\theta\, d\theta} \qquad (4.43)$$

Substituting $\frac{pE}{kT} = x$ and $\cos\theta = y$, $dy = -\sin\theta\, d\theta$.

$$\therefore \langle p\rangle = \frac{p\int_{+1}^{-1} y e^{xy}\, dy}{\int_{+1}^{-1} e^{xy}\, dy}$$

$$\text{or} \quad \frac{\langle p\rangle}{p} = \frac{e^x + e^{-x}}{e^x - e^{-x}} - \frac{1}{x} = \coth x - \frac{1}{x} \qquad (4.44)$$

$$\text{or} \quad \frac{\langle p\rangle}{p} = L(x), \qquad (4.45)$$

where $L(x) = \coth x - \frac{1}{x}$ is the *Langevin function.*

The variation of $L(x)$ with x is shown in Fig. 4.9.

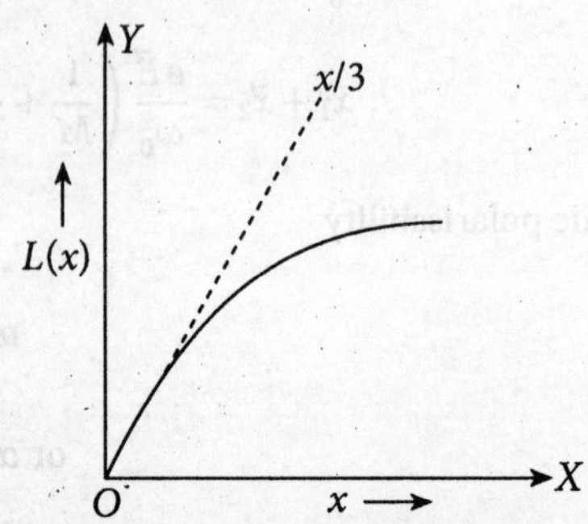

Fig. 4.9: Variation of $L(x)$ with x.

If E is not too large and T is not too small, i.e., $(pE/kT)=x \ll 1$, then

$$L(x)=\frac{x}{3}=\frac{pE}{3kT}. \tag{4.46}$$

Then from (4.45),

$$\langle p\rangle = p\left(\frac{pE}{3kT}\right)=\frac{p^2E}{3kT}. \tag{4.47}$$

The dipolar polarisation is

$$\vec{P}=N\langle p\rangle=\frac{Np^2E}{3kT}. \tag{4.48}$$

∴ the dipolar (or oriental) polarisability is

$$\alpha_d=\frac{\langle p\rangle}{E}=\frac{p^2}{3kT}. \tag{4.49}$$

Thus *dipolar polarisability is inversely proportional to the absolute temperature of the dielectric.*

4.9.4 Space charge polarisability

The polarisability arising due to the accumulation of charges at the electrodes or at the interfaces in a multiphase material is called *space charge polarisability* (α_s). In presence of an applied electric field the ions diffuse over appreciable distances giving rise to redistribution of charges in the dielectric medium.

The space charge polarisation is the slowest process as it involves the diffusion of ions over several inter-atomic distances.

4.10 Langevin-Debye Equation

The total polarisability of a dielectric is given by ,

$$\alpha=\alpha_e+\alpha_i+\alpha_d+\alpha_s=\alpha_e+\alpha_i+\alpha_d. \tag{4.50}$$

[Since α_s is very small as compared to other polarisabilities]

α decreases with increase in frequency.

$$\therefore \alpha=\frac{e^2/m}{\omega_0^2-\omega^2}+\frac{e^2}{\omega_0^2}\left(\frac{1}{M}+\frac{1}{m}\right)+\frac{p^2}{3kT}.$$

Hence total polarisation is given by

$$P=N\alpha E=NE\left[\frac{e^2/m}{\omega_0^2-\omega^2}+\frac{e^2}{\omega_0^2}\left(\frac{1}{M}+\frac{1}{m}\right)+\frac{p^2}{3kT}\right]. \tag{4.51}$$

(4.51) is called *Langevin-Debye equation.*

4.11 Clausius-Mossotti Relation

We consider a piece of dielectric of cubic structure. As in a dielectric there is no ions and permanent dipole, the ionic polarisability (α_i) and oriental polarisability (α_d) are zero, i.e., $\alpha_i=\alpha_d=0$, $\therefore \alpha=\alpha_e$.

Polarisation

$$P = N\alpha_e E_{\text{loc}} = N\alpha_e\left(E + \frac{P}{3\epsilon_0}\right)\left[\begin{array}{l}\text{where } N \text{ is the number of}\\ \text{molecules per unit volume.}\end{array}\right]$$

$$\text{or } P\left[1 - \frac{N\alpha_e}{3\epsilon_0}\right] = N\alpha_e E \text{ or } P = N\alpha_e E\Big/\left[1 - \frac{N\alpha_e}{3\epsilon_0}\right]. \tag{4.52}$$

Again, since $D = P + \epsilon_0 E$.

$$\therefore\ P = D - \epsilon_0 E \text{ or } \frac{P}{E} = \frac{D}{E} - \epsilon_0 = \epsilon - \epsilon_0 = \epsilon_0\epsilon_r - \epsilon_0 \text{ or } P = E\epsilon_0(\epsilon_r - 1). \tag{4.53}$$

From (4.52) and (4.53), we get

$$E\epsilon_0(\epsilon_r - 1) = N\alpha_e E\Big/\left(1 - \frac{N\alpha_e}{3\epsilon_0}\right)$$

$$\text{or } 1 - \frac{N\alpha_e}{3\epsilon_0} = \frac{N\alpha_e}{\epsilon_0(\epsilon_r - 1)} \text{ or } 1 = \frac{N\alpha_e}{3\epsilon_0}\left[1 + \frac{3}{\epsilon_r - 1}\right]$$

$$\text{or } \frac{N\alpha_e}{3\epsilon_0} = \frac{1}{1 + \frac{3}{\epsilon_r - 1}} = \frac{\epsilon_r - 1}{\epsilon_r + 2} \text{ or } \frac{\epsilon_r - 1}{\epsilon_r + 2} = \frac{N\alpha_e}{3\epsilon_0}. \tag{4.54}$$

(4.54) is known as the *Clausius-Mossotti relation*. It relates dielectric constant to atomic polarisability provided the condition of cubic symmetry holds good. Measuring ϵ_r we can find out α_e.

In a more general form it is given by

$$\frac{\epsilon_r - 1}{\epsilon_r + 2} = \frac{1}{3\epsilon_0}\sum_j N_j\alpha_j. \tag{4.55}$$

4.11.1 Debye equation

If N_A and ρ be the molecular weight and density of the dielectric respectively, then molar volume is N_A/ρ.
$\therefore$ the molar polarisation is

$$\left(\frac{\epsilon_r - 1}{\epsilon_r + 2}\right)\frac{M_A}{\rho} = \frac{N\alpha}{3\epsilon_0}\left(\frac{M_A}{\rho}\right)$$

$$\text{or } \left(\frac{\epsilon_r - 1}{\epsilon_r + 2}\right)\cdot\frac{M_A}{\rho} = \frac{N_A}{3\epsilon_0}(\alpha_e + \alpha_i + \alpha_d)\left[\because \frac{NM_A}{\rho} = N_A\right]$$

$$\text{or } \left(\frac{\epsilon_r - 1}{\epsilon_r + 2}\right)\frac{M_A}{\rho} = \frac{N_A}{3\epsilon_0}(\alpha' + \alpha_d). \tag{4.56}$$

(4.57) is known as the *Debye equation*. (4.57) can be written as

$$\left(\frac{\epsilon_r - 1}{\epsilon_r + 2}\right)\frac{M_A}{\rho} = \frac{N_A}{3\epsilon_0}\alpha' + \frac{N_A p^2}{9\epsilon_0 kT} = a + \frac{m}{kT}\left[\because \alpha_d = \frac{p^2}{3kT}\right], \tag{4.57}$$

where $a = \dfrac{N_A}{3\epsilon_0}\alpha'$ and $m = \dfrac{N_A p^2}{9\epsilon_0 k}$.

Now the plot of

$$\left(\frac{\epsilon_r - 1}{\epsilon_r + 2}\right)\frac{M_A}{\rho}$$

against $1/T$ will be a straight line of slope m as shown in Fig. 4.10. Knowing the slope m from the graph we can determine the dipole moment using the equation

$$p = 3\left[\sqrt{\frac{mk\epsilon_0}{N_A}}\right]. \tag{4.58}$$

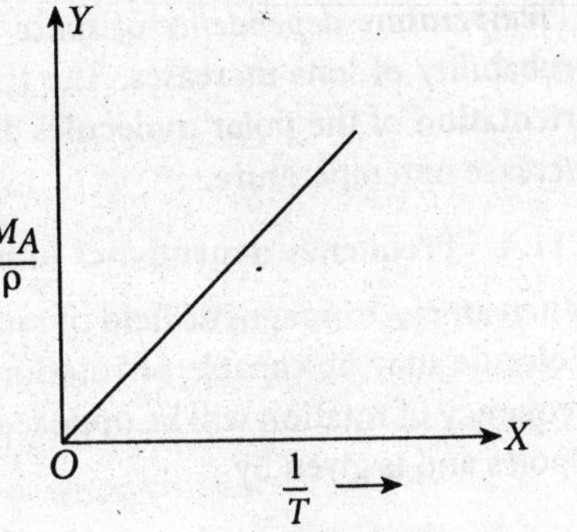

Fig. 4.10: Plot of $\left(\frac{\varepsilon_r-1}{\varepsilon_r+2}\right)\frac{M_A}{\rho}$ against $\frac{1}{T}$.

4.11.2 Refractive index

We know that the velocity of light in vacuum is $c = \frac{1}{\sqrt{\mu_0\epsilon_0}}$ and that in a dielectric medium, $v = \frac{1}{\mu\epsilon}$, where μ_0, ϵ_0 and μ, ϵ are respectively the permeability and permittivity of vacuum and the medium.

$$\therefore\ v = \frac{1}{\sqrt{\mu_r\mu_0\epsilon_r\epsilon_o}}.$$

But refractive index

$$n = \frac{\text{Velocity of light in vacuum } (c)}{\text{Velocity of light in the medium } (v)} = \frac{1}{\sqrt{\mu_0\epsilon_0}}\Big/\frac{1}{\sqrt{\mu_r\mu_0\epsilon_r\epsilon_0}} = \sqrt{\mu_r\epsilon_r}.$$

Now from (4.54),

$$\frac{n^2-1}{n^2+2} = \frac{N\alpha_e}{3\epsilon_0}\quad [\text{as } \sqrt{\epsilon_r} = n]. \tag{4.59}$$

(4.59) is known as the *Lorentz-Lorentz relation.*

4.11.3 Effect of temperature on dielectric constant

(4.39) and (4.41) show that the electronic polarisability (α_e) and ionic polarisability (α_i) are independent of temperature for normal temperature range. But oriental polarisation and space charge polarisation are affected by temperature (T).

Temperature dependence of oriental polarisability. With the increase of temperature (i.e., thermal energy) the permanent dipoles are randomly oriented and their tendency to orient along the applied field $(\vec{E})$ reduces. This causes a decrease in the dielectric constant with the increase in temperature.

Putting $\alpha_d = \frac{p^2}{3kT}$ in equation $\frac{\epsilon_r - 1}{\epsilon_r + 2} = \frac{N\alpha}{3\epsilon_0}$ in place of α, we get

$$\frac{\epsilon_r - 1}{\epsilon_r + 2} = \frac{N\alpha_d}{3\epsilon_0} = \frac{Np^2}{9\epsilon_0 kT} = \frac{T_c}{T}\left[\text{where } T_c = \frac{Np^2}{9\epsilon_0 k}\right] \text{ or } \epsilon_r(T - T_c) = 3T_c + (T - T_c)$$

$$\text{or } \epsilon_r = 1 + \frac{3T_c}{T - T_c}. \tag{4.60}$$

(4.60) shows that *spontaneous polarisation* sets in at $T = T_c$ even for $\vec{E} = 0$. T_c is called the *critical temperature* for the dielectric.

Temperature dependence of space charge polarisability. With the increase of temperature the diffusion probability of ions increases. The thermal energy also aids in overcoming the activation barrier for the orientation of the polar molecules along the applied field $(\vec{E})$. Hence the value of ϵ_r increases with the increase of temperature.

4.11.4 Frequency dependence of polarisation

When an electromagnetic field of radio frequency (ω) region be applied to a dipolar dielectric, the dipolar molecule may be capable of rotation in response to the oscillations of the alternating electric field. The frequency of rotation will be opposed by dissipative forces resulting a characteristic relaxation time for the dipoles and is given by

$$\tau = \frac{1}{\omega_0},$$

where ω_0 is the *resonant frequency* at which maximum power absorption by the oscillator takes place.

If the frequency (ω) of the applied field be very much smaller than resonant frequency, i.e., $\omega \ll \omega_0$, the dipoles being in phase with the field no energy is absorbed. For $\omega \gg \omega_0$, the dipole cannot respond to the field and again there is no absorption of energy. The contribution of the dipoles to the permittivity is the maximum at low frequencies and minimum at high frequencies.

The time dependence of polarisability may be expressed as

$$P(t) = P_0 e^{-t/\tau}, \tag{4.61}$$

where τ is the *relaxation time*.

Now, the frequency dependence of polarisation is expressed as

$$P(\omega) = \int_0^\infty P_0 e^{-t/\tau} e^{-i\omega t}\, dt \text{ or } P(\omega) = \frac{P_0}{\omega_0 + i\omega} \tag{4.62}$$

with $1/\tau = \omega_0$.

Since P depends directly on the permittivity ϵ, we write

$$\epsilon(\omega) = \frac{A}{\omega_0 + i\omega} + B = \frac{A}{\omega_0\left[1 + i\frac{\omega}{\omega_0}\right]} + B, \tag{4.63}$$

where A and B are constants.

The static permittivity ϵ_s corresponds to $\omega = 0$.

$$\therefore \epsilon(0) + \frac{A}{\omega_0} + B. \tag{4.64}$$

At $\omega > \omega_0$, $\epsilon = \epsilon_\infty$.

$$\therefore \epsilon_\infty = B. \tag{4.65}$$

From (4.64) and (4.65),

$$\epsilon_s = \frac{A}{\omega_0} + \epsilon_\infty \text{ or } A = \omega_0(\epsilon_s - \epsilon_\infty). \tag{4.66}$$

Substituting the values of A and B in (4.63), we get

$$\epsilon(\omega) = \frac{\epsilon_s - \epsilon_\infty}{1 + i\frac{\omega}{\omega_0}} + \epsilon_\infty \tag{4.67}$$

$$\text{or } \epsilon(\omega) = \frac{(\epsilon_s - \epsilon_\infty)\left(1 - i\frac{\omega}{\omega_0}\right)}{\left[1 + \left(\frac{\omega}{\omega_0}\right)^2\right]} + \epsilon_\infty \tag{4.68}$$

$$\text{or } \epsilon(\omega) = \epsilon'(\omega) - i\epsilon''(\omega), \tag{4.69}$$

where

$$\epsilon'(\omega) = \epsilon_\infty + \frac{(\epsilon_s - \epsilon_\infty)}{\left[1 + \left(\frac{\omega}{\omega_0}\right)^2\right]} \tag{4.70}$$

$$\text{and } \epsilon''(\omega) = \frac{\left(\frac{\omega}{\omega_0}\right)(\epsilon_s - \epsilon_\infty)}{\left[1 + \left(\frac{\omega}{\omega_0}\right)^2\right]}. \tag{4.71}$$

$\epsilon''(\omega)$ is the imaginary part of ϵ which is responsible for the absorption of energy. $\epsilon''(\omega)$ becomes maximum at $\omega = \omega_0$ and becomes zero for $\omega > \omega_0$ and $\omega < \omega_0$. The real part of $\epsilon(\omega)$, i.e., ϵ' falls off for $\omega < \omega_0$ to the static dielectric constant ϵ_s. ϵ' decreases as ω_0 is approached and reaches halfway between ϵ_s and ϵ_∞ at $\omega = \omega_0$. The variations of ϵ' with $\ln\omega$ and ϵ'' with $\ln\omega$ are represented in Figs 4.11 and 4.12 respectively.

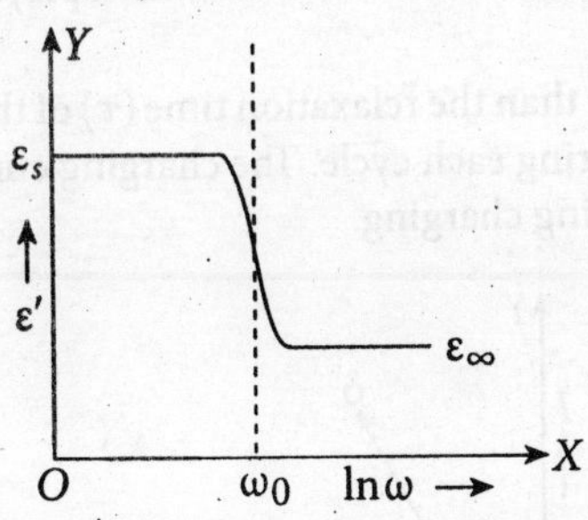

Fig. 4.11: Variation of ε' with $\ln\omega$.

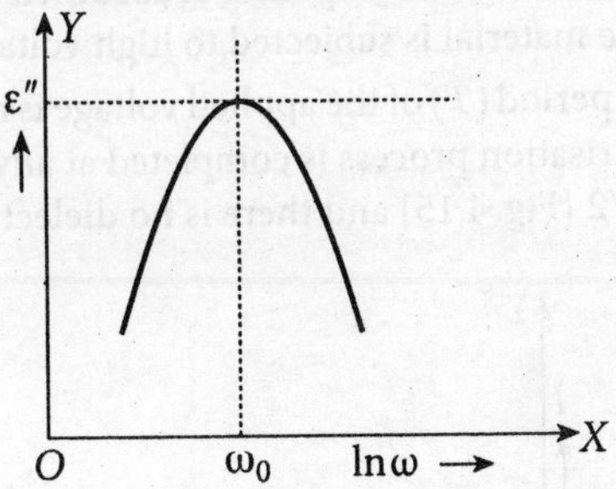

Fig. 4.12: Variation of ε'' with $\ln\omega$.

In a dielectric there are several absorption mechanism like oriental molecular and electronic processes. For each process there are peaks for ϵ'' and reduction of ϵ' with frequency ω.

The electronic absorption process occurs very quickly in presence of an electric field and electronic polarisation occurs at UV and VIS region of the *electromagnetic spectrum*. The ionic polarisation occurs in IR region as the ionic motion is several thousand times slower than electronic motion. Oriental polarisation gives rise to absorption peaks in UHF and microwave frequencies.

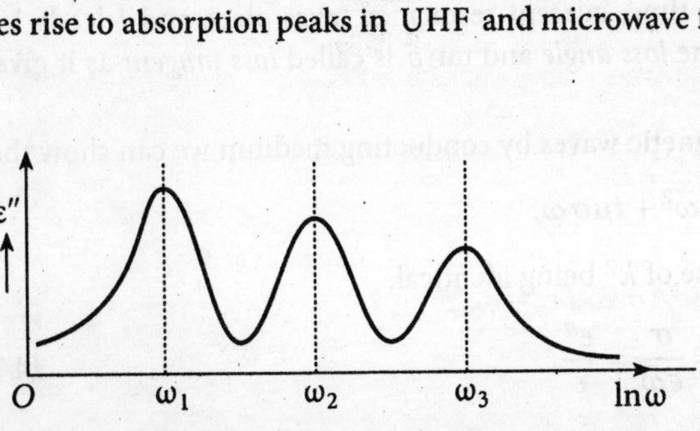

Fig. 4.13: Variation of ε'' with $\ln\omega$.

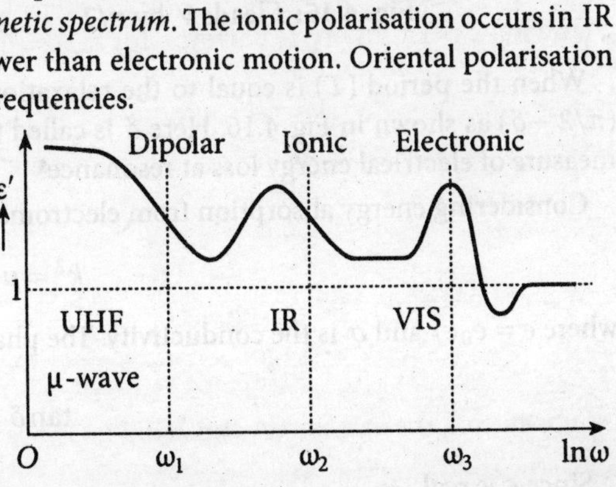

Fig. 4.14: Variation of ε' with $\ln\omega$.

The variation of ϵ'' and ϵ' with $\ln \omega$ are depicted in Figs 4.13 and 4.14 respectively.

Table 4.1: Dielectric strength of some materials

Material	*Dielectric strength* (kV/mm)	*Permittivity*	$\tan\delta$
Bakellite	22	6	0.02
Cotton	4	5	0.02
Ebonite	25	2.8	0.01
Glass	8	6.5	0.004
Rubber	17	4	0.016
Dry paper	7	3	0.005
Paraffin wax	8	2.2	0.0003
Porcelain	15	6	0.002

4.12 Dielectric Loss

If a dielectric is subjected to ac voltage, electrical energy is absorbed by it which is subsequently dissipated in the form of heat. This dissipation of absorbed energy is known as *dielectric loss* which plays an important role when the material is subjected to high voltage.

If the time period (T) of the applied voltage is much larger than the relaxation time (τ) of the polarisation process, polarisation process is completed at any instant during each cycle. The charging current leads the voltage by $\pi/2$ [Fig. 4.15] and there is no dielectric loss during charging.

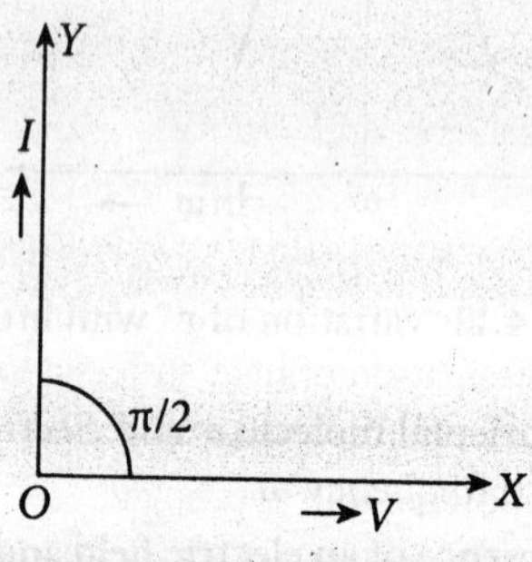

Fig. 4.15: *I* leads *V* by π/2.

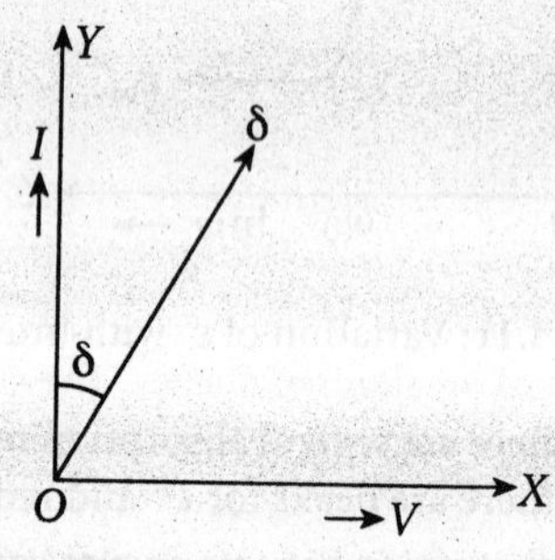

Fig. 4.16: *I* leads *V* by (π/2 – δ).

When the period (T) is equal to the relaxation time, current resonance takes place and I leads V by $(\pi/2 - \delta)$ as shown in Fig. 4.16. Here δ is called the *loss angle* and $\tan\delta$ is called *loss tangent* as it gives a measure of electrical energy loss at resonance.

Considering energy absorption from electromagnetic waves by conducting medium we can show that

$$k^2 = \mu\epsilon\omega^2 + i\mu\sigma\omega,$$

where $\epsilon = \epsilon_0\epsilon_r$ and σ is the conductivity. The phase of k^2 being identical,

$$\tan\delta = \frac{\sigma}{\epsilon\omega} = \frac{\epsilon''}{\epsilon}. \tag{4.72}$$

Since ϵ is real,

$$\epsilon'' = \frac{\sigma}{\omega}. \tag{4.73}$$

Now, power loss per unit volume of the conducting medium is given by

$$W = \sigma E^2 = \epsilon'' \omega E^2 \; [\because \epsilon'' = \sigma/\omega].$$

Again, $\epsilon'' = \epsilon' \tan\delta$,

$$\therefore \; W = \omega E^2 \epsilon' \tan\delta. \qquad (4.74)$$

The power factor is $\sin\delta = \tan\delta$. As dielectric losses are proportional to E^2, these are to important while designing high voltage cables and capacitors. The power loss dependent on $\tan\delta$ as long as E, ω and ϵ' are constants. *Power loss is thus depends on frequency ω and applied field intensity E.*

In radio frequency region, the dielectric loses are due to dipole rotation or ions jumping from one equilibrium position to another. In lower frequencies the losses are due to electrical resistivity. In the optical region the losses are due to optical absorption. The electronic polarisation, however, is free from power loss. In the infrared region the dielectric losses arise due to *ionic vibration.*

4.13 Dielectric Breakdown

Dielectric breakdown is a process in which a dielectric losses its resistivity and permits very large current to flow through it.

The mechanism responsible for dielectric breakdown are as follows:

1. *Intrinsic breakdown.* Breakdown voltage per unit thickness is called *dielectric strength.* Under the action of high electric field, large number of valence electrons in the valence band jump to the conduction band across a large forbidden gap giving rise to a large current. This is called *intrinsic breakdown* or *zener breakdown.*

   If the electric field is $\sim 10^6$ V/cm, the electrons collide with high velocity with other atoms and molecules when covalent bonds are broken. Electrons and holes are created in this process and number of charge carriers increases with time leading to high current. This is *Avalanche breakdown.*
2. *Thermal breakdown.* If the rate of generation of heat in a dielectric is greater than the rate of loss of heat, thermal breakdown occurs. The surrounding of dissipation of heat suffers a local melting due to high temperature when the region becomes conductive and breakdown occurs.
3. *Discharge breakdown.* Discharge breakdown is of the following four types, viz.,
   (a) *External breakdown.* It is caused due to *corona discharge.* Moisture or dust on the dielectric surface are also responsible for external breakdown.
   (b) *Internal breakdown.* It occurs when the dielectric contains *occluded* gas bubbles. The ionisation energy of the gases being smaller than the solid, the gas becomes ionised first. The gaseous atoms then bombard the solid dielectric releasing a large number of electrons and breakdown occurs even at low voltages.
   (c) *Defect breakdown.* If the material has defects like porosity, crack, then impurities like dust or moisture are collected leading breakdown.
   (d) *Electrochemical breakdown.* The mobility of ions in the material increases with the rise of temperature, and electrochemical reaction sets in. With the increase of mobility of the ions, the leakage current increases leading to breakdown.

4.14 Applications of Dielectric Materials

Following are some of the important applications of dielectric materials:

1. *In electric iron.* Mica or asbestos insulators are used to prevent the flow of current to the outer part of the iron.
2. *In transformer.* Varnished cotton is used as insulators in transformers, generators and motors.
3. *In electric distribution lines.* Porcelain structures are used as insulators in electric distribution lines.
4. *In electric heater.* In order to insulate the outer body and avoid short circuit ceramic beads are used in electric heater.
5. *In electric wiring.* The copper or aluminium wires used in electric wiring are insulated from one another by rubber or plastic coating.

4.15 Answers of Some Important Questions

4.1 What are dielectrics? What are polar and non-polar dielectrics?

Ans. See Arts 4.1 and 4.2 (P. 169)

4.2 What do you mean by dielectric polarisation? Define dipole moment and polarisability.

Ans. When a dielectric is placed in an external electric field, the centre of positive charge is displaced along the applied field direction and the centre of negative charge is displaced in the opposite direction. Thus a dipole is produced. *The process of producing electric dipoles which are oriented along the field direction is called polarisation in dielectrics.*

The product of one of the charges and their separating distance is called the *dipole moment.* If the centres of $+q$ and $-q$ charges be separated by a distance d, then the dipole moment is $\vec{p} = q\vec{d}$. The dipole moment is a vector quantity and is directed from the negative to positive charge.

If the strength of the applied electric field be increased, then the dipole moment also increases. The induced dipole moment is proportional to the applied field $\vec{E}$, i.e., $\vec{p} \propto \vec{E}$ or $\vec{p} = \alpha\vec{E}$, where α is a constant of proportionality and is known as the *polarisability of the material.* It is defined as the induced dipole moment per unit applied electric field.

4.3 State the different processes of polarisation.

Ans. In presence of a dc electric field $\vec{E}$, a dielectric is polarised by the following four processes: (a) Electric polarisation (b) Ionic polarisation (c) Dipolar or oriental polarisation and (d) Space charge polarisation.

The contribution of a particular process towards polarisation depends on the nature of the dielectric and the frequency of the applied field if the applied field be an ac field.

4.4 What do you mean by electronic polarisability? Write down an expression for the electronic polarisability.

Ans. When a dielectric is placed in an electric field, the positively charged nucleus and the electrons of an atom are displaced in opposite direction producing electronic polarisation. If $\vec{p}_e$ be the dipole moment produced in each atom and $\vec{E}$, the applied electric field, then $\vec{p}_e = \alpha_e \vec{E}$, where α_e is called the *electronic polarisability.*

$\therefore \alpha_e = \vec{p}_e / \vec{E}$.

In other words, *electronic polarisability is defined as the dipole moment produced in each atom per unit applied field.*

4.5 What is ionic polarisability? On what factors does it depend?

Ans. If an electric field be applied to an ionic solid, the anions and cations get displaced in opposite directions. This phenomenon is called ionic *polarisation.* The ionic polarisability (α_i) is defined as the resultant dipole moment (p) per unit applied electric field, i.e.,

$$\alpha_i = \frac{p}{E} = \frac{e^2}{\omega_0^2}\left(\frac{1}{M} + \frac{1}{m}\right),$$

where m, M are the masses of the positive and negative ions respectively.
Thus ionic polarisability α_i depends (i) inversely on the square of the natural frequency (ω_0) of the ionic molecule and (ii) on the reduced mass $\left(\frac{1}{M}+\frac{1}{m}\right)^{-1}$ of the system.

4.6 Find a relation between ionic polarisability and relative permittivity.

Ans. If N be the number of atoms per unit volume of the dielectric, then electronic polarisation is given by $P_e = Np_e = N\alpha_e E$.
Again, $P_e = \epsilon_0 E(\epsilon_r - 1)$.

$$\therefore\ N\alpha_e E = \epsilon_0 E(\epsilon_r - 1) \text{ or } \alpha_e = \frac{\epsilon_0(\epsilon_r - 1)}{N}.$$

This is the relation between the electronic polarisability and relative permittivity of the material.

4.7 What is oriental polarisability? How does it depend on the temperature of the dielectric?

Ans. When an electric field is applied to a polar molecules (molecules having permanent dipole moment) they tend to align themselves in the direction of applied field. The polarisation due to such alignment of the polar molecules along the applied field is called oriental on dipolar polarisation. Oriental polarisation is given by

$$P_d = N\vec{p} = Np^2E/3kT = N\alpha_d E \text{ or } \alpha_d = \frac{P_d}{NE} = \frac{p^2}{3kT}.$$

Thus oriental polarisability is defined as the polarisation per unit volume per unit applied field. α_d is inversely proportional to temperature of the material.

4.8 What is local field? Write down its expression.

Ans. When a dielectric is placed in an external field $\vec{E}_0$, dielectric polarisation takes place and a polarising field $\vec{E}_p$ is established. The net electric field at the site of an atom or molecule is significantly different from $\vec{E}_0$ and is called the *local field*. The local field of a dielectric placed in an external electric field $(\vec{E})$ is given by

$$\vec{E}_{\text{loc}} = \vec{E}_0 + \vec{E}_p + \frac{\vec{P}}{3\epsilon_0} = \vec{E} + \frac{P}{3\epsilon_0}.$$

This field is responsible for the polarisation of each atom or molecule of a solid having cubic symmetry. So $\vec{E}_{\text{loc}}$ apart from macroscopic field $\vec{E}$ contains field due to polarisation of the remaining molecules. $\vec{E}_{\text{loc}}$ depends on the shape of the material.

4.9 Find a relation between relative permittivity and susceptibility of a dielectric.

Ans. The electric displacement vector $\vec{D}$ of an isotropic medium is given by $\vec{D} = \epsilon_0\epsilon_r\vec{E} = \epsilon_0\vec{E} + \vec{P}$, where $\epsilon_r = \epsilon/\epsilon_0$ is the *relative permittivity* or *dielectric constant*.

$$\therefore\ \epsilon_r = \frac{\epsilon_0\vec{E} + \vec{P}}{\epsilon_0\vec{E}} = 1 + \frac{\vec{P}}{\epsilon_0\vec{E}}.$$

But $\frac{\vec{P}}{\epsilon_0\vec{E}}$ is the susceptibility (χ_e) of the dielectric.
$\therefore\ \epsilon_r = 1 + \chi_e$.

Thus larger the value of $\vec{P}/\vec{E}$ greater is the value of ϵ_r which gives a measure of polarisation of the material.

4.10 What do you understand by static dielectric constant of gases?

Ans. The molecules of a gas are widely separated from one another and hence the interactions between them is negligible. The field experienced by a dipole is, therefore, the externally applied field $(\vec{E}_0)$. The total polarisation $\vec{P}$ may, therefore, be expressed as $\vec{P}=\vec{P}_e+\vec{P}_i+\vec{P}_d=N(\alpha_e+\alpha_i+\alpha_d)\vec{E}_0$, where α_e, α_i and α_d are the electronic ionic, and dipolar polarisabilities respectively and N is the number of molecules per unit volume.

The total polarisability of the gas is

$$\alpha=\alpha_e+\alpha_i+\alpha_d=\alpha_e+\alpha_i+\frac{p^2}{3kT}.$$

This is called the *Langevin-Debye equation.*

Again, the polarisation $\vec{P}$ is given by

$$\vec{P}=\epsilon_0(\epsilon_r-1)\vec{E}_0.$$

$$\therefore\ \epsilon_0(\epsilon_r-1)=N\left(\alpha_e+\alpha_i+\frac{p^2}{3kT}\right).$$

This shows that $\epsilon_0(\epsilon_r-1)$ depends on temperature T of the gas. The plot of $\epsilon_0(\epsilon_r-1)$ against $\frac{1}{T}$ is shown in the Fig. 4.17, which is a straight line of slope $Np^2/3kT$. Thus finding slope of the curve, p can be determined and hence the molecular structure can be known.

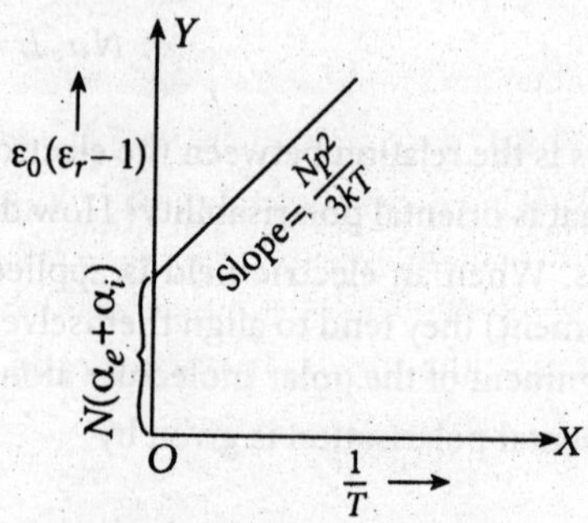

Fig. 4.17: Variation of $\varepsilon_0(\varepsilon_{r-1})$ against $1/T$.

4.11 Starting from Clausius-Mossotti equation explain the origin of spontaneous polarisation.

Ans. If ϵ_r be the relative permeability and α, the polarisability of a dielectric material, then Clausius-Mossotti relation is given by

$$\frac{\epsilon_r-1}{\epsilon_r+2}=\frac{N\alpha}{3\epsilon_0},$$

where N is the number of atoms per unit volume and $\alpha=\alpha_e+\alpha_i+\alpha_d+\alpha_s$.

Putting $\alpha_d=\dfrac{Np^2}{9\epsilon_0 kT}$ and the contributions of α_e, α_i and α_s, we get

$$\frac{\epsilon_r-1}{\epsilon_r+2}=\frac{Np^2}{9\epsilon_0 kT}=\frac{T_c}{T},\ \text{where } T_c=\frac{Np^2}{9\epsilon_0 k}$$

$$\text{or } \epsilon T-T=\epsilon T_c+2T_c \text{ or } \epsilon(T-T_c)=3T_c+(T-T_c)$$

$$\text{or } \epsilon=1+\frac{3T}{T-T_c}.$$

From the above equation we note that $\epsilon\to\infty$ for $T=T_c$, even for $E=0$. T_c is called the *critical temperature.* This phenomenon is called *spontaneous polarisation.*

Thus *the occurrence of the phenomena of dielectric polarisation even for zero applied field is called spontaneous polarisation.* Spontaneous polarisation is the cause of *ferroelectricity.*

4.12 What is polarisation catastrophe?

Ans. From Clausius-Mossotti relation,

$$\frac{\epsilon_r - 1}{\epsilon_{r-2}} = \frac{1}{3\epsilon_0} \sum_j N_j \alpha_j,$$

where α_j and N_j are the polarisability and number of ions of j type per unit volume respectively.

$$\therefore\ \epsilon_r = \frac{1 + \frac{2}{3\epsilon_0} \sum_j n_j \alpha_j}{1 - \frac{1}{3\epsilon_0} \sum_j N_j \alpha_j}.$$

For $\sum N_j \alpha_j = 3\epsilon_0$, $\epsilon_r \to \infty$ even for $E = 0$.

The condition expressed by $\sum_j N_j \alpha_j = 3\epsilon_0$ is called *polarisation catastrophe* and the phenomenon is called the *spontaneous polarisation*.

4.13 What is *dielectric loss* and the *loss tangent*?

Ans. If a dielectric material is subjected to an ac voltage, electric energy is absorbed by the material and is dissipated in the form of heat. This dissipation of energy is called the *dielectric loss*. The high voltage applications cause a serious effect as it is associated with heat generation and heat dissipation. If ω be the angular frequency of an ac voltage E then dielectric loss per unit volume is

$$W = \omega E^2 \epsilon' \tan\delta,$$

where ϵ' is the real part of ϵ and δ is the loss angle. The factor $\tan\delta$ is called the *tangent loss*.

4.14 Mention the different mechanisms of dielectric breakdown.

Ans. The different mechanisms for the dielectric breakdown are as follows:

(a) Intrinsic breakdown (b) Thermal breakdown
(c) Discharge breakdown (d) Electrochemical breakdown and
(e) Defect breakdown.

4.15 What are the essential qualities of a good insulating material?

Ans. The essential qualities of a good insulator are as follows:

(a) *Electrical*

(i) The dielectric material should have high electrical resistivity and high dielectric strength to withstand high voltage.

(ii) The dielectric loss must be minimum.

(iii) Some liquid and gas dielectrics must have the property of cooling, e.g., hydrogen, helium and transformer oil.

(b) *Mechanical*

(i) Insulators used in electric machines must have sufficient mechanical strength to withstand vibration.

(ii) Certain mechanical properties of insulators are necessary depending on the field of applicability.

(c) *Chemical*

(i) The insulators should have the capacity to resist oil, gas fumes, acids and alkalis.

(ii) The insulators must be waterproof.

4.16 Solved Problems

4.1 The relative permittivity and refractive index of a dielectric material are 4.94 and 2.69 respectively. Find the ratio between electronic and ionic polarisability of the material.

Solution. The formula,

$$\frac{\epsilon_r - 1}{\epsilon_r + 2} = \frac{N\alpha}{3\epsilon_0},$$

where $\alpha = \alpha_e + \alpha_i + \alpha_d + \alpha_s \simeq \alpha_e + \alpha_i$ [$\because \alpha_d$ and α_s are negligible]

$$\therefore \frac{\epsilon_r - 1}{\epsilon_r + 2} = \frac{N(\alpha_e + \alpha_i)}{3\epsilon_0} \text{ or } \frac{4.94 - 1}{4.94 + 2} = \frac{N(\alpha_e + \alpha_i)}{3\epsilon_0} \quad [\because \epsilon_r = 4.94]$$

$$\text{or } \frac{N(\alpha_e + \alpha_i)}{3\epsilon_0} = 0.5677.$$

Again, $\epsilon_r = n^2$.

$$\therefore \frac{n^2 - 1}{n^2 + 2} = \frac{N\alpha_e}{3\epsilon_0} \text{ or } \frac{N\alpha_e}{3\epsilon_0} = \frac{2.69 - 1}{2.69 + 2} \; [\because n = 2.69] = 0.36034.$$

Hence

$$\frac{N(\alpha_e + \alpha_i)}{N\alpha_e} = \frac{0.5677}{0.36034} \text{ or } 1 + \frac{\alpha_i}{\alpha_e} = 1.575 \text{ or } \frac{\alpha_i}{\alpha_e} = 0.575.$$

$$\therefore \frac{\alpha_e}{\alpha_i} = \mathbf{1.739}.$$

4.2 The crystal of sodium chloride has static dielectric constant of 5.6 and optical index of refraction 1.5. Calculate the percentage contribution of ionic polarisability. **[HPU 1995]**

Solution. From *Clausius–Mossotti* relation,

$$\frac{\epsilon_r - 1}{\epsilon_r + 2} = \frac{N(\alpha_e + \alpha_i)}{3\epsilon_0},$$

where α_e and α_i are the electronic and ionic polarisability respectively.

At optical frequencies, the above equation takes the form:

$$\frac{n^2 - 1}{n^2 + 2} = \frac{N\alpha_e}{3\epsilon_0},$$

n is the refractive index of the crystal.

$$\therefore \left(\frac{n^2 - 1}{n^2 + 2}\right) \Big/ \left(\frac{\epsilon_r - 1}{\epsilon_r + 2}\right) = \frac{\alpha_e}{\alpha_e + \alpha_i}.$$

Hence, the percentage contribution of ionic polarisability is

$$\left(\frac{\alpha_i}{\alpha_e + \alpha_i}\right) \times 100 = \left(1 - \frac{\alpha_e}{\alpha_e + \alpha_i}\right) \times 100 = \left[1 - \left(\frac{n^2 - 1}{n^2 + 2}\right)\left(\frac{\epsilon_r + 2}{\epsilon_r - 1}\right)\right] \times 100$$

$$= \left[1 - \frac{(1.5)^2 - 1}{(1.5)^2 + 2} \times \frac{5.6 + 2}{5.6 - 1}\right] \times 100 = \mathbf{51.4\%}.$$

4.3 **Helium gas contains 2.7×10^{25} atoms/m^3 and dielectric constant of He atom at NTD is 1.0000684. Calculate the electronic polarisability of He atom.**

Solution. Electronic polarisability $\alpha_e = \dfrac{\epsilon_0(\epsilon_r - 1)}{N}$.

Here, $\epsilon_0 = 8.854\times10^{-12}$ F/m, $\epsilon_r = 1.0000684$, $N = 2.7\times10^{25}$ atoms/m^3

$$\therefore \alpha_e = \frac{8.854\times10^{-12}\times(1.0000684-1)}{2.7\times10^{25}}$$

$$= \frac{8.854\times10^{-12}\times(1.0000684-1)}{2.7\times10^{25}} = \mathbf{2.243\times10^{-41}\ Fm^2}.$$

4.4 **The polarisability of ammonium molecule is 2.42×10^{-39} C^{2}m/N and 1.74×10^{-39} C^{2}m/N at 309 K and 448 K respectively. Calculate the oriental polarisability for each of the mentioned temperatures.**

Solution. The total molecular polarisability of a polar molecule is given by

$$\alpha = \alpha_e + \alpha_i + \alpha_d = \alpha_e + \alpha_i + \frac{p^2}{3kT},$$

where α_e, α_i and α_d are the electronic, ionic and oriental polarisability respectively.

For $T = 309$ K,

$$2.42\times10^{-39} = \alpha_e + \alpha_i + \alpha_d = \alpha_e + \alpha_i + \frac{p^2}{3\times1.38\times10^{-38}\times309}$$

and for $T = 448$ K,

$$1.74\times10^{-39} = \alpha_e + \alpha_i + \frac{p^2}{3\times1.38\times10^{-38}\times448}.$$

$$\therefore \frac{2.42\times10^{-39} - (\alpha_e+\alpha_i)}{1.74\times10^{-39} - (\alpha_e+\alpha_i)} = \frac{448}{309}.$$

$$\therefore \alpha_e + \alpha_i = \mathbf{0.228\times10^{-39}\ C^2m/N.}$$

Now α_e at 309 K is $= 2.42\times10^{-39} - 0.228\times10^{-39} = \mathbf{2.192\times10^{-39}\ C^2m/N}$.

and α_e at 448 K is $= 1.74\times10^{-39} - 0.228\times10^{-39} = \mathbf{1.512\times10^{-39}\ C^2m/N}$.

4.5 **For sulphur, the relative dielectric constant at 27 K is 3.75. If the internal field constant $\gamma = \frac{1}{3}$, calculate the electronic polarisability of sulphur. [Density at 27 °C is 2050 kg/m^3 and At. wt. = 32].**

Solution. The number of atoms of sulphur per m^3 is

$$= \frac{\text{Avogadro number}\times\text{density}}{\text{Atomic weight}} = \frac{6.02\times10^{26}\times2050}{32} = 385.66\times10^{26}\ \text{atoms/m}^3.$$

Now internal field $E_i = E + \dfrac{\gamma P}{\epsilon_0} = E + \dfrac{P}{3\epsilon_0}$ $[\because \gamma = 1/3]$

$$\therefore \frac{\epsilon_r - 1}{\epsilon_r + 2} = \frac{N\alpha_e}{3\epsilon_0} \text{ or } \alpha_e = \left(\frac{\epsilon_r-1}{\epsilon_r+2}\right)\times\frac{3\epsilon_0}{N}$$

$$= \left(\frac{3.75-1}{3.75+2}\right)\times\left(\frac{3\times8.55\times10^{-12}}{385.66\times10^{26}}\right) = \left(\frac{2.75}{5.75}\right)\times\left(\frac{25.65}{385.66}\right)\times10^{-38}$$

$$= \mathbf{3.181\times10^{-40}\ F/m^2}.$$

4.6 The dielectric constant of Si is 12. The length of the edge of its cubic unit cell is 5.43 Å. Find the electrical polarisability of Si atoms.

Solution. From *Clausius-Mossotti* relation, $\frac{\epsilon_r - 1}{\epsilon_r + 2} = \frac{N\alpha_e}{3\epsilon_0}$.

For Silicon, there are 8 atoms per unit cell of volume $(5.43 \times 10^{-10})^3$ cubic metre.

The number of atoms per unit volume is

$$N = \frac{8}{(5.43 \times 10^{-10})^3} = 4.997 \times 10^{28}/\text{m}^3.$$

Putting $\epsilon_r = 12$ in the *Clausius-Mossotti* relation, we get

$$\frac{11}{13} = \frac{4.997 \times 10^{28} \times \alpha_e}{3 \times 8.854 \times 10^{-12}} \quad [\because \epsilon_0 = 8.854 \times 10^{-12}\,\text{F/m}]$$

$$\text{or } \alpha_e = \frac{11 \times 3 \times 8.854}{13 \times 4.997 \times 10^{40}} = \mathbf{4.5 \times 10^{-40}\ F.m^2}.$$

4.7 A solid dielectric has electronic polarisability of 10^{-40} Fm^2. If the internal electric field be a Lorentz field, what is the dielectric constant of the material? [Given: density of the dielectric material = 3×10^{28} atoms/m^3]

Solution. The relative dielectric constant of the material is given by

$$\epsilon_r - 1 = \frac{N\alpha_e}{\epsilon_0} \text{ or } \epsilon_r = 1 + \frac{N\alpha_e}{\epsilon_0} = 1 + \frac{(3 \times 10^{28}) \times 10^{-40}}{8.854 \times 10^{-12}} = \mathbf{1.339},$$

where the density of the material $N = 3 \times 10^{28}$ atoms/m^3 and the dielectric constant of free space $\epsilon_0 = 8.854 \times 10^{-12}$ F/m^3.

EXERCISES

4(A) Objective Type Questions

Choose the correct alternatives.

4.1 If ϵ and ϵ_0 be the permittivity of a material and of free space respectively, then the relative permittivity (or dielectric constant) of the material is

(a) $\epsilon_r = \epsilon_0/\epsilon$
(b) $\epsilon_r = \epsilon_0\epsilon$
(c) $\epsilon_r = \epsilon/\epsilon_0$
(d) $\epsilon_r = 1/\epsilon\epsilon_0$.

4.2 In polar molecules in absence of an electric field the centre of gravity of the positive and negative charges

(a) coincides
(b) does not coincide
(c) almost coincide
(d) none of these.

4.3 The non-polar in absence of an electric field molecules have a

(a) dipole moment
(b) zero dipole moment
(c) transient dipole moment
(d) none of these.

4.4 Temperature dependence of polarisability is observed in case of

(a) α_3
(b) α_i
(c) α_d
(d) α_s.

4.5 At moderate temperature (T) the electronic polarisability (α_e)

(a) depends linearly on T
(b) is independent of T
(c) inversely dependent on T
(d) none of these.

4.6 The unit of dipole moment per unit volume is

(a) coulomb-metre
(b) metre/coulomb
(c) coulomb
(d) coulomb/metre^2.

4.7 Select the correct relation:

(a) $D=\epsilon_0(\epsilon_r-1)E$
(b) $E=\epsilon_0(\epsilon_r-1)P$
(c) $P=\epsilon_0(\epsilon_r-1)E$
(d) $\epsilon_r=(\chi-1)$.

4.8 Ionic polarisability (α_i)

(a) increases with T
(b) decreases with T
(c) independ of T
(d) none of these.

4.9 Lorentz field is expressed as

(a) $E_i=E+\dfrac{P}{3\epsilon_0}$
(b) $E_i=E+\dfrac{3P}{\epsilon_0}$
(c) $E_i=E-\dfrac{P}{3\epsilon_0}$
(d) $E_i=\dfrac{3P}{\epsilon_0 E}$.

4.10 If α_s and T be the space charge polarisability and temperature of a dielectric, then

(a) α_s increases with T
(b) α_s decreases with T
(c) α_s is independent of T
(d) none of these.

4.11 The dipolar polarisability α_d is equal to

(a) μ^2/kT
(b) $\mu^2/3kT$
(c) $3kT/\mu^2$
(d) $\mu^2 kT$.

4.12 At optical frequencies $\left(\sim 10^{15}\text{ Hz}\right)$ the dielectric constant (ϵ_r) depends on the refractive index (n) as

(a) $\epsilon_r \propto n$
(b) $\epsilon_r \propto n^2$
(c) $\epsilon_r \propto \dfrac{1}{n}$
(d) $\epsilon_r \propto \dfrac{1}{n^2}$.

4.13 The relation $\dfrac{N\alpha}{3\epsilon_0}=\dfrac{\epsilon_r-1}{\epsilon_r+2}$ is called

(a) Einstein relation
(b) Lorentz relation
(c) Clausius-Mossotti relation
(d) none of these.

4.14 Dielectrics are basically

(a) conductors (c) insulators
(b) semiconductors (d) superconductors.

4.15 Which is the correct relation?

(a) $\epsilon_r = 1 - \frac{3T_c}{T - T_c}$ (c) $\epsilon_r = 1 + \frac{3T_c}{T - T_c}$
(b) $\epsilon_r = 1 - \frac{3T_c}{T + T_c}$ (d) $\epsilon_r = 1 + \frac{3T}{T - T_c}$.

4.16 The time-dependence of polarisability is expressed as

(a) $P(t) = P_0 e^{-t/\tau}$ (c) $P(t) = P_0 e^{\tau/t}$
(b) $P(t) = P_0 e^{t/\tau}$ (d) $P(t) = P_0 e^{-\tau/t}$.

4.17 The relaxation time (τ) and resonant frequency (ω_0) are related by

(a) $\tau\omega_0 = 1$ (c) $\tau\omega_0 = \infty$
(b) $\tau\omega_0 = 0$ (d) $\tau + \omega_0 = 1$.

4.18 In a dipolar dielectric, in absence of electric field, the dipoles are

(a) parallel (c) randomly oriented
(b) antiparallel (d) none of these.

4.19 Which is the correct relation of ionic polarisability (α_i) and the angular frequency of a molccule (ω_0)?

(a) $\alpha_i = \frac{e^2}{\omega_0^2}\left(\frac{1}{M} + \frac{1}{m}\right)$ (c) $\alpha_i = e\omega_0^2\left(\frac{1}{\mu} + \frac{1}{m}\right)$
(b) $\alpha_i = \frac{\omega_0^2}{e^2}\left(\frac{1}{M} + \frac{1}{m}\right)$ (d) $\alpha_i = \omega_0^2\left(\frac{1}{M} + \frac{1}{m}\right)$.

4.20 Space charge polarisation occurs at frequencies

(a) $\sim 10^{15}$ Hz (c) $\sim 10^{13}$ H
(b) $\sim 10^{3}$ Hz (d) $\sim 10^{10}$ Hz.

4.21 The condition of spontaneous polarisation is

(a) $\sum_j N_j \alpha_j = 3\epsilon_0$ (c) $\sum_j N_j \alpha_j = \epsilon_0$
(b) $\sum_j N_j \alpha_j = 2\epsilon_0$ (d) $\sum_j N_j \alpha_j = 6\epsilon_0$.

4.22 Spontaneous polarisation takes place due to

(a) atoms (c) permanent dipoles
(b) free electrons (d) ions.

4.23 Dielectric loss in a dielectric is proportional to

(a) ω^2 (c) ω
(b) $\frac{1}{\omega}$ (d) $\frac{1}{\omega^2}$.

4.24 The temperature coefficient of permittivity is given by

(a) $T_{c\epsilon} = \frac{1}{\epsilon}\frac{d\epsilon}{dT}$

(b) $T_{cE} = \frac{d\epsilon}{dT}$

(c) $T_{c\epsilon} = \frac{dT}{d\epsilon}\epsilon$

(d) $T_{cE} = \epsilon d\epsilon dT$.

ANSWERS

4.1 (c)	4.4 (c)	4.7 (c)	4.10 (a)	4.13 (c)	4.16 (a)	4.19 (a)	4.22 (c)
4.2 (b)	4.5 (b)	4.8 (c)	4.11 (b)	4.14 (c)	4.17 (a)	4.20 (b)	4.23 (c)
4.3 (b)	4.6 (d)	4.9 (b)	4.12 (b)	4.15 (c)	4.18 (c)	4.21 (a)	4.24 (a).

4(B) Short Answer Type Questions

4.1 What is a dielectric? State what is the local field in a dielectric and give its decomposition explaining the significance of each part. [BUH 1993]

4.2 What are polar and non-polar dielectrics? How would you distinguish them?

4.3 What do you mean by dielectric susceptibility? Explain its physical significance.

4.4 Define polarisation and polarisability of a dielectric material. What are the different sources of polarisability? [BUH 2000]

4.5 What is oriental polarisability? Derive an expression for the oriental polarisability of a gas of polar molecule. [BUH 1997]

4.6 What is electronic polarisability? What is the dependence of electronic polarisability of an atom on the radius of the atom? [BUH 2004]

4.7 What is dipolar polarisability? Obtain an expression for dipolar polarisability of a dielectric at moderate temperature. [BUH 2001]

4.8 Explain how the polarisability of an atom may be determined? Deduce the Clausius-Mossotti relation. [BUH 2003]

4.9 Derive Clausius-Mossotti relation between polarisability and the dielectric constant of a solid. [BUH 2003]

4.10 State the different contributions to the local polarisability of a dielectric. Deduce the Clausius-Mossotti formula for an isotropic dielectric. [BUH 1995]

4.11 What is oriental polarisability? How does it differ from atomic polarisability? [BUH 1994]

4.12 Explain what is meant by space charge polarisation. Does it depend on temperature?

4.13 Discuss frequency dependence of polarisation.

4.14 What are dielectric loss and loss tangent? Find an expression for the dielectric loss.

4.15 What is dielectric breakdown? What factors are responsible for it?

4.16 What is polarisation catastrophe?

4(C) Numerical Problems

4.1 If the susceptibility of CO_2 is 0.985×10^{-3}, calculate the total polarisability of CO_2 if its density is 1.977 kg/m. $\left(Ans.\ 3.24 \times 10^{-40}\ Fm^2\right)$

4.2 If the refractive index and dielectric constant of water are 1.33 and 8.1 respectively, then calculate the percentage of ionic polarisability. (*Ans.* 71%)

4.3 NaCl crystal has a static dielectric constant $\epsilon_r = 5.6$ and refractive index $n = 1.5$. Calculate the percentage contribution of the ionic polarisability.

4.4 The structure of $BaTiO_3$ transforms from cubic to tetragonal below the curie point with lattice parameters $c = 4.03$ and $a = 3.98$. The displacement of titanium ion is opposite to that of oxygen ions. The magnitude of displacement is 0.06 for titanium, 0.06 for oxygen ions on the side faces and 0.08 for oxygen ions on the top and bottom faces of the cube. Neglecting the displacement of barium ions calculate the polarisation of a $BaTiO_3$ crystal. (*Ans.* $0.16/\text{cm}^2$)

4.5 The bakelite is found to have the real part of its complex relative dielectric constant at 4.36 with a loss tangent of 2.8×10^{-2} at frequency 1 MHz. Calculate the complex polarisability of the material assuming Lorentz field [$N = 4 \times 10^{28}/\text{m}^3$]. (*Ans.* $(3.5 - i0.06) \times 10^{-40}$)

4.6 Calculate the value of the radius of electron orbit of helium and hence the permittivity for 2.7×10^{25} atoms/m^3. [Given electronic polarisability of helium $= 0.18 \times 10^{-40}$ Fm^3]

(*Ans.* $5.46 \times 10^{-11}; 1.000057776$)

4.7 The static dielectric constant of water is 8.125 and its refractive index is 1.33. Calculate the percentage contribution of ionic polarisability. [BUH 2004]
[**Hints.** cf. Solved Problem no. 4.2.]

4.8 Calculate the electronic polarisability of Ar atom having radius 10^{-10} m. (*Ans.* 1.11×10^{-40} Fm^2)

4(D) Broad Answer Type Questions

4.1 (a) What is a dielectric? What do you mean by dielectric polarisation? Distinguish metal and semiconductor from a dielectric.
(b) What are the various components of the local electric field at an atom in a crystal? Obtain the Lorentz relation for the local electric field. When is this relation valid? [BUH 1999]

4.2 (a) Distinguish between polar and non-polar dielectrics. Give some examples of the above two types of dielectric solids.
(b) What are the different types of polarisability?
(c) Discuss the dependence of the electronic polarisability of an atom on the radius of the atom. [BUH 2004]

4.3 (a) Define electric polarisation and the local field. Explain how the local field may be expressed as a sum of different components. Find the Lorentz field.
(b) What are the different factors which contribute to polarisation in a solid? Explain how would you determine the polarisability of an atom. Derive the formula you would use. [BUH 1997]
(c) What is Lorentz-Lorentz relation?

4.4 (a) Define polarisation and polarisability of a dielectric material. Derive Clausius-Mossotti relation between polarisability and dielectric constant of a solid. What are the different sources of polarisability. [BUH 2003]
(b) Distinguish carefully between the following terms: (i) the applied electric field, (ii) the macroscopic electric field.
(c) the local field in a dielectric.

4.5 (a) What is oriental polarisability? How does it differ from atomic polarisability?

(b) Deduce an expression for the oriental polarisability per molecule of a gas of polar substance at temperature T. [Assume that $pE/kT \ll 1$, where p is the dipole moment of the polar molecule and E is the magnitude of the applied electric field]. [BUH 1994]

(c) Discuss the temperature dependence of different types of polarisabilities.

4.6 (a) What is a dielectric? State what is local field in a dielectric and give its decomposition explaining the significance of each part.

(b) State the different contributions to the total polarisability of a dielectric material.

(c) Discuss the Clausius-Mossotti formula for an isotropic dielectric. [BUH 1995]

4.7 (a) Deduce an expression for the electronic polarisability of an atom on the basis of classical theory. [BUH 1996]

(b) Discuss the temperature dependence of oriental polarisability. What is spontaneous polarisation?

(c) What do you mean by dielectric loss and dielectric breakdown?

4.8 (a) Give a schematic sketch of the variation of the total polarisability of a dielectric as a function of frequency, explaining the physical origin of the various contributions and the relevant frequency ranges. [JNTU 2004]

(b) (i) Explain the ionic polarisability of a molecule and its frequency dependence.

(ii) Define piezoelectric effect. Discuss some of the important applications of the piezoelectrics. [JNTU 2003]

4.9 Write notes on:

(a) Relaxation time of polarisation

(b) Internal field of solids [JNTU 2003]

(c) Behaviour of ferroelectric crystals

(d) Debye equation.

Chapter 5

Magnetic Properties of Solids

5.1 Introduction

The magnetic properties of solids were discovered by ancient people of 800 BC. They found that a mineral called *magnetite* which is an oxide of iron (Fe_3O_4) possessed this property and the ancient mariners used the material for finding out direction of a ship on sea. In view of modern concept all materials, viz., metal, semiconductors and insulators exhibit magnetic properties to different extent.

The origin of magnetism in material is the magnetic moment of the magnetic dipole in the same. Materials which can be magnetised by the application of an external magnetic field are called *magnetic materials.* The important magnetic materials in view of practical applications are *diamagnetic, paramagnetic, ferromagnetic, antiferromagnetic and ferrimagnetic.*

5.2 Terms and Definitions in Magnetism

(a) Magnetic Permeability. Magnetic permeability of a material is defined as the ratio of magnetic induction to the magnetic field strength, i.e.,

$$\mu = \frac{B}{H}, \tag{5.1}$$

where B and H are respectively the *magnetic induction* and *magnetic field* at any point in the material.

The permeability of free space is denoted by μ_0 and has value $4\pi \times 10^{-7}$ H/m in SI units.

The ratio of μ/μ_0 is called the *relative permeability* (μ_r) of the solid, i.e.,

$$\mu_r = \frac{\mu}{\mu_0}. \tag{5.2}$$

(b) Magnetisation. A magnetic material when placed in a magnetic field gets magnetised.

The magnetic moment developed per unit volume in the material is called the intensity of magnetisation or magnetisation denoted by M. The unit of M is A/m.

If H be the applied magnetic field, then

$$\chi = \frac{M}{H} \tag{5.3}$$

is called the *magnetic susceptibility* and is defined as:

The magnetisation developed in the material per unit applied magnetic field. Since M and H are measured in A/m, χ is a *dimensionless quantity.*

From (5.1),

$$B = \mu H = \mu_0 \mu_r H \quad \left[\because \mu_r = \frac{\mu}{\mu_0}\right]$$

$$= \mu_0 \mu_r H + \mu_0 H - \mu_0 H = \mu_0 H + \mu_0 H(\mu_r - 1) = \mu_0 H + \mu_0 M,$$

where $M = H(\mu_r - 1)$,

$$\text{or } B = \mu_0(H + M). \tag{5.4}$$

$\mu_0 H$ and $\mu_0 M$ represent magnetic induction due to external field and due to the magnetisation respectively.

$$\text{or } \mu_0 = \frac{B}{H+M}.$$

∴ the relative permeability

$$\mu_r = \frac{\mu}{\mu_0} = \frac{B/H}{B/(H+M)} = \frac{H+M}{H} = 1 + \frac{M}{H} \text{ or } \mu_r = 1 + \chi \quad \left[\because \chi = \frac{M}{H}\right]. \tag{5.5}$$

In free space, i.e., in absence of material medium, $M = 0$, $\chi = 0$, $\mu = \mu_0$ and hence $\mu_r = 1$.

∴ from $B = \mu H = \mu_0 \mu_r H$, we get

$$B = \mu_0 H. \tag{5.6}$$

The unit of B in emu is gauss and in SI unit tesla, where 1 tesla = 10^4 gauss.

5.3 Classification of Magnetic Materials

In the atom the presence of permanent magnetic dipole and the interaction between them is responsible for the origin of different types of magnetic materials. The permanent dipole arises due to the *orbital and spin motion of the electron as well as the spins of the nucleus.* The electron orbit behaves as a current carrying coil and produces magnetic effect. For the production of permanent dipole moment the spins of unpaired valence electrons are mainly responsible. The orientation of the magnetic moments in the material may lead to a zero or non-zero resultant magnetic moments giving rise to various types of magnetism as follows: (i) *Diamagnetism* (ii) *Paramagnetism* (iii) *Ferromagnetism* (iv) *Antiferromagnetism* (v) *Ferrimagnetism.*

(i) *Diamagnetism.* Diamagnetic solids do not contain permanent magnetic moments, the presence of small non-zero magnetic moment is due to the orbital motion of the electrons and directed oppositely to the applied field. The diamagnetic effect is quite small.

Example. *Copper, Gold, Silver,* He, Ne, *Benzene, Naphthalene, etc.*

(ii) *Paramagnetism.* Paramagnetism arises due to the presence of permanent atomic or electronic magnetic moments. Paramagnetic effect is quite weak but unlike diamagnetism, the magnetic moment is aligned along the direction of the applied field.

Example. *All atoms and molecules with unfilled outer shells like Ti, etc., salts of transition elements, rare-earth metals and actinide series containing elements like* Cr^{3+}, Dy^{3+}, *etc. and metals like* Al.

(iii) *Ferromagnetism.* Ferromagnetic effect is very strong and arises due to the alignments of the neighbouring magnetic moments along the same direction in absence of magnetic field below *Curie temperature.* Above Curie temperature the Ferromagnetic materials behave as paramagnetic.

Example. Fe, Co, Ni.

(iv) *Antiferromagnetism.* In antiferromagnetism the adjacent moments being equal and opposite cancel each other completely and thus the total value over a finite volume is zero. Above *Néel temperature* antiferromagnetism vanishes and the material behaves as paramagnetic.

Example. MnO, NiO, MnTe, etc.

(v) *Ferrimagnetism.* In Ferrimagnetic materials, unlike antiferromagnetic, the number of neighbouring moments are unequal and hence though these are opposite, complete cancellation of moments do not take place. Hence, there is a net magnetic moment.

The magnetic ordering of different kinds of magnetic materials is depicted in Fig. 5.1. The ferro- and ferrimagnetic materials are consisting of *domains* of completely magnetised regions. It is interesting to note that *above certain critical temperature the distinction of the various magnetic materials disappears and all magnetic materials become paramagnetic.*

Example. Fe_3O_4, *Nickel ferrite and Yttrium iron garnet.*

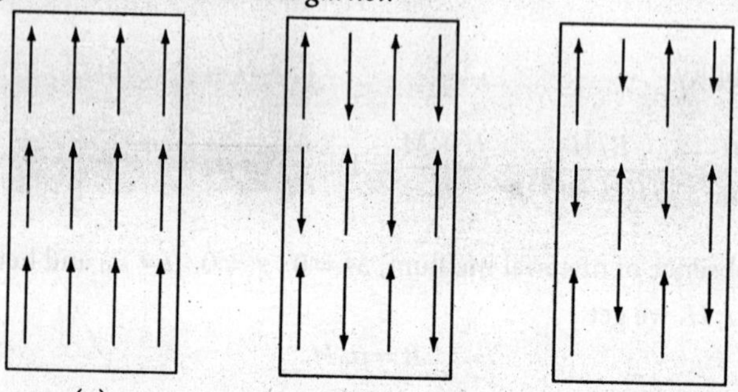

(a) (b) (c)

Fig. 5.1: Various types of magnetic ordering. Each arrow indicates a magnetic moment and its orientation.
(a) Ferromagnetic (b) Antiferromagnetic (c) Ferrimagnetic.

5.4 The Quantum Numbers

The magnetism in materials arises due to the orbital motion of the electron, the spin motion of the electron and the nuclear spin. To understand the origin of the magnetism, specification of the electron in the atom is essential which could be done by the following four quantum numbers.

1. *The Principal Quantum Number* (n). Corresponding to electronic energy levels K, L, M, N the principal quantum number of the electron $n = 1, 2, 3, 4, \ldots$. It determines the size of the electron orbit and the number of electron in a given shell with principal quantum number n, i.e., $N = n^2$.
2. *The Orbital Quantum Number* (l). For each value of n the orbital quantum number is $l = 0$ to $(n-1)$. Each value of l represents a sub-shell in the orbit. For $l = 0$, the sub-shell is called s sub-shell. Similarly, $p, d, f, \ldots$ sub-shells are for $l = 1$, $l = 2$, $l = 3$ and so on. The number of electrons in each sub-shell is $(2l+1)$ and the total angular momentum for a given l is

$$L_l = \hbar[l(l+1)]^2 \text{ with } \hbar = \frac{h}{2\pi}. \tag{5.7}$$

3. *The Orbital Magnetic Quantum Number* (m_l). An atom placed in an external magnetic field precesses about the field direction as axis and can have only specific orientations in space. The projection of L_l along the field direction is given by

$$L_{l,B} = L_l \cos\alpha, \tag{5.8}$$

where B is the magnetic field and α is the angle as shown in Fig. 5.2. Now $L_{l,B}$ can take only integral values in the multiples of $\hbar$, i.e.,

$$L_{l,B} = m_l \hbar = m_l \frac{h}{2\pi}. \tag{5.9}$$

m_l is called the *orbital magnetic quantum number* and can have values

$$m_l = -l, -(l-1), \ldots, 0, \ldots, (l-1), l$$
$$\text{or } m_l = 0 \pm 1, \pm 2, \ldots, \pm l. \tag{5.10}$$

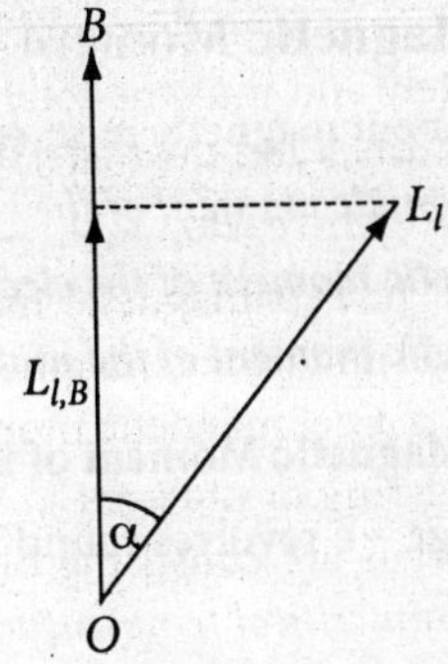

Fig. 5.2: Projection of L_l and the field direction (B).

Figs 5.3(a) and (b) show the possible orientations of L_l for an electron in p and d states respectively.

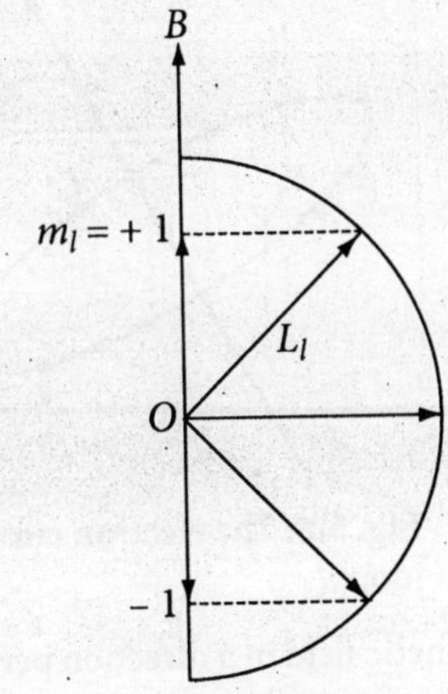

(a) p-state

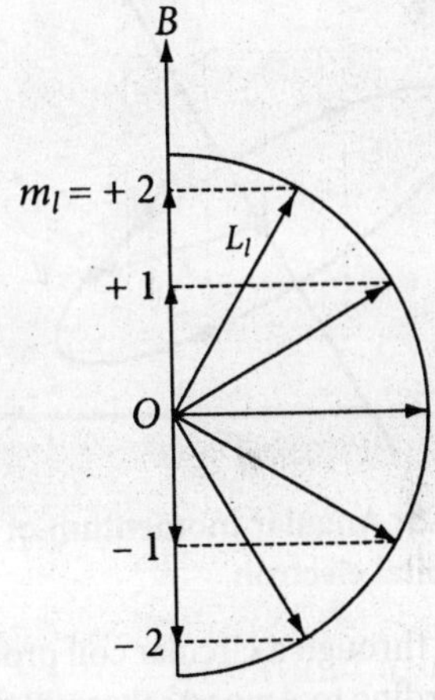

(b) d-state

Fig. 5.3: Possible orientations of L_l defined by l in magnetic field B.

4. *The spin magnetic quantum number* (m_s). The spin angular momentum arises due to the spin motion of the electron and has two values corresponding to spin magnetic quantum number, $m_s = \pm\frac{1}{2}$, i.e.,

$$S = m_s \hbar = m_s \frac{h}{2\pi}. \tag{5.11}$$

For clockwise or anticlockwise spin motion, the spin angular momentum points either upward or downward direction respectively as shown in Fig. 5.4.

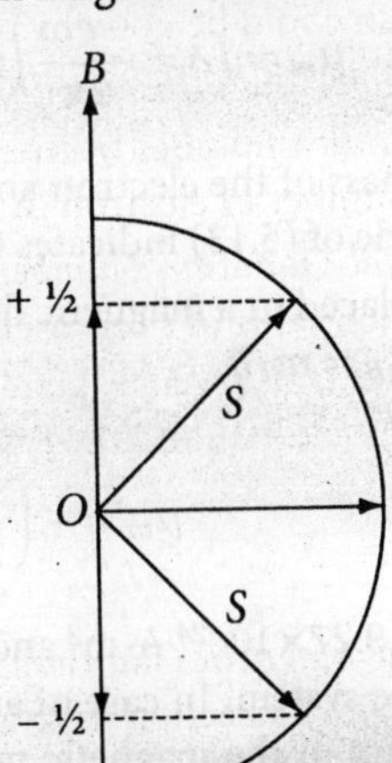

Fig. 5.4: Spin quantisation of the electron.

5.5 Origin of Magnetic Moment

The origins of permanent magnetic moment in an atom are as follows:

1. *The orbital magnetic moment of the electron arising due to its orbital motion around the nucleus.*
2. *The spin magnetic moment of the electron due to its spin motion about its own axis.*
3. *The spin magnetic moment of the nucleus due to spin motion about its own axis.*

5.5.1 The Orbital Magnetic Moment of the Electrons

If an electron of charge $-e$ revolves round a nucleus in a circular orbit of radius r, then the current in the orbit is

$$I = \frac{-e}{T} = \frac{-e}{2\pi/\omega} = -\frac{e\omega}{2\pi}, \tag{5.12}$$

ω is the *angular velocity* of the electron.

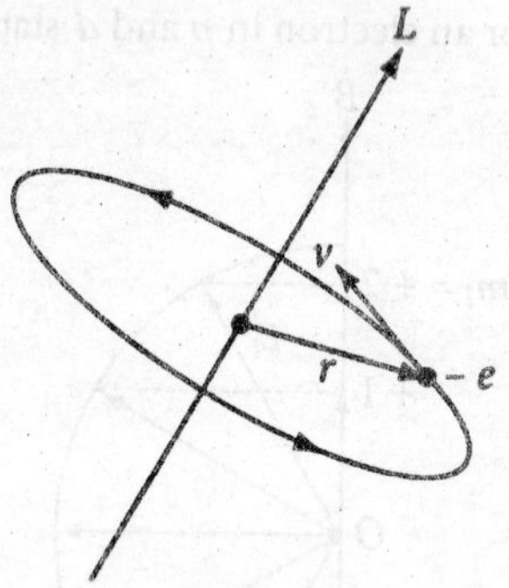

Fig. 5.5: Angular momentum of an orbital electron.

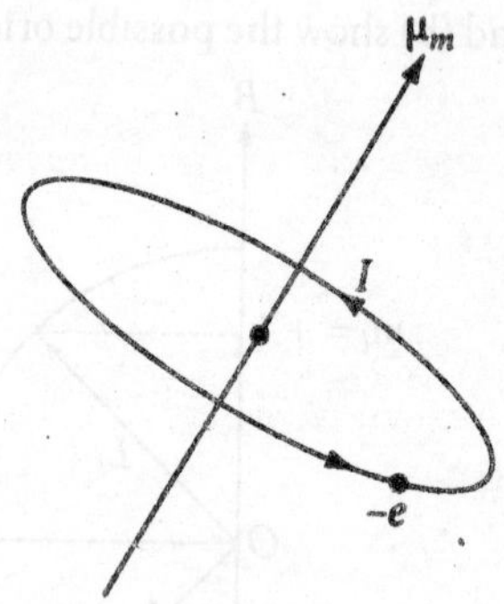

Fig. 5.6: The electron current loop.

The current flowing through a circular coil produces a magnetic field in a direction perpendicular to the plane of the coil. According to *Ampere's theorem* the current carrying coil is equivalent to a magnetic dipole of moment [Fig. 5.6],

$$\mu_m = IA = \frac{-e\omega}{2\pi}\left(\pi r^2\right) = -\frac{e\omega r^2}{2} = -\frac{e}{2m}\left(m\omega r^2\right) = -\left(\frac{e}{2m}\right)L, \tag{5.13}$$

where m is the mass of the electron and $L = m\omega r^2$ is its *orbital angular momentum*. The negative sign in the right-hand side of (5.13) indicates that μ_m is *anti-parallel to L.*

If the atom is placed in a magnetic field, then the possible orientation of the angular momentum vector (L) is given by $L_{l,B} = m_l \hbar$.

∴ from (5.13),

$$\mu_m = -\left(\frac{e}{2m}\right) m_l \hbar = -\left(\frac{e\hbar}{2m}\right) m_l = -\mu_B m_l, \tag{5.14}$$

where $\mu_B = \frac{e\hbar}{2m} = 9.27\times10^{-24}$ A-m^2 and is called the *Bohr magneton*. It is also the accepted unit of magnetic moment in atomic system. In case of an atom having many electrons, the total orbital magnetic moment is the algebraic sum of the magnetic moments of individual electrons. The magnetic moment (μ_m) of a completely filled shell is zero but for an atom with partially filled shells will have non-zero orbital magnetic moment (μ_l).

Hund's Rule

Since the filled electron shells do not contribute to the *magnetic moment* of the atom, it arises from the partially filled shells. This is in accordance with the *Pauli's principle,* that no two electrons can have all the quantum numbers n, l, m_l and m_s identical. For the *ground state* of such an atom, *Hund's rule* for the *incomplete shells* may be stated as follows:

1. *The spins of the electron add to give the maximum possible values of S consistent with the Pauli's principle.*
2. *The orbital moments combine to give a maximum value of L that is consistent with* (1) above.
3. *For an incomplete shells,*

$J = L - S$, when the shell is *less than the half* occupied and

$J = L + S$, when the shell is *more than the half* occupied.

Illustration of the application of Hund's Rule

Example 1. CO^{2+} ion in a $3d^7$ state ($l = 2$)

$$\begin{array}{lcccccccccccc} m_s & \rightarrow & \frac{1}{2} & \frac{1}{2} & \frac{1}{2} & \frac{1}{2} & \frac{1}{2} & -\frac{1}{2} & -\frac{1}{2} & -\frac{1}{2} & -\frac{1}{2} & -\frac{1}{2} \\ m_l & \rightarrow & 2 & 1 & 0 & -1 & -2 & 2 & 1 & 0 & -1 & -2 \end{array}$$

Similarly, for the seven electrons from the left, $S = 3/2; 2S+1 = 2\times\frac{3}{2}+1 = 4; L = 3$.

Example 2. Dy^{3+} ion in a $4f^9$ state ($l = 3$)

$$\begin{array}{lcccccccccccccccc} m_s & = & \frac{1}{2} & \frac{1}{2} & \frac{1}{2} & \frac{1}{2} & \frac{1}{2} & \frac{1}{2} & \frac{1}{2} & -\frac{1}{2} & -\frac{1}{2} & -\frac{1}{2} & -\frac{1}{2} & -\frac{1}{2} & -\frac{1}{2} & -\frac{1}{2} \\ m_l & = & 3 & 2 & 1 & 0 & -1 & -2 & -3 & 3 & 2 & 1 & 0 & -1 & -2 & -3 \end{array}$$

Similarly, if the nine electrons from the left, $S = 5/2; 2S+1 = 2\times\frac{5}{2}+1 = 6; L = 5$.

Combination of $\vec{L}$ and $\vec{S}$

As the electron is negatively charged $\vec{\mu}_l$ and $\vec{\mu}_s$ are opposite in direction to $\vec{L}$ and $\vec{S}$ respectively. The resultant vector $\vec{J}$ of $\vec{L}$ and $\vec{S}$ is given by $\vec{J} = \vec{L} + \vec{S}$.

Under the action of an external magnetic field $\vec{B}$, $\vec{L}$ tends to precess round $\vec{J}$ with *Larmour frequency* $\vec{\omega}_L$. $\vec{S}$ will precess round $\vec{J}$ with frequency $2\vec{\omega}_L$. If $\vec{B}$ is small, then the coupling between $\vec{L}$ and $\vec{S}$ will be strong *R-S* coupling and $\vec{J}$ will be maintained and will precess around $\bar{B}$ with a compromise frequency. As $\vec{L}, \vec{S}, \vec{\mu}_l$ and $\vec{\mu}_s$ all precess around $\vec{J}$, their normal components to $\vec{J}$ will average to zero. Under such conditions, the resultant moment $\vec{\mu}_j$ will be the sum of their components parallel to $\vec{J}$.

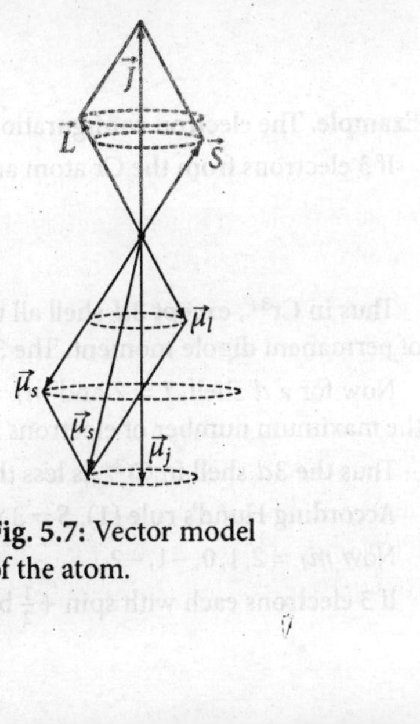

Fig. 5.7: Vector model of the atom.

Now, the component of μ_l parallel to $\vec{J}$ is

$$\mu_l \cos\left(\vec{L}, \vec{J}\right) = L\frac{e}{2m}\cos\left(\vec{L}, \vec{J}\right). \tag{5.15}$$

Component of $\vec{\mu}_s$ parallel to $\vec{J}$ is

$$S\times\frac{2e}{m}\cos\left(\vec{S},\vec{J}\right). \tag{5.16}$$

$$\therefore\ \mu_j=\left[L\cos\left(\vec{L},\vec{J}\right)+2S\cos\left(\vec{S},\vec{J}\right)\right]\frac{e}{2m}. \tag{5.17}$$

From Fig. 5.8,

$$S^2=L^2+J^2-2LJ\cos\left(\vec{L},\vec{J}\right) \tag{5.18}$$

$$\text{or } L\cos\left(\vec{L},\vec{J}\right)=\frac{J^2+L^2-S^2}{2J}. \tag{5.19}$$

Similarly,

$$S\cos\left(\vec{S},\vec{J}\right)=\frac{J^2-L^2+S^2}{2J}. \tag{5.20}$$

Substituting (5.19) and (5.20) in (5.17), we get

$$\mu_j=\left[1+\frac{J^2-L^2+S^2}{2J^2}\right]J\frac{e}{2m}=\left[1+\frac{J(J+1)+S(S+1)-L(L+1)}{2J(J+1)}\right]\sqrt{J(J+1)}\frac{3\hbar}{2m}$$

$$=g\mu_B\sqrt{J(J+1)}, \tag{5.21}$$

where $\mu_B=\frac{e\hbar}{2m}$ is the Bohr magneton and

$$g=\frac{1+J(J+1)+S(S+1)-L(L+S)}{2J(J+1)}, \tag{5.22}$$

where g is the *Landé g-factor*.

When $\vec{B}=0$, all the atoms have the same value of μ_j. But when $\vec{B}$ is applied the energy depends on the statistical occupation of the $(2J+1)$ sub-shells and all the atoms will have M_J value different from μ_j. The magnetic energy of a given atom is then

$$E_M=\vec{\mu}_j\cdot\vec{B}=-g\mu_B M_J B. \tag{5.23}$$

Example. The electron configuration of Cr atom ($Z=24$) is $1s^2 2s^2 2p^6 3s^2 3p^6 3d^5 4s^1$.

If 3 electrons from the Cr atom are removed, it becomes C_r^{3+} whose electronic configuration is

$$1s^2 2s^2 2p^6 3s^2 3p^6 3d^3.$$

Thus in Cr^{3+}, except $3d$ shell all the shells are filled and hence the $3d$ shell is responsible for the origin of permanent dipole moment. The $3d$ shell has 3 electrons.

Now for a d shell, $l=2$ and $m_l=2,1,0,-1,-2$, each can accommodate 2 electrons with $m_s=\pm\frac{1}{2}$. So the maximum number of electrons in the $3d$ shell is $5\times2=10$.

Thus the $3d$ shell in Cr^{3+} is less than half occupied (as the atomic number $Z=24$).

According Hund's rule (1), $S=3\times\frac{1}{2}=\frac{3}{2}$ as there are three electrons in the $3d$ shell.

Now $m_l=2,1,0,-1,-2$.

If 3 electrons each with spin $+\frac{1}{2}$ be placed in the first three of these, then $L=2+1+0=3$.

Hence, according to *Hund's rule* (3),

$$J = L - S = 3 - \tfrac{3}{2} = \tfrac{3}{2} \quad \left[\because L = 3 \text{ and } S = \tfrac{3}{2}\right]$$

$\therefore$ the *Landé splitting factor* is

$$g = 1 + \frac{J(J+1) + S(S+1) - L(L+1)}{2J(J+1)} = 1 + \frac{\frac{3}{2}\left(\frac{3}{2}+1\right) + \frac{3}{2}\left(\frac{3}{2}+1\right) - 3(3+1)}{2\cdot\frac{3}{2}\left(\frac{3}{2}+1\right)}$$

$$= 1 + \frac{\frac{3}{2}\times\frac{5}{2} + \frac{3}{2}\times\frac{5}{2} - 3\times 4}{3\times\frac{5}{2}} = 1 + \frac{\frac{15}{4} + \frac{15}{4} - 12}{\frac{15}{2}} = 0.4.$$

5.5.2 Spin Angular Momentum of the Electrons

The spin magnetic moment of the electron is

$$\mu_s = -\left(\frac{e}{2m}\right)S = -\left(\frac{e}{2m}\right)\times m_s\hbar \quad [\because S = m_s\hbar]$$

$$= \pm\frac{1}{2}\left(\frac{e}{2m}\right)\hbar \quad \left[\because m_s = \pm\frac{1}{2}\right]$$

$$= \pm\frac{1}{2}\mu_B. \tag{5.24}$$

Thus the magnitude of spin magnetic moment is half of Bohr magneton μ_B.

The magnetic moment component due to electron spin along the field direction is

$$\mu_{sz} = g\left(\frac{e}{2m}\right)m_s\hbar = g\left(\frac{e\hbar}{2m}\right)m_s, \tag{5.25}$$

where g is called the *Landé splitting factor* or *spectroscopic splitting factor* and is given by

$$g = 1 + \frac{J(J+1) + S(S+1) - L(L+1)}{2J(J+1)}, \tag{5.26}$$

where S is the spin angular momentum, L is the orbital angular momentum and J is the total angular momentum of the electron.

For $m_s = \frac{1}{2}$,

$$\mu_{sz} = g\frac{e\hbar}{4m} = g\left(\frac{e}{2m}\right)\frac{\hbar}{2}.$$

For the electron spin, $g = 2.0023$.

$\therefore$ $M_{sz} = \mu_B$, i.e., the dipole moment due to electron spin is equal to *Bohr magneton* in the direction of the applied field B.

$\therefore$ the total angular momentum of the atom is

$$\mu = gJ\mu_B, \tag{5.27}$$

where $J\mu_B = \left(\frac{e}{2m}\right)\frac{\hbar}{2}$.

From the knowledge of g and $J\mu_B$ we can find the total magnetic moment of the atom.

5.5.3 Why g is called *Splitting Factor*?

Let an electron with spin $\frac{1}{2}$ but no angular momentum be subjected to an external magnetic field $\vec{B}$. Then for $m_s = \frac{1}{2}$, $|\mu_{sz}| = \mu_B$, the direction being anti-parallel to $\vec{B}$. But for $m_s = -\frac{1}{2}$, $|\mu_{sz}| = \mu_B$, the direction being parallel to $\vec{B}$. The PE of magnetic dipole of moment $\vec{\mu}$ when placed in a magnetic field $\vec{B}$ is $-\vec{\mu}\cdot\vec{B}$ and the energy level of the spin $\frac{1}{2}$ electron is split up as shown in Fig. 5.8.

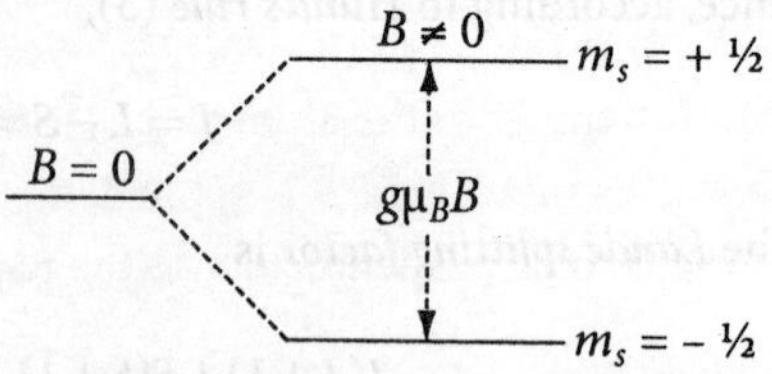

Fig. 5.8: Splitting of energy level of the spin ½ electron for $B = 0$.

The separation of the split-up energy levels is

$$\Delta E = 2|\mu_{sz}|B = g\left(\frac{e}{2m}\right)\hbar B = g\mu_B B \quad \left[\because \frac{e\hbar}{2m} = \mu_B\right].$$

So the split-up of the energy levels is determined by g and hence the name *splitting factor.*

5.5.4 Nuclear Spin Magnetic Moment

Due to its spin motion, the nucleus is associated with a magnetic moment and is called the *nuclear magnetic moment.* The nuclear magnetic moment is expressed in terms of nuclear magneton as

$$\mu_n = \frac{e\hbar}{2M_p} = 5.05 \times 10^{-29} \text{ A-m}^2, \tag{5.28}$$

where M_p is the mass of a proton.

5.6 Diamagnetism

In an atom the electrons revolve in closed orbits around the nucleus. The electron orbit may be considered as a current carrying closed coil producing magnetic field perpendicular to its plane. The electrons in an atom revolve in discrete orbits whose planes are not parallel to each other and direction of rotation of electrons is not the same. That is why *the vector sum of the magnetic moments due to all the revolving electrons becomes zero.*

If the electric circuit is subjected to a changing magnetic flux, an induced current is set up in it which, according *Lenz's law,* opposes the change in flux. The induced magnetic field in the coil opposes the applied field and produces *diamagnetism.* Diamagnetism is produced in all materials but it is suppressed by the stronger *paramagnetic, Ferromagnetic effects.*

5.6.1 Langevin's Classical Theory of Diamagnetism

We consider an electron revolving around the nucleus in a circular orbit with frequency ω_0 in absence of any external magnetic field. Then the centrifugal force $\frac{mv^2}{\rho}$ is balanced by the *Coulomb force* of attraction between the nucleus and the electron.

$$\therefore \frac{mv^2}{\rho} = \frac{Ze^2}{4\pi\varepsilon_0\rho^2} \text{ or } m\rho\omega_0^2 = \frac{Ze^2}{4\pi\varepsilon_0\rho^2}, \; [\because v = \rho\omega_0] \tag{5.29}$$

$$\therefore \omega = \sqrt{\frac{Ze}{4\pi m\varepsilon_0\rho^3}}. \tag{5.30}$$

Here m is the mass of the electron, ρ is the radius of the electron orbit, e is the electronic charge, Z is the atomic number of the atom and ε_0 is the permittivity of the free space.

If now a magnetic field $\vec{B}$ is applied, the additional *Lorentz force*, $\vec{F} = e\left(\vec{v}\times\vec{B}\right)$ will act on the electron. Assuming $\vec{B}$ to be perpendicular to the plane of the orbit, the magnitude of this force is

$$F = e\rho\omega B, \tag{5.31}$$

where ω is the frequency of the electron in presence of the field.

Then

$$m\omega^2\rho = \frac{Ze^2}{4\pi\varepsilon_0\rho^2} - e\rho\omega B = m\omega_0^2\rho - e\rho\omega B\left[\because \frac{Ze^2}{4\pi\varepsilon_0\rho^2} = m\rho\omega_0^2\right]. \tag{5.32}$$

The minus sign with the Lorentz force is due to the negative charge of the electron.

Thus from (5.32),

$$\omega^2 + \frac{eB}{m}\omega - \omega_0^2 = 0 \text{ or } \omega = \frac{1}{2}\left[-\frac{eB}{m} \pm \sqrt{\left(\frac{eB}{m}\right)^2 + 4\omega_0^2}\right] = \pm\sqrt{\omega_0^2 + \left(\frac{eB}{2m}\right)^2} - \frac{eB}{2m}.$$

For $\frac{eB}{2m} \ll \omega_0$, because $\omega_0 \simeq 10^{14}$ to 10^{15} rad/s and $B \simeq 1$ Wb/m²,

$$\omega \simeq \pm\omega_0 - \frac{eB}{2m}. \tag{5.33}$$

(5.33) shows that the frequency of the electron changes by $\left(\frac{eB}{2m}\right)$ in presence of the magnetic field B. The $\pm$ sign in (5.33) signifies that *the electrons with orbital moments parallel to the field are slowed down, whereas with moments anti-parallel to the field are speeded up*. This is called *Larmor theorem*. The quantity $\frac{eB}{2m} = \omega_L$ is called the *Larmor frequency*.

This change in frequency gives rise to an additional current,

$$I = (-e)\left(\frac{1}{2\pi}\frac{eB}{2m}\right) = -\frac{e^2B}{4\pi m}. \tag{5.34}$$

The magnetic moment associated with the current loop having one electron is

$$\vec{\mu}_e = I\times\vec{a} = \frac{e^2\vec{B}}{4\pi m}\times\pi\rho^2 = -\frac{e^2\vec{B}}{4m}\rho^2, \tag{5.35}$$

where $a = \pi\rho^2$ is the area of the current loop of radius ρ.

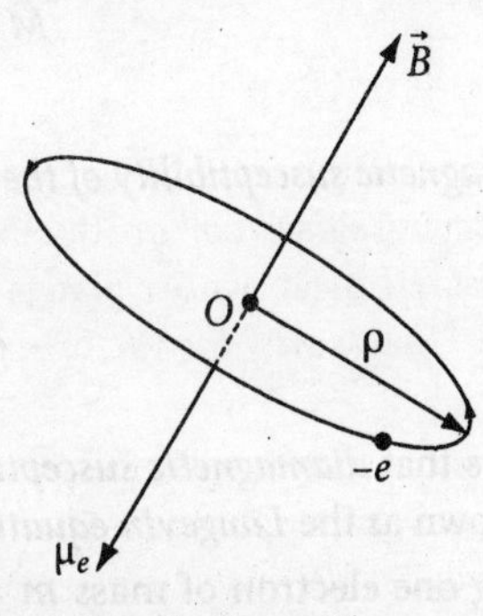

Fig. 5.9: μ_e is directed opposite to the magnetic field $\vec{B}$.

The significance of the negative sign on the right side of (5.35) is that, the induced magnetic moment μ_e is directed opposite to the magnetic field as shown in Fig. 5.9.

If the electron orbit is not perpendicular to the magnetic field $\vec{B}$, then it precesses about $\vec{B}$ with frequency ω_L. In this case the induced magnetic moment μ_e will have a component opposite to the direction of $\vec{B}$. If the orbits of the electrons of an atom are randomly oriented, then the total induced magnetic moment is given by

$$\mu_a = -\frac{e^2\vec{B}}{4m}\sum_{i=1}^{n'}\left\langle\rho_i^2\right\rangle, \tag{5.36}$$

where n' = the total number of electrons in the atom.

If O be the origin of a Cartesian coordinate system OX, OY and OZ coinciding with the centre of the circular orbits and $\vec{B}$ be along the Z-axis, then

$$\langle \rho^2 \rangle = \langle x^2 \rangle + \langle y^2 \rangle, \tag{5.37}$$

where x, y, z represent the coordinates of a point on the orbit of radius ρ.

The *mean squared distance* of the electrons from the nucleus is

$$\langle r^2 \rangle = \langle x^2 \rangle + \langle y^2 \rangle + \langle z^2 \rangle. \tag{5.38}$$

If the charge distribution is assumed to be *spherically symmetric*, then

$$\langle x^2 \rangle = \langle y^2 \rangle = \langle z^2 \rangle$$
$$\therefore \langle r^2 \rangle = \tfrac{3}{2} \langle \rho^2 \rangle. \tag{5.39}$$

$\therefore$ (5.36) becomes

$$\vec{\mu}_a = -\frac{e^2 \vec{B}}{6m} \sum_{i=1}^{n'} \langle r^2 \rangle = -\frac{\mu_0 e^2 \vec{H}}{6m} \sum_{i=1}^{n'} \langle r^2 \rangle \; [\because \vec{B} = \mu_0 \vec{H}]. \tag{5.40}$$

If the magnetic solid contains N atoms/volume and the number of electrons in each atom is Z, then the magnetic moment per unit volume is

$$\vec{M} = -N\vec{\mu}_a = -\frac{N\mu_0 e^2 \vec{H}}{6m} \sum_{i=1}^{n'} \langle r^2 \rangle. \tag{5.41}$$

The *diamagnetic susceptibility of the material* is

$$\chi_{\text{dia}} = \frac{M}{H} = -\frac{N\mu_0 e^2}{6m} \sum_{i=1}^{n'} \langle r^2 \rangle. \tag{5.42}$$

(5.42) shows that *diamagnetic susceptibility is negative and is independent on temperature of the material.* (5.42) is known as the *Langevin equation* for diamagnetism.

Assuming one electron of mass $m = 9.1 \times 10^{-31}$ kg, $\langle r^2 \rangle = 0.1\,\text{mm}^2$, $N = 5 \times 10^{28}/\text{m}^3$ and $e = 1.6 \times 10^{-19}$ C we get

$$\chi_{\text{dia}} = \frac{-(5 \times 10^{28}) \times (4\pi \times 10^{-7}) \times (1.6 \times 10^{-19})^2 \times (0.1 \times 10^{-9})^2}{6 \times (9.1 \times 10^{-31})} = \mathbf{-3 \times 10^{-6}}. \tag{5.43}$$

The experimental value of χ_{dia} for copper at room temperature is -9×10^{-6}. This shows that there is a close agreement between theoretical and experimental values of χ_{dia}.

The *molar susceptibility* $\chi_{\text{dia}}^{(\text{molar})}$ may be obtained by multiplying χ_{dia} by the volume of a mole (N_A/N), $N_A = 6.022 \times 10^{23}$ is the *Avogadro number.*

So

$$\chi_{\text{dia}}^{(\text{molar})} = -\frac{N_A}{N} \times \frac{N\mu_0 e^2}{6m} \sum_{i=1}^{n'} \langle r^2 \rangle = -\frac{N_A \mu_0 e^2}{6m} \sum_{i=1}^{n'} \langle r^2 \rangle \tag{5.44}$$

The *atomic radius is defined* as

$$\langle\rho^2\rangle_a = \frac{1}{n'}\sum_{i=1}^{n'}\langle r^2\rangle, \quad \therefore \chi_{\text{dia}}^{(\text{molar})} = -\frac{n'e^2 N_A\mu_0}{6m}\langle\rho^2\rangle_a. \tag{5.45}$$

In case of He-*atom*, $n'=2$,

$$\chi_{\text{dia}}^{(\text{molar})} = -1.9\times10^{-6}\,\text{cm}^3/\text{mol}.$$

$$\therefore \sqrt{\langle\rho^2\rangle_a} = 1.64\times10^{-9}\,\text{cm}$$

is the *radius of the* He-*atom*.

Conclusions. Following conclusions can be drawn from the *Langevin's theory of diamagnetism*:

1. Diamagnetic susceptibility χ_{dia} is independent of temperature of the solid as the orbital areas of the electrons are independent of temperature.
2. Since $\chi_{\text{dia}} \propto \langle\rho^2\rangle_a$, the outer electrons of the atoms contribute in a large scale.
3. The diamagnetic susceptibility increases with the increase in number of atoms.

It is to be noted that, an atom having permanent dipole moment will also possess the diamagnetic property. The permanent magnetic moment in this case tends to rotate into the direction of $\vec{B}$ and the paramagnetic effect results and the diamagnetic effect disappears.

Apart from Cu, Au, Ge, Si, NaCl, etc. *inert gases* are also diamagnetic as they have completely filled electron shells in their atoms.

5.6.2 Quantum Theory of Diamagnetism

The Hamiltonian $\mathcal{H}$ of an electron changes to $\mathcal{H}+\mathcal{H}'$ in presence of a magnetic induction $\vec{B}$, where $\mathcal{H}'$ represents a small perturbation.

Using *quantum mechanics* we can show that

$$\mathcal{H}' = -\frac{ie\hbar}{2m}\left(\nabla\cdot\vec{A}+2\vec{A}\cdot\nabla\right)+\frac{e^2}{2m}A^2, \tag{5.46}$$

where $\vec{A}$ is the *magnetic vector potential* given by $\vec{B}=\nabla\times\vec{A}$.

If $\vec{B}$ is uniform, then it can be expressed as

$$\vec{A} = \tfrac{1}{2}\vec{B}\times\vec{r} = \tfrac{1}{2}\left[\hat{i}\left(B_y z - B_z y\right)+\hat{j}(B_z x - B_x z)+\hat{k}\left(B_x y - B_y x\right)\right]. \tag{5.47}$$

If $\vec{B}$ is along the Z-direction, then $B_x = B_y = 0$; $B_z = B$.

$$\therefore A_x = -\tfrac{1}{2}yB,\ A_y = \tfrac{1}{2}xB \text{ and } A_z = 0.$$

Also

$$\nabla\cdot\vec{A} = \frac{\partial A_x}{\partial x}+\frac{\partial A_y}{\partial y}+\frac{\partial A_z}{\partial z} = 0 \tag{5.48}$$

$$\text{and } \vec{A}\cdot\nabla = A_x\frac{\partial}{\partial x}+A_y\frac{\partial}{\partial y}+A_z\frac{\partial}{\partial z} = \tfrac{1}{2}B\left(x\frac{\partial}{\partial x}-y\frac{\partial}{\partial y}\right). \tag{5.49}$$

Now (5.46) becomes

$$\mathcal{H}' = -\frac{ie\hbar B}{2m}\left(x\frac{\partial}{\partial x}-y\frac{\partial}{\partial y}\right)+\frac{e^2B^2}{8m}\left(x^2+y^2\right). \tag{5.50}$$

The *quantum mechanical angular momentum operator* is

$$\vec{L} = -i\hbar\vec{r}\times\nabla$$

$$\text{or } \hat{i}L_x + \hat{j}L_y + \hat{k}L_z = -i\hbar\left[\hat{i}\left(y\frac{\partial}{\partial z} - z\frac{\partial}{\partial y}\right) + \hat{j}\left(z\frac{\partial}{\partial x} - x\frac{\partial}{\partial z}\right) + \hat{k}\left(x\frac{\partial}{\partial y} - y\frac{\partial}{\partial x}\right)\right]. \quad (5.51)$$

$$\therefore \quad L_z = k\left(x\frac{\partial}{\partial y} - y\frac{\partial}{\partial x}\right). \quad (5.52)$$

∴ from (5.50) and (5.52) it is clear that the first term on the right-hand side of (5.50) is proportional to L_z.

If now N be the number of atoms per unit volume and n', the number of electrons in each atom, then the second term of (5.50) becomes

$$\frac{Ne^2B^2}{8m}\sum_{i=1}^{n'}\left(x_i^2 + y_i^2\right) = \frac{Nn'e^2B^2}{8m}\left\langle\rho^2\right\rangle \quad (5.53)$$

where $\langle\rho^2\rangle$ is the mean squared radius of the projection of the orbit on a plane perpendicular to the magnetic field $\vec{B}$. If $\vec{B}$ induces a dipole moment in the specimen, the corresponding energy term should contain B^2. Hence, $\frac{Nn'e^2B^2}{8m}\langle\rho^2\rangle$ is the energy term associated with the diamagnetism of the material. But the diamagnetic energy is given by $\frac{1}{2}\vec{M}\cdot\vec{B} = -\frac{1}{2\mu_0}\chi_{\text{dia}}B^2$.

$$\therefore \frac{Nn'e^2B^2}{8m}\left\langle\rho^2\right\rangle = -\frac{1}{2\mu_0}\chi_{\text{dia}}B^2 \text{ or } \chi_{\text{dia}} = -\frac{Nn'\mu_0e^2}{4m}\left\langle\rho^2\right\rangle \quad (5.54)$$

$$\text{or } \chi_{\text{dia}} = -\frac{Nn'\mu_0e^2}{6m}\left\langle r^2\right\rangle \quad (5.55)$$

$$\left[\because \left\langle r^2\right\rangle = \tfrac{3}{2}\left\langle\rho^2\right\rangle\right]$$

Here $\langle r^2\rangle$ is the mean squared distance of the electron from the nucleus. Determining $\langle r^2\rangle$ we can calculate the value of χ_{dia} of the material. In quantum mechanical treatment $\langle r^2\rangle$ has different significance from that in classical treatment.

The experimental value of χ_{dia} agrees fairly with theoretical result specially for light elements and rare gases. For He-atom $\chi_{\text{dia}} = -1.9\times10^{-6}$ indicating that *diamagnetism is a very weak effect.*

5.7 Paramagnetism

The paramagnetic property of the materials arise due to the *permanent magnetic moments* of the constituent atoms or molecules. These magnetic moments orient randomly and the net magnetic moment becomes zero. But when a magnetic field is applied these moments orient themselves along the field direction and a net magnetisation along the field direction is produced.

The susceptibility of the paramagnetic materials is small but positive and depends on the temperature of the materials. In paramagnetic materials the permanent magnetic moments arise from (i) *the orbital motion of the electrons* (ii) *the spin and intrinsic motion of the electrons* and (iii) *the spin magnetic moment of the nucleus.*

Some examples of paramagnetic materials and their susceptibilities at room temperature are shown in Table 5.1.

Table 5.1

Material	$\chi = \mu_r - 1$	*Material*	$\chi = \mu_r - 1$
Oxygen	0.19×10^{-5}	$CuSO_4$, H_2O	2.0×10^{-3}
Platinum	35.5×10^{-5}	$MnSO_4$	3.6×10^{-3}
Air	0.035×10^{5}	Fe_2O_3	1.4×10^{-3}
Nitrogen	0.0012×10^{-5}	$FeCl_2$	3.7×10^{-3}
Tungsten	7.5×10^{-5}	$FeSO_4$	2.8×10^{-3}
CoO	585×10^{-5}	$NiSO_4$	1.2×10^{-3}
$CrCl_2$	1.5×10^{-3}		
Cr_2O_3	1.7×10^{-3}		

5.7.1 Langevin's Classical Theory of Paramagnetism

Langevin's *classical theory* of paramagnetism is based on the following assumptions:

1. *The permanent dipoles of the paramagnetic material may have all possible orientations.*
2. *The dipoles are so separated that their mutual interaction force may be neglected.*

If a magnetic dipole of material $\vec{\mu}$ be placed in a magnetic field $\vec{B}$ [Fig. 5.10], then *torque* experienced by the dipole is given by

$$\vec{\tau} = \vec{\mu} \times \vec{B}. \tag{5.56}$$

This torque tends to establish an order by rotating the dipole in the direction of $\vec{B}$. This action is opposed by thermal agitation energy and tends to establish disorder by randomising the orientations of the dipoles.

The PE of the dipole is

$$E = -\vec{\mu} \cdot \vec{B} = -\mu B \cos\theta, \tag{5.57}$$

where θ is the angle between $\vec{\mu}$ and $\vec{B}$.

According to *Maxwell-Boltzmann statistics,* the probability of finding the dipole at the angle θ is $p(\theta)$ given by

$$p(\theta) \propto e^{-(\vec{\mu}\cdot\vec{B})/kT} \text{ or } p(\theta) \propto e^{-\mu B \cos\theta/kT},$$

where k is the *Boltzmann constant* and T is the absolute temperature of the material.

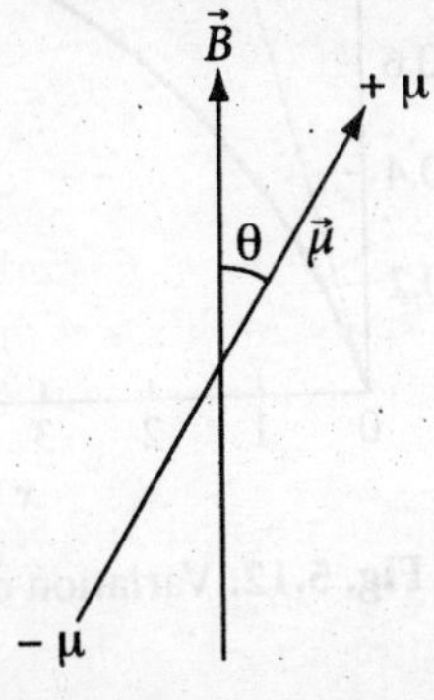

Fig. 5.10: Magnetic dipole in magnetic field $\vec{B}$.

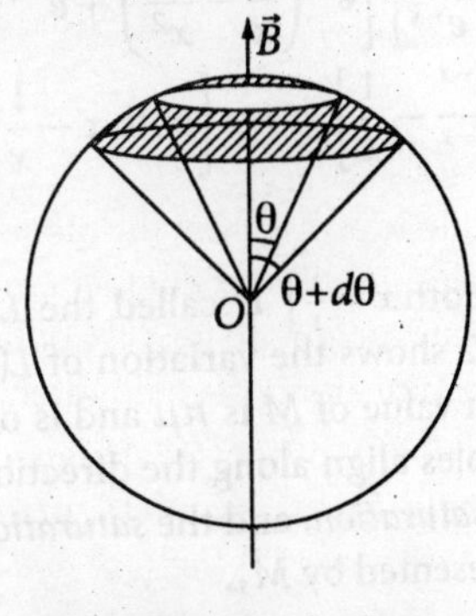

Fig. 5.11: Dipoles lying between θ and (θ + dθ).

If $n(\theta)\,d\theta$ be the number of dipole moments in a direction between θ and $\theta + d\theta$, then

$$n(\theta)\,d\theta \propto p(\theta)\,d\Omega,$$

where $d\Omega$ is the *solid angle* between two hollow cones with semi-vertical angles θ and $\theta + d\theta$ [Fig. 5.11].

$$\therefore n(\theta)\,d\theta = 2\pi A \sin\theta\, e^{-\frac{\mu B\cos\theta}{kT}}\, d\theta, \tag{5.58}$$

where $d\Omega = 2\pi \sin\theta\, d\theta$ and A is a constant of proportionality

$$\text{or } n(\theta)\,d\theta = C_1 e^{-x\cos\theta} \sin\theta\, d\theta, \tag{5.59}$$

where $C_1 = 2\pi A$ and $x = \frac{\mu B}{kT}$.

If n be the number of dipole moments per unit volume, then

$$n = \int_0^{\pi} n(\theta)\,d\theta = C_1 \int_0^{\pi} e^{x\cos\theta} \sin\theta\, d\theta = C_1 \int_{+1}^{-1} e^{-xy}\, dy = \frac{C_1}{x}(e^x - e^{-x})$$

$$\text{or } C_1 = \frac{xn}{e^x - e^{-x}}. \tag{5.60}$$

Each of the $n(\theta)\,d\theta$ dipoles has a component of magnetic moment $\mu\cos\theta$ in the direction of the applied field $\vec{B}$, whereas the component perpendicular to $\vec{B}$ cancels each other. Hence, the total magnetic moment of the n dipoles, i.e., the *magnetisation* (M) is

$$M = \int_0^{\pi} (\mu\cos\theta)\, n(\theta)\,d\theta = \int_0^{\pi} \mu\cos\theta\, C_1 e^{x\cos\theta} \sin\theta\, d\theta$$

$$= -\mu C_1 \int_1^{-1} y e^{xy}\, dy = -\mu C_1 \left[e^x \left(\frac{1}{x} - \frac{1}{x^2} \right) + e^{-x} \left(\frac{1}{x} + \frac{1}{x^2} \right) \right]. \tag{5.61}$$

Substituting the value of C_1 from (5.60) in (5.61), we get

$$M = -\mu \times \frac{xn}{(e^x - e^{-x})} \left[e^x \left(\frac{1}{x} - \frac{1}{x^2} \right) + e^{-x} \left(\frac{1}{x} + \frac{1}{x^2} \right) \right]$$

$$= \mu n \left[\frac{e^x + e^{-x}}{e^x - e^{-x}} - \frac{1}{x} \right] = n\mu \left[\coth x - \frac{1}{x} \right]$$

$$= n\mu L(x), \tag{5.62}$$

where $L(x) = \left[\coth x - \frac{1}{x}\right]$ is called the *Langevin function.* Fig. 5.12 shows the variation of $L(x)$ with x. The maximum value of M is $n\mu$ and is obtained when all the dipoles align along the direction of $\vec{B}$. This is *magnetic saturation,* and the *saturation magnetisation* is represented by M_s.

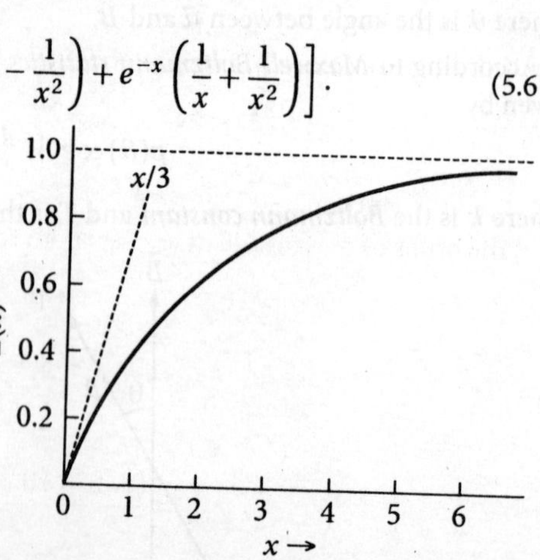

Fig. 5.12: Variation of $L(x)$ with x.

$$\therefore M = M_s L(x). \tag{5.63}$$

Now the following *two* cases arise:

Case I. For $x \ll 1$, i.e., at *normal field strength* and *ordinary temperature*,

$$L(x) = \coth x - \frac{1}{x} = \frac{1 + \frac{x^2}{2!} + \frac{x^4}{4!} + \ldots}{x + \frac{x^3}{3!} + \frac{x^5}{5!} + \ldots} - \frac{1}{x} = \left(1 + \frac{x^2}{2!}\right)\frac{1}{x}\left(1 - \frac{x^2}{3!}\right) - \frac{1}{x}$$

$$\simeq \frac{1}{x}\left(1 + \frac{x^2}{2} - \frac{x^2}{6}\right) - \frac{1}{x} = \frac{x}{3}. \tag{5.64}$$

Thus the variation of $L(x)$ with x near the origin is linear having slope 1/3. Thus

$$L(x)\bigg|_{x=0} = \frac{x}{3} = \frac{\mu B}{3kT}. \tag{5.65}$$

From (5.62),

$$M \simeq \frac{n\mu^2 B}{3kT}. \tag{5.66}$$

The *paramagnetic susceptibility* is given by

$$\chi_{\text{para}} = \frac{\mu_0 M}{B} = \frac{\mu_0 n \mu^2}{3kT}. \tag{5.67}$$

This shows that $\chi_{\text{para}} \propto \frac{1}{T}$.

Putting $\frac{\mu_0 n \mu^2}{3k} = C$, we get

$$\chi_{\text{para}} = \frac{C}{T}, \tag{5.68}$$

where C is a constant, called the *Curie constant*. Expression (5.68) is called the *Curie law* and holds good for $\mu B \ll kT$.

The *molar susceptibility* can be obtained by replacing n by *Avogadro number* N_A in (5.67), i.e.,

$$\chi_{\text{para}}^{(\text{molar})} = \frac{\mu_0 N_A \mu^2}{3kT}. \tag{5.69}$$

The order of magnitude of χ_{para} at room temperature can be estimated by taking

$$\mu = \mu_B = 9.27 \times 10^{-24}\ \text{A-m}^2$$
$$n = 5 \times 10^{23}/\text{m}^3$$
$$T = 300\,\text{K}.$$

$$\therefore\ \chi_{\text{para}} = \frac{(9.27 \times 10^{-24}) \times (1.257 \times 10^{-6}) \times (5 \times 10^{28})}{3 \times (1.38 \times 10^{-23}) \times 300} \simeq \mathbf{10^{-3}}.$$

We have obtained $\chi_{\text{dia}} \sim 10^{-5}$.

$$\therefore\ \frac{\chi_{\text{para}}}{\chi_{\text{dia}}} = \frac{10^{-3}}{10^{-5}} = \mathbf{10^2}. \tag{5.70}$$

Thus *paramagnetic susceptibility is roughly hundred times greater than diamagnetic susceptibility.* Table 5.1 shows the paramagnetic susceptibilities of different materials at room temperature.

5.7.2 Quantum Theory of Paramagnetism

According to the *classical theory* of paramagnetism proposed by *Langevin*, the permanent magnetic dipole moment of a given atom or ion can rotate freely and *orient in any direction* to the applied magnetic field. But in view of *quantum theory*, the magnetic moments, on the contrary, can orient along *certain discrete orientations* w.r.t. the magnetic field. Since the magnetic moments $\vec{\mu}$ and its component μ_Z are quantised in the direction of the applied field, they cannot have arbitrary values.

If $\vec{\mu}$ and $\vec{J}$ be the magnetic moment and total angular momentum of an atom (or ion) respectively, then

$$\vec{\mu} = -g\mu_B \vec{J}, \tag{5.71}$$

where $\mu_B = \frac{e\hbar}{2m}$ is the *Bohr magneton* in SI *system*, g is the *Landé splitting factor* having value 2, if $\vec{J}$ is due to orbital motion and electron spin only, i.e.,

$$g = 1 + \frac{J(J+1) + S(S+1) - L(L+1)}{2J(J+1)}, \tag{5.72}$$

where S and L represent respectively spin and angular momentum.

Now the possible orientations of $\vec{\mu}$ along the field direction is given by

$$\mu_Z = -g\mu_B m_J, \tag{5.73}$$

where $m_J = -J, -J+1, \ldots, J-1, J$ is the magnetic quantum number associated with $\vec{J}$. Thus for each value of J, m_J can have $(2J+1)$ values. This means $\vec{\mu}$ *can have* $(2J+1)$ *orientations relative to the applied field* $\vec{B}$.

The PE of the dipole in presence of $\vec{B}$ is

$$E = -\vec{\mu}\cdot\vec{B} = g\mu_B \vec{J}\cdot\vec{B} = g\mu_B BJ\cos\theta = g\mu_B Bm_J = m_J g\mu_B B. \tag{5.74}$$

According to *Maxwell-Boltzmann distribution law*, the number of dipoles having energy E at temperature T is proportional to

$$e^{-E/kT}, \text{ i.e., } e^{\frac{-m_J g\mu_B B}{kT}}.$$

If N be the number of atoms, the magnetisation in the direction of $\vec{B}$ is given by

$$M = N \times (\text{statistical average of the magnetic moment component along } \vec{B})$$

$$= N \sum_{m_J=-J}^{+J} \frac{\mu_Z e^{\frac{-m_J g\mu_B B}{kT}}}{e^{-\frac{m_J g\mu_B B}{KT}}} = N\frac{\sum_{-J}^{+J} -m_J g\mu_B e^{-\frac{m_J g\mu_B B}{kT}}}{\sum_{-J}^{+J} e^{-\frac{m_J g\mu_B B}{kT}}}. \tag{5.75}$$

Case I. For *normal values of* $\vec{B}$ and at *ordinary temperature* $\frac{m_J g\mu_B B}{kT} \ll 1$.

$\therefore$ (5.75) reduces to

$$M = N\frac{g\mu_B \sum_{-J}^{+J} -m_J\left(1 - \frac{m_J g\mu_B B}{kT}\right)}{\sum_{-J}^{+J}\left(1 - \frac{m_J g\mu_B B}{kT}\right)}.$$

But

$$\sum_{-J}^{+J} m_j = 0 \text{ and } \sum_{-J}^{+J} m_J^2 = 2\sum_{m_J=0}^{J} m_J^2 = \frac{J(J+1)(2J+1)}{3}.$$

$$\therefore\ M = N\frac{\frac{g^2\mu_B^2 B}{kT}\frac{J(J+1)(2J+1)}{3}}{(2J+1)} = N\frac{g^2\mu_B^2 B}{3kT}J(J+1). \tag{5.76}$$

Hence, the paramagnetic susceptibility is

$$\chi_{\text{para}} = \frac{M}{H} = \frac{\mu_0 M}{B} = \frac{\mu_0 N\mu_B^2}{3kT}g^2 J(J+1) \tag{5.77}$$

$$\text{or } \chi_{\text{para}} = \frac{\mu_0 N p_{\text{eff}}^2 \mu_B^2}{3kT}, \tag{5.78}$$

$$\text{or } \chi_{\text{para}} = \frac{C}{T}, \text{ where } C = \mu_0 N p_{\text{eff}}^2 \mu_B^2/3k, \tag{5.79}$$

where

$$p_{\text{eff}} \text{ being the effective number of Bohr magneton} = g\sqrt{J(J+1)}. \tag{5.80}$$

Effective magneton number for trivalent rare-earth ions (near room temperature) are shown in Table 5.2.

Table 5.2

Ion	*Electronic Configuration*	*Spectroscopic ground state*	p_{eff} *(calculated)* $= g\sqrt{J(J+1)}$	$p_{\text{effective}}$ *(experimental)*
Ce^{3+}	$4f^1 5s^2 p^6$	$^3F_{5/2}$	2.54	2.4
Pr^{3+}	$4f^2 5s^2 p^6$	$^3H_4$	3.58	3.5
Nd^{3+}	$4f^3 5s^2 p^6$	$^4I_{9/2}$	3.62	3.5
Pm^{3+}	$4f^4 5s^2 p^6$	$^5I_4$	2.68	—
Sm^{3+}	$4f^5 5s^2 p^6$	$^6H_{5/2}$	0.84	1.5
Eu^{3+}	$4f^6 5s^2 p^6$	$^7F_0$	0	3.4
Gd^{3+}	$4f^7 5s^2 p^6$	$^8S_{7/2}$	7.9	8.0
Tb^{3+}	$4f^8 5s^2 p^6$	$^7F_6$	9.72	9.5
Dy^{3+}	$4f^9 5s^2 p^6$	$^6H_{15/2}$	10.63	16.6
Ho^{3+}	$4f^{10} 5s^2 p^6$	$^5I_8$	10.6	10.4
Er^{3+}	$4f^{11} 5s^2 p^6$	$^4I_{15/2}$	9.59	9.5
Tm^{3+}	$4f^{12} 5s^2 p^6$	$^3H_6$	7.57	7.3
Y^{3+}	$4f^{13} 5s^2 p^6$	$^2F_{7/2}$	4.54	4.5

Case II. If the magnetic field $\vec{B}$ is strong and the temperature T of the material is low, then after some algebraic calculation it can be shown from (5.75) that

$$M = NgJ\mu_B B_J(x), \tag{5.81}$$

where $x = \frac{gJ\mu_B B}{kt}$ and $B_J(x)$ is the *Brillouin function* given by

$$B_J(x) = \frac{2J+1}{2J}\coth\left(\frac{2J+1}{2J}\right)x - \frac{1}{2J}\coth\left(\frac{x}{2J}\right) \tag{5.82}$$

(a) If $x \ll 1$, $\coth \simeq \frac{1}{x} + \frac{x}{3}$;

$$\therefore\ B_J(x) = \frac{x(J+1)}{3J}.$$

Then

$$\chi_{\text{para}} = \frac{\mu_0 M}{B} = \frac{\mu_0 N J(J+1) g^2 \mu_B^2}{3kT} \text{ or } \chi_{\text{para}} = \frac{\mu_0 N p_{\text{eff}} \mu_B^2}{3kT}. \tag{5.83}$$

This is the same expression for χ_{para} as (5.78).

(b) If $x \gg 1$, i.e., when temperature is low and field $\vec{B}$ is strong, $\coth x \simeq 1$.

$\therefore\ B_J(x) \simeq 1$.

(5.77) then takes the form:

$$M = N g J \mu_B. \tag{5.84}$$

(5.84) represents the state, where all the dipoles are aligned along the direction of $\vec{B}$. This is the *state of magnetic saturation and takes place at low temperature and strong magnetic field.* Under this limiting condition (5.84) becomes analogous to *Langevin* (5.62).

For large number of allowed orientations of a magnetic dipoles, i.e., $J \to \infty$, we have

$$\coth \frac{x}{2J} \to \frac{2J}{x} \text{ and } \coth\left(1 + \frac{1}{2J}\right) x \to \coth x.$$

$$\therefore\ B_J(x) \to \left(\coth x - \frac{1}{x}\right) = L(x). \tag{5.85}$$

(5.85) indicates that for $J \to \infty$, the quantum mechanical value of M approaches that due to classical value. For other values of J the two results differ widely.

Discussion

The paramagnetic susceptibility given by (5.83) is found to depend on the following factors:

1. χ_{para} is inversely proportional to temperature T of the material. For example, the value of χ_{para} calculated from (5.83) is $\sim 10^{-7}$ at room temperature. But at 1 K, χ_{para} becomes $\sim 10^{-5}$, i.e., hundred times greater.
2. χ_{para} is proportional to the number of atoms N and the strength of the magnetic field $\vec{B}$.
3. χ_{para} is proportional to the effective number of Bohr magneton (p_{eff}).
4. (5.83) is unable to account the experimental results on susceptibility for the ions of iron group. This is due to the presence of crystal field which cannot be ignored compared to the applied field.

5.7.3 Pauli's Theory of Paramagnetism

Substances like metals, free atoms or ions with a partially filled inner shell, e.g., *transition elements* and atoms, molecules and lattice defects possessing an odd number of electrons, e.g., sodium atoms, gaseous nitric oxide, F-centres in alkali halides exhibit paramagnetism. *Classical theory* fails to account the paramagnetic susceptibility of the conduction electrons since it is expected that the magnetisation of most normal non-ferromagnetic metals is independent of the temperature of the specimen.

Pauli applied *Fermi-Dirac distribution* and obtained satisfactory result which agrees with experimental observations.

According to *Pauli's exclusion principle* two electrons with opposite spin can occupy the same region in $\vec{k}$ space. So a graph of total density of states versus the KE of two components with

$$g(E) = \frac{V}{2\pi^2}\left(\frac{2m}{\hbar^2}\right)^{3/2} E^{1/2}. \tag{5.86}$$

will be of the form as shown in Fig. 5.13(a).

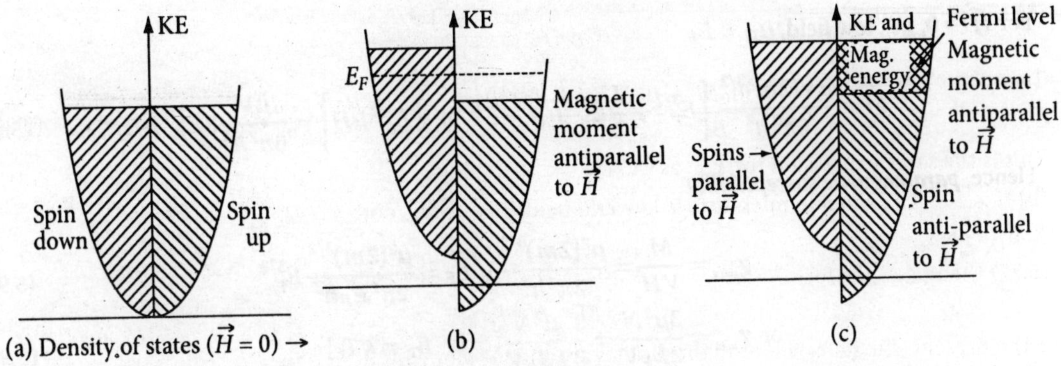

Fig. 5.13: Free electron model to describe paramagnetism in presence of $\vec{H}$.

In Fig. 5.13(a) states with spin up and spin down are similarly occupied when the magnetic field $\vec{H}=0$. Figs 5.13(b) and (c) show the occupancy when $\vec{H}$ is applied. The cross-hatched area represents the excess up-spin electrons.

The energy of a Fermi particle subjected to $\vec{H}$ is

$$E=\frac{p^2}{2m}-\mu H, \tag{5.87}$$

when spin is parallel to $\vec{H}$. For anti-parallel spin

$$E=\frac{p^2}{2m}+\mu H, \tag{5.88}$$

where μ is the *intrinsic moment of the particle.*

All the energy levels up to *Fermi level* are filled at $T=0$. Then the KE of the first group will be from 0 to $\left(E_p+\mu H\right)$ and those of second group will be from 0 to $\left(E_p-\mu H\right)$. Some of the electrons, with anti-parallel spin having magnetic moments, parallel to $\vec{H}$, will flip over and orient parallel to $\vec{H}$ to maintain common electrochemical potential of the system. This will give rise to paramagnetism.

Paramagnetic Susceptibility

If N^+ and N^- be the particles in the two groups respectively, then

$$N^+=\frac{V}{4\pi^2\hbar^3}(2m)^{3/2}\int_0^{E_F+\mu H}E^{1/2}\,dE \tag{5.89}$$

$$=\frac{V}{4\pi^2\hbar^3}(2m)^{3/2}\left(E_F+\mu H\right)^{3/2}$$

$$=\frac{V}{6\pi^2\hbar^3}(2m)^{3/2}\left(E_F+\mu H\right)^{3/2}. \tag{5.90}$$

Likewise

$$N^-=\frac{V}{6\pi^2\hbar^3}(2m)^{3/2}\left(E_F-\mu H\right)^{3/2}. \tag{5.91}$$

∴ the net magnetic moment is

$$M=\mu\left(N^+-N^-\right)=\frac{\mu V(2m)^{3/2}}{6\pi^2\hbar^3}\left[\left(E_F+\mu H\right)^{3/2}-\left(E_F-\mu H\right)^{3/2}\right]. \tag{5.92}$$

For $H \to 0$, i.e., low field, $\mu_H < E_F$

$$\therefore M = \frac{\mu V(2m)^{3/2}}{6\pi^2\hbar^2}\left[E_F^{3/2} + \frac{3}{2}E_F^{1/2}\mu H - E_F^{3/2} + \frac{3}{2}E_F^{1/2}\mu H\right] = \frac{\mu V}{6\pi^2\hbar^3}3E_F^{1/2}\mu H. \tag{5.93}$$

Hence, *paramagnetic susceptibility*

$$\chi_{\text{para}} = \frac{M}{VH} = \frac{\mu^2(2m)^{3/2}}{2\pi^2\hbar^3}\sqrt{E_F} = \frac{\mu^2(2m)^{3/2}}{2\pi^2 E_F\hbar^3}E_F^{3/2} \tag{5.94}$$

$$\text{or } \chi_{\text{para}} = \frac{3\mu^2 N}{2E_F V} = \frac{3}{2}\frac{\mu^2 N}{kT_F V}, \quad [\because E_F = kT_F] \tag{5.95}$$

where $E_F = \frac{\hbar^2}{2m}\left(\frac{3\pi^2 N}{V}\right)^{2/3}$

T_F is the *Fermi temperature,* which is constant for a given metal and hence χ_{para} is independent of temperature (T). Except some deviation, where diamagnetic susceptibility contributes, the *Pauli's theory of paramagnetism agrees closely with experimental results.*

5.7.4 Paragmagnetism of Ionic Solids

The transition group elements (rare-earth group) having *incomplete 4f shell* and the iron group elements having *incomplete 3d shell* show paramagnetic properties. The paramagnetic susceptibility (χ_{para}) [(5.78)] with J values obtained by *Hund's rule,* i.e.,

$J = L - S$, for less than half-filled shells
$J = 0$, for just half-filled shells
and $J = L + S$, for more than half-filled shells

is in excellent agreement with experimental results for many rare-earth group of ions.

But for iron group ions, the χ_{para} values obtained from (5.78) differ widely from the experimental values. Why?

The $3d$ shell in iron group ions is the outermost shell and is exposed to *Crystal Field* due to the neighbouring ions. The crystal field destroys the rotational symmetry of such crystals; as a result the total angular momentum differs from the proper quantum number. Also the average values of L_Z is reduced to zero. This effect is known as the *quenching of the orbital angular momentum L.*

Hence, putting $J = 0 + S = S$, we get from (5.77),

$$\chi_{\text{para}} = \frac{\mu_0 N\mu_B^2}{3kT}g^2 S(S+1). \tag{5.96}$$

This values of χ_{para} is in good agreement with experiment values for iron group.

Example. χ_{para} for Mn^{3+} ion is zero according to (5.77) but according to (5.96), χ_{para} with $S = 2$ is in good agreement with experimental results.

Table 5.3 shows the calculated values of *effective magneton number* for Iron Group Ions.

Table 5.3: Calculated values of effective magneton number for iron group ions.

Iron Group Ions	*Electronic Configuration*	*Basic level*	p_{eff} *(exptl)*	p_{eff} *(calculated)* $p_{\text{eff}} = g\sqrt{J(J+1)}$	p_{eff} *(calculated)* $p_{\text{eff}} = 2\sqrt{S(S+1)}$	p_{eff} *(calculated)* $p_{\text{eff}} = [L(L+1)+4S(S+1)]^{1/2}$
Ti^{3+}, V^{4+}	$3d^1$	$^2D_{3/2}$	1.8	1.55	1.73	3.01
V^{2+}	$3d^2$	$^3F_2$	2.8	1.63	2.83	4.49
$Cr^{3+}V^{2+}$	$3d^3$	$^4F_{3/2}$	3.8	0.77	3.87	5.20
$Mn^{3+}Cr^{3+}$	$3d^4$	$^5D_0$	4.9	0	4.90	5.50
$Fe^{3+}Mn^{2+}$	$3d^5$	$^6S_{1/2}$	5.9	5.92	5.92	5.92
Fe^{2+}	$3d^6$	$^5D_4$	5.4	6.70	4.90	5.50
CO^{2+}	$3d^7$	$^4F_{9/2}$	4.8	6.63	3.87	5.21
Ni^{2+}	$3d^8$	$^3F_4$	3.2	5.59	2.87	4.49
Cu^{2+}	$3d^9$	$^2D_{5/2}$	1.9	3.55	1.73	3.01

5.7.5 Cooling by Adiabatic Demagnetisation

Temperature as low as 0.3 K can be attained by *evaporating liquid helium under low pressure.* But *P. Debye* and *W.F. Giaugue* suggested that, *adiabatic demagnetisation of paramagnetic salt* can produce temperature ~ 0.001 K and lower than this value. The principle of the method is that *at fixed temperature the entropy of a system of magnetic moments is lowered by the application of an external magnetic field.*

The variation of entropy of a system of magnetic moments is shown in Fig. 5.14. The curve *OCA* represents the variation of entropy (*S*) with temperature (*T*) in absence of the magnetic field $\vec{B}$. If at temperature T_1, $\vec{B}$ be applied on the paramagnetic material, the magnetic dipoles will tend to align along $\vec{B}$. The spin system will become more orderly and hence the entropy of the system will decrease from *A* to B_1 along the path *AB*. If now the paramagnetic specimen is isolated from its surroundings and the magnetic field $\vec{B}$ is withdrawn, then the system will suffer an adiabatic transformation and goes from B_1 to *C* along B_1C with no change in entropy (i.e., $\Delta S = 0$). The temperature of the system at *C* is T_1, where $T_1 < T_2$. By repeating the process the temperature of the specimen can be reduced to ~ 0.001 K.

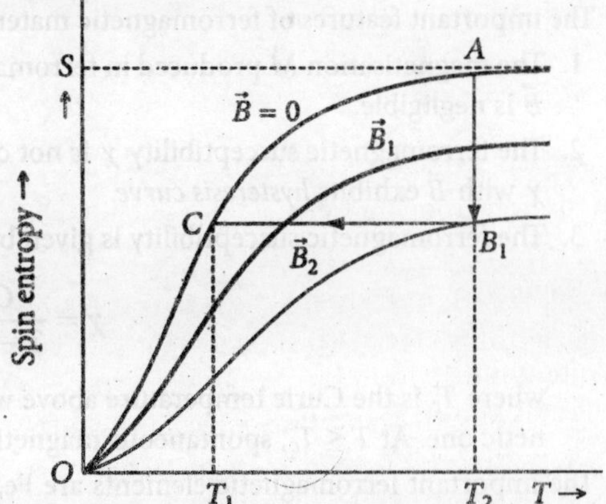

Fig. 5.14: Entropy variation with temperature without and with magnetic field.

From *Maxwell's thermodynamical relations* we can show that the entropy of the system in presence of external magnetic field $\vec{B}$ is

$$S = S_0 + \int_0^B \left(\frac{\partial M}{\partial T}\right)_B dB, \tag{5.97}$$

where *M* is the magnetisation and S_0 is the entropy at 0 K.

Now from (5.76),

$$M = \frac{Ng^2\mu_B^2 B}{3kT} J(J+1).$$

$$\therefore \quad \frac{\partial M}{\partial T} = -\frac{Ng^2\mu_B^2 B}{3kT^2} J(J+1).$$

$$\therefore \quad \int_0^B \left(\frac{\partial M}{\partial T}\right)_B dB = -\frac{Ng^2\mu_B^2 J(J+1)}{3kT^2} \int_0^B B\,dB$$

$$= -\frac{Ng^2\mu_B^2 B^2}{6kT^2} J(J+1). \tag{5.98}$$

Substituting (5.98) in (5.97), we get

$$S = S_0 - \frac{Ng^2\mu_B^2 B^2}{6kT} J(J+1). \tag{5.99}$$

This shows that the entropy S of the system at T_1 with magnetic field B will be lower than the entropy S_0 at T_1 in absence of magnetic field (i.e., $S < S_0$) and will correspond to S_0 at lower temperature T_2. This is obtained by *adiabatic change* along B_1C (i.e., $\Delta S = 0$) in absence of the magnetic field. Hence, by *adiabatic demagnetisation* temperature below 1 K can be attained.

5.8 Ferromagnetism

The ferromagnetism also arises due to the presence of permanent magnetic dipole moment like paramagnetism. But in ferromagnetism *the adjacent magnetic dipoles are oriented in a particular direction even if the external magnetic field be zero.* Hence ferromagnetic materials exhibit magnetic moment in absence of magnetic field — a phenomenon called *spontaneous magnetisation.*

The important features of ferromagnetic materials are as follow:

1. The magnetisation $\vec{M}$ produced in ferromagnetic materials is high even if the applied magnetic field $\vec{B}$ is negligible.
2. The ferromagnetic susceptibility χ is not constant but varies with the applied field. The variation of χ with $\vec{B}$ exhibits *hysteresis curve.*
3. The ferromagnetic susceptibility is given by

$$\chi = \frac{C}{T - T_c}, \text{ [Curie-Weiss law]}$$

where T_c is the Curie temperature above which the ferromagnetic substance behaves as a paramagnetic one. At $T \leq T_c$, spontaneous magnetisation occurs.

The important ferromagnetic elements are Fe, Ni, Co, Gd and Dy. A number of alloys and oxides like MnBi, MnAs, CrO_2, etc. also exhibit ferromagnetic property.

5.8.1 Weiss Theory of Ferromagnetism

The fundamental assumptions on which the *Weiss' theory* is based are the following:

1. A ferromagnetic specimen of macroscopic dimension contains a large number of small regions called *domains* which are spontaneously magnetised. The magnetisation of the whole material is determined by the vector sum of the magnetic moments of individual domains.
2. The spontaneous magnetisation is due to the existence of *molecular field* $\vec{B}_1$ within the material which tends to align the atomic dipoles. The field B_1 is quite strong being ~ 1000 T and is assumed to be proportional to the magnetisation of each domain, i.e.,

$$\vec{B}_1 = \lambda \vec{M}. \tag{5.100}$$

- The field $\vec{B}_1$ arising out of the exchange interaction is known as the *Weiss field* and λ is a constant called *Weiss-field constant* and is independent of temperature.

The effective magnetic field on an atom or ion is given by

$$\vec{B}_{\text{eff}} = \vec{B} + \vec{B}_1 = \vec{B} + \lambda\vec{M}. \tag{5.101}$$

As $\vec{B}, \vec{B}_1, \vec{M}$ and $\vec{B}_{\text{eff}}$ are all in the same direction,

$$B_{\text{eff}} = B + \lambda M. \tag{5.102}$$

If N be the number of atoms per unit volume of the ferromagnetic material, then as in case of paramagnetism (5.81), the magnetisation

$$M = NgJ\mu_B B_J(x), \tag{5.103}$$

where

$$B_J(x) = \frac{2J+1}{2J}\coth\left(\frac{2J+1}{2J}\right)x - \frac{1}{2J}\coth\left(\frac{x}{2J}\right) \tag{5.104}$$

$$\text{and } x = \frac{gJ\mu_B B_{\text{eff}}}{kT} = \frac{gJ\mu_B}{kT}(B + \lambda M). \tag{5.105}$$

For *spontaneous magnetisation,* $B = 0$.

$$\therefore x = \frac{gJ\mu_B\lambda M}{kT} \tag{5.106}$$

$$\text{or } M(T) = \frac{xkT}{\lambda gJ\mu_B}. \tag{5.107}$$

For $T \to 0$, $x \to \infty$ and $B_J(x) \to 1$. The magnetic moments of each domain align along the field and magnetisation M under this condition becomes *saturated magnetisation* $M_s(0)$, given by

$$M_s(0) = NgJ\mu_B. \tag{5.108}$$

From (5.107) and (5.108) we get

$$\frac{M(T)}{M_s(0)} = \frac{xkT}{\lambda N g^2 J^2 \mu_B^2}. \tag{5.109}$$

Also from (5.81) and (5.108), we get

$$\frac{M(T)}{M_s(0)} = B_J(x). \tag{5.110}$$

The simultaneous solutions of (5.109) and (5.110) can be obtained by plotting $\frac{M}{Ms(0)}$ against x for $T > T_c$, $T = T_c$ and $T < T_c$ which is a straight line for different values of T. But the plot of $\frac{M}{M_s(0)}$ against x according to (5.109) for different T values gives a curve as shown in Fig. 5.15. The point of intersection of the straight line and the curve of Fig. 5.15 gives the solution for $M(T)$.

Discussion

1. *At critical temperature or Curie Temperature* (T_c). From Fig. 5.15 it is evident that at T_c the straight line given by (5.109) is tangent to the curve given by (5.110) at the origin O.

The straight line has slope given by

$$\frac{M(T)}{M_s(0)} = \frac{kT}{\lambda N g^2 J^2 \mu_B^2} x. \qquad (5.111)$$

At $T = T_c$,

$$\frac{\left(\frac{M(T)}{M_s(0)}\right)}{x} = \frac{kT_c}{\lambda N g^2 J^2 \mu_B^2}. \qquad (5.112)$$

Again, for $x \ll 1$,

$$B_J(x) = \frac{J(J+1)}{3J} x \text{ [from (5.104)]}.$$

So the tangent to the curve at the origin has a slope

$$\frac{B_J(x)}{x} = \frac{J(J+1)}{3J}. \qquad (5.113)$$

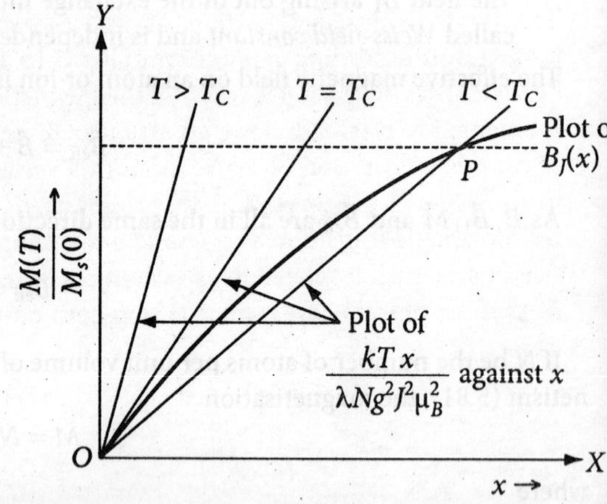

Fig. 5.15: Graphical solution of the simultaneous equations (5.109) and (5.110). The point P determines the spontaneous magnetisation $M_s(T)$ at a given temperature. For $T > T_C$, no solution exists.

Equating the two slopes from (5.112) and (5.113), we get

$$\frac{kT_c}{\lambda N g^2 J^2 \mu_B^2} = \frac{(J+1)}{3J} \text{ or } T_c = \frac{\lambda N g^2 J^2 \mu_B^2}{3k} J(J+1) = \frac{N\mu_j^2 \lambda}{3k}, \qquad (5.114)$$

where

$$\mu_j = \text{the total magnetic moment per atom} = g_j \mu_B \sqrt{J(J+1)}. \qquad (5.115)$$

(5.114) shows that, *T_c is proportional to the Weiss field constant λ.*

2. *Below Curie Temperature (T_c).* At $T < T_c$, the straight line given by (5.109) and the curve given by (5.110) intersects at $x = 0$ and at P. The intersection at $x = 0$ represents *unstable magnetisation,* whereas the intersection at P represents non-zero value of M for $\vec{B} = 0$ and corresponds to *spontaneous magnetisation.*

   Substituting

$$\frac{k}{\lambda N g^2 J^2 \mu_B^2} = \frac{1}{T_c} \frac{(J+1)}{3J}$$

   in (5.111), we get

$$\frac{M(T)}{M_s(0)} = \frac{T}{T_c}\left(\frac{J+1}{3J}\right) x. \qquad (5.116)$$

   (5.116) gives the maximum possible values of $M_s(0)$ and shows that $M_s(0)$ is inversely proportional to temperature (T).

   Practically, a specimen of ferromagnetic at $T < T_c$ is not magnetised in absence of external magnetic field ($\vec{B}$). But it becomes magnetised only after the application of $\vec{B}$. *This is due to the fact that though each domain is spontaneously magnetised, these are so arranged to form closed chains within the material and the magnetic effect of one cancels that due to the other. Hence, the material becomes unmagnetised in absence of external field.*

The spontaneous magnetisation (M_s) varies with temperature according to the curve of Fig. 5.16. It becomes maximum at 0 K when all the magnetic moments are oriented along a particular direction under the action of *Weiss field*. This variation of M_s with T helps to classify paramagnetic-ferromagnetic transition as a *second order phase transition* which is related to an order parameter M_s (in the present case) which is non-zero only at $T < T_c$.

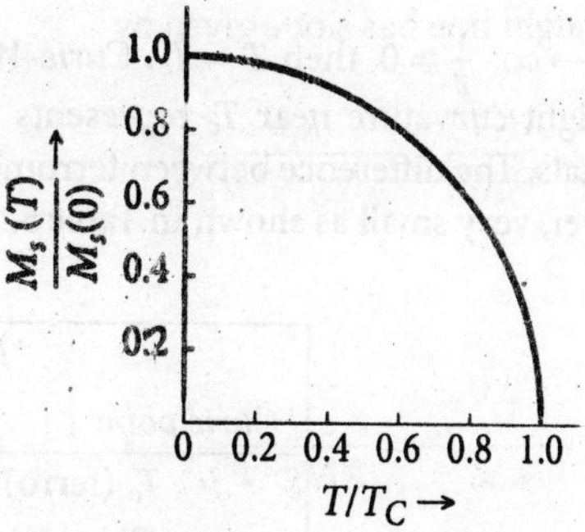

Fig. 5.16: Spontaneous magnetisation versus temperature at $T < T_C$.

3. *Above Curie Temperature* ($T > T_C$). Above Curie temperature, spontaneous magnetisation is not possible. An applied magnetic field can produce only magnetisation in this case. In this region ($T > T_c$), we find from (5.105) that $x \ll 1$.

$$\therefore\ B_J(x) \simeq \left(\frac{J+1}{3J}\right) x \text{ [from (5.104)]}$$

(5.103) and (5.105) reduce to

$$M = \frac{Ng\mu_B(J+1)x}{3}\left[x = \frac{gJ\mu_B(B+\lambda M)}{kT}\right]. \tag{5.117}$$

$$\therefore\ M = \frac{Ng^2\mu_B^2 J(J+1)}{3kT}(B+\lambda M).$$

Then susceptibility,

$$\chi = \frac{\mu_0 M}{B} = \frac{\mu_0 T_c/\lambda}{T - T_c} \text{ or } \chi = \frac{C}{T - T_c}, \tag{5.118}$$

where

$$C = \frac{\mu_0 T_c}{\lambda} \tag{5.119}$$

$$\text{and } T_c = \frac{\lambda N g^2 \mu_B^2 J(J+1)}{3k}. \tag{5.120}$$

(5.118) is known as the *Curie-Weiss law*. Curie-Weiss law describes the temperature dependence of susceptibility in the paramagnetic region with satisfaction, if temperature is well above the Curie temperature.

From (5.118)

$$\frac{1}{\chi} = \frac{T}{C} - \frac{T_c}{C}.$$

Thus the graph between $\frac{1}{\chi}$ versus T is a straight line as shown in Fig. 5.17.

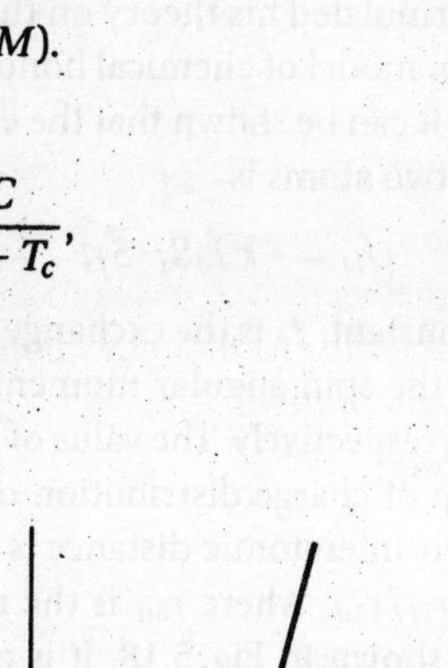

Fig. 5.17: Distinction between ferromagnetic and paramagnetic Curie points.

For $\chi \to \infty$, $\frac{1}{\chi}=0$, then $T=T_C$. *Curie-Weiss law* agrees quite accurately with experimental results. The slight curvature near T_C represents the distinction between ferromagnetic and paramagnetic materials. The difference between ferromagnetic Curie point T_C and paramagnetic Curie point T_C' is, however, very small as shown in Table 5.4.

Table 5.4

Material → / *Curie point* ↓	Fe	CO	Ni
T_C (ferro) (K)	1043	1393	631
T_C' (para) (K)	1093	14.28	650

The *Weiss' theory*, however, fails to provide the explanation of the origin of *Weiss field* until *Heisenberg* proposed his *quantum exchange interpretation* (1928) for the origin of Weiss field in ferromagnetism.

5.8.2 Origin of Weiss Molecular Field (Heisenberg's Exchange Interaction)

Weiss field calculated on the basis of magnetic dipole-dipole interaction is only of the order of 10^3 G, whereas the actual field strength are observed to be to the order of 10^7 G. Hence, Weiss field cannot be simply due to magnetic dipole-dipole interaction.

Heisenberg in 1928 first proposed that the large Weiss field is due to *quantum mechanical exchange interaction* between the electron spin. The exchange interaction is due to the *Pauli's exclusion principle.* According to this principle, *the spatial charge distribution is disturbed due to any change in the relative orientation of the two spins, and interaction between two atoms. The extent of overlap of the wave functions of two neighbouring atoms and the relative orientation of the electron spin determine the strength of the exchange interaction.* Hence, exchange interaction is *electrostatic and non-magnetic in nature.*

Heisenberg formulated his theory on the analogy of *Heitler-London* model of chemical bonding. In view of this theory, it can be shown that the exchange energy between two atoms is

$$U_{ij}=-kJ_e\vec{S}_i\cdot\vec{S}_j, \qquad (5.121)$$

where k is a constant, J_e is the exchange integral and $\vec{S}_i$ and $\vec{S}_j$ are the spin angular momenta of the ith and jth atoms respectively. The value of U_{ij} depends on the overlap of charge distribution of two atoms and hence their interatomic distance is r_{ij}. The plot of J_e against r_{ij}/r_{3d}, where r_{3d} is the radius of the 3d orbital, is shown in Fig. 5.18. It is evident from this figure that J_e is positive for large interatomic distances and negative for $\left(r_{ij}/r_{3d}\right)<3$.

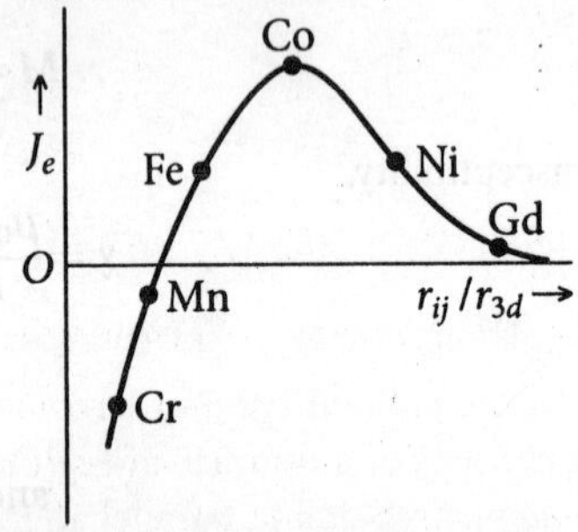

Fig. 5.18: Variation of exchange interaction J_e with r_{ij}/r_{3d}.

Note. *It was conclusively proved through the gyromagnetic experiment conducted by Einstein, de Haas and Barnett that ferromagnetism is due to spin moments and not orbited moments.*

For J_e value positive, the interaction energy U_{ij} is negative and hence minimum for parallel arrangement of spin and is more stable than anti-parallel spins. The atom under this condition exhibits ferromagnetism. In view of this, Fe, CO, Ni and Gd are ferromagnetic. But $\left(r_{ij}/r_{3d}\right)$ is less than 3 for Cr and Mn and hence these are not ferromagnetic.

Two materials with $\left(r_{ij}/r_{3d}\right)<3$ can be combined to a compound for which $\left(r_{ij}/r_{3d}\right)>3$ and thus a ferromagnetic body can be produced.

Examples. Mn–As, Cu–Mn and Mn–Sb alloys are ferromagnetic as r_{ij}/r_{3d} for them is greater than 3, although for the individual elements $\left(r_{ij}/r_{3d}\right) < 3$. *This is the greatest triumph for solid state theory that by proper cooking of the molecules we can fabricate materials of any desired characteristics.*

Now, if J_e be constant for all the neighbouring pairs, then considering the effect of neighbouring pairs only, we have the energy of the ith atom as

$$U_i = -k J_e \sum_j \vec{S}_i \cdot \vec{S}_j. \tag{5.122}$$

$\sum_j$ represents the summation over the nearest neighbours of the ith atom. If the number of nearest neighbours be Z, then

$$U_i = -kZ J_e \left(S_{xi}\left\langle S_{xj}\right\rangle + S_{yi}\left\langle S_{yj}\right\rangle + S_{zi}\left\langle S_{zj}\right\rangle\right), \tag{5.123}$$

where the instantaneous values of neighbouring spins have been replaced by their time averages. Considering magnetisation M along Z-axis only,

$$\left\langle S_{xj}\right\rangle = \left\langle S_{yj}\right\rangle = 0 \text{ and } \left\langle S_{zj}\right\rangle = \frac{M}{g\mu_B N}. \tag{5.124}$$

$$\therefore\ U_i = -\frac{kZJ_eS_{zi}M}{gN\mu_B} \tag{5.125}$$

$$\text{or } U_i = -gS_{zi}\mu_B B_1, \tag{5.126}$$

$$\text{where } B_1 = \frac{kZJ_e}{Ng^2\mu_B^2}M \tag{5.127}$$

$$\text{or } B_1 = \lambda M, \tag{5.128}$$

$$\text{where } \frac{kZJ_e}{Ng^2\mu_B^2} = \lambda \text{ is a constant.} \tag{5.129}$$

Thus, from (5.128) the *Weiss field* B_1 *is proportional to magnetisation* M.

5.8.3 Domain Theory of Ferromagnetism

To explain the phenomenon of ferromagnetism *Weiss* proposed that a virgin specimen of ferromagnetic material consists of a number of region (~ 10^{-6} m or larger) called *domains*. These domains are *spontaneously magnetised* due to parallel alignment of all spin magnetic moments in a particular direction. The direction of alignments of the spin magnetic moments varies from domain to domain and as such the resultant magnetic moment of the specimen may be zero or nearly zero. In presence of an external magnetic field, the random domains align in the following ways:

1. *By the motion of domain walls.* The size of the domains in which the magnetic moments subtend the minimum angle w.r.t. the external field $\vec{B}$ increases with the increase of B [Fig. 5.19(b)]. The domain in which the magnetic moments are oriented at larger angles w.r.t. $\vec{B}$ suffers a decrease in size with the increase of $\vec{B}$ [Fig. 5.19(c)]. This type of change in domain size leads to a diminution in the magnetic energy, producing more stable configuration. The process is known as *displacement process of magnetisation* as the changes in domain size are produced due to the displacement of *domain boundaries.*
2. *By the rotation of domains.* If the applied magnetic field $\vec{B}$ be strong, the rotation of the direction of the magnetic moment towards $\vec{B}$ occurs as shown in Fig. 5.19(c). When the magnetic moment of the domain turns fully in the direction of $\vec{B}$, *technical saturation* occurs.

3. *By the Para-Process.* If $\vec{B}$ be further increased, then the magnetisation M increases by a small amount. We assume that in the spontaneously magnetised domains the spin moments of all the atoms are aligned along $\vec{B}$. But this occurs strictly at $T = 0$ K. At higher temperature complete magnetic saturation in a domain cannot be achieved due to thermal agitation and some of the spin moments may align anti-parallel to $\vec{B}$. At higher magnetic field these magnetic moments may be made to align parallel to $\vec{B}$, whereby magnetisation increases slightly beyond technical saturation. *This is the para-process.*

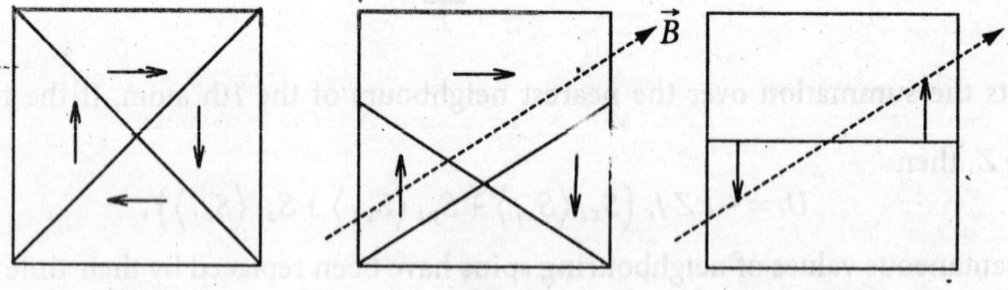

Fig. 5.19: (a) Ferromagnetic domain, (b) Domain magnetised by wall moment, (c) Domain magnetised by rotation.

5.8.4 Hysteresis

Hysteresis in ferromagnetic materials is referred to the lag of magnetisation behind the magnetising field. Ferromagnetic materials show hysteresis below the ferromagnetic Curie temperature (T_C). The phenomenon of hysteresis can be well understood by the concept of *domain.* The increase of resultant magnetic moment of a ferromagnetic specimen under the action of an applied magnetic field (B) is due to (i) *the motion of domain walls* and (ii) *the rotation of the domains.*

If a *weak magnetic field B* be applied, then the domains whose spin magnetic moments are aligned along B and in the easy direction of magnetisation, grow in size at the expense of less favourably oriented domains due to the motion of the domain walls. After the removal of B, the domains reverse back to the original size. This reversible domain wall displacement is indicated by the portion OA of the magnetisation curve [Fig. 5.20].

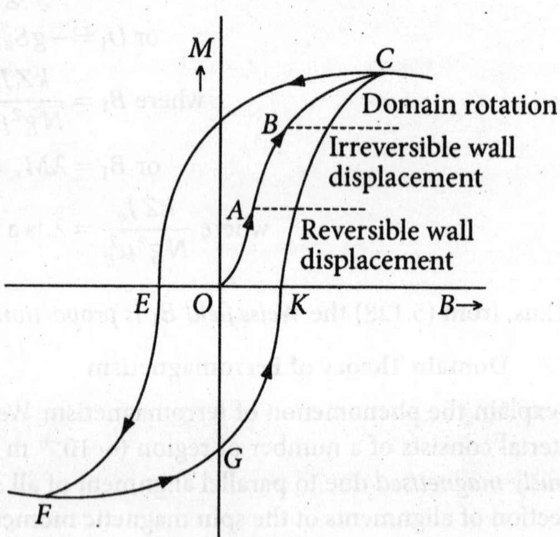

Fig. 5.20: Hysteresis loop of a ferromagnetic material.

If the *applied field B be stronger,* then the movement of domain walls becomes irreversible as indicated by the portion AB of the magnetisation curve. This irreversible domain wall movement is responsible for the phenomenon of hysteresis.

The point B represents the situation, where all the domains are aligned along their easy directions. If the value of B be still further increased, then the rotation of the domains takes place in a direction away from the easy direction and thereby an *isotropic energy* is stored in the specimen. At the completion of rotation of the domains, *magnetic saturation* occurs as represented by the point C of the curve. If now the applied field

B be withdrawn, the specimen tends to attain the original configuration by the motion of domain walls. But this tendency is opposed by the *impurities, lattice imperfection*, etc. present in it. So more energy is required to overcome the opposite action. This is equivalent to say that, a negative field, called the *coercive force*, is required to reduce the magnetisation to zero. This energy is actually a loss and increases the temperature of the specimen.

Hence *hysteresis is the loss of energy in taking a ferromagnetic body through a complete cycle of magnetisation and is proportional to the area enclosed by the hysteresis loop.*

Calculation of Hysteresis Loss

The magnetic induction B and the intensity of magnetisation are related by $B = \mu_0(H + \mu)$, where H is the applied magnetic field and μ_0 is the magnetic permeability of the free space.

$\therefore\ dB = \mu\delta(dH + M) = \mu_0 dH + \mu_0 dM.$

Now, multiplying both sides of the above equation and integrating over a complete cycle of magnetisation, we get

$$\oint H\,dB = \mu_0 \oint H\,dH + \mu_0 \oint H\,dM = \mu_0 \oint H\,dM.$$

$\because \oint H\,dH = 0$ as the curve between H against dH is a straight line.

$$\therefore\ \mu_0 \oint H\,dM = \oint H\,dB.$$

$\mu_0 \oint H\,dM$ represents the loss of energy per unit volume per cycle of hysteresis and $\oint H\,dB$ is the area of the hysteresis loop.

Therefore, loss of energy per unit volume of the specimen per cycle is

$$\oint H\,dB = \text{area of } B\text{-}H \text{ loop}.$$

The energy thus lost is dissipated in the form of heat in the specimen.

Importance of Hysteresis Curve

The hysteresis curves are important in the (i) determination of the quality of magnetic material, (ii) their selection for particular purpose, e.g., for making permanent magnet, electromagnet or core of transformers and dynamos and (iii) it provides information regarding retentivity, coercivity, permeability, susceptibility and energy loss per cycle of magnetisation.

5.8.5 Hard and Soft Magnetic Materials

Magnetic materials based on the area of their hysteresis loop are classified as Hard and Soft magnetic materials. Table 5.5 shows a comparative study of these materials with their examples and important field of applications.

Table 5.5: Distinction between Hard and Soft magnetic materials

Hard Magnetic Materials	*Soft Magnetic Materials*
1. These materials have large area of the hysteresis loop and hence large hysteresis loss.	1. The hysteresis loss is small as the area of hysteresis loop is less.
2. Here the impurities, imperfections in the lattice oppose the domain wall movement making the process irreversible.	2. Here the domain wall movement is easier. The change of magnetisation even for low field is large.
3. These materials cannot be easily magnetised or demagnetised due to large coercivity and retentivity.	3. These materials can be easily magnetised or demagnetised due to small coercivity and retentivity.
4. In these materials impurities and crystal imperfections are causing more mechanical strain and, magnetostatics energy is large.	4. The magnetostatics energy in these materials is small due to the absence of irregularities.
5. The values of permeability and susceptibility are small.	5. Permeability and susceptibility of these materials are large.
6. *Examples.* (a) Fe–Ni–Al alloys with certain amount of Co (Alnico). (b) Cu–Ni–Fe alloys (cunif) (c) Cu–Ni–Co alloys (cunico)	6. *Examples.* (a) Fe–Si alloys (b) Ferrites (c) Garnets
7. Used for production of permanent magnets which are used in magnetic detectors, flux meters, magnetic separators, etc.	7. Used in electromagnets, which are used in electromagnetic machinery and transformer's cores, matrix storage of computers, etc.

5.9 Antiferromagnetism

Heisenberg showed that, *quantum mechanical exchange interactions* produce large internal fields and the neighbouring magnetic dipoles tend to align in the same direction [Fig. 5.21(a)]. It can also be proved quantum mechanically that, for small separation of the interacting dipoles, the dipole moments of the neighbouring atoms has a tendency of anti-parallel alignments [Fig. 5.21(b)] by the action of the exchange forces. This occurs below a certain temperature T_N, called the *Néel temperature,* and the corresponding material is known *as antiferromagnetic material.*

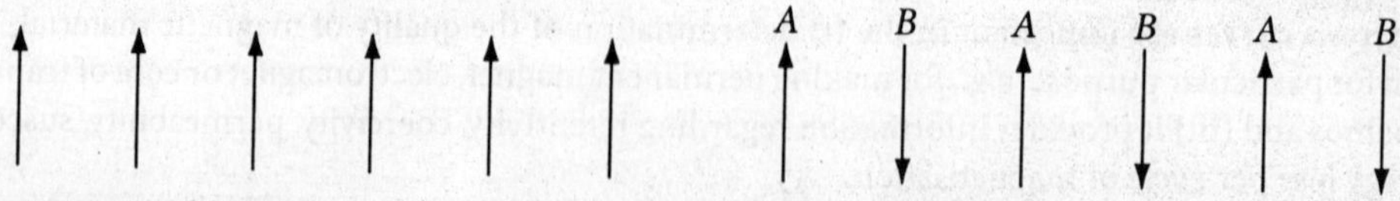

Fig. 5.21: (a) Ferromagnetism produced by parallel spin moments of neighbouring dipoles.

Fig. 5.21: (b) Neighbouring antiparallel spin moments of *A* and *B* sublattices in antiferromagnetism.

5.9.1 Theory of Antiferromagnetism

We consider that, an antiferromagnetic material consists of two sub-lattices *A* and *B* so that an atom of one type has all the nearest neighbours of the opposite type. In antiferromagnetic materials besides nearest neighbour negative *AB* interactions, nearest neighbour *AA* (or *BB*) weak negative interaction is also present.

The internal fields at A- and B-sites can be expressed as

$$\left.\begin{aligned} B_{mA} &= -\alpha M_A - \beta M_B \\ B_{mB} &= -\beta M_A - \alpha M_B \end{aligned}\right\}, \tag{5.130}$$

where α and β are *Weiss field constants for AA (or BB) and AB interaction respectively* and M_A and M_B are the corresponding magnetisation.

If B be the *external field,* then

$$\left.\begin{aligned} B_A &= B - \alpha M_A - \beta M_B \\ \text{and } B_B &= B - \beta M_A - \alpha M_B \end{aligned}\right\}. \tag{5.131}$$

Now we consider the situation in two different cases:

Case I. $T > T_N$, *i.e., above Néel temperature* (T_N).

If the temperature of the material be above Néel temperature, it behaves as a paramagnetic. The magnetisation of sub-lattices A and B can be written as

$$\left.\begin{aligned} M_A &= \frac{N\mu^2}{3kT} B_A \\ M_B &= \frac{N\mu^2}{3kT} B_B \end{aligned}\right\}, \tag{5.132}$$

by (5.66), where $\mu^2 = \mu_B^2 g^2 J(J+1)$.

Here we have assumed $N_A = N_B = N$.

$\therefore$ the total magnetisation is given by

$$M = M_A + M_B = \frac{N\mu^2}{3kT} B_A + \frac{N\mu^2}{3kT} B_B = \frac{N\mu^2}{3kT}(B_A + B_B) = \frac{N\mu^2}{3kT}[2B - (\alpha+\beta)(M_A + M_B)].$$

$\because$ from (5.131),

$$(B_A + B_B) = [2B - (\alpha+\beta)(M_A+M_B)] = \frac{N\mu^2}{3kT}[2B - (\alpha+\beta)M] \quad [\because M_A + M_B = M]$$

$$\text{or } M\left[1 + \frac{N\mu^2}{3kT}(\alpha+\beta)\right] = \frac{N\mu^2}{3kT} 2B \tag{5.133}$$

$$\text{or } M = \frac{\frac{2}{3}\frac{N\mu^2 B}{k}}{T + \frac{N\mu^2}{3k}(\alpha+\beta)}. \tag{5.134}$$

$\therefore$ the susceptibility of the antiferromagnetic material can be expressed as

$$\chi = \frac{\mu_0 M}{B} = \frac{2\mu_0 N\mu^2/3k}{T + \frac{N\mu^2(\alpha+\beta)}{3k}} \text{ or } \chi = \frac{C}{T+\theta}, \tag{5.135}$$

where

$$C = \frac{2\mu_0 N\mu^2}{3k} \tag{5.136}$$

$$\text{and } \theta = \frac{N\mu^2(\alpha+\beta)}{3k}, \tag{5.137}$$

where θ is called the *asymptotic Curie temperature.*

(5.135) is comparable with $\chi = \frac{C}{T-\theta}$ for ferromagnetic materials above *Curie temperature* except (5.135) contains $(T+\theta)$ instead of $(T-\theta)$ [or $(T-T_c)$]. The *Curie constant* C in (5.135) is twice the value of Curie constant for individual A and B lattice. To illustrate the difference of paramagnetic, ferromagnetic and antiferromagnetic we draw curves of χ against T in the above three cases [Fig. 5.22].

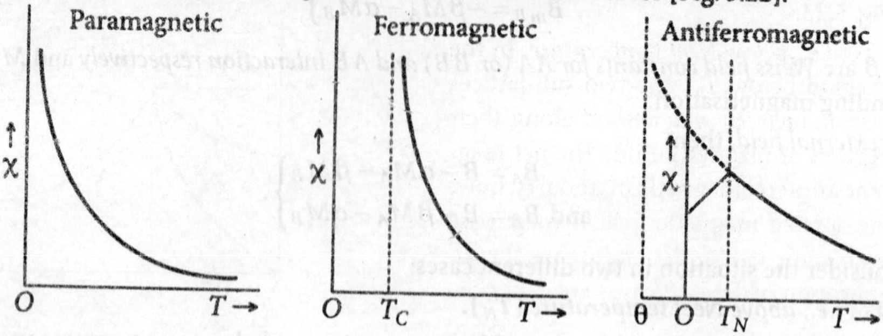

Curie law: $\chi = C/T$

Curie-Wiess law:

$\chi = C/(T - T_C)$, $[T > T_C]$ $\chi = C/(T + \theta)$, $[T > T_N]$

Fig. 5.22

At $T = T_N$, the *Néel temperature*, in absence of the applied field B, we have from (5.132)

$$M_A = -\left[N\mu^2/3kT_N\right](\alpha M_A + \beta M_B)$$

$$\because B_A = -(\alpha M_A + \beta M_B) \text{ [from (5.132) for } B = 0]$$

$$\text{or } M_A\left[1 + \frac{N\mu^2\alpha}{3kT_N}\right] + \left[\frac{N\mu^2\beta}{3kT_N}\right] M_B = 0. \quad (5.138)$$

Similarly,

$$M_B\left[1 + \frac{N\mu^2\alpha}{3kT_N}\right] + \left[\frac{N\mu^2\beta}{3kT_N}\right] M_A = 0. \quad (5.139)$$

The above equations have the trivial solution for $M_A = M_B = 0$. This means, there is no spontaneous magnetisation for above *Néel temperature.* For spontaneous magnetisation to occur there should exist a solution of the simultaneous equations for a finite value of M_A and M_B. This is possible if

$$\begin{vmatrix} \left(1 + \frac{C\alpha}{2T_N}\right) & \frac{C\beta}{2T_N} \\ \frac{C\beta}{2T_N} & \left(1 + \frac{C\alpha}{2T_N}\right) \end{vmatrix} = 0$$

$$\text{or } \left(1 + \frac{C\alpha}{2T_N}\right)^2 - \left[\frac{C^2\beta^2}{4T_N^2}\right] = 0$$

$$\text{or } 1 + \frac{C^2\alpha^2}{4T_N^2} + \frac{C\alpha}{T_N} - \frac{C^2\beta^2}{4T_N} = 0$$

$$\text{or } 4T_N^2 + C^2\alpha^2 = 4C\alpha T_N - C^2\beta^2 = 0$$

$$\text{or } T_N^2 + C\alpha T_N + \frac{C^2}{4}(\alpha^2 - \beta^2) = 0$$

$$\text{or } T_N = \frac{-C\alpha \pm \sqrt{C^2\alpha^2 - C^2(\alpha^2 - \beta^2)}}{2} = \frac{-C\alpha \pm C\beta}{2}$$

$$\text{or } T_N = \frac{C(\beta - \alpha)}{2}. \quad (5.140)$$

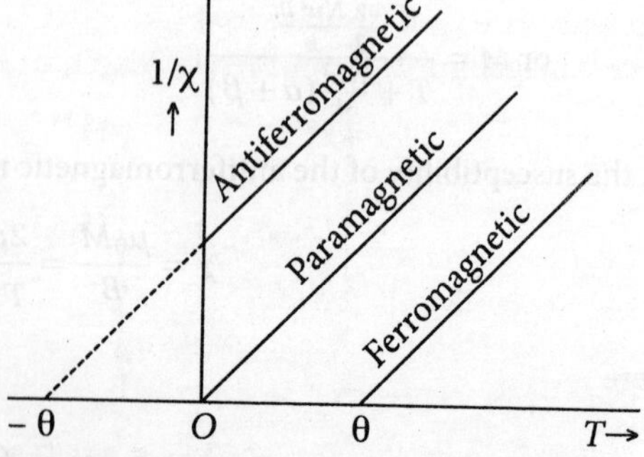

Fig. 5.23: 1/χ versus T plot for para-ferro- and antiferromagnetic materials below critical temperature.

(5.140) shows that the *Néel temperature* T_N increases as AB interaction (β) is stronger but T_N decreases as AA or BB interaction (α) is stronger.

The difference between para-, ferro- and antiferromagnetic materials above critical temperature is represented in Fig. 5.23.

Case II. $T < T_N$, *i.e., below Néel temperature.* In the temperature region below T_N, the two sub-lattices A and B are *spontaneously magnetised* along their preferred direction of magnetisation. The net magnetisation of the antiferromagnetic material is, however, zero. The applied magnetic field $\vec{B}$ magnetise the specimen to some extent depending on the direction of application of $\vec{B}$. Two cases are of special interest:

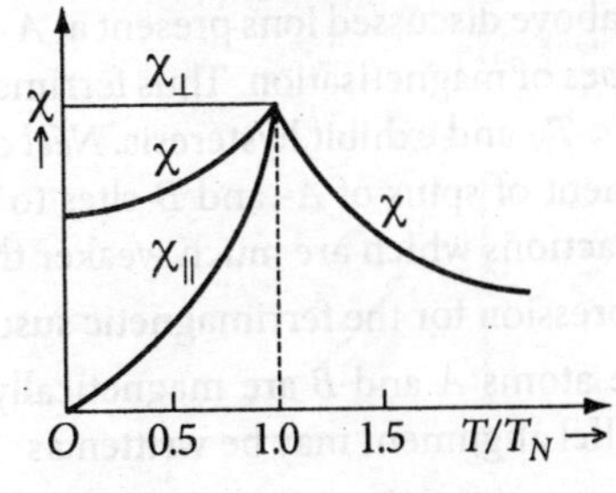

Fig. 5.24: Variation of χ with T/T_N for an antiferromagnetic material.

1. When $\vec{B}$ is *perpendicular* to the natural spin direction.
2. When $\vec{B}$ is *parallel* to the natural spin direction.

The antiferromagnetic susceptibility in the above two cases were separately determined and the average of the two is found out for the net susceptibility, i.e., $\chi = \frac{1}{2}\left(\chi_\perp + \chi_\parallel\right)$, where $\chi_\perp$ and $\chi_\parallel$ refer respectively to perpendicular and parallel directions of $\vec{B}$. $\chi_\perp$ is found to be independent of temperature. But $\chi_\parallel$ is found to decrease continuously with temperature and $\chi_\parallel \to 0$ as $T \to 0$. This implies that for $T > T_N$, the internal field is very strong which restricts the change of orientation of spins of one sub-lattice to the orientation of spins of the other sub-lattice even in presence of $\vec{B}$. Fig. 5.24 shows the variation of $\chi_\perp, \chi_\parallel$ and χ with temperature, where $\chi_\perp = \chi_\parallel = \chi$ at $T = T_N$.

5.10 Ferrimagnetism

Ferrimagnetism is identical with antiferromagnetism, where the opposing magnetic moments of two sets of atoms in the sub-lattices are of different magnitude resulting a large net magnetisation. Ferrites which are basically the oxides of various metal elements are the examples of ferromagnetic. The chemical formula for the ferrite may be written as XY_2Z_4, where X is a divalent negative ion, Z is mostly divalent oxygen ion O^{2-} and Y is Fe^{3+}ion. Chemical composition of ferrite may be written as $M_e^{2+}Fe_2^{3+}O_4^{2-}$, where M_e^{2+} represents variety of divalent ions like F_e^{2+}, CO^{2+}, Mn^{2+}, Cd^{2+}, Mg^{2+}, etc. The most common example of ferrite is magnetite or ferrous ferrite, Fe_3O_4 or FeO, Fe_2O_3.

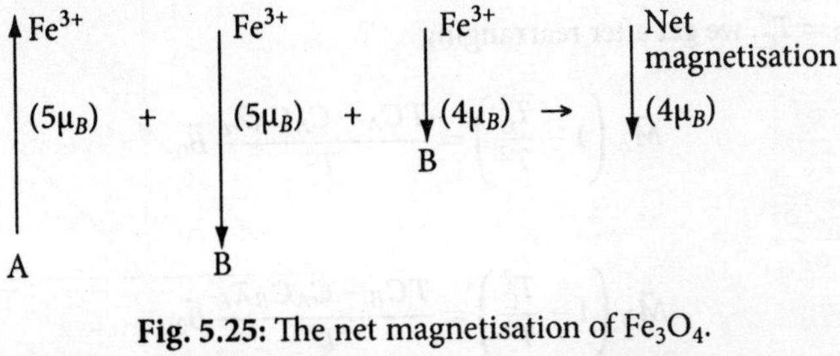

Fig. 5.25: The net magnetisation of Fe_3O_4.

The state of ferric ions Fe^{3+} with spin is $S = 5/2$ and zero orbital moment. The contribution to saturation magnetic moment is $5\mu_B$. Similarly, for Fe^{2+} ferrous ions the spin state is $S = 2$ and the contribution to magnetic moment is $4\mu_B$ in addition to any residual orbital magnetic moment, where μ_B is the

Bohr magneton. If all the ions of $Fe^{2+}Fe^{3+}O_4^{2-}$ are aligned in parallel, then the magnetic moment per mole $=(5\times2+2\times2)\mu_B=14\mu_B$. But the observed value is $4.1\mu_B$. This discrepancy can be explained if the two Fe^{3+} ions are anti-parallel so that their net moment is zero. The observed value of magnetic moment will then be due to Fe^{2+} ions having value $4\mu_B$ [Fig. 5.25].

In the above discussed ions present at A- and B-sites constitute A and B sub-lattices respectively with opposite types of magnetisation. Thus ferrimagnetic like ferromagnetic possesses spontaneous magnetisation below $T<T_N$ and exhibit hysteresis. *Néel* described the ferrimagnetic property by attributing anti-parallel arrangement of spins of A- and B-sites to the negative AB interaction. Apart from that negative AA and BB interactions which are much weaker than AB interaction are also present.

An expression for the ferrimagnetic susceptibility can be obtained as follows:

Let the atoms A and B are magnetically not equivalent. Then the magnetic moments of A and B for anti-parallel alignment may be written as

$$\vec{B}_A=\vec{B}_0-\lambda_A\vec{M}_B$$
$$\text{and } \vec{B}_B=\vec{B}_0-\lambda_B\vec{M}_A,$$

where $\vec{B}_0$ is the applied magnetic field and λ_A and λ_B are constants independent of temperature. $\vec{B}_A$ and $\vec{B}_B$ are the total magnetic fields for the atoms A and B respectively.

According to *Curie's law* above, $T>T_N$.

$$\frac{C_A}{T}=\frac{\vec{M}_A}{\vec{B}_0-\lambda_A\vec{M}_B},\quad \therefore \vec{M}_A=\frac{C_A}{T}\left(\vec{B}_0-\lambda_A\vec{M}_B\right). \tag{5.141}$$

Similarly,

$$\vec{M}_B=\frac{C_B}{T}\left(\vec{B}_0-\lambda_B\vec{M}_A\right), \tag{5.142}$$

where

$$C_A=N_A\left(\frac{e\hbar g_A}{2m}\right)^2\frac{J_A(J_A+1)}{3k} \text{ and } C_B=N_B\left(\frac{e\hbar g_B}{2m}\right)\frac{J_B(J_B+1)}{3k}.$$

Using the value of M_B from (5.142) in (5.141), we get

$$\vec{M}_A=\frac{C_A}{T}B_0-\frac{C_A\lambda_A}{T}\left[\frac{C_B}{T}\left(\vec{B}_0-\lambda_B\vec{M}_A\right)\right]=\left(\frac{TC_A-C_AC_B\lambda_A}{T^2}\right)\vec{B}_0+\frac{C_AC_B\lambda_A\lambda_B}{T^2}\vec{M}_A.$$

Putting $C_AC_B\lambda_A\lambda_B=T_C^2$, we get after rearranging

$$\vec{M}_A\left(1-\frac{T_C^2}{T^2}\right)=\frac{TC_A-C_AC_B\lambda_A}{T^2}\vec{B}_0. \tag{5.143}$$

Similarly,

$$\vec{M}_B\left(1-\frac{T_C^2}{T^2}\right)=\frac{TC_B-C_AC_B\lambda_B}{T^2}\vec{B}_0. \tag{5.144}$$

Adding (5.143) and (5.144), we get

$$(M_A+M_B)\left(T^2-T_C^2\right)=[T(C_A+C_B)-C_AC_B(\lambda_A+\lambda_B)]\,\vec{B}_0.$$

Now the ferrimagnetic susceptibility

$$\chi = \frac{\mu_0(M_A + M_B)}{B_0} = \frac{\mu_0[T(C_A + C_B) - C_A C_B(\lambda_A + \lambda_B)]}{(T + T_C)(T - T_C)}. \tag{5.145}$$

$\therefore$ $\chi \to \infty$ for $T \to T_C$. *This implies that for $T < T_N$, there is a spontaneous magnetisation.*

5.10.1 Ferrites and Their Applications

Ferrite forms an important class of *ferrimagnets* having large magnetisation at room temperature and high resistivity. The net magnetisation due to the *antiferromagnetic interaction* is zero as the sites *A* and *B* are equal and anti-parallel. In special cases the spins in these two sites do not cancel each other and we get a class of substances called *ferrimagnets* or ferrite exhibiting spontaneous magnetisation below *Néel temperature.*

Ferrites have spinel structure same as the structure of $MgAl_2O_4$. In this structure the oxygen atoms are closely packed in face-centred lattices into the interstices of which the metal ions are distributed. In a unit cell of *magnetite,* there are 8 molecules with 32 divalent oxygen ions, 16 trivalent iron ions and 8 divalent iron ions per unit cell. Due to the arrangement of oxygen atoms in face-centred structure, there are 8 tetrahedral voids (*A*-sites) and 16 octahedral voids (*B*-sites). In the *spinel structure* of $MgAl_2O_4$, the *divalent ions* occupy the sites *A* and the *trivalent ions* sites *B*. But magnetite (the natural ferrite) has an inverse spinel structure. Here the $8Fe^{3+}$ ions occupy all the *A*-sites and the other $8Fe^{3+}$ ions half the number of *B*-sites. The remaining *B*-sites are filled by Fe^{++} ions. In *nickel ferrite,* the divalent nickel ions occupy half of the *B*-sites and the remaining *B*-sites as well as the *A*-sites are filled by trivalent iron ions. As several ion-combinations in *A*- and *B*-sites are possible, a large variety of ferrite may exist.

As has been discussed earlier in magnetite, Fe_3O_4 (or, FeO, Fe_2O_3) the Fe^{3+} ions have spin state with spin $S = \frac{5}{2}$ and zero orbital moment. Thus each ion contributes $5\mu_B$ to the saturation magnetic moment, whereas the Fe^{2+} ions have spin 2 and the contribution to the magnetic moment is $4\mu_B$. Hence, the effective number of Bohr magnetons per Fe_3O_4 unit is $(2\times5+4)\mu_B = 14\mu_B$, if all the spins are parallel. The observed magnetic moment is $4.1\mu_B$, μ_B being *Bohr magneton.* This discrepancy between theory and experiment, can, however, be resolved if we assume that the moments of two Fe^{3+} ions are anti-parallel so that their net magnetic moment is zero. Then the magnetic moment of Fe_3O_4 will be due to the Fe^{2+} ions which is $4\mu_B$. Thus the discrepancy is explained.

Mixed ferrites like Mn-Zn ferrite (Ferroxcube IV) have great technical importance under the trade name *Ferroxcube.* Mixferrite have higher saturation magnetisation than simple ferrite.

Application of Ferrites

Ferrites have great technical importance as follows:

1. Ferrites are insulators having resistivity $10^{10}\,\Omega$-m and do not have eddy current loss problem as in iron specially at high frequencies. For this reason ferrite are used in *high frequency transformers* and *high frequency generators.*
2. The ferrite are used as core computer memories because of their low coercive force and short switching time.
3. Gyromagnetic properties are used in developing *isolators* and *circulators* having application in radar communication.
4. Ferrites are used in microwave devices based on *Faraday effect* in a ferromagnetic medium.

5.10.2 Ferrites and Their Properties

Ferrites are ferrimagnetic materials which are prepared from electrically non-conducting materials. Magnetite $\left[Fe^{2+}O^{2-}\cdot Fe^{3+}, O_3^2\right]$ is a common ferrite which forms a class of mixed oxides [MO, Fe_2O_3, where M stands for a divalent material like Mn, Fe, Ni, Cu, Zn].

Ferrites have the following general electrical and magnetic properties:

1. They are mechanically *hard, brittle* and difficult to work with machine.
2. They are prepared using *ceramic method.*
3. Although they show similar hysteresis curve, domain structure and motion similar to ferromagnets, the coercive force and saturation magnetisation are smaller.
4. Curie temperature for ferrite varies from 100°C to several hundred degrees centigrade.
5. Ferrites have resistivity ($\sim 10^{12}\,\Omega$-cm) and hence are insulators without eddy current loss.
6. The permeability is high (500–1000) for mixed Ni–Zn ferrite ($Ni_{0.36}Zn_{0.64}Fe_2O_3$) at low frequencies. At high frequencies ≥ 300 MHz, permeability is ~ 10.
7. They have *low dielectric loss* as microwave dielectric constant is ~ 10 to 12.
8. Saturation magnetisation and Curie temperature of some ferrimagnetic (ferrite) are shown in Table 5.6.

Table 5.6

Ferrites	*Saturation magnetisation* (Wb/m²)	*Curie temperature* (°C)
Fe_2O_3	0.60	585
$MnFe_2O_4$	0.49	300
$NiFe_2O_4$	0.34	585
$MgFe_2O_4$	0.15	440

Table 5.7: Distinction Between Magnetic Materials

Type	*Magnitude and sign of* χ	*Temperature dependence*	*Remarks*	*Examples*
Diamagnetism	Small, negative ($\sim 10^{-6}$)	Independent	No permanent dipoles, small magnetic effects.	Organic materials, light elements, Alkali earth, Bismuth.
Paramagnetism	Small, positive	Independent	Permanent dipoles randomly oriented, in absence of B net magnetisation in any given direction is zero.	Alkali and transition metals.
	large, positive	$\chi = \frac{C}{T-\theta}$		Rare-earth elements.
Ferromagnetism	Very large, positive	For $T > \theta_C$, $\chi = \frac{C}{T-\theta}$. For $T < \theta_C$, $\chi \to$ complex, $\theta_C \to$ Curie temperature	Due to large internal field, permanent dipoles align in the same direction. A large spontaneous magnetisation results in absence of $\vec{B}$.	Some transition and rare-earth metals

Type	*Magnitude and sign of χ*	*Temperature dependence*	*Remarks*	*Examples*
Antiferro-magnetism	Small, positive	When $T > T_N$, $\chi = \frac{C}{T+\theta}$; when $T < T_N$, $\chi \propto T$; $T_N \to$ Néel temperature	Due to negative molecular field, magnetic ions on lattice site–A align anti-parallel to those on lattice site–B. Also $M_A = M_B$.	Salts of transition elements
Ferrimagnetism	Very large, positive	For $T > T_N$, $\chi = \frac{C}{T \pm \theta}$. For $T < T_N$; $\chi \to$ complex	Special case of antiferromagnetism. Opposed moments are of different magnitudes and a large magnetisation occurs.	Ferrites

5.11 Answers of Some Important Questions

5.1 What is a magnetic material? What are its different classes?

Ans. Any material that can be magnetised by the application of an external magnetic field is called a *magnetic material*.

The different types of magnetic materials are:

(a) Diamagnetic materials, viz., Cu, Si, He, Au, etc.

(b) Paramagnetic materials, viz., O, N, $FeCl_2$, etc.

(c) Ferromagnetic materials, viz., Fe, Co, Ni, etc.

(d) Antiferromagnetic materials, viz., MnO, FeO, NiO, etc.

(e) Ferimagnetic materials, viz., Fe_3O_4, $MgAl_2O_4$, etc.

5.2 Where are the origins of magnetic moment in an atom?

Ans. The origins of permanent magnetic moments in an atom are as follows:

(a) The orbital magnetic moment of the electron arising out of the orbital motion of the electron around the nucleus.

(b) The spin magnetic moment of the electron due to its orbital motion around the nucleus.

(c) The spin magnetic moment of the nucleus about its own axis.

5.3 What is a Bohr magneton?

Ans. Bohr magneton is the unit of magnetic movements of atomic system.

If *e*, *m* be the charge and mass of an orbiting electron in the first Bohr orbit, then its magnetic moment is *Bohr magneton* given by

$$\mu_B = \frac{e\hbar}{2m} = 9.27 \times 10^{-24} \text{ A-m}^2,$$

where $\hbar = \frac{h}{2\pi}$, *h* is the *Planck's constant.*

5.4 What is Landé splitting factor? Why is it so called?

Ans. The energy level of an electron splits up by the action of an external magnetic field B by an amount $\Delta E = g\mu_B B$, since the split up energy ΔE of the energy level of the electron is determined by the factor g and hence the name *splitting factor*. The concept was introduced for the first time by spectroscopist *Landé* and hence the name *Landé splitting factor.*

5.5 What is Hund's rule? Explain with example.

Ans. See Art. 5.5.2.

5.6 Discuss the origin of Diamagnetism. Does diamagnetic susceptibility depend on temperature?

Ans. In an atom the electrons revolve in closed orbits around the nucleus. The electron orbit may be considered as current carrying closed coil producing magnetic field perpendicular to its plane. The electrons in atom revolve in discrete orbits whose planes are not parallel to each other and the direction of rotation of the electrons is not the same. This is why the vector sum of the magnetic moments due to all the revolving electrons on an average becomes zero.

If the electron circuit be subjected to a changing magnetic flux, an induced current is stet up in it, which according to *Lenz's law,* opposes the change in flux. The induced magnetic field in the coil opposes the applied field and produces diamagnetism. Diamagnetism is produced in all materials but it is suppressed by the stronger paramagnetic, ferromagnetic effects.

Diamagnetic susceptibility is of the order of -10^{-16} (very small) and independent of temperature.

5.7 What is the origin of paramagnetism? Discuss the temperature dependence of paramagnetic susceptibility.

Ans. The paramagnetism originates from the permanent magnetic moment of its constituent atoms or molecules. The completely filled electronic shells contribute to diamagnetism. But as the magnetic moments due to orbital and spin motions are self-neutralising, they cannot give rise to paramagnetism. Paramagnetism originates due to some of the incomplete electronic shells in atoms or molecules in which permanent dipole moment exists.

The paramagnetic susceptibility χ_{para} is small but positive and depends on temperature of material. The temperature dependence of χ_{para} is expressed by Curie law which holds good for $\mu_B \ll kT$:

$$\chi_{\text{para}} = \frac{C}{T}, \text{ i.e., } \chi_{\text{para}} \propto \frac{1}{T},$$

where C is the Curie constant and T is the temperature. The order of χ_{para} at room temperature is $\sim 10^{-5}$.

5.8 What is magneto-caloric effect? Discuss how temperature as low as 10^{-3} K can be produced.

Ans. The process of production of low temperature of the order of 10 K by the adiabatic demagnetisation is called the *magneto-caloric effect.*

For the answer of the second part of the question, see Art. 5.7.5 (P. 219).

5.9 Discuss the origin of ferromagnetism and point out its important features.

Ans. See Art. 5.8 (P. 220).

5.10 How does internal field (Weiss-field) arise in ferromagnetic crystals?

Ans. According to Weiss a ferromagnetic material of macroscopic dimension contains a number of small regions called *domain* which are spontaneously magnetised.

In a domain, the spontaneous magnetisation arises due to the alignment of the magnetic moments of the adjacent atoms in the same direction. The alignments of all the magnetic moments in a domain is caused due to the *quantum mechanical exchange interaction* of the electrons which is electrostatic in nature. This exchange interaction is called *Weiss field* or *internal field* which arises due to the overlap of the wave functions of two neighbouring atoms. As the electrons are indistinguishable they belong to both the neighbouring atoms and as such the symmetry or antisymmetry of the wave function strongly influences the energy of the system. The nature and extent of exchange interaction depends on the exchange symmetry between the spin and extent of overlap of the wave functions.

The internal field $\vec{B}_1$ is given by $\vec{B}_1 = \lambda\vec{M}$, where λ is Weiss constant and $\vec{M}$ is the magnetisation. So $\vec{B}_1 \propto \vec{M}$. $\vec{B}_1$ is generally very strong being ~ 1000 T.

5.11 What is Curie-Weiss law? Discuss the significance of Curie temperature.

Ans. The Curie-Weiss law for ferromagnetic material is given by $\chi = \frac{C}{T-T_C}$, where C is Curie constant and T_C is the Curie Temperature. The above law shows that for

(a) $T > T_C$, χ is positive and the material behaves as paramagnetic.

(b) $T = T_C$, the transition from ferromagnetism to paramagnetism occurs as χ is ∞.

(c) $T < T_C$, χ is negative, *spontaneous magnetisation* takes place.

Thus Curie temperature T_C is that temperature above which the spontaneous magnetisation vanishes. It separates the disorder paramagnetic state at $T < T_C$ from the order ferromagnetic phase at $T > T_C$ of the material.

5.12 Discuss the variation of spontaneous magnetisation with temperature for ferromagnetic materials.

Ans. We know that spontaneous magnetisation arises, due to the interaction between the neighbouring atoms which tend to align their spins in the same direction. Magnetisation of the material becomes maximum when all spins are aligned parallel to a particular direction. This long range alignment takes place due to the strong nearest neighbour interactions and the continuity of the crystal. This alignment of spins is, however, obstructed by thermal vibration of the atoms of crystal. Hence, maximum magnetisation occurs only when thermal vibration is stopped, i.e., at 0 K. If the temperature T is increased, magnetisation is decreased due to thermal vibration and becomes zero at Curie temperature (θ). The variation of spontaneous magnetisation $M_S(T)$ with temperature is shown in Fig. 5.26.

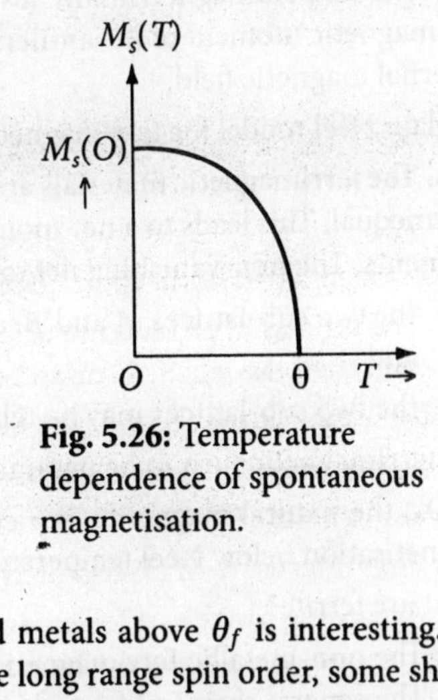

Fig. 5.26: Temperature dependence of spontaneous magnetisation.

At $T > \theta$, the ferromagnetic material behaves like paramagnetic having susceptibility

$$\chi = \frac{C}{T-\theta},$$

which is the Curie-Weiss law. Here C is the Curie constant and θ is the paramagnetic Curie temperature.

Now the ferromagnetic Curie temperature θ_f is less than paramagnetic Curie temperature ($\theta_f < \theta$) and $(\theta - \theta_f)$ is small. These elements have large value of χ above θ_f and they do not obey *Curie-Weiss law*. Fe, Co, etc. obey *Curie-Weiss law* but above θ_f, χ deviates from exact agreement of Curie-Weiss' law. The behaviour of the aforesaid metals above θ_f is interesting. Because though the thermal motion of the atoms has destroyed the long range spin order, some short range spin order still exists giving rise to magnetisation.

5.13 How the domain structure can be experimentally demonstrated?

Ans. The domain structure proposed by Weiss in ferromagnetic can be demonstrated by microphotographs of domain boundaries by *magnetic powder pattern*. In this method a drop of colloidal suspension of finely divided ferromagnetic material (e.g., magnetite) is placed on the surface of a ferromagnetic crystal. It is observed through microscope that the colloidal particles become strongly concentrated about certain well-defined lines, which are the boundaries between the domains that are magnetised in different directions. This is due to the fact, the strong magnetic field in the vicinity of the colloidal particles, attracts the magnetic particles of the solution.

5.14 What is magnetostriction? How does the magnetostriction energy originate?

Ans. A ferromagnetic material subjected to external magnetic field undergoes a change in dimension. This phenomenon is known as *magnetostriction*. The magnitude of deformation is different for different direction of the crystal but independent of the direction of the applied magnetic field. The dimension may increase or decrease depending on the nature of the material. As for example, the length of a nickel rod undergoes a decrease, whereas that for a permalloy undergoes a decrease due to the application of a magnetic field. If an iron rod be placed in an alternating magnetic field, the rod vibrates with twice the frequency of the applied field. The amplitude of vibration assumes the maximum value when the frequency of the field be varied to the natural frequency of the rod. Magnetostriction is successfully applied in ultrasonic generators.

The energy due to the mechanical stresses generated by the rotation of the domains is called the *magnetostriction energy*.

5.15 Discuss the origin of antiferromagnetism.

Ans. In ferromagnetism, the quantum mechanical exchange forces produce large internal field and the neighbouring magnetic dipoles tend to align in the same direction [Fig. 5.19(a)]. It can also be quantum mechanically proved that for small separation of the interacting magnetic dipoles, the dipole moments of the neighbouring atoms have a tendency of anti-parallel alignments [Fig. 5.19(b)] by the action of the exchange forces. This occurs below a certain temperature T_N called the *Néel temperature* and the corresponding material is called *antiferromagnetic material.* MnO, FeO, CoO and MnS are some of the examples of antiferromagnetic materials. Above Néel temperature these materials exhibit paramagnetism. In absence of external magnetic field and below Néel temperature the net magnetic moment of the antiferromagnetic is zero. Magnetisation develops by the application of external magnetic field.

5.16 Explain Néel model for ferrimagnetism?

Ans. The ferrimagnetic materials are defined as antiferromagnetic in which the sub-lattice moments are unequal. This leads to a net spontaneous magnetisation in the direction of the greater sub-lattice moments. The non-vanishing net magnetisation is attributed to the following two main cases:

(a) the two sub-lattices, A and B, can be occupied by different types and number of magnetic ions and

(b) the two sub-lattices may be related to two different crystallographic sites.

The ferrimagnetic turn to be paramagnetic if $T > T_N$, T_N being the Néel temperature.

Fe_3O_4, the natural magnetite, is a common example of ferrite. These materials show spontaneous magnetisation below Néel temperature.

5.17 What are ferrite?

Ans. The non-metallic ferrimagnetic materials are called *ferrite*. They form class of solid oxides of iron. The general chemical formula is $Me^{2+}Fe^{3+}O_4$, where Me^{2+} is a divalent metal ion like Mn, Ni, Co, Mg, Cu, etc. Mixed ferrite are called *Feroxcube,* e.g., Mn Zn.

Ferrites have important electrical and electronic applications.

5.18 Give the units of the magnetic quantities.

Ans. The following table shows the units of various magnetic quantities:

Table 5.8

Name of the magnetic quantities	*emu-cgs units*	*Rationalised MKS (SI) units*	*Conversion*
Magnetic Induction (B)	gauss = $g^{1/2}/cm^{1/2}$-s	Wb/m² = kg/(s-C) = tesla (T) = kg/(s-C)	1 Wb/m² = 10^4 gauss
Magnetic field strength (H)	oersted = $g^{1/2}/(cm^{1/2}$-s)	ampere/metre	1 A/m = $\frac{4\pi}{10^3}$ Oe
Magnetisation (M)	maxwell/cm² = $g^{1/2}/(cm^{1/2}$-s)	ampere/metre	1 A/m = $\frac{4\pi}{10^3}$ maxwell/cm²
Magnetic flex (ϕ)	maxwell ($cm^{3/2} - g^{1/2}$)/s	Wb = (kg-m²)/s-C = V-s	1 Wb = 10^8 maxwell
Susceptibility (χ)	Unit-less	Unit-less	$\chi_{SI} = 4\pi\chi_{cgs}$
Permeability (μ_r)	Unit-less	Unit-less	Same value

1 gauss = 10^{-4} T (tesla)

5.12 Solved Problems

5.1 A paramagnetic material has a magnetic field intensity of 10^4 A/m. If the susceptibility of the material at room temperature is 3.7×10^{-3}, calculate the magnetisation and flux density of the material.

Solution. The susceptibility $\chi = \frac{M}{H}$ and flux density $B = \mu_0(M+H)$.

$\therefore M = \chi H = (3.7 \times 10^{-3}) \times 10^4 = \mathbf{3.7\,A\text{-}m^{-1}}$.

Flux density $B = \mu_0(M+H) = (4\pi \times 10^{-7}) \times 10037 = \mathbf{0.0126\,Wb/m^2}$.

5.2 The magnetic field strength of a piece of metal is 10^6 A/m. Calculate the magnetisation and flux density of the material. [Given: $\chi = 1.5 \times 10^{-3}$]

Solution. The magnetisation of the metal.

$M = \chi H = (1.5 \times 10^{-3}) \times 10^6 = \mathbf{1.5 \times 10^3\ A/m}$.

Flux density $B = \mu_0(M+H) = 4\pi \times 10^{-7}\,(1.5 \times 10^3 + 10^6) = \mathbf{1.257\,T}$

5.3 Calculate the diamagnetic susceptibility of atomic hydrogen in the ground state at STP. [Assume the mean square distance of electronic charge distribution of atomic hydrogen from the nucleus $\langle r^2 \rangle = 3a_0^2$, a_0 being the radius of the first Bohr orbit of hydrogen.]

Solution. Diamagnetic susceptibility of atomic hydrogen

$$\chi_{\text{dia}} = -\frac{N\mu_0 Ze^2}{6m}\langle r^2 \rangle = -\frac{N\mu_0 Ze^2}{2m}a_0^2.$$

Here $N = \frac{6.023 \times 10^{26}}{2.24 \times 10^{-2}} = 2.69 \times 10^{28}\ \text{m}^{-3}, Z = 1, a_0 = 0.46 \times 10^{-10}$.

$$\therefore \chi_{\text{dia}} = -\frac{(2.69 \times 10^{28}) \times (4\pi \times 10^{-7}) \times (1.6 \times 10^{-19})^2 \times (0.46 \times 10^{-10})^2}{2 \times 9.1 \times 10^{-31}} = \mathbf{-1.01 \times 10^{-6}}.$$

5.4 An atom with $L = 2$ and 0 spin angular momentum is placed in a uniform magnatic field of induction 2 Wb/m². Calculate the rate of precession of the resultant magnetic moment. Also calculate the corresponding linear frequency and the radius of an electron in the ground state.

Solution. The rate of precession is the Larmor frequency given by $\omega_L = g_L\left(\frac{eB}{2m}\right)$, where

$$g_J = \frac{J(J+1) + S(S+1) - L(L+1)}{2J(J+1)} + 1.$$

Here $L=2,\ S=0,\ J=L+S=2.$

$$\therefore\ g_J=g_L=\frac{2\times3+0-2\times3}{2\times2\times3}+1=1.$$

So

$$\omega_L=\frac{1\times\left(1.6\times10^{-19}\right)\times2}{2\times\left(9.1\times10^{-31}\right)}=\mathbf{1.76\times10^{11}\ rad/s.}$$

The corresponding linear frequency

$$\nu_L=\frac{\omega_L}{2\pi}=\frac{1.76\times10^{11}}{2\times3.14}=\mathbf{2.8\times10^{10}\ Hz.}$$

The mean radius of an electron in the atom in the ground state (1s) is

$$\langle r\rangle=\frac{\hbar^2}{me^2}=\mathbf{0.53\times10^{-8}\ cm.}$$

5.5 If $N=27\times10^{23}/\text{cm}^3$ be the atomic density of helium gas and $e^2/mc^2=2.8\times10^{-13}$ cm, calculate the susceptibility of the He-atom. $[\langle r\rangle=0.53\times10^{-8}$ cm.]

Solution. The susceptibility of the He-atom is

$$\chi=-2\left(\frac{Ne^2}{6mc^2}\right)\langle r\rangle^2\ [\because \text{He is diamagnetic and there are two electrons per atom}]$$

$$\text{or } \chi=\frac{-\left(27\times10^{23}\right)\times\left(2.8\times10^{-13}\right)\times\left(0.053\times10^{-8}\right)^2}{3}$$

$$=\mathbf{-0.71\times10^{-7}\ (emu).}$$

5.6 If the magnetisation and flux density of a magnetic material be 3200 A/m and 0.005 Wb/m². Calculate the relative permeability of the material.

Solution. Magnetic induction $B=\mu_0(M+H)$, where M and H are the magnetisation and magnetising field respectively.

Here $B=0.005\ \text{Wb/m}^2$, $\mu_0=4\pi\times10^7$ and $M=3200$ A/m.

$$\therefore\ 0.005=\left(4\pi\times10^7\right)(3200+H)\ \text{or}\ H=\frac{0.005}{4\pi\times10^7}-3200.$$

$$\therefore\ H=\mathbf{780.9\ A/m.}$$

Again, the relative permeability μ_r is given by

$$\mu_r-1=\frac{M}{H}=\frac{3200}{780.9}=4.1\ \text{or}\ \mu_r=\mathbf{5.1.}$$

5.7 Calculate the change in magnetic moment of a circulating electron in an applied field of 2 T acting perpendicular to the plane of the orbit. $[r=5.29\times10^{-11}$ m for the radius of the orbit] [JNTU May 2004]

Solution. The induced magnetic moment in the orbit of the electron is

$$\mu_i=\left(\frac{e^2}{4m}\right)r^2B.$$

Here $r = 5.29 \times 10^{-11}$ m, $B = 2$T, electronic charge $e = 1.602 \times 10^{-19}$ C and $m = 9.108 \times 10^{-31}$ kg.

$$\therefore \mu_i = \frac{(1.602 \times 10^{-19})^2 \times (5.29 \times 10^{-11})^2 \times 2}{4 \times 9.108 \times 10^{-31}} = \mathbf{3.943 \times 10^{-29}\ Am^2}.$$

5.8 **A paramagnetic system of electronic spin dipole moment is placed in an external field of 10^5 A/m. (a) What is the average magnetic moment per dipole at 300 K? (b) What is the fractional number of spins parallel to the external field?**

Solution. (a) The average magnetic moment per dipole at the temperature T is

$$\langle \mu_m \rangle = \frac{M}{N} = \frac{\mu_0 H \mu_B^2}{kT}.$$

Here $\mu_0 = 4\pi \times 10^{-7}$ henry/m, $H = 10^5$ A/m, $\mu_B = 9.27 \times 10^{-24}$ Am^2, $k = 1.38 \times 10^{-23}$ and $T = 300$ K.

$$\therefore \langle \mu_m \rangle = \frac{(4\pi \times 10^{-7}) \times 10^5 \times (9.27 \times 10^{-24})^2}{(1.38 \times 10^{-23}) \times 300} = \mathbf{2.6 \times 10^{-27}\ Am^2}.$$

(b) If N be the total number and N_p the number of parallel spins, the fractional number of spins which are parallel to the field is

$$\frac{N_p}{N} = \frac{e^{(\mu_0 \mu_B H/kT)}}{[e^{(\mu_0 \mu_B H/kT)} + e^{-(\mu_0 \mu_B H/kT)}]} = \frac{e^{(2.8 \times 10^{-4})}}{\left[e^{(2.8 \times 10^{-4})} + e^{-(2.8 \times 10^{-4})}\right]} = 0.5.$$

$$\therefore \frac{N_p}{N} = \mathbf{0.5}.$$

5.9 **A paramagnetic material has 10^{28} atoms/m^3. Its susceptibility at 350 K is 2.8×10^{-4}. Calculate the susceptibility at 300 K.** [JNTU Dec. 2004]

Solution. If C be the Curie constant and T, the temperature of the paramagnetic material, then its susceptibility is given by

$$\chi = \frac{C}{T}.$$

$\therefore$ χ_1 and χ_2 be the susceptibilities at temperatures T_1 and T_2 respectively, then

$$\frac{\chi_1}{\chi_2} = \frac{T_2}{T_1}.$$

Here $\chi_1 = 2.8 \times 10^{-4}$, $T_1 = 350$ K, $T_2 = 300$ K, $\chi_2 = ?$

$$\therefore \chi_2 = \frac{\chi_1 T_1}{T_2} = \frac{(2.8 \times 10^{-4}) \times 350}{300} = \mathbf{3.267 \times 10^{-4}}.$$

5.10 **Assuming iron has a body-centred cubic structure with an edge length of 2.87 Å, calculate the average number of Bohr magnetons which contributes to the saturation magnetisation per atom. [Saturation magnetisation of iron = 1.75×10^6 A/m].**

Solution. If n be the total number of atoms per unit volume and μ, the dipole moment per atom, then saturation magnetisation is $M_s = n\mu$.

Now

$$n = \frac{2}{(2.87 \times 10^{-10})^3}.$$

Since there are two atoms per unit cell in a body-centred cubic crystal having volume of each cubic cell $V = (2.87 \times 10^{-10})^3\ m^3$.

$$\therefore\ n = \frac{2}{(2.87 \times 10^{-10})^3} = \frac{2 \times 10^{30}}{(2.87)^3}\ m^{-3}$$

$$\text{and } \mu = \frac{M_s}{n} = \frac{(1.75 \times 10^6) \times (2.87)^3}{(2 \times 10^{30}) \times (9.27 \times 10^{-24})}\ \text{Bohr magneton} = 2.23.$$

[$\because$ 1 Bohr magneton $\mu_B = 9.27 \times 10^{-24}\ Am^2$]

$\therefore$ the average number of Bohr magneton which contributes to the saturation magnetisation per atom is $\mu = 2.23$.

5.11 Gadolinium is ferromagnetic below 16°C with average magnetic moment per atom 7.1 Bohr magneton. (a) What is the magnetic moment per gram? (b) What is the value of saturation magnetisation? Given that the atomic weight of gadolinium is 157.26 and its density is $7.8 \times 10^3\ kg/m^3$?

Solution. (a) We know that 157.26 kg of gadolinium contains $= 6.025 \times 10^{26}$ atoms.

$$\therefore\ 7.8 \times 10^3 \text{ kg of gadolinium contains} = \frac{7800 \times 6.025 \times 10^{26}}{157.26} \text{ atoms.}$$

So 1 kg of gadolinium contains $= 298.836 \times 10^{26}$ atoms.

1 gm of the element contains $= 298.836 \times 10^{23}$ atoms.

$\therefore$ magnetic moment per gram

$$= (298.836 \times 10^{23}) \times 7.1 \text{ Bohr magneton}$$
$$= (298.836 \times 10^{23}) \times 7.1 \times 9.27 \times 10^{-24}\ Am^2$$
$$= 1966.849\ Am^2.$$

Hence, saturation magnetisation

$$M_s = N\mu_0\mu_m = (298.836 \times 10^{26}) \times (4\pi \times 10^{-7}) \times (9.27 \times 10^{-24})\ Wb/m^2$$
$$= 2.4726\ Wb/m^2.$$

5.12 Find the order of magnitude of the exchange integral of a ferromagnet having Curie temperature 727°C. What is the internal field? [Given: $\mu_B = 9.3 \times 10^{-21}$ ergs/gauss]

Solution. The exchange integral is given by

$$J = \frac{k\,T_c}{3},$$

where the Boltzmann constant $k = 1.4 \times 10^{-16}$ ergs/K

$$T_c = 273 + 727 = 1000\ K$$

$$\therefore\ J = \frac{(1.4 \times 10^{-16}) \times 1000}{3} = 4.66 \times 10^{-14} \text{ ergs.}$$

Again, internal field in the ferromagnet is

$$H=\frac{kT_c}{\mu_B}=\frac{(1.4\times10^{-16})\times1000}{9.3\times10^{-21}}=\mathbf{1.5\times10^7\ gauss.}$$

5.13 The outer electronic configuration of Dy^{3+} ion is $4f^9 6s^0$. Calculate the magnetic susceptibility for a salt containing one kg-mole of Dy^{3+} ions at 300 K.

Solution. Unpaired electrons in Dy^{3+} is 5.

Applying Hund's rules,

$$S=5\times\frac{1}{2}=\frac{5}{2}, L=3+2+1+0-1=5, \text{ and } J=L+S=5+\frac{5}{2}=\frac{15}{2}.$$

$$\therefore\ g=1+\frac{J(J+1)+S(S+1)-L(L+1)}{2J(J+1)}=1+\frac{\left(\frac{15}{2}\times\frac{17}{2}\right)+(5\times6)-5\times6}{2\times\frac{15}{2}\times\frac{17}{2}}.$$

$$=1+0.33=1.33.$$

The susceptibility of the material is

$$\chi=\frac{\mu_0 N\mu_B^2}{3kT}g^2 J(J+1)$$

$$\text{or } \chi=\frac{(4\pi\times10^{-7})\times(6.023\times10^{26})\times(9.27\times10^{-24})^2\times(1.33)^2}{3\times(1.38\times10^{-23})\times300}\times\frac{15}{2}\times\frac{17}{2}=\mathbf{5.9\times10^{-4}},$$

where $N=6.023\times10^{26}$ for 1 kg mole of the salt, $\mu_B=(9.27\times10^{-24})\ \text{Am}^2$ and $T=300\,\text{K}$.

5.14 Calculate the internal field for iron having Curie temperatures 1043 K and effective magnetic moment of 2.2 Bohr magneton.

Solution. If B_1 be the internal field and M the magnetisation, then $B_1=\lambda M$, λ is a constant of proportionality given by $\lambda=\frac{\mu_0 T_c}{C}$,

$$\text{where } C=\frac{\mu_0 g^2 N\mu_B^2 S(S+1)}{3k} \text{ with } g=2 \text{ and } M=N\mu_m.$$

Here T_c is the Curie temperature and C is the Curie constant. The other symbols have their usual significances.

The magnetic moment of an iron atom is $=2.2\mu_B$. So the magnetic moment $\mu_m=g\mu_B[S(S+1)]^{1/2}$ or, $2.2\mu_B=g\mu_B[S(S+1)]^{1/2}$ or, $1.1=[S(S+1)]^{1/2}$ [$\because g=2$].

$\therefore\ S(S+1)=(1.1)^2$.

Hence

$$B_1=\frac{\mu_0 T_c M}{C}=\frac{\mu_0 T_c M\times3k}{\mu_0 g^2 N\mu_B^2 S(S+1)}=\frac{\mu_0 T_c(N\mu_m)\times3k}{\mu_0 g^2 N\mu_B^2 S(S+1)}$$

$$=\frac{3kT_c(2.2\mu_B)}{g^2 S(S+1)\mu_B^2}\quad[\because \mu_m=2.2\mu_B]$$

$$=\frac{3\times(1.38\times10^{-23})\times2.2\times(9.27\times10^{-24})}{4\times(9.27\times10^{-24})^2\times(1.1)^2}\quad[\because \mu_B=9.27\times10^{-24}\ \text{Am}^2]$$

$$=2117\ \text{T}.$$

5.15 The adiabatic demagnetisation of gadolinium sulphate from an initial magnetic field of 6200 gauss to zero yields a final temperature of 0.346 K. What is the initial temperature of the specimen. [Given: Curie constant $C = 7.85\,\text{cm}^2$ deg/(gm-ion); specific heat constant $= 0.32\,R$].

Solution. If T_1 and T_2 be the initial and final temperatures of the specimen, then for adiabatic demagnetisation

$$\left(\frac{T_1}{T_2}\right)^2 = 1 + \frac{H_1^2 \times \text{Curie const}}{\text{specific constant}},$$

where H_1 is the initial field $= 6200$ gauss.

$$\therefore \quad \left(\frac{T_1}{T_2}\right)^2 = 1 + \frac{(6200)^2 \times (7.85)}{(1.38 \times 10^{-16})^2 \times (6.02 \times 10^{23}) \times 0.32}$$

$$= 1 + \frac{6200^2 \times 7.85}{1.38 \times 6.02 \times 0.32 \times 10^7} = 11.3 + 1 = 12.3$$

$$\text{or } T_1^2 = 12.3 \times T^2 = 12.3 \times (0.346)^2 \text{ or } T_1 = \mathbf{1.213\ K}.$$

5.16 Calculate (a) the intrinsic flux density near 0 K and (b) the ratio of magnetisation at 300 K in presence of an external field of 1 mT to the spontaneous magnetisation at 0 K of a ferromagnetic material. [Given: $J = \frac{3}{2}$ and $g = 2$; Curie temperature $T_c = 125$ K]

Solution. (a) The intrinsic flux density B_1 is given by

$$B_1 = \lambda M = \lambda M_s(0) = \lambda N g J \mu_B \quad [\because M_s(0) = N g J \mu_B]$$

Again, from

$$T_c = \frac{\lambda N g^2 \mu_B^2 J(J+1)}{3k} = \frac{B_1 g \mu_B (J+1)}{3k},$$

$$B_1 = \frac{3kT_c}{g\mu_B(J+1)} = \frac{3 \times (1.38 \times 10^{-23}) \times 125}{2 \times (9.27 \times 10^{-24}) \times \frac{5}{2}} = \mathbf{111.6\ T}$$

From

$$\chi = \frac{\mu_0 M}{B} = \frac{C}{T - T_c}, \quad \text{[Ref. (5.1), (5.2), (5.3) and (5.118)]}$$

we have the magnetisation

$$M = \frac{CB}{\mu_0(T - T_c)} = \frac{N g^2 \mu_B^2 J(J+1) B}{3k(T - T_c)} \quad \left[\because C = \frac{\mu_0 T_c}{\lambda} \text{ and } T_c = \frac{\lambda N g^2 \mu_B^2 J(J+1)}{3k}\right]$$

$$\therefore \quad \frac{M}{M_s(0)} = \frac{g\mu_B B(J+1)}{3k(T - T_c)} \quad [\because M_s(0) = N g J \mu_B]$$

$$\text{or} \quad \frac{M}{M_s(0)} = \frac{2 \times (9.27 \times 10^{-24}) \times 10^{-3} \times (5/2)}{3 \times (1.38 \times 10^{-23}) \times (300 - 125)} = \mathbf{6.4 \times 10^{-6}}.$$

5.17 Show that for europium oxide the paramagnetism at 300 K is much smaller than ferromagnetism at 0 K for an applied field 0.01 T. [Given: The europium ion has $g = 2$, the Curie temperature $= 70$ K and $J = 7/2$]

Solution. If M and B be magnetisation and magnetic field respectively, then the susceptibility of the specimen is

$$\chi = \frac{\mu_0 M}{B} = \frac{C}{T - T_c} = \frac{\mu_0 N g^2 \mu_B^2 J(J+1)}{3k(T - T_c)}, \quad \therefore M = \frac{N g^2 \mu_B^2 J(J+1) B}{3k(T - T_c)}.$$

At $T \to 0$, $M(0) = Ng\mu_B J$.

$$\therefore \frac{M(300)}{M(0)} = \frac{g\mu_B(J+1)B}{3k(T-T_c)} = \frac{2\times(9.27\times10^{-24})\times(4.5)\times(0.01)}{3\times(1.38\times10^{-23})\times(300-70)} = \frac{2\times9.27\times4.5\times0.01}{3\times1.38\times2300}.$$

$$\text{or } \frac{M(300)}{M(0)} = \mathbf{8.76\times10^{-5}}.$$

The specimen behaves as paramagnetic and ferromagnetic at 300 K and 0 K respectively. Since $M(300) = (8.76\times10^{-5})\,M(0)$, therefore, the paramagnetic magnetisation is $\sim 10^{-5}$ smaller than the ferromagnetic magnetisation. Hence paramagnetism is much smaller than ferromagnetism.

5.18 The hysteresis loop of a transfer has an area 2500 ergs/cm³. Calculate the loss of energy per hour at 50 Hz frequency. Density of iron is 7.5 gm-cm^{-3} and weight is 10 kg. [KU 1995]

Solution. Here the volume of the iron of the transformer $= \dfrac{\text{mass of iron}}{\text{density of iron}} = \dfrac{10\times10^3}{7.5}$ cm³.

The area of the hysteresis loop of a transformer iron = 2500 ergs/cm³.

∴ the energy loss per unit volume per cycle = 2500 ergs and the loss of energy per unit volume per hour = $2500\times(50\times60\times60)$ ergs.

Hence, the total loss of energy of the transformer per hour

$$= 2500\times(50\times60\times60)\times\frac{10\times10^3}{7.5}\ \text{ergs} = \mathbf{6\times10^{11}}\ \text{ergs}.$$

EXERCISES

5(A) Objective Type Questions

Choose the correct alternatives.

5.1 The chemical formula for magnetite is

(a) Fe_2O_3
(b) FeO
(c) Fe_3O_4
(d) $Fe(OH)_2$.

5.2 The origin of magnetism in material is the presence of

(a) electric dipole
(b) magnetic dipole
(c) magnetic monopole
(d) electric monopole.

5.3 All materials exhibit

(a) ferromagnetism
(b) ferrimagnetism
(c) paramagnetism
(d) diamagnetism.

5.4 Diamagnetic susceptibility is

(a) small and positive **(c) large and positive**

(b) small and negative **(d) large and negative.**

5.5 If a diamagnetic material is subjected to an external magnetic field the induced dipole moment acts along the

(a) applied field **(c) perpendicular to the applied field**

(b) opposite to the applied field **(d) none of these.**

5.6 Bohr magneton is the magnetic moment of

(a) an electron spin **(c) an electron orbital motion**

(b) a nuclear spin **(d) none of these.**

5.7 If the susceptibility of a material is independent of temperature, then it is

(a) ferromagnetic (c) ferrimagnetic

(b) paramagnetic (d) diamagnetic.

5.8 The materials in which the permanent dipoles are aligned due to bonding forces are called

(a) diamagnetic (c) ferromagnetic

(b) paramagnetic (d) ferrimagnetic.

5.9 The paramagnetic susceptibility varies as

(a) T (c) $\frac{1}{T}$

(b) T^2 (d) $\frac{1}{T^2}$.

5.10 The induced dipole moment (μ_i) of an orbiting electron in presence of an external magnetic field (B) is

(a) $\frac{e^2 r^2 B}{4m}$ (c) $\frac{e^2 B m}{4\pi r^2}$

(b) $\frac{e^2 B}{4m r^2}$ (d) $\frac{e^2 B}{4\pi r^2}$.

5.11 In a ferrimagnetic material the number of atoms with opposite spins is

(a) equal (c) zero

(b) unequal (d) none of these.

5.12 Interaction between the neighbouring dipoles is negligible in

(a) antiferromagnetic (c) diamagnetic

(b) ferrimagnetic (d) paramagnetic.

5.13 The relative permeability is related to magnetic susceptibility by

(a) $\mu_r = 1 - x$ (c) $\mu_r = x - 1$

(b) $\mu_r = \frac{1}{x}$ (d) $\mu_r = x + 1$.

5.14 The ferromagnetic susceptibility is given by

(a) $\chi = \frac{C}{T + T_c}$ (c) $\chi = \frac{C}{T - T_c}$

(b) $\chi = C(T + T_c)$ (d) $\chi = \frac{CT}{T + T_c}$.

5.15 The paramagnetic-ferromagnetic transition is an example of
(a) first order phase transition
(b) second order phase transition
(c) no phase transition
(d) neither of these.

5.16 If T_c and T'_c represent respectively ferromagnetic and paramagnetic Curie temperatures, then
(a) $T_c = T'_c$
(b) $T_c > T'_c$
(c) $T_c < T'_c$
(d) $T_c = \frac{1}{T'_c}$.

5.17 The experiment which conclusively proves that ferromagnetism is due to spin magnetic moments and not due to orbital magnetic moment is
(a) Davisson-Germer experiment
(b) Stern-Gerlach experiment
(c) Gyromagnetic experiment by Einstein, de Haas and Barnett.

5.18 Magnetically saturated ferrite can
(a) produce low eddy current
(b) interact with electromagnetic waves
(c) cause large hysteresis loss
(d) interact with ultrasonic waves.

5.19 The susceptibility of antiferromagnetic is given by
(a) $\chi = \frac{C}{T+\theta}$
(b) $\chi = C(T+\theta)$
(c) $\chi = \frac{C}{T-\theta}$
(d) $\chi = \frac{C}{\theta - T_c}$.

5.20 For simple ferrite, the chemical formula is
(a) $Me^{2+}Fe_2^{3+}O_4^{3+}$
(b) $Me^{3+}Fe^{4+}O_4^{2+}$
(c) $Me^{4+}Fe^{2+}O_4^{3+}$
(d) none of these.

5.21 Ferrimagnetic materials show spontaneous magnetisation for
(a) $T > T_N$
(b) $T = T_N$
(c) $T < T_N$
(d) $T = 1/T_N$.

5.22 If χ and T be the antiferromagnetic susceptibility and temperature of the specimen respectively, then
(a) χ increases with T
(b) χ decreases with T
(c) χ is independent of T
(d) all are true.

5.23 The temperature at which antiferro- to paramagnetism take place is called
(a) Debye temperature
(b) Curie-Weiss temperature
(c) Néel temperature
(d) Curie temperature.

5.24 The striking feature of antiferromagnetic is
(a) magnetic domains
(b) hysteresis loss
(c) decrease of susceptibility with increase of temperature
(d) sharp maximum in the χ versus T curve.

5.25 Materials which can be readily magnetised in either direction are called
(a) soft magnetic materials
(b) hard magnetic materials
(c) low hysteresis loss materials
(d) high hysteresis loss materials.

5.26 The dimension of ferromagnetic domain is of the order of
(a) 10^6 m
(b) 10^{-9} m
(c) 10^{-6} m
(d) 10^{-3} cm.

5.27 The susceptibility of diamagnetic material is about
(a) 10^5
(b) -10^{-5}
(c) 10^7
(d) 10^{-5}.

5.28 The material used to make permanent magnet is
(a) carbon steel
(b) Alnico V
(c) Platinum cobalt
(d) all of the three.

5.29 The ferromagnetic Curie temperature of iron is
(a) 631 K
(b) 1043 K
(c) 900 K
(d) 1428 K.

5.30 The spontaneous magnetisation of ferromagnetic at Curie temperature is
(a) 1
(b) 0
(c) ∞
(d) none of these.

5.31 The permanent magnetic moment in ferromagnetic arises due to
(a) partially filled electronic shells
(b) completely filled electronic shells
(c) innermost electronic shells
(d) none of these.

ANSWERS

5.1 (c)	5.5 (b)	5.9 (c)	5.13 (d)	5.17 (c)	5.21 (c)	5.25 (a)	5.29 (b)
5.2 (b)	5.6 (a)	5.10 (a)	5.14 (c)	5.18 (b)	5.22 (a)	5.26 (c)	5.30 (b)
5.3 (d)	5.7 (d)	5.11 (b)	5.15 (b)	5.19 (a)	5.23 (c)	5.27 (b)	5.31 (a).
5.4 (b)	5.8 (c)	5.12 (d)	5.16 (c)	5.20 (a)	5.24 (d)	5.28 (d)	

5(B) Short Answer Type Questions

5.1 What is the origin of magnetism in materials? How does it depend on temperature? [JNTU May 2003]

5.2 Discuss how diamagnetism, paramagnetism, ferromagnetism, antiferromagnetism and ferrimagnetism appear in materials?

5.3 Write down the names of three kinds of magnetic materials and discuss their differences in respect of magnetic susceptibilities. [BUH 2006]

5.4 Give a comparative study of dia-, para- and ferromagnetic materials. Give some examples of each kind. [BUH 2002]

5.5 Show that diamagnetic susceptibility increases with the number of atoms per unit volume but is independent of temperature. [BUH 2002]

5.6 Why does a paramagnetic sample display greater magnetisation when cooled? Why is it not so in diamagnetism? [BUH 2001]

5.7 Define magnetisation and show that $B = \mu_0(H + M)$. [JNTU 2004]

5.8 In hydrogen atom an electron e revolves round the nucleus at a distance of r metre with an angular velocity ω radians. Obtain expression for magnetic moment associated with it due to its orbital motion. [JNTU May 2003, Dec. '04]

5.9 Explain the detailed classification of magnetic materials on the basis of electron spin. [JNTU Jan. 2003]

5.10 What is Bohr magneton? How is it related to magnetic moment of electron?

5.11 What is a ferromagnet? How do you account for the alignment of spins in such systems? [BUH 1998]

5.12 Point out the essential difference between the classical theory and quantum theory of paramagnetism.

5.13 What is spontaneous magnetisation? How does it vary with temperature of the material?

5.14 State Curie-Weiss law and discuss its application to ferromagnetic materials.

5.15 Explain how the exchange interaction helps to explain the phenomenon of iron group of elements.

5.16 What is hysteresis? What is the significance of hysteresis loop?

5.17 How does susceptibility vary with temperature in case of antiferromagnetic materials?

5.18 Define Curie temperature and Néel temperature. Point out the essential difference between the two.

5.19 What is internal field (or Weiss field) in ferromagnetic? Find an expression for it.

5.20 How can the domain structure be experimentally demonstrated?

5.21 What is ferromagnetism? What are ferrite? Give some applications of ferrite.

5.22 Explain the *B-H* loop with the help of domain theory.

5.23 How ferrite are superior to ferromagnetic materials? [JNTU June 2005]

5.24 Explain the difference of ferro-, antiferro- and ferrimagnetic materials. [JNTU May 2003]

5.25 Explain clearly the difference between hard and soft magnetic materials. What are mixed ferrite? Mention their uses. [JNTU June 2005]

5.26 Write the importance of hard magnetic materials in engineering applications. [JNTU Jan. 2003]

5(C) Numerical Problems

5.1 A paramagnetic system of electronic spin magnetic dipole moment is placed in an external field of 10^5 A/m. Calculate the average magnetic moment per dipole at 300 K. (*Ans.* 2.6×10^{-27} A/m^2)

5.2 Given the diamagnetic susceptibility of He to be -1.9×10^{-6} cm/mole. Estimate the radius of the He atom. [BUH 1999]

(*Ans.* 0.0164 nm)

5.3 The earth has a magnetic dipole moment of 6.4×10^{21} Am^2. What current would have to be set up in a single turn of wire going around the earth at its magnetic equator, if we wish to set up such a dipole? [The mean radius of the earth $= 6.4 \times 10^6$ m] (*Ans.* 5×10^7 A)

5.4 Determine the flux density and magnetisation in Fe_2O_3 when placed in a magnetic field of strength 1.2×10^6 A/m. [Given: the susceptibility of Fe_2O_3 at 300 K is 1.4×10^{-3}] (*Ans.* 1.51 T; 1.68×10^3 A/m)

5.5 Calculate p_{eff} of Nd^{3+} with electronic configuration [Xe]$4f^3$. [BUH 2004]

5.6 Calculate p_{eff} for the following ions in the Hund ground state:

$$Pr^{3+}\ [4f^3 5s^2 p^6; {}^4I_{9/2}] \text{ and } Ni^{2+}\ [3d^8; 3F_4].$$ [BUH 2000]

5.7 Calculate the Landé g-factor and the total magnetic moment of an atom in the ${}^8S_{7/2}$ state. [BUH 1999]

5.8 State Hund's rule. Find the ground states of the following ions:

(a) Gd^{3+} ([Xe]$4f^7$)

(b) Ti^{3+} ([Ar]3d^1)

(c) Co^{2+} ([Ar]3d^7)

[BUH 1997]

5.9 **Compute the internal field for iron having Curie temperature 1043 K and an effective moment of $2.2\mu_B$ per atom.**

(Ans. 2100 T)

5.10 **What magnetic field must be applied to a salt containing Cu^{2+} ions at 1 K so that 90% of the ions are in the ground state. [Cu^{2+} has 9 electrons in the 3d shell]**

(Ans. 3 T)

5.11 **Compute the magnetic susceptibility for a salt containing one kg-mole of Cr^{2+} ions at 300 K. [The electronic configuration of Cr^{2+} is $3d^4 4s^0$]**

(Ans. 1.25×10^{-4})

5.12 **The magnetic intensity of a ferric oxide piece is 10^6 A/m. If the susceptibility of the material at room temperature is 1.5×10^{-3}, calculate the flux density and magnetisation of the material.**

(Ans. 1.259 T and 1500 A/m)

5.13 **Anhydrous nickel chloride ($NiCl_2$) has a paramagnetic susceptibility of 78.5×10^{-8} kg^{-1} at 20°C. Using simplified Langevin formula calculate the magnetic moment of the Ni^{2+} ion in Bohr magneton. [Atomic weight of Ni is 58.7 and of chlorine 35.5]**

(Ans. 4.36 Bohr magneton)

5.14 **A system of electron spins is placed in a magnetic field of 2 Wb/m^2 at temperature T. If the number of spins parallel to the field be twice as large as the number of anti-parallel spins, calculate T.**

(Ans. $T = 3.8$ K)

5.15 **A paramagnetic material is subjected to a homogeneous field of 10^6 A/m at 37°C. Calculate the average magnetic moment along the field direction per spin in Bohr magneton.**

(Ans. 2.72×10^{-3} Bohr magneton/spin)

5.16 **Calculate the theoretical saturation magnetisation in pure iron if iron has BCC structure with lattice parameter $a = 0.286$ nm and $\mu_B = 9.27 \times 10^{-24}$ Am^2.**

(Ans. 1.7×10^5 A/m^2)

5.17 **A typical magnetic field achievable with an electromagnet of iron core is 1 T. Compare the magnetic interaction energy of an electron spin magnetic dipole moment with kT at room temperature.**

(Ans. 2.3×10^{-3})

5.18 **A ferromagnetic material with $J = \frac{3}{2}$ and $g = 2$ has a transition temperature $T_c = 200$ K. Estimate the internal field near 0 K.**

[BUH 1995]

5(D) Broad Answer Type Questions

5.1 Deduce an expression for the molar diamagnetic susceptibility of atoms. Does the diamagnetic susceptibility vary with temperature? Give reasons for your answer. [BUH 1994]

5.2 (a) Calculate the magnetic moments of the following: (a) Na^+ (b) Fe^{3+} ([Ar]3d^5) (b) Ce^{3+} ([Xe]4f^1).

(b) Find an expression for the susceptibility of a ferromagnet in the paramagnetic phase. [BUH 2005]

5.3 (a) Deduce Langevin's formula for the molar diamagnetic susceptibility arising from the core electrons in an atom.

(b) How would you determine whether a material is dia-, para- or ferromagnetic? Derive an expression or the molar magnetic susceptibility of a paramagnetic compound, where each molecule contains one ion with total angular momentum $\hbar\sqrt{(J(J+1)}$. [BUH 1996]

5.4 (a) Define effective Bohr magneton number. Write the expressions for the molar susceptibility of the compounds of Gd^{3+}, Ti^{3+} and Co^{2+}, where a molecule contains only one paramagnetic ion.

(b) Derive the expression for the susceptibility of a ferromagnetic material much above the transition temperature (T_c). [BUH 1997]

5.5 What do you mean by ferromagnetism and the ferromagnetic domains? Discuss the Weiss' theory of ferromagnetism and explain how magnetic susceptibility varies with temperature. [BUH 2008]

5.6 Find an expression for paramagnetic susceptibility from quantum theory. Under what condition does it reduce to classical expression of Langevin? [CUH 2005]

5.7 (a) Discuss the characteristics of dia-, para- and ferromagnetic substances. How would you distinguish between them experimentally? [CUH 1990]

(b) Name an element which exhibits paramagnetism. Explain why inert gases do not show paramagnetism. [CUH 1997]

5.8 (a) Discuss Langevin's classical theory on paramagnetism and its physical significance. [BUH 2006]

(b) Derive Curie's law of paramagnetism from Langevin's theory.

5.9 (a) Give a comparative study of dia-, para- and ferromagnetic materials. Give some examples of each kind of material.

(b) Show that diamagnetic susceptibility increases with number of atoms per unit volume but is independent of temperature. [BUH 2002]

5.10 (a) Deduce Langevin's formula for the molar diamagnetic susceptibility. What is the diamagnetic susceptibility of a perfect diamagnet?

(b) Discuss briefly the Weiss' theory of ferromagnetism. [BUH 1999]

5.11 What is meant by ferromagnetic materials? Give some examples. What are their characteristics? Give an account of Weiss' theory of ferromagnetism. Discuss the temperature variation of saturation magnetisation. [BUH 2001]

5.12 (a) Give an account of the quantum theory of paramagnetism and explain how it removes the shortcomings of Langevin's theory.

(b) Give an account of domain theory of hysteresis.

(c) Is the permeability of a ferromagnetic material independent of magnetic field? If not, is it more for lower or higher fields? [BUH 2000]

5.13 (a) Describe the Heisenberg's exchange interaction. How does it explain ferromagnetism?

(b) Relate exchange integral to the Weiss constant and the ferromagnetic Curie temperature.

5.14 What is ferromagnetism? Discuss the Weiss-Field theory of ferromagnetism and explain how magnetic susceptibility varies with temperature. [BUH 2003]

5.15 (a) Draw the *B-H* curve for a ferromagnetic material and identify the retentivity and the coercive field on the curve. [JNTU 2005]

(b) Explain ferromagnetism and antiferromagnetism. [JNTU 2004]

5.16 (a) Deduce Langevin's formula for the molar diamagnetic susceptibility. What is the diamagnetic susceptibility of a perfect diamagnet?

(b) Calculate the diamagnetic susceptibility of atomic hydrogen in the ground state at STP using the wave function

$$\psi(r)=\frac{1}{\left(\pi a_0^3\right)^{1/2}}\,e^{-r/a_0},$$

where $a_0(=0.053\text{ nm})$ is the atomic radius. [BUH 2004]

[Hints. We have the wave function for the ground state of hydrogen atom

$$\psi(r)=\frac{1}{\left(\pi a_0^3\right)^{1/2}}\,e^{-r/a_0}.$$

Now the mean square distance of the electronic charge distribution from the nucleus is

$$\langle r^2 \rangle = \int_0^\infty \psi^* r^2 \psi \, d\tau = 4\pi \int_0^\infty \psi^* r^2 \psi r^2 \, dr = \frac{4\pi}{\pi a_0^3} \int_0^\infty r^4 e^{-2r/a_0} \, dr.$$

Let $-\frac{2r}{a_0} = x$.

$$\therefore r = -\frac{a_0}{2} x \text{ and } dr = -\frac{a_0}{2} dx, \therefore \langle r^2 \rangle = \frac{4}{a_0^3} \left(\frac{a_0^4}{16} \right) \left(-\frac{a_0}{2} \right) \int_0^\infty x^4 e^{-x} \, dx.$$

But

$$\int_0^\infty x^4 e^{-x} \, dx = 24, \therefore \langle r^2 \rangle = 3a_0.$$

Now the susceptibility of the diamagnet is

$$\chi_{\text{dia}} = -\frac{N\mu_0 Z e^2}{6m} \langle r^2 \rangle = -\frac{N\mu_0 Z e^2}{2m} a_0^2.$$

But here

$$N = \frac{6.023 \times 10^{26}}{2.24 \times 10^{-2}} = 2.69 \times 10^{28} \text{ m}^{-3}.$$

Here $Z = 1$

and $a_0 = 0.053 \text{ nm} = 0.53 \times 10^{-10}$ m.

$$\therefore \chi_{\text{dia}} = -\frac{(2.69 \times 10^{28}) \times (4\pi \times 10^{-7}) \times (1.6 \times 10^{-19})^2 \times (0.53 \times 10^{-10})^2}{2 \times 9.1 \times 10^{-31}}$$

$$= -\frac{2.69 \times (4 \times 3.14) \times (1.6 \times 1.6) \times (0.53 \times 0.53)}{2 \times 9.1} \times 10^{-6}$$

$$= -\frac{2.69 \times 12.56 \times 2.56 \times 0.28}{18.2} \times 10^{-6} = \mathbf{-1.33 \times 10^{-6}} \text{ (SI units)]}$$

5.17 **(a) What is the difference between diamagnetic and paramagnetic atoms? Describe a method for measuring magnetic susceptibility of a powder sample. Derive the expression you use.**

(b) What is Curie law of magnetic susceptibility?

Derive the susceptibility expression for a paramagnetic compound having one Mn^{2+} ([Ar]$3d^5$) molecule per molecule.

(c) What is a ferromagnet? How do you account for the alignment of spins in such a system?

[BUH 1998]

5.18 **(a) Describe the two-sub-lattice model to explain antiferromagnetism.**

(b) How does this model account for the difference between the Néel temperature (T_N) and Curie-Weiss temperature (θ)?

5.19 **(a) Discuss the phenomenon of antiferromagnetism.**

(b) Show that the susceptibility of antiferromagnetic materials is inversely proportional to ($T + T_N$), where T_N is the Néel temperature.

5.20 (a) Explain Ferrimagnetism.
(b) What are ferrite? Discuss their applications.

5.21 (a) Give an account of Néel theory of antiferromagnetism and show how the ferromagnetic behaviour of ferrite can be explained from Néel theory.
(b) Describe the structure of ferrite. How is the magnetic moment of ferrite molecule calculated?

MODULE 4

Chapter 6

Crystal Structure

6.1 Introduction

In nature, matter exists in the *solid, liquid* and *gaseous states.* The atoms and molecules in a solid are bound to one another with strong forces of attraction and as such solids have definite *shape* and *size.* A solid deforms elastically under small shear stresses. The branch of physics which deals with the properties of solid is known as *Solid State Physics* or *Physics of the Condensed Matter.*

Solids are classified broadly into two classes, viz., *crystalline* and *non-crystalline* (or *amorphous*).

In *crystalline solids,* the atoms are arranged in periodic manner in all three directions. The crystalline state being the low energy state, most of the solids prefer to be in crystalline state. This is because energy released during the formation of ordered structure is more than that in disordered structure. The crystalline solids are again subdivided into *single crystal* and *microcrystalline* (or *polycrystalline*). The periodicity of atoms extends throughout the volume of the matter in single crystal-like diamond, quartz, etc. In polycrystalline material, there is an aggregate of so many small *crystallites* oriented randomly and separated by *grain boundaries.* The crystallines are called *grains.*

In *amorphous solids,* the atoms or molecules are randomly arranged and the solid exhibits short range order. Amorphous solids are formed when the atoms do not get sufficient time to be arranged in periodic manner. *Glass* is an example of amorphous solid.

The sketch in Fig. 6.1 shows the classification of solids.

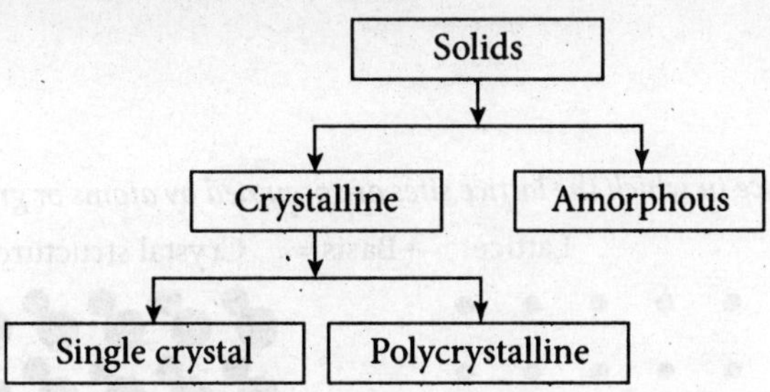

Fig. 6.1: Classification of solids.

The physical properties of solids depend on the surroundings of atoms, ions or molecules forming the solid. As the amorphous solids have no directional properties they are known as *isotropic* substances. Whereas a crystal has different periodic arrangements in all the three directions, its physical properties vary with direction and, therefore, the crystal is called *anisotropic* substance. The branch of science dealing with the study of forms and physical properties of crystalline solid is called *crystallography.*

6.1.1 Properties of Crystalline Solids

Crystalline solids like calcium carbonate, ammonium chloride, quartz, etc. have the following specific properties:

1. A crystalline solid has definite melting point.
2. A crystalline material has a definite geometrical shape and size.
3. Crystalline solids are anisotropic.
4. The strength of all the bonds between the ions, atoms or molecules being the same, their chemical composition is uniform.
5. The crystalline solids have definite colour (some are transparent) in visible light.

6.1.2 Properties of Amorphous Solids

Amorphous solids like glass, etc. have the following properties:

1. Amorphous solids have no definite shape.
2. They are isotropic in nature.
3. Their chemical composition is not uniform as the bonds between the atoms are not of equal strength.
4. Their melting point is not fixed due to the variation of bond strength.

6.2 Space Lattice (or Point Lattice)

An infinite regular and periodic array of points in three dimensions in which every point has surroundings identical to that of every other point in the array is called a space lattice.

The term *identical surroundings* means that the lattice appears to have the same view when looked from a point $P(\vec{r})$ as from another point $Q(\vec{r}\,')$ in the lattice with reference to an arbitrary origin O, Fig. 6.2.

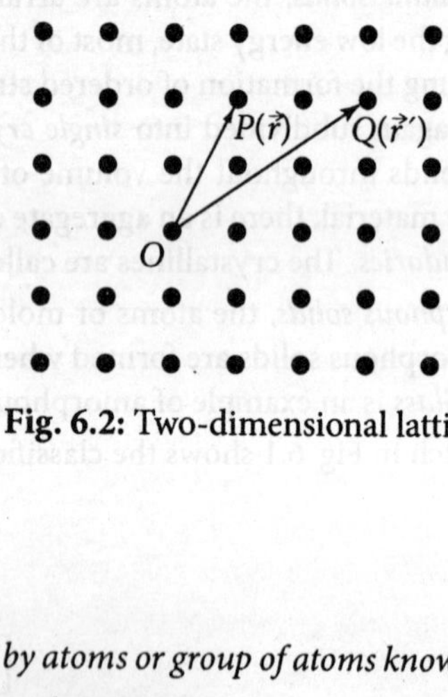

Fig. 6.2: Two-dimensional lattice.

6.3 Crystal Lattice

It is defined as a space lattice in which the lattice sites are occupied by atoms or group of atoms known as basis.

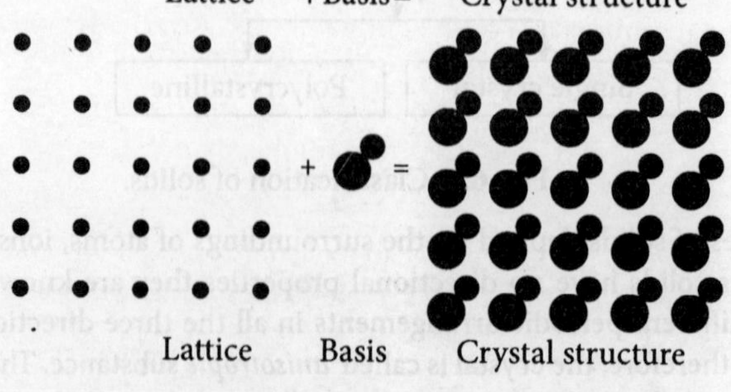

Fig. 6.3: Lattice, Basis and Crystal structure.

6.4 Basis

A basis is defined as a cluster of atoms, ions or molecules identical in composition, arrangement and orientation. **The periodic repetition of the basis in three dimensions forms the structure of the crystal.**

The simple relation between lattice, basis and crystal structure can be represented as in Fig. 6.3.

6.5 Symmetry Considerations

The operation which leaves the crystal and its environment invariant is called *symmetry operation.* External symmetry of most of the crystal shows that, most of them possess *geometrical symmetry.*

The following sketch shows the classification of symmetry operations.

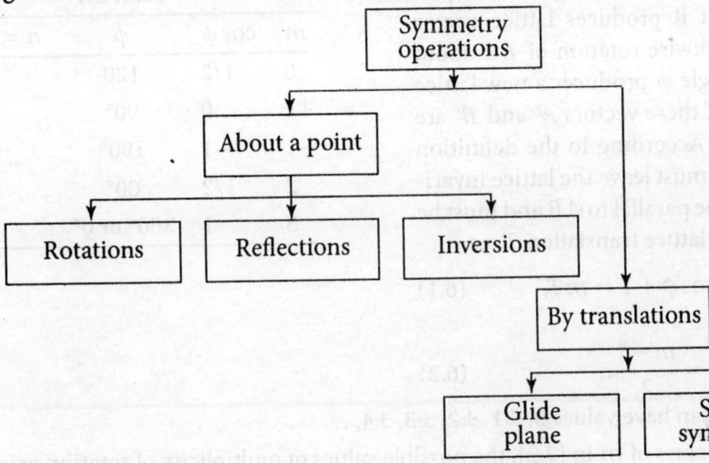

Fig. 6.4: Symmetry operations.

6.6 Point Group Symmetry

The symmetry operations performed about a point or a line are called *point group symmetry.* The different point group symmetries are as follows: (i) Rotation, (ii) Reflection, (iii) Inversion, (iv) Rotation-inversion axis, (v) Combination of symmetry elements.

(i) *Rotational symmetry.* If after rotation of a body about an axis by some angle ϕ [Fig. 6.5], the body appears as it did prior to rotation, i.e., the body is left *invariant* as a result of rotation, then the body is said to possess rotational symmetry. For the body to be left invariant by the rotation, the rotation angle ϕ must be an integral sub-multiple of 2π, i.e., $\phi_n = \frac{2\pi}{n}$ where n is an integer and is called the *multiplicity* of the rotation axis. A point has multiplicity of rotation axis $n = 1$, a square has $n = 4$, a hexagon has $n = 6$, etc.

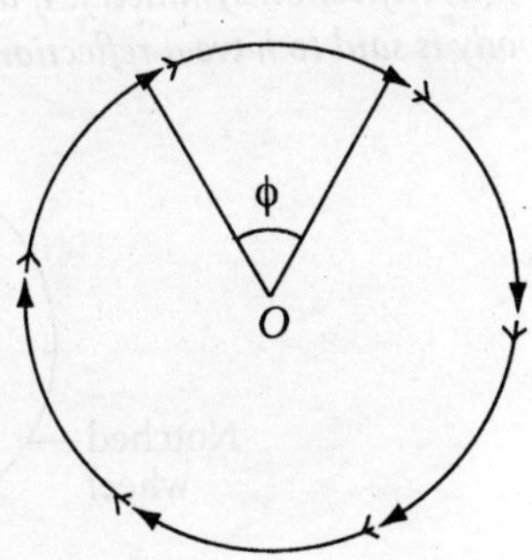

Fig. 6.5: Rotation through an angle ϕ.

Absence of five-fold symmetry axis. A crystal can have only 1-, 2-, 3-, 4- and 6-fold axes of rotational symmetry. ***It can be proved that 5-fold rotational symmetry axis is impossible.***

Proof. We consider a lattice row $OABC$ as shown in Fig. 6.6. Let the lattice has n-fold rotation axis and $\vec{T}$ is the lattice translation vector. The rotation axis passes through the lattice point A and is perpendicular to the plane of the paper. As all the lattice points are equivalent, there must be a similar axis through the tip of B.

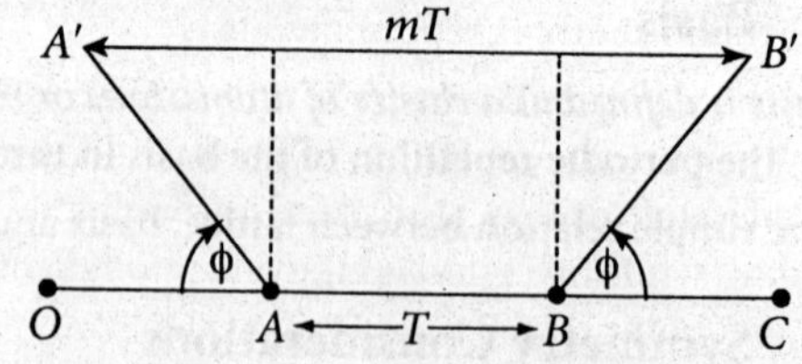

Fig. 6.6: Proof of non-existence of 5-fold symmetry axis of rotation.

Let AO be rotated through an angle ϕ, where $\phi = 2\pi/n$ about A so that it produces lattice vector AA'. Similarly, anticlockwise rotation of BC about B through the same angle ϕ produces a new lattice vector BB'. The tips of these vectors A' and B' are the new lattice points. According to the definition of rotation operation it must leave the lattice invariant. Hence, $A'B'$ must be parallel to AB and must be integral multiple of the lattice translation T, i.e.,

$$A'B' = 2T\cos\phi + T = mT, \tag{6.1}$$

where

$$\cos\phi = \frac{m-1}{2}. \tag{6.2}$$

Table 6.1

m	$\cos\phi$	ϕ	$n = 360/\phi$
0	−1/2	120	3
+1	0	90°	4
−1	−1	180°	2
2	1/2	60°	6
3	−1	360° or 0°	1

Since $|\cos\phi| \leq 1$, m can have values $0, \pm1, \pm2, \pm3, \pm4, \ldots$

Now using different values of m in (2.2), the possible values of multiplicity of rotation axis (n) will be as shown in Table 6.1.

Thus, a crystal can have only 1-, 2-, 3-, 4- and 5-fold rotational axes of symmetry and cannot have 5-*fold axis* of rotation symmetry.

(ii) *Reflection symmetry. If a body is left unchanged in every way after being reflection at a plane, then the body is said to have a reflection symmetry.*

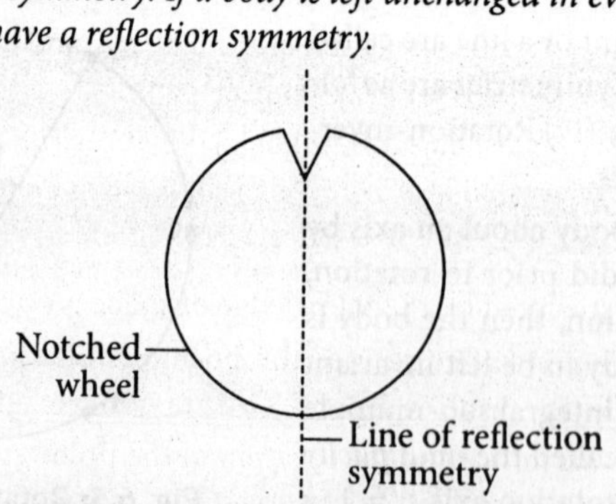

Fig. 6.7: (a) Reflection symmetry using a notched wheel.

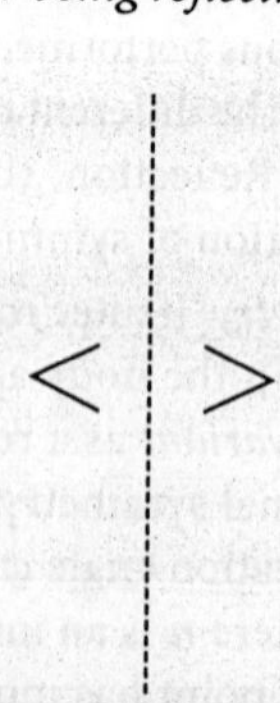

Fig. 6.7: (b) Reflection across a line.

The reflection symmetry is illustrated in Fig. 6.7, where the reflection plane passes through the centre of a *notched* wheel. The symbol used to represent reflection symmetry is m.

(iii) *Inversion. If for every lattice point represented by the position vector $\vec{r}$ there are corresponding lattice points at the position $-\vec{r}$, then the crystal structure is said to possess inversion symmetry.*

An inversion centre is represented by I.

Inversion symmetry operation is similar to Reflection symmetry operation, the only difference is that reflection occurs at a plane through the lattice point, while inversion is equivalent to reflection through a point [Fig. 6.8].

(iv) *Rotation-inversion axis. If a crystal is brought into self-coincidence by rotation followed by an inversion about a lattice point through which rotation axis passes, then the crystal structure is said to possess a rotation-inversion axis of symmetry.*

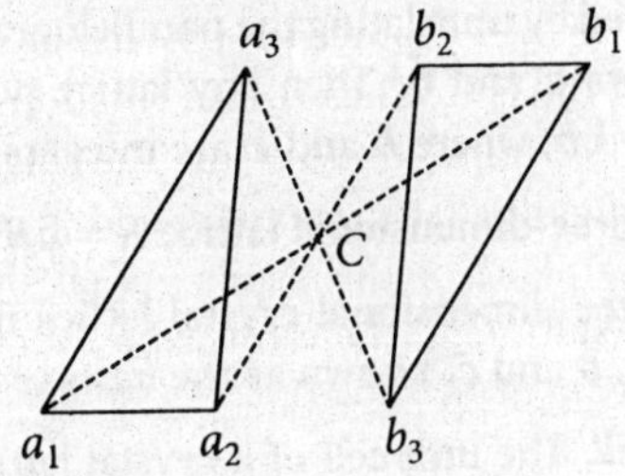

Fig. 6.8: Points a_1, a_2, a_3 of a triangle is reflected through the point C and producing the new points b_1, b_2, b_3.

A crystal structure can possess 1-, 2-, 3-, 4- and 6-fold rotation inversion axes denoted by $\bar{1}$, $\bar{2}(\equiv m)$, $\bar{3}$, $\bar{4}$, $\bar{6}(\equiv 3)$ plus normal reflection planes. A *rotation inversion operation* is known as *improper rotation* and the corresponding axis as *improper axis*.

(v) *Combination of symmetry elements. Point group.* Symmetry elements like rotation, reflection and inversion can be combined provided they are *compatible.* Different crystal elements possess different symmetries. *The various combinations of allowed rotation and reflection operations is called point group.* There are 42 point groups.

Note. For a detailed discussion on *Point group symmetry*, the reader is referred to *Introduction to Solids*, Azaroff, Tata McGraw Hill (1997).

6.7 Translation Symmetry of Crystals

(i) *Glide plane.* The combination of a mirror plane with simultaneous translation operation in a crystal is known as Glide plane. The glide plane in a crystal is always *parallel* to the mirror plane.

(ii) *Screw axis.* The rotation axis coupled with the translation parallel to the rotation axis gives rise to screw axis.

The symbol used for screw axis is n_m which is performed by rotation of $\theta = 2\pi/n_m$ and a translation of m/n times the translation vector parallel to the rotation axis in a crystal.

(iii) *Space group.* The combination of *point group symmetry* and *translation symmetry elements*, if they are compatible, give rise to *space group symmetry.* There are 230 *space groups* exhibited by crystals.

Characteristics of Space Group

1. The symmetry of a crystal structure can be specified by the knowledge of space group.
2. The space group, through its symmetry elements, determines the positions of equivalent points within the unit cell.
3. The space group is characterised by the *Bravais lattice* and the location of the point group and other symmetry elements.
4. For a particular space group, if only one point in a cell is occupied, then it is necessary that all equivalent points be occupied by identical atoms or molecules.

6.8 Some Crystallographic Terms

We consider a two-dimensional crystal lattice as shown in Fig. 6.9, where the parallelogram $ABCD$ defined by the vectors $\vec{a}$ and $\vec{b}$ may be taken as a *unit cell* of the crystal. It is evident that the whole crystal may be reproduced by translating the parallelogram $ABCD$ along $\vec{a}$ and $\vec{b}$ direction merely by suitably multiplying the vectors $\vec{a}$ and $\vec{b}$. Then any lattice point in the crystal may be represented by a vector $\vec{R}$ such that, $\vec{R} = h\vec{a} + k\vec{b}$, where h and k are integers.

For a three-dimensional lattice $\vec{R} = h\vec{a} + k\vec{b} + l\vec{c}$.

The three-dimensional crystal lattice may be reproduced by repeated translation of the non-coplanar vectors $\vec{a}$, $\vec{b}$ and $\vec{c}$, known as the *basis vectors.*

(i) *Unit cell.* The unit cell of a crystal lattice is defined as *the region of the crystal represented by the basis vectors $\vec{a}$, $\vec{b}$ and $\vec{c}$ such that a translation of this region by any integral multiple of these vectors will lead to a similar region of the crystal.*

It is also defined as *the fundamental elementary pattern of minimum of atoms, molecules or group of molecules which represent fully all the characteristics of the crystal.*

The choice of a unit cell is not unique.

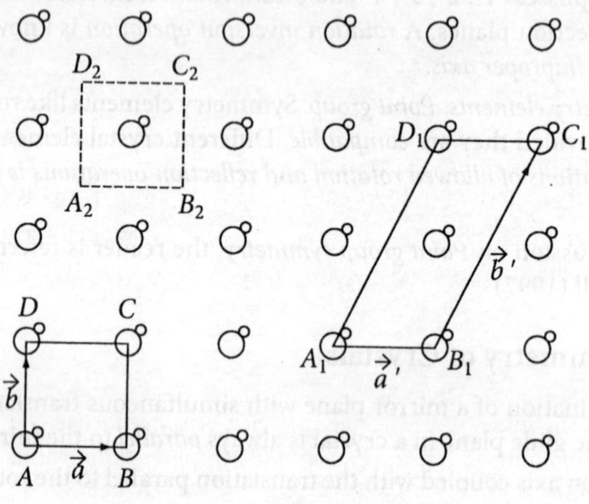

Fig. 6.9: Unit cells in two-dimensional crystal lattice.

(ii) *Basis vectors.* The three non-coplanar vectors $\vec{a}$, $\vec{b}$ and $\vec{c}$ used to define a unit cell are called the *basis vectors.*

(iii) *Primitive unit cell. The unit cell occupying the smallest volume in the crystal lattice is termed as a primitive unit cell.* The unit cells $ABCD$, $A_1 B_1 C_1 D_1$, $A_2 B_2 C_2 D_2$ are all primitive unit cells [Fig. 6.10].

(iv) *Primitive basis vectors. The three non-coplanar vectors $\vec{a}$, $\vec{b}$ and $\vec{c}$ which define a primitive unit cell are called primitive basis vectors.*

$\vec{a}'$, $\vec{b}'$, $\vec{c}'$ are also primitive basis vectors for the primitive unit cell, $A_1 B_1 C_1 D_1$.

The primitive basis vectors $\vec{a}$, $\vec{b}$, $\vec{c}$ and interfacial angles α, β, γ are called the *basic lattice parameters* as they determine the form and actual size of the *unit cell.*

Difference between Unit Cell and Primitive Cell

A unit cell, in addition to an atom at its each corner, may or may not contain additional atom or atoms [Fig. 6.10 (a)].

But a primitive unit cell contains only one atom at its each corner [Fig. 6.10 (b)].

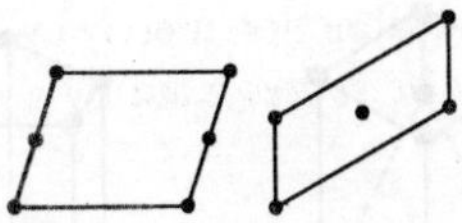

Fig. 6.10: (a) Unit cells.

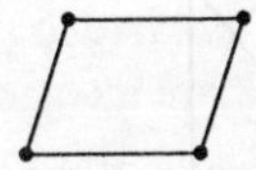

Fig. 6.10: (b) Primitive unit cells.

From above we may conclude that *all primitive cells are unit cells but all unit cells may or may not be primitive cells.*

Lattice Constants

The length a, b and c specifying the size of a unit cell are called *lattice constants.* For a cubic crystal $a = b = c$ and a is the *lattice constant.*

Wigner-Seitz Primitive Cell

Wigner-Seitz primitive unit cell may be constructed as follows:

(a) Connect a given lattice point to all the nearby lattice points.

(b) Draw normals at the *midpoints* of the lines connecting the lattice points as shown in Fig. 6.11.

The smallest volume enclosed by the normal planes is the *Wigner-Seitz primitive unit cell.* The volume of this cell is $v_0 = |\vec{a} \cdot \vec{b} \times \vec{c}|$.

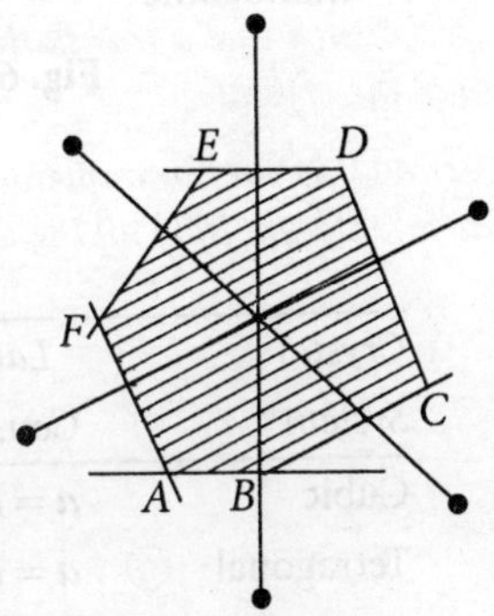

Fig. 6.11: Wigner-Seitz primitive unit cell (*ABCDEF*).

6.9 Bravais Lattice

Bravais in 1884 proposed the concept of the *space lattice. A space lattice is a mathematical concept which represents infinite number of points in space such that the arrangement of points about a given point is similar to that about any other point.*

There are 14 distinguishable ways in which points can be arranged in three-dimensional space. Therefore, Bravais lattice are 14 *in number* as shown in Figs 2.12 (i) to 2.12 (xiv) and belong to 7 *crystal system* [Table 6.2]. The 14 Bravais lattice and their names are listed in Fig. 6.12.

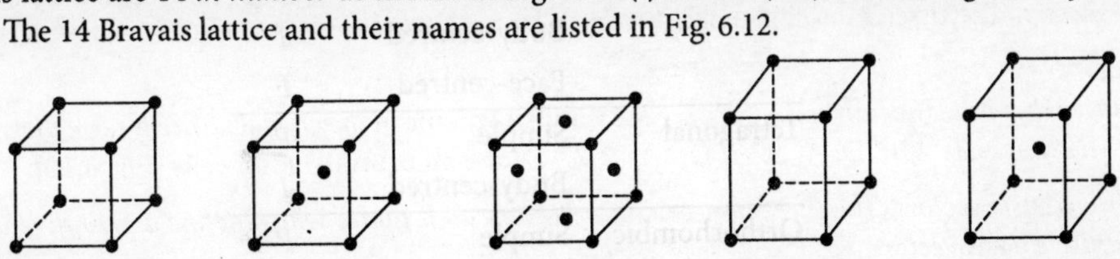

(i) Simple cubic (SC) (ii) Body-centred cubic (BCC) (iii) Face-centred cubic (FCC) (iv) Simple Tetragonal (ST) (v) Body-centred Tetragonal (BCT)

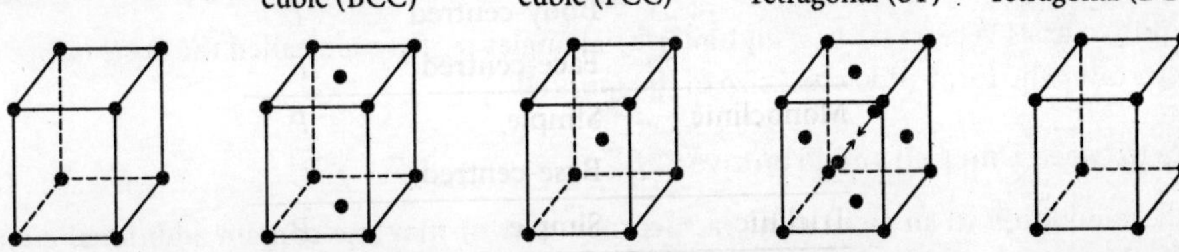

(vi) Simple Orthorhombic (SOR) (vii) Base-centred Orthorhombic (viii) Body-centred Orthorhombic (ix) Face-centred Orthorhombic (x) Simple monoclinic

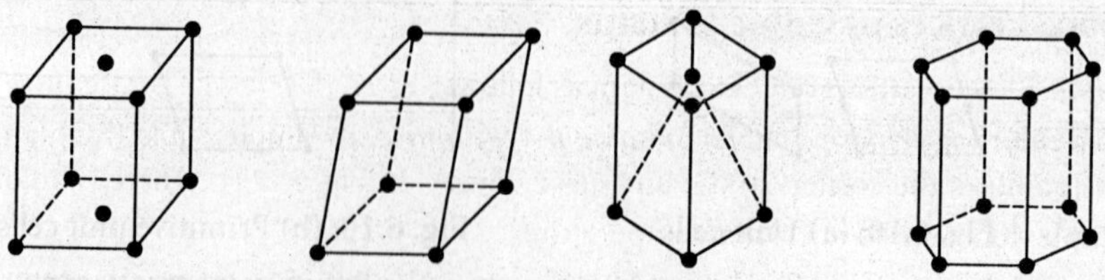

(xi) Base-centred monoclinic (xii) Triclinic (xiii) Trigonal (xiv) Hexagonal

Fig. 6.12: The Bravais lattices in 3 dimensions.

Table 6.2: Crystal Systems

Crystal System	*Lattice Constants*	α, β, γ	*Example*
Cubic	$a=b=c$	$\alpha=\beta=\gamma=90°$	$NaCl, CaF_2, NaClO_2$
Tetragonal	$a=b\neq c$	$\alpha=\beta=\gamma=90°$	$NiSO_4, SnO_2$
Orthorhombic	$a\neq b\neq c$	$\alpha=\beta=\gamma=90°$	$KNO_3, BaSO_4$, $MgSO_4$
Monoclinic	$a\neq b\neq c$	$\alpha=\beta=90°, \gamma\neq 90°$	Na_2SO_4, $FeSO_4$
Triclinic	$a\neq b\neq c$	$\alpha\neq\beta\neq\gamma\neq 90°$	$CuSO_4, K_2Cr_2O_7$
Trigonal	$a=b=c$	$\alpha=\beta=\gamma\neq 90°$	$CaSO_4$
Hexagonal	$a=b\neq c$	$\alpha=\beta=90°, \gamma=120°$	SiO_2, AgI

Table 6.3: Bravais lattices and crystal systems

Crystal systems	*Bravais lattices*	*Symbols*
Cubic	Simple	*P*
	Body-centred	*I*
	Face-centred	*F*
Tetragonal	Simple	*P*
	Body-centred	*I*
Orthorhombic	Simple	*P*
	Base-centred	*C*
	Body-centred	*I*
	Face-centred	*F*
Monoclinic	Simple	*P*
	Base-centred	*C*
Triclinic	Simple	*P*
Trigonal	Simple	*P*
Hexagonal	Simple	*P*

6.10 Space Lattices of Cubic Systems

There are *three* Bravais lattices of cubic system as follows:

1. *Simple cubic (or primitive cubic) or cubic p-type lattice*: In simple cubic (SC) lattice, there are eight lattice points at the corners of the unit cell as shown in Fig. 6.12 (i). There is no other lattice point.
   Example: KCl and Po at certain temperature.
2. *Face-centred cubic or cubic F lattice*: In addition to the lattice points at the eight corners it contains lattice points at the centre of each face [Fig. 6.12 (ii)]. There is an extra lattice point at the centre of each face of the face-centred cubic (FCC) lattice unit cell.
   Example: Cu, Ni, Au, Ag crystals.
3. *Body-centred cubic or cubic I lattice*: In body-centred cubic (BCC) lattice, there is one additional lattice point situated at the body centre besides eight lattice points situated at the eight corners as shown in Fig. 6.12 (iii). Thus, there is an extra lattice point at the body centre of the unit cell.
   Example: Li, Na, K, Cr crystals.

6.10.1 Calculation of Number of Atoms per Unit Cell

1. *Simple cubic (SC) lattice*: Every lattice point in SC lattice situated at the corner of the unit cell is shared by eight unit cells and hence its contribution to each cell is $\frac{1}{8}$.
   Total contribution of eight corners of the SC lattice is $\frac{1}{8} \times 8 = 1$. Thus, for an SC lattice there is one lattice point per unit cell.
2. *Face-centred cubic (FCC) lattice*: There are eight lattice points at the eight corners of the FCC lattice and contribution of these eight lattice points to the unit cells is $\frac{1}{8} \times 8 = 1$.
   Moreover, there are six-faces of the unit cell each atom of which belongs to two unit cells. So, the contribution of these six lattice points to the unit cell is $\frac{1}{2} \times 6 = 3$.
   ∴ total number of lattice points per FCC cell is $1 + 3 = 4$.
3. *Body-centred cubic (BCC) lattice*: Here there are eight lattice points at the eight corners of the unit cell and each is a member of eight unit cells. In addition, there is one lattice point at the body centre. So, the total number of lattice points per BCC unit cell is $\frac{1}{8} \times 8 + 1 = 2$.

6.10.2 Coordination Number

For a particular lattice, *coordination number* (CN) *is defined as the number of equivalent nearest neighbours to a particular atom in the crystal lattice.*

6.10.3 Nearest Neighbour Distance

It is defined as the distance between the centres of two nearest neighbouring atoms in the crystal.

6.10.4 Atomic Radius (r)

Atomic radius is defined as half of the distance between nearest neighbour atoms in the crystal.

Here the atoms are supposed to be spherical in size and any two nearest neighbouring atoms touch each other.

1. SC *lattice*: In Fig. 6.13 AB represents the radius r of an atom and $r = a/2$.
   The nearest neighbour distance $d = 2r = 2\frac{a}{2} = a$.

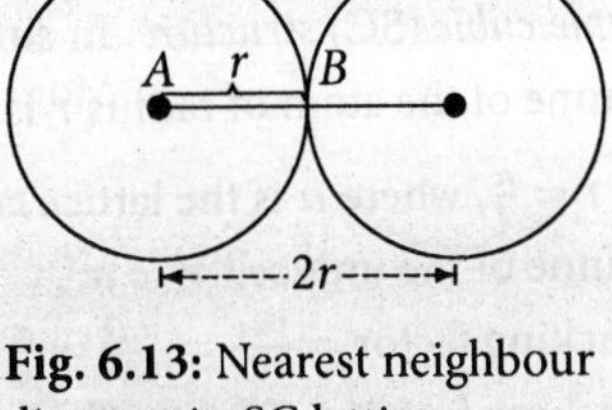

Fig. 6.13: Nearest neighbour distance in SC lattice.

2. BCC *lattice*: In Fig. 6.15 a BCC unit cell is shown, where $AB^2 = a^2 + a^2$.

$\therefore AB = \sqrt{2}a$ and $AC^2 = \left(\sqrt{2}a\right)^2 + a^2 = 3a^2$.

$\therefore AC = \sqrt{3}a$.

$\therefore$ atomic radius $r = \frac{\sqrt{3}a}{4}$ $\left(\because 4r = \sqrt{3}a\right)$.

So, the atomic radius of a BCC crystal is $r = \frac{\sqrt{3}a}{4}$.

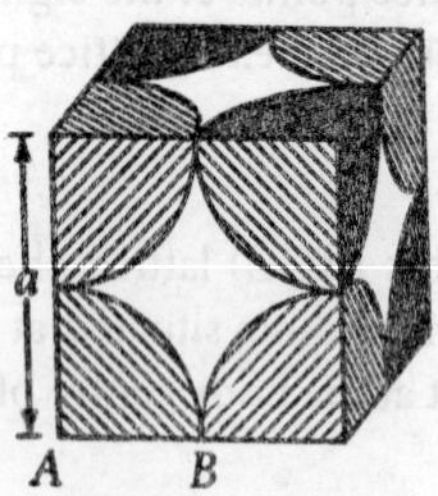

Fig. 6.14: Geometry of an SC unit cell.

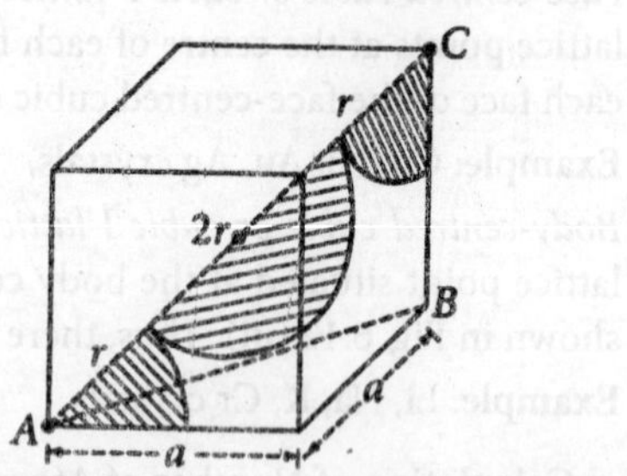

Fig. 6.15: Geometry of a BCC unit cell.

The nearest neighbour distance,

$$d = 2r = \frac{\sqrt{3}}{2}a.$$

3. FCC *lattice*: Fig. 6.16 shows an FCC unit cell whose face diagonal is AC.

Face diagonal, $AC = \sqrt{2}a = 4r$.

$$\therefore \text{atomic radius } r = \frac{\sqrt{2}a}{4} \text{ or } r = \frac{a}{2\sqrt{2}}.$$

The nearest neighbour distance,

$$d = 2r = \frac{a}{\sqrt{2}}.$$

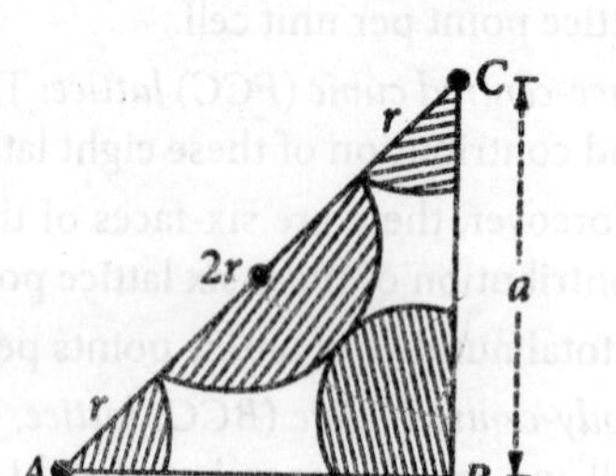

Fig. 6.16: Face diagonal of an FCC unit cell.

6.10.5 Atomic Packing Factor

The atomic packing factor of a unit cell is defined as the ratio of the volume occupied by the atoms in the unit cell to the total volume of the unit cell.

If there are n atoms each of volume v, and V is the volume of the unit cell, then Packing Factor (PF) $= \frac{nv}{V}$. Sometimes packing factor (PF) is expressed in %, then % PF = PF × 100.

Atomic packing factor is sometimes called *packing fraction*.

Calculation of Packing Factor

1. *Simple cubic* (SC) *structure*: In simple cubic structure there is one atom in the unit cell.

Volume of the atom of radius r is $v = \frac{4}{3}\pi r^3$.

But $r = \frac{a}{2}$, where a is the lattice constant of SC structure. $\therefore v = \frac{4}{3}\pi\left(\frac{a}{2}\right)^3 = \frac{\pi a^3}{6}$.

Volume of the unit cell $V = a^3$.

$\therefore$ packing factor $= \frac{1 \times v}{V} = \frac{\pi a^3}{6a^3} = \frac{\pi}{6} = 0.523$.

$\therefore$ packing fraction $= \frac{\pi}{6} \times 100 = 0.523 \times 100 = 52.3\%$.

The structure is thus loosely packed. Example of this structure is polonium.

2. *Body-centred cubic* (BCC) *structure*: Here the number of atoms per unit cell is 2.

$\therefore$ volume of these two atoms $= 2 \times \frac{4}{3}\pi r^3$.

But atomic radius $r = \frac{\sqrt{3}}{4}a$.

$\therefore$ total volume occupied by the two atoms is $2 \times \frac{4}{3}\pi\left(\frac{\sqrt{3}}{4}a\right)^3 = \frac{\sqrt{3}\pi}{8}a^3$.

Volume of the unit cell $V = a^3$.

$\therefore$ packing factor (PF) $= \frac{\sqrt{3}\pi a^3}{8 \times a^3} = \frac{\sqrt{3}\pi}{8} = 0.68$.

$\therefore$ packing fraction $= \frac{\sqrt{3}\pi}{8} \times 100 = 0.68 \times 100 = 68\%$.

Examples of this type of structure are tungsten, sodium, iron and chromium.

3. *Face-centred cubic* (FCC) *structure*: Here the number of atoms per unit cell = 4.

$\therefore$ volume of 4 atoms $= 4 \times v = 4 \times \frac{4}{3}\pi r^3$.

But atomic radius $r = \frac{\sqrt{2}}{4}a$.

$\therefore$ total volume occupied by the 4 atoms is $4 \times \frac{4}{3}\pi\left(\frac{\sqrt{2}}{4}a\right)^3 = \frac{\sqrt{2}\pi}{6}a^3$ and volume of the unit cell $V = a^3$.

$\therefore$ packing factor (PF) $= \frac{\sqrt{2}\pi a^3}{6 \times a^3} = \frac{\sqrt{2}\pi}{6} = 0.74$.

So, packing fraction $= \frac{\sqrt{2}\pi}{6} \times 100 = 0.74 \times 100 = 74\%$.

So, FCC structure has highest packing fraction compared to SC and BCC structure. Examples of this type of structure are most of the metals like copper, aluminium, lead and silver.

Table 6.4 shows some crystal structure with examples.

Table 6.4: Some crystal structure with examples

Parameters	*Simple Cubic* (SC)	*Body-Centred Cubic* (BCC)	*Face-Centred Cubic* (FCC)	*Hexagonal Close Packed Structure* (HCP)	*Diamond Cube* (DC)
Atomic radius	$a/2$	$\frac{\sqrt{3}a}{4}$	$\frac{a}{2\sqrt{2}}$	$\frac{a}{2}$	$\frac{\sqrt{3}a}{8}$
Nearest neighbour distance	a	$\frac{\sqrt{3}a}{2}$	$\frac{a}{\sqrt{2}}$	a	$\frac{\sqrt{3}a}{4}$
Coordination number	6	8	12	12	4
Volume of the unit cell	a^3	a^3	a^3	$\frac{3\sqrt{3}}{2}a^2c$	a^3
No. of atoms per unit cell	1	2	4	6	8
Packing factor	0.52	0.68	0.74	0.74	0.34
Example	Polonium	Sodium, Lithium and Chromium	Aluminium, Copper, Silver and Lead	Magnesium, Zinc and Cadmium	Germanium, Silicon and Diamond

6.11 Calculation of Lattice Constant

We consider a cubic crystal of lattice constant a. Then volume of the unit cell $= a^3$.

If ρ be the density of the crystal, then mass in each unit cell of the crystal $= \rho a^3$.

6.12 Some Other Crystal Structures

6.12.1 Hexagonal Closed Packed (HCP) Structure

The HCP structure is a closest packing structure having packing fraction close to that of FCC space lattice as shown in Fig. 6.17. The unit cell of HCP structure contains one atom at each corner, one atom at the centre of the hexagonal face, in addition to three atoms within the body of the unit cell. An atom in the unit cell touches three atoms in the layers above and below and six atoms in its plane situated at the corners of the hexagon.

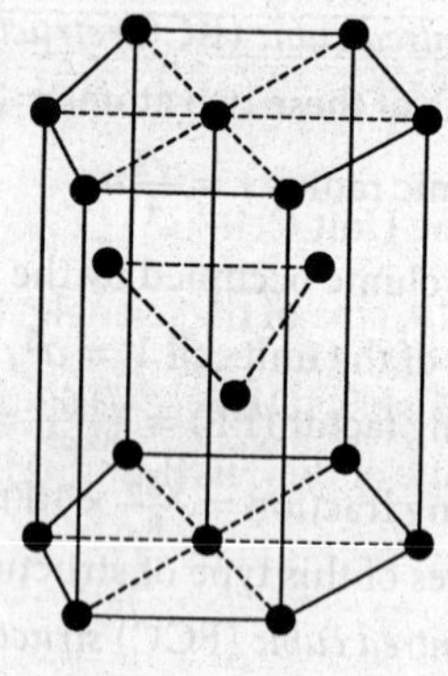

Fig. 6.17: Unit cell of HCP structure.

If r be the radius of each atom and a the inter-atomic distance, then $2r = a$. Also the coordination number is 12.

Three atoms are within the body of the cell having full contribution to the cell. The top layer contains seven atoms, each corner atom is shared by surrounding hexagon cells. The centre atom is shared by two adjacent cells.

Therefore, in a unit cell the total number of atoms $= 3/2 + 3/2 + 3 = 6$.

c/a Ratio for an Ideal HCP Structure

There are three atoms at the bottom and the top faces in a horizontal plane at $c/2$ (where c is the height of the cell) from the ortho-centres of alternate equivalent triangles at the top or base of the cell. These three atoms rest on the three corner atoms of the triangle as shown in Fig. 6.18. From the triangle ABY of Fig. 6.19, we have

$$\left.\begin{aligned}\cos 30° = \frac{AY}{AB}, \therefore\ AY = \frac{a\sqrt{3}}{2}\\ \text{and } AZ^2 = AX^2 + ZX^2\end{aligned}\right\}. \tag{6.3}$$

From the triangle AXZ,

$$AX = \frac{2}{3}AY = \frac{2a\sqrt{3}}{6} = \frac{a}{\sqrt{3}}.$$

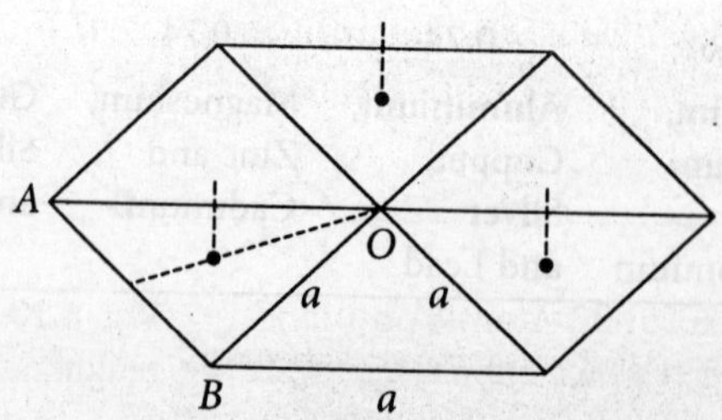

Fig. 6.18: Ideal HCP structure.

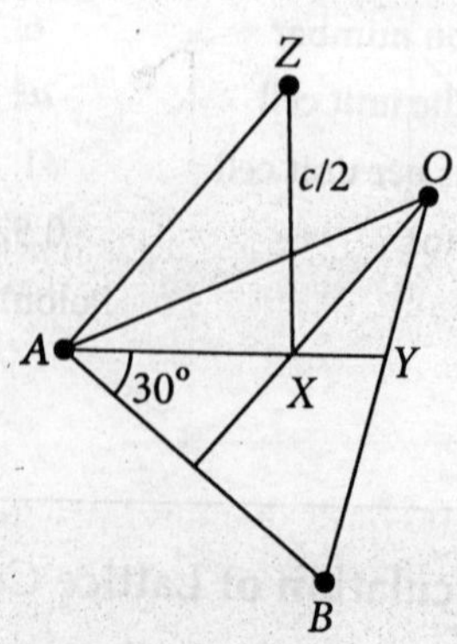

Fig. 6.19

$\therefore$ from the second equation of (2.3)

$$a^2 = \frac{a^2}{3} + \frac{c^2}{4} \text{ and } \frac{c^2}{4} = a^2 - \frac{a^2}{3} = \frac{2}{3}a^2 \text{ or } \frac{c^2}{a^2} = \frac{8}{3} \text{ or } \frac{c}{a} = \sqrt{\frac{8}{3}}. \tag{6.4}$$

Volume of the Unit Cell

From Fig. 6.19, area of the triangle $AOB = \frac{1}{2}BO \times AY = \frac{1}{2}a\left[a\frac{\sqrt{3}}{2}\right]$.

$\therefore$ base area of $\Delta AOB = \frac{6\times a^2\sqrt{3}}{4} = \frac{3}{2}\sqrt{3}\,(a^2)$.

Now volume of the HCP unit cell

$$V = \frac{3}{2}\sqrt{3}(a^2 \times c) = \frac{3\sqrt{3}a^2 \times c}{2}. \tag{6.5}$$

Packing Fraction

We have for the HCP structure, the nearest neighbour distance, $2r = a$, lattice parameter, $a = 2r$, number of atoms in the unit cell = 6,

volume of all the atoms per unit cell, $v = 6 \times \frac{4}{3}\pi r^3 = \frac{24\pi r^3}{3} = \pi a^3$,

volume of the unit cell, $V = \frac{3\sqrt{3}a^2c}{2}$,

coordination number, $CN = 12$, and $\frac{c}{a} = \sqrt{\frac{8}{3}}$.

$\therefore$ density of packing fraction

$$\text{PF} = \frac{v}{V} = \frac{\pi a^3 \times 2}{3\sqrt{3}a^2c} = \left(\frac{2\pi}{3\sqrt{3}}\right)\left(\frac{a}{c}\right) = \left(\frac{2\pi}{3\sqrt{3}}\right)\left(\frac{3}{8}\right)^{1/2} = \frac{2\pi}{3\sqrt{3}} \times \frac{\sqrt{3}}{2\sqrt{2}}$$

$$\text{or PF} = \frac{\pi}{3\sqrt{2}} = 0.74 \text{ or } 74\%. \tag{6.6}$$

6.12.2 Diamond Cubic Structure

The unit cell of diamond is similar to that of germanium and silicon as shown in Fig. 6.20 (a). The space lattice is FCC with a basis of two carbon atoms at the lattice points. The projection of atoms in the cubic

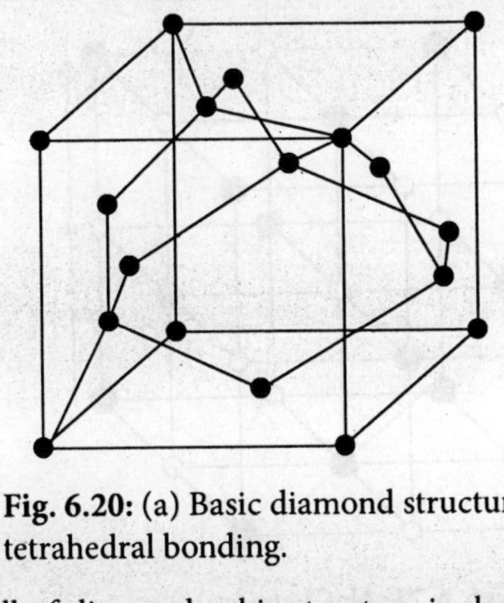

Fig. 6.20: (a) Basic diamond structure tetrahedral bonding.

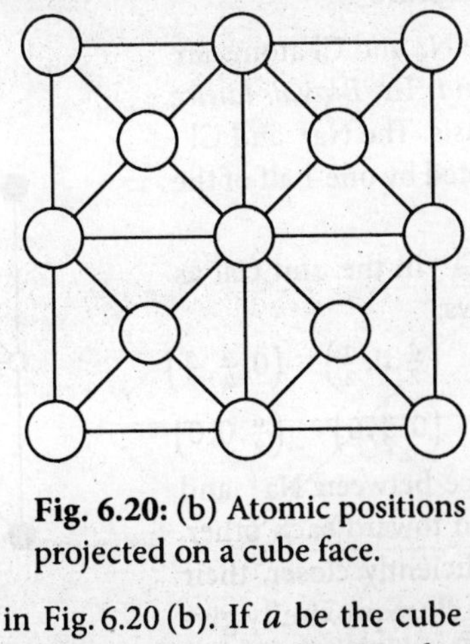

Fig. 6.20: (b) Atomic positions projected on a cube face.

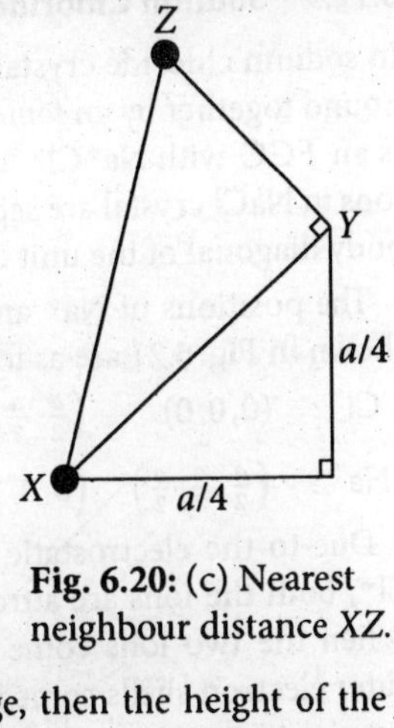

Fig. 6.20: (c) Nearest neighbour distance XZ.

cell of diamond cubic structure is shown in Fig. 6.20 (b). If a be the cube edge, then the height of the atoms above the base is as represented in the figure. The points at O and $a/2$ are on the face lattice and those at $a/2, 3a/4$ are on a similar lattice displaced along the body diagonal by 1/4 of its length. Thus, the diamond lattice is a combination of two inter-penetrating FCC sub-lattices along the body diagonal displaced relative to each other by $a/4$.

The coordination number being 4, the diamond cubic structure is loosely packed. There are eight corner atoms along with six face-centred atoms and four more atoms are presented in the unit cell. Again, each face-centred atom is shared by two unit cells and the corner atoms are shared by eight adjacent unit cells. Hence, the number of atoms per unit cell $= \left(\frac{1}{2}\times 6\right) + \left(\frac{1}{8}\times 8\right) + 4 = 8$.

Packing Fraction

From Fig. 6.20 (c),

$$XY^2 = \left(\frac{a}{4}\right)^2 + \left(\frac{a}{4}\right)^2 = \frac{a^2}{8}.$$

Also

$$XZ^2 = XY^2 + YZ^2 = \frac{a^2}{8} + \frac{a^2}{16} = \frac{3a^2}{16}.$$

But $XZ = 2r$.

$\therefore$ nearest neighbour distance

$$2r = \left(\frac{3a^2}{16}\right)^{1/2} \quad \text{or } (2r)^2 = \frac{3a^2}{16}.$$

Hence, lattice constant, $a = \frac{8r}{\sqrt{3}}$.

Volume of all atoms in the unit cell, $v = 8 \times \frac{4}{3}\pi r^3$.

Volume of the unit cell, $V = a^3$.

$\therefore$ packing fraction

$$\text{PF} = \frac{v}{V} = \frac{\frac{32\pi r^3}{3}}{\left(\frac{8r}{\sqrt{3}}\right)^3} = \frac{32 \times \pi r^3 \times 3\sqrt{3}}{8 \times 8 \times 8 \times r^3} = \frac{\pi\sqrt{3}}{2 \times 8} = 0.34.$$

$\therefore$ PF = 34%.

6.12.3 Sodium Chloride Structure

In sodium chloride crystal the Na and Cl atoms are bound together by an *ionic bond*. The *Bravais lattice* is an FCC with Na^+Cl^- as basis. The Na^+ and Cl^- ions in NaCl crystal are separated by one-half of the body diagonal of the unit cell.

The positions of Na^+ and Cl^- in the unit cell as shown in Fig. 6.21 are as follows:

Cl^-: $(0,0,0)$ $\left(\frac{a}{2},\frac{a}{2},0\right)$ $\left(\frac{a}{2},0,\frac{a}{2}\right)$ $\left(0,\frac{a}{2},\frac{a}{2}\right)$

Na^+: $\left(\frac{a}{2},\frac{a}{2},\frac{a}{2}\right)$ $\left(0,0,\frac{a}{2}\right)$ $\left(0,\frac{a}{2},0\right)$ $\left(\frac{a}{2},0,0\right)$

Due to the electrostatic force between Na^+ and Cl^-, both the ions are attracted toward each other. When the two ions come sufficiently closer, their outer electron shells come into close proximity giving rise to a strong repulsive force. Equilibrium is attained when the force of attraction and repulsion become equal.

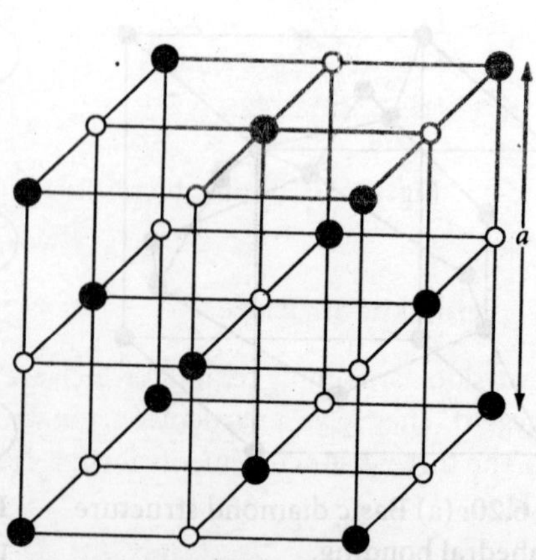

Fig. 6.21: NaCl structure

○ → Na^+ ● → Cl^-

The coordination number of Na^+ or Cl^- in NaCl crystal is 6. The ionic radius of Cl^- is 1.18 Å whereas that of Na^+ is 0.92 Å. The number of molecules in a unit cell of NaCl is 4.

The crystals having the same structure as NaCl are KCl, KBr, MgO, AgBr, etc.

6.12.4 Cesium Chloride Structure

The space lattice of cesium chloride is simple cubic as shown in Fig. 6.22. The ionic radii of both Cs^+ and Cl^- are approximately equal. The basis comprises of Cs^+-ion at (0,0,0) and Cl^--ion at $(a/2, a/2, a/2)$. Each of Cs^+ and Cl^- is situated at the centre of a cube of ions of electrically opposite in nature.

The lattice points of CsCl consist of two interpenetrating simple cubic lattice, one occupied by Cs^+-ion and the other by Cl^--ion. So, the coordination number is 8.

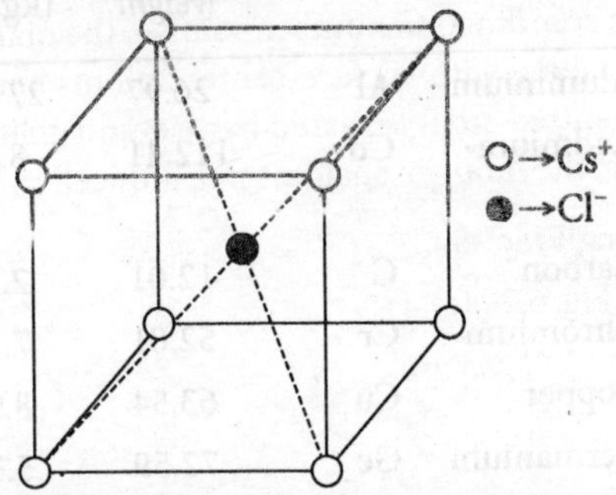

Fig. 6.22: CsCl structure.

6.12.5 Zinc Blend (ZnS) Structure

The structure of ZnS is identical to diamond structure except that the two interpenetrating sub-lattices are occupied by two different elements. In this structure the Zn-atoms are placed on one FCC lattice and S-atoms on the other [Fig. 6.23]. The ZnS structure does not possess centre of inversion whereas diamond does.

Other similar structures are CdS, AgI, ZnSe and SiC, $InSb_2$, GaAs and CuCl.

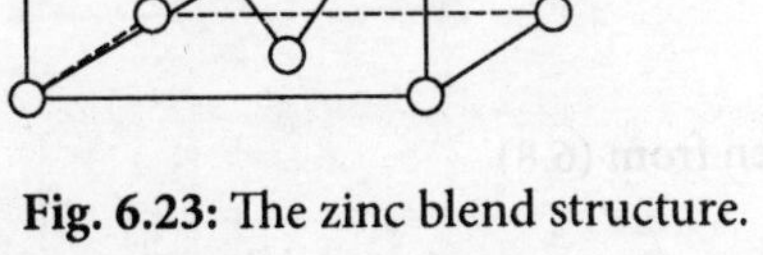

Fig. 6.23: The zinc blend structure.

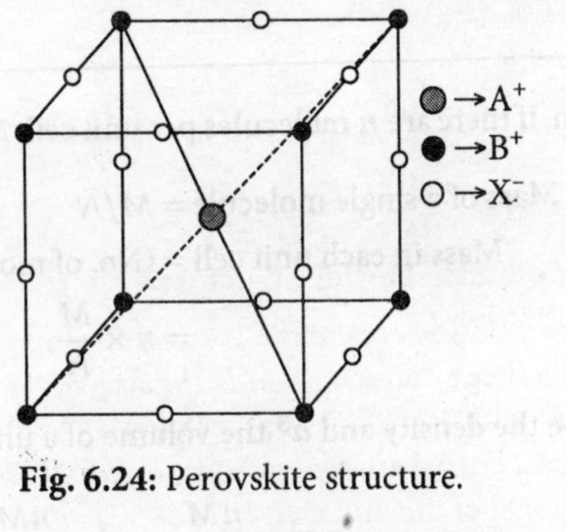

Fig. 6.24: Perovskite structure.

6.12.6 Perovskite Structure

The perovskites are simply ceramics which are solids comprising of metallic elements and non-metals (usually oxygen). They have a particular arrangement of atoms in them. The general formula for perovskite is ABX_3. The three distinct elements form a cube and present in the ratio $A : B : X_3 = 1 : 1 : 3$ as shown in Fig. 6.24.

In this solid, A and B are metallic cations and X is non-metallic axioms. The A-cation is largest in size and occupies at the centre of the cube; B-cations occupy the eight corners. The non-metal X-axions occupy the mid-points of each edge of the cube.

Table 6.5: Crystal structure and some properties of representative elements

Elements	*Symbol*	*Atomic weight*	*Density* (kg/m³)	*Crystal structure*	*Lattice constant at* 20°C (nm)	*Approximate atomic radius*	*Melting Point* (°C)
Aluminium	Al	26.97	27	FCC	0.4049	0.1428	660.2
Cadmium	Cd	112.41	8.65	HCP	$a = 0.2979$ $c = 0.5617$	0.1489	320.9
Carbon	C	12.01	2.22	DC	0.3564	0.0770	3700
Chromium	Cr	52.01	7.19	BCC	0.2885	0.1249	1890
Copper	Cu	63.54	8.96	FCC	0.3615	0.1278	1083
Germanium	Ge	72.59	5.36	DC	0.5658	0.1224	937
Iron (α)	Fe	55.85	7.87	BCC	0.2866	0.1238	1539
Lead	Pb	207.21	11.34	FCC	0.4949	0.1750	327
Magnesium	Mg	24.32	1.74	HCP	$a = 0.3209$ $c = 0.5210$	0.1594	650
Potassium	K	39.10	0.86	BCC	0.5339	0.2312	63.7
Silicon	Si	28.09	2.34	DC	0.5428	0.1176	1430
Silver	Ag	107.88	10.49	FCC	0.4086	0.1444	960.5
Sodium	Na	22.99	0.97	BCC	0.4281	0.1857	97.8
Zinc	Zn	65.38	7.13	HCP	$a = 0.2664$ $c = 0.4945$	0.1332	419.5

Again, if there are n molecules per unit cell, M, the molecular weight and N, the Avogadro number, then

$$\text{Mass of a single molecule} = M/N \tag{6.7}$$

Mass in each unit cell = (No. of molecules in a unit cell) × (mass of a single molecule)

$$= n \times \frac{M}{N}. \tag{6.8}$$

If ρ be the density and a^3 the volume of a unit cell, then from (6.8)

$$\rho a^3 = \frac{nM}{N} \text{ or } a^3 = \frac{nM}{\rho N} \text{ or lattice constant } a = \left(\frac{nM}{\rho N}\right)^{1/3}. \tag{6.9}$$

Thus, if n, M, ρ and N are known, the value of the lattice constant a of a crystal can be determined from (6.9).

6.13 Lattice Planes and Miller Indices

A crystal lattice may be considered as an aggregate of a set of parallel equidistant planes passing through the lattice points. These set of parallel planes are called *lattice planes* which can be chosen in a number of ways like (1), (2), (3) and (4) as shown in Fig. 6.25. *Miller* discovered a method to designate a plane in a crystal using three numbers h, k and l, known as *Miller indices*.

Miller indices are the three smallest possible integers which have the same ratios as the reciprocals of the intercepts of the plane concerned on the three axes.

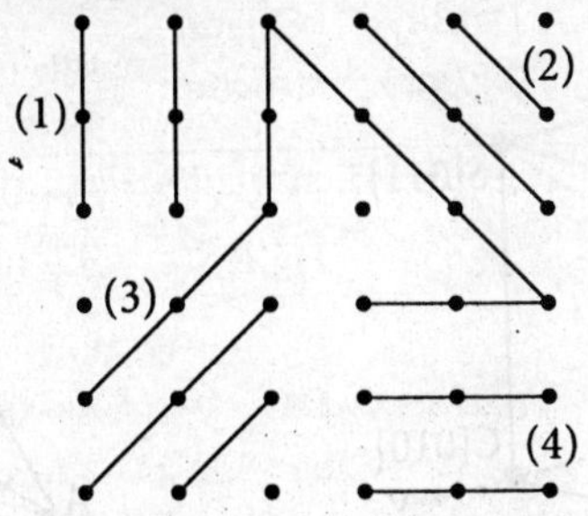

Fig. 6.25: Set of lattice planes.

6.13.1 Crystallographic Directions

In a crystal, the concentration of atoms in different direction is different which leads to the anisotropy of it. For this reason it is necessary to locate different directions and planes for crystallographic analysis.

Definition. *The crystallographic direction in a crystal is defined as line joining any two points in the lattice.*

In two dimensions, the crystallographic direction is expressed as a line passing through a point $A(x,y)$ and any other lattice point taken as the origin $O(0,0)$ as shown in Fig. 6.26, *the coordinates x and y being whole numbers.*

In Fig. 6.26, the directions of the point A and B with O as the origin are $(6,5)$ and $(7,3)$ respectively.

In *three dimensions,* the crystallographic direction of any point A having coordinates (x,y,z), where x,y,z are whole numbers, is conventionally denoted by putting x,y,z in brackets, i.e., $[x,y,z]$.

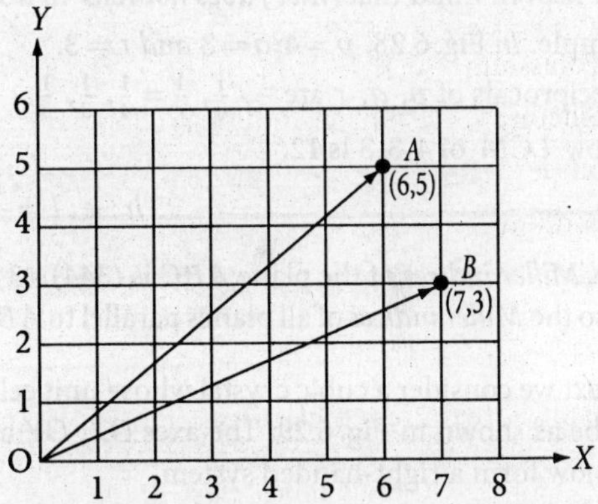

Fig. 6.26: Representation of crystallographic directions in two dimensions.

The crystallographic directions of the corner points of a unit cube are shown in the table below:

Points	$A(100)$	$B(110)$	$C(010)$	$P(001)$	$Q(101)$	$R(111)$	$S(011)$
Direction	[100]	[110]	[010]	[001]	[101]	[110]	[011]

6.13.2 Rules to find the Miller Indices of a Lattice Plane

1. First, find the intercepts of the plane on the three crystal axes (OX, OY, OZ). Let these be pa, qb and rc, where a,b,c are the corresponding *primitives* and p,q,r are the *intercepts* along $\vec{a}$, $\vec{b}$ and $\vec{c}$ respectively [Fig. 6.28].
2. Take the reciprocals of the numbers p,q,r in the form $\frac{1}{p}, \frac{1}{q}, \frac{1}{r}$.
3. Find the smallest possible integers h,k,l by multiplying the reciprocals by the LCM of p,q,r, such that $h:k:l = \frac{1}{p}:\frac{1}{q}:\frac{1}{r}$.

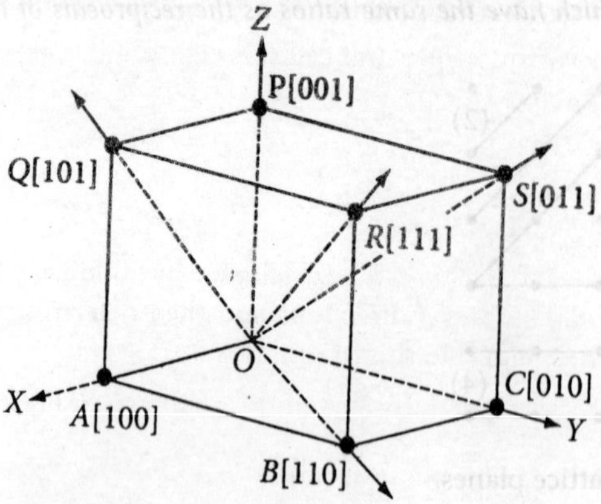

Fig. 6.27: Representation of crystallographic directions in three dimensions.

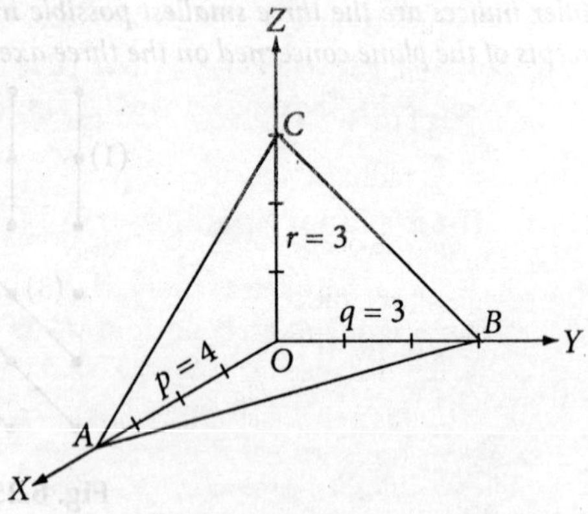

Fig. 6.28: A lattice plane specified by Miller indices (344).

The numbers h, k, l are the *Miller indices*, the corresponding lattice plane is represented by (hkl). It is to be kept in mind that (hkl) *does not refer to a single plane instead a set of parallel planes.*

Example. *In* Fig. 6.28, $p = 4, q = 3$ *and* $r = 3$.

Reciprocals of p, q, r are $\frac{1}{p}, \frac{1}{q}, \frac{1}{r} = \frac{1}{4}, \frac{1}{3}, \frac{1}{3}$.

Now LCM of 4,3,3 is 12.

$$\therefore h:k:l = \frac{12}{4} : \frac{12}{3} : \frac{12}{3} = 3:4:4. \quad (6.10)$$

So, *Miller indices* of the plane ABC is (344). (344) is also the *Miller indices* of all planes parallel to ABC.

Next we consider a cubic crystal whose unit cell is a cube as shown in Fig. 6.29. The axes OX, OY and OZ now form a right-handed system.

The plane $ABCD$ of this cube has an intercept OD on the X-axis but it is parallel to the Y- and Z-axes, i.e., the intercepts on the Y- and Z-axes are each *infinity.* If OD is taken as unity, then the *Miller indices* are the reciprocals of $1, \infty, \infty$, i.e., they are (100).

Similarly, the *Miller indices* of the plane $GADE$ are (110) and those for the plane FAG are (111) [Fig. 6.29].

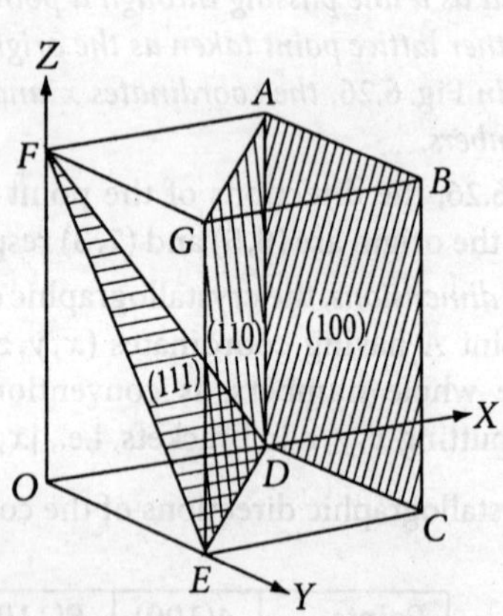

Fig. 6.29: Miller indices of lattice planes in a cubic lattice.

6.13.3 Important Features of Miller Indices

Following are the important features of the *Miller indices:*

1. Any plane parallel to one of the coordinate axes has intercept of infinity and index zero for that axis.
2. *Miller indices* define a set of parallel planes and not a particular plane.
3. All equally spaced parallel planes have the same index number (hkl).
4. When the intercept of a plane on any axis is negative a bar is put on the corresponding Miller index [e.g., $(2\bar{3}4)$].

5. If a plane passes through the origin it is defined in terms of parallel plane having non-zero intercept.
6. The complete set of crystallographically equivalent directions of (hkl) is a member and is expressed by enclosing the direction indices in single bracket like $\langle hkl \rangle$.

6.14 Inter-planar Spacing

The spacing d_{hkl} between a family of parallel adjacent planes (hkl) can be calculated by taking any lattice point as an origin and erecting coordinate axes in the $\vec{a}$, $\vec{b}$ and $\vec{c}$ directions and then determining the perpendicular distance between the origin and the plane nearest to this origin.

The intercepts of the (hkl)-plane on $\vec{a}, \vec{b}$ and $\vec{c}$ axes are respectively $\frac{a}{h}$, $\frac{b}{k}$ and $\frac{c}{l}$ as shown in Fig. 6.30.

From Fig. 6.30,

$$d_{hkl} = OP = \frac{a}{h}\cos\alpha = \frac{b}{k}\cos\beta = \frac{c}{l}\cos\gamma, \quad (6.11)$$

where α, β and γ are the angles between the normal to the plane and the $\vec{a}$, $\vec{b}$ and $\vec{c}$ axes respectively.

If $\hat{n}$ be the unit vector normal to the plane, then

$$\hat{n}\cdot\vec{a} = a\cos\alpha$$

$$\hat{n}\cdot\vec{b} = b\cos\beta$$

$$\hat{n}\cdot\vec{c} = c\cos\gamma.$$

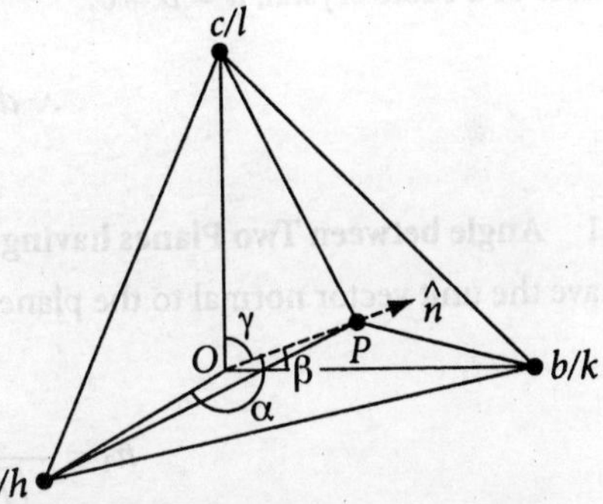

Fig. 6.30: Direction angles of the normal to the (hkl)-plane.

$\therefore$ from (6.11),

$$d_{hkl} = \frac{\hat{n}\cdot\vec{a}}{h} = \frac{\hat{n}\cdot\vec{b}}{k} = \frac{\hat{n}\cdot\vec{c}}{l}. \quad (6.12)$$

Now in an orthogonal lattice, taking X-, Y- and Z-axes along $\vec{a}$-, $\vec{b}$- and $\vec{c}$-directions respectively, the equation to the (hkl)-plane having intercepts $\frac{a}{h}$, $\frac{b}{k}$ and $\frac{c}{l}$ is

$$\frac{x}{a/h} = \frac{y}{b/k} = \frac{z}{c/l} = 1 \text{ or } \frac{hx}{a} + \frac{ky}{b} + \frac{lz}{c} = 1 \text{ or } \phi(x,y,z) = 1 \text{ (constant).} \quad (6.13)$$

$$\therefore \nabla\phi = \hat{i}\frac{\partial\phi}{\partial x} + \hat{j}\frac{\partial\phi}{\partial y} + \hat{k}\frac{\partial\phi}{\partial z} = \hat{i}\frac{h}{a} + \hat{j}\frac{k}{b} + \hat{k}\frac{l}{c} = |\nabla\phi|\hat{n}, \quad (6.14)$$

where $\hat{n}$ is the unit vector along the normal to the (hkl) plane and

$$|\nabla\phi| = \sqrt{(h/a)^2 + (k/b)^2 + (l/c)^2}.$$

$$\therefore \hat{n} = \frac{\nabla\phi}{|\nabla\phi|} = \frac{\hat{i}\left(\frac{h}{a}\right) + \hat{j}\left(\frac{k}{b}\right) + \hat{k}\left(\frac{l}{c}\right)}{\sqrt{(h/a)^2 + (k/b)^2 + (l/c)^2}}.$$

Now from (6.12),

$$d_{hkl}=\frac{\hat{n}\cdot\vec{a}}{h}=\frac{1}{h}\left[\frac{\hat{\imath}\left(\frac{h}{a}\right)+\hat{\jmath}\left(\frac{k}{b}\right)+\hat{k}\left(\frac{l}{c}\right)}{\sqrt{(h/a)^2+(k/b)^2+(l/c)^2}}\cdot\hat{\imath}a\right]=\frac{1}{h}\left[\frac{\frac{h}{a}a}{\sqrt{(h/a)^2+(k/b)^2+(l/c)^2}}\right],$$

$$\therefore d_{hkl}=\frac{1}{\sqrt{(h/a)^2+(k/b)^2+(l/c)^2}},\tag{6.15}$$

where $\hat{\imath},\hat{\jmath}$ and $\hat{k}$ are the usual triad of unit vectors along X-, Y- and Z-directions respectively.

In case of a cubic crystal, $a=b=c$,

$$\therefore d_{hkl}=\frac{a}{\sqrt{h^2+k^2+l^2}}.\tag{6.16}$$

6.14.1 Angle between Two Planes having Miller Indices $(h_1k_1l_1)$ and $(h_2k_2l_2)$

We have the unit vector normal to the plane $(h_1k_1l_1)$. Then

$$\hat{n}_1=\frac{\hat{\imath}\frac{h_1}{a}+\hat{\jmath}\frac{k_1}{b}+\hat{k}\frac{l_1}{c}}{\sqrt{\left(\frac{h_1}{a}\right)^2+\left(\frac{k_1}{b}\right)^2+\left(\frac{l_1}{c}\right)^2}}.$$

For a cubic system, $a=b=c$.

$$\therefore \hat{n}_1=\frac{(\hat{\imath}h_1+\hat{\jmath}k_1+\hat{k}l_1)}{\sqrt{h_1^2+k_1^2+l_1^2}}.$$

Similarly, the unit vector normal to the plane (h_2,k_2,l_2) of a cubic system is

$$\hat{n}_2=\frac{(\hat{\imath}h_2+\hat{\jmath}k_2+\hat{k}l_2)}{\sqrt{h_2^2+k_2^2+l_2^2}}.$$

Now

$$\hat{n}_1\cdot\hat{n}_2=|\hat{n}_1||\hat{n}_2|\cos\theta \text{ [where } \theta \text{ is the angle between the planes]}$$

$$=\frac{\hat{\imath}h_1+\hat{\jmath}k_1+\hat{k}l_1}{\sqrt{h_1^2+k_1^2+l_1^2}}\cdot\frac{\hat{\imath}h_2+\hat{\jmath}k_2+\hat{k}l_2}{\sqrt{h_2^2+k_2^2+l_2^2}}=\frac{h_1h_2+k_1k_2+l_1l_2}{\sqrt{(h_1^2+k_1^2+l_1^2)(h_2^2+k_2^2+l_2^2)}}.$$

$$\therefore \theta=\cos^{-1}\left[\frac{h_1h_2+k_1k_2+l_1l_2}{\sqrt{(h_1^2+k_1^2+l_1^2)(h_2^2+k_2^2+l_2^2)}}\right].\tag{6.17}$$

6.14.2 Inter-Planar Spacing between Lattice Planes of Cubic Lattice

1. **SC lattice:** Here the unit cell is a cube of side a and there is only one lattice point at each corner of the cube.

The inter-planar spacing of (hkl)-planes for such lattice is

$$d_{hkl} = \frac{a}{\sqrt{h^2+k^2+l^2}}.$$

The miller indices of the three family of planes are

(100), (110) and (111).

For (100)-planes,

$$d_{100} = \frac{a}{\sqrt{1^2+0^2+0^2}} = a.$$

For (110)-planes,

$$d_{110} = \frac{a}{\sqrt{1^2+1^2+0^2}} = \frac{a}{\sqrt{2}}.$$

For (111)-planes,

$$d_{111} = \frac{a}{\sqrt{1^2+1^2+1^2}} = \frac{a}{\sqrt{3}}.$$

2. BCC lattice: In this cubic lattice apart from one atom at each corner, there is an additional atom at the body centre.
   The family of the lattice planes are represented by (100), (110) and (111). For (100)-planes,

   $$d_{100} = \frac{a}{2},$$

   since there is an additional plane (100) just half-way between (100)-planes.
   For (110)-planes,

   $$d_{110} = \frac{a}{\sqrt{2}},$$

   as the (110)-plane of a BCC lattice is similar to (110) plane of an SC lattice.
   For (111)-planes,
   In a BCC lattice, the plane (111) is just midway between (111)-planes of an SC lattice.

   $$\therefore d_{111} = \frac{1}{2}\frac{a}{\sqrt{3}}.$$

3. FCC lattice: For (100)-plane,

   $$d_{100} = \frac{a}{2},$$

   since there is an additional plane (100) half-way between (100)-planes of an SC lattice.
   For (110)-planes,

   $$d_{110} = \frac{1}{2}\frac{a}{\sqrt{2}},$$

   since there is an additional plane (110) midway and parallel to the first set between (110)-planes of an SC lattice.
   For (111)-planes,

   $$d_{111} = \frac{a}{\sqrt{3}}$$

   as the (111)-planes of an SC lattice pass through the centres of the faces of an FCC lattice.

6.14.3 Comparison of SC, BCC, FCC Structures

Table 6.6

Properties	*SC*	*BCC*	*FCC*
Position of atoms	Corners of the cube	Corners and body centre of the cube	Corners and face centres of the cube
Volume of the unit cell	a^3	a^3	a^3
Number of atoms per unit cell	1	2	4
Coordination number (CN)	6	8	12
Nearest neighbour distance ($2r$)	a	$\frac{a\sqrt{3}}{2}$	$\frac{a}{\sqrt{2}}$
Atomic radius	$\frac{a}{2}$	$\frac{a\sqrt{3}}{4}$	$\frac{a}{2\sqrt{2}}$
Atomic packing fraction (APF)	0.52	0.68	0.74
Example	KI	NaCl, KCl	$AlCl_2$, $CuSO_4$

6.15 Answers of Some Important Questions

6.1 What do you mean by single crystal and microcrystalline (or polycrystallites)?

Ans. If in a solid the periodicity extends throughout the volume of the matter, then the solid is a single crystal, e.g., diamond, quartz, etc.

If there is an aggregate of so many small crystallites oriented at random and separated by *grain boundaries*, then the corresponding solid is called a *microcrystalline*. The crystallites are called *grains*.

6.2 Amorphous solids are isotropic but crystals are anisotropic. Why?

Ans. Amorphous solids have no directional properties, hence they are isotropic. But crystals have different periodic arrangements in all three directions which lead to the variation of its physical properties with direction. That is why a crystal is *anisotropic*.

6.3 All primitive cells are unit cells but all unit cells may or may not be primitive cell. — Justify.

Ans. In a primitive cell there is one atom at each corner of the unit cell, i.e., the number of atoms per unit cell is one. But in a unit cell there may or may not be additional atom or atoms in addition to an atom at each corner.

Hence, all primitive cells are unit cells but all unit cells may or may not be primitive cell.

6.4 What are the basic lattice parameters?

Ans. In a unit cell, the intercepts a, b and c made by three adjacent faces with the crystallographic axes are called *primitives* or *lattice constants*. The angles α, β and γ between the crystallographic axes are known as *interfacial angles*. The quantities a, b, c and α, β, γ are the *basic lattice parameters*. These determine the form and actual size of the unit cell.

6.5 What do you mean by space lattice?

Ans. A space lattice is a mathematical concept which represents infinite number of points in space such that the arrangement of points in space about a given point is similar to that about any other point.

6.6 What are Bravais lattice?

Ans. The points may be arranged in 14 distinguishable ways in three-dimensional space maintaining perfect periodicity. These 14 space lattices are called *Bravais lattice* and belong to 7-crystal system.

6.7 Define space lattice and crystal lattice.

Ans. A *space lattice* is an infinite array of points in three dimensions in which every point has surroundings identical to that of every other point in that array.

A *crystal lattice* is a space lattice in which the lattice sites are occupied by *atoms* or *cluster of atoms.*

6.8 What do you mean by lattice constant?

Ans. In a lattice, a unit cell has the length of the edge as well as the angles between the edges are of definite values. Each edge of the unit cell is the distance between the atoms of the same kind and is known as the *lattice constant.*

6.9 Define coordination number.

Ans. Coordination number (CN) is defined as the number of nearest neighbours to a particular atom in the crystal lattice.

6.10 What is atomic packing factor?

Ans. The atomic packing factor of a unit cell is defined as the ratio of volume occupied by the atoms in the unit cell to the total volume of the unit cell.

If there are n atoms each of volume v, and V is the volume of the unit cell, then

$$\text{Packing Factor (or Fraction) (PF)} = \frac{nv}{V}.$$

6.11 Find an expression for the density of crystalline material in terms of lattice constant for a cubic lattice.

Ans. Let ρ be the density of a cubic crystalline material of lattice constant a. If M_A be the atomic weight, then the volume of one mole of the substance is $v_M = \frac{M_A}{\rho}$. v_M is also the volume of Avogadro number (N_A) of atoms.

$\therefore$ the volume of a single atom $v = \frac{M_A}{N_A\rho}$ and volume of n number of atoms $v_n = nv = \frac{nM_A}{N_A\rho}$.

But the effective volume of n atoms is equal to the volume of the unit cell a^3.

So

$$a^3 = \frac{nM_A}{N_A\rho} \text{ or } \rho = \frac{nM_A}{a^3 N_A}$$

6.12 What are Miller indices?

Ans. Miller indices are the three smallest possible integers (hkl) which have the same ratios as the reciprocals of the intercepts of the plane concerned on the three axes. A particular set of values of (hkl) represents a family of parallel crystal planes.

6.13 What are the specialities of the Miller indices?

Ans. The Miller indices have the following specialities:

(a) Miller indices (hkl) represent a set of parallel planes and not a particular plane.

(b) If the intercepts are at infinity, the corresponding Miller indices are zero. If a plane passes through the origin, it is defined in terms of parallel plane having non-zero intercept.

(c) The Miller indices (hkl) have the same value if the planes are equispaced and parallel.

(d) The Miller index $\bar{h}$ represents that the plane cuts the axis on the negative side of the origin.

6.14 What are crystallographic directions?

Ans. The crystallographic direction is a line joining any two points of the space lattice relative to specific point taken as the origin.

6.15 If $(h_1k_1l_1)$ and $(h_2k_2l_2)$ be the Miller indices of two planes, then what is the angle between them?

Ans. The angle between the two lattice planes with *Miller indices* $(h_1 k_1 l_1)$ and $(h_2 k_2 l_2)$ is given by

$$\theta = \cos^{-1}\left[\frac{h_1 h_2 + k_1 k_2 + l_1 l_2}{\sqrt{\left(h_1^2+k_1^2+l_1^2\right)\left(h_2^2+k_2^2+l_2^2\right)}}\right].$$

6.16 Find the angle between the plane (111) and plane direction [111].

Ans. The (111)-plane in a cubic unit cell is represented by *PAC*. *PA* is on the (111)-plane and the direction of *PA* is represented by $[10\bar{1}]$. The angle between the direction $[10\bar{1}]$ and plane [111] is the angle between the plane (111) and the direction [111].

If θ be the angle between (111) and [111], then

$$\theta = \cos^{-1}\frac{h_1 h_2 + k_1 k_2 + l_1 l_2}{\sqrt{\left(h_1^2+k_1^2+l_1^2\right)\left(h_2^2+k_2^2+l_2^2\right)}}.$$

Here $h_1 = 1 \quad h_2 = 1$
$k_1 = 0$ and $k_2 = 1$
$l_1 = -1 \quad l_2 = 1.$

$$\therefore \theta = \cos^{-1}\frac{1\times 1 + 1\times 0 + (-1\times 1)}{\sqrt{1^2+1^2+(-1)^2}\sqrt{(1^2)+(0)^2+(1)^2}} = 0.$$

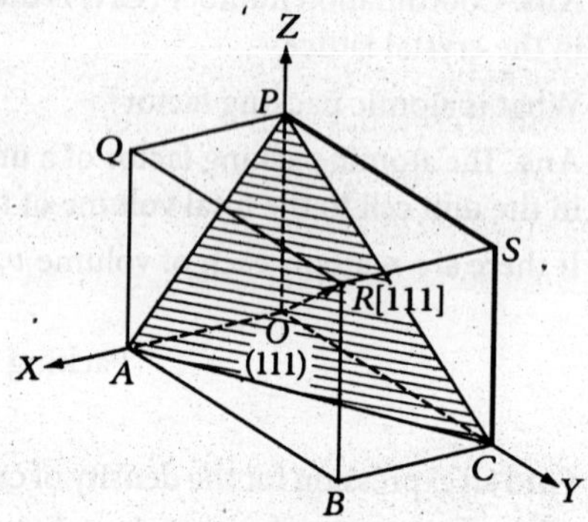

Fig. 6.31

Note. If a plane and the direction are represented by same value of h, k, l, then they are always perpendicular to each other.

6.16 Solved Problems

6.1 Copper has FCC structure and the atomic radius is 0.1278 nm. Calculate the density and the inter-planar spacing for (321) planes. Take the atomic weight of copper as 63.5. **[WBUT 2005]**

Solution. For a cubic crystal having lattice constant d and number of atoms $n(=4$ for FCC lattice)

$$a = \left(\frac{nM_A}{\rho N}\right)^{1/3},$$

where M_A, ρ and N are respectively the atomic mass, density and Avogadro number.

Again, for an FCC crystal, $a = \frac{4}{\sqrt{2}}r$, where $r = 0.1278\,\text{nm} = 0.1278\times 10^{-9}$ m is the atomic radius of the crystal.

$$\text{Now } \rho = \frac{nM_A}{Na^3} = \frac{4\times 63.5}{6.023\times 10^{23}\times\left(\frac{4\times 0.1278\times 10^{-9}}{\sqrt{2}}\right)^3}\ \text{gm/cm}^3 = \mathbf{8.926\,gm/cm^3}.$$

$$\text{and } d_{hkl} = \text{inter-planar spacing for (321)-plane} = \frac{a}{\sqrt{(h^2+k^2+l^2)}} = \frac{\frac{4\times 0.1278\times 10^{-9}}{\sqrt{2}}}{\sqrt{3^2+2^2+1^2}}\ \text{m}$$

$$= 0.0966\times 10^{-9}\ \text{m} = \mathbf{0.0966\ nm}.$$

6.2 A cubic crystal has lattice constant 4.3 Å and density 963 kg/m^3. If its atomic weight is 630, then what type of cubic unit cell does it form?

Solution. Here edge of the unit cell, $a = 4.3$ Å $= 4.3 \times 10^{-10}$ m.
Density of the material of the crystal, $\rho = 963\,\text{kg-m}^3$.
Atomic weight, $M_A = 23$ kg/k-mol.
Then

$$n = \frac{a^3 \rho N}{M_A},$$

where n = the number of atoms per unit cell and N = the Avogadro number.

$$\therefore\ n = \frac{(4.3 \times 10^{-10}) \times 963 \times 6.02 \times 10^{26}}{23} = 2.$$

We know that in a body-centred cubic (BCC) crystal the number of atoms per unit cell is 2. So, the atomic crystal is of BCC type.

6.3 NaCl crystal has FCC structure and is of density 2.18 gm/cm³. Calculate the distance between the two adjacent atoms.

Solution. Molecular weight of NaCl

$$M = 23 + 35.5 = 58.5.$$

$$\text{Mass of a NaCl molecule } m = \frac{\text{molecular weight } (M)}{\text{Avogadro number } (N)} = \frac{58.5\,\text{gm-mol}}{6.023 \times 10^{23}\ (\text{mol}^{-1})}$$

$$= 9.713 \times 10^{-23}\ \text{gm}.$$

$$\text{Number of molecules per unit volume} = \frac{\text{mass of unit volume of NaCl } (\rho)}{\text{mass of a NaCl molecule } (m)}$$

$$= \frac{2.18}{9.713 \times 10^{-23}}$$

$$= 2.244 \times 10^{22}\ \text{molecule/cm}^3.$$

Now number of atoms per unit volume of NaCl $= \mathbf{2 \times 2.244 \times 10^{22}}$ ($\because$ NaCl is diatomic).
If a and n represent respectively the distance between adjacent atoms and number of atoms along the edge of 1 cm cube, then the length of the edge of a unit cube $= na$.
Volume of the unit cube $= n^3 a^3 = 1\ \text{cm}^3$.

$$\text{or } (4.49 \times 10^{22})(a)^3 = 1 \quad [\because n^3 = 4.49 \times 10^{22}]$$

$$\therefore\ a = \left(\frac{1}{4.49 \times 10^{22}}\right)^{1/3} = 2.81 \times 10^{-8}\ \text{cm} = 2.81\ \text{Å}.$$

6.4 Calculate the lattice constant of rock-salt (NaCl) from the following data:
Molecular weight by NaCl = 58.45 kg-mol
Density of NaCl $= 2.17 \times 10^3$ kg-m³
Avogadro number $N = 6.023 \times 10^{23}$/gm-mol $= 6.023 \times 10^{26}$/kg-mol. **[BUH 1996; CUH 2004]**

Solution. The number of lattice point (n) per unit cell of NaCl = 4 (as NaCl is an FCC crystal)
Now lattice constant

$$a = \left(\frac{4M}{\rho N}\right)^{1/3} = \left[\frac{4 \times 58.45}{2.17 \times 10^3 \times 6.023 \times 10^{26}}\right]^{1/3} \text{m} = 5.64\ \text{Å}.$$

6.5 KCl is of FCC crystal structure having density $1.98 \times 10^3\,kg/m^3$. If its molecular weight is 74.6, calculate (i) the distance from one atom to the next atom of same kind (ii) the distance between adjacent atoms.

[cf. BUH 1995]

Solution. If a be the lattice constant of the crystal, then

$$a = \left(\frac{nM_A}{\rho N}\right)^{1/3},$$

where n = number of ions in a unit cell = 4
M_A = molecular weight = 74.6 kg/k-mol
ρ = density = 963 kg/m^3
and N = Avogadro number = 6.02×10^{26} atoms/kg-mol.

$$\therefore\ a = \left(\frac{4 \times 74.6}{1.98 \times 10^3 \times 6.023 \times 10^{26}}\right)^{1/3} \simeq 6.3\,\text{Å}.$$

Thus, distance between two atoms of the same kind is $a = 6.3$ Å.
Hence, the distance between two adjacent atoms is

$$\frac{a}{2} = 3.15\,\text{Å}.$$

6.6 Calculate the lattice constant of potassium bromide. Given, density of potassium bromide = $2.7 \times 10^3\,kg/m^3$ and it belongs to FCC lattice. Molecular weight of potassium bromide = 199 and Avogadro No. = 6.02×10^{23}/gm-mol = 6.02×10^{26}/kg-mol. **[CUH 1992, cf. 91]**

Solution. Mass of a unit cell = $\dfrac{4 \times 119}{6.02 \times 10^{26}}$ kg.

[$\because$ KBr is an FCC crystal having 4 atoms/cell.

Again, mass of a unit cell of KBr can be written as $a^3 \times 2.7 \times 10^3$ kg [a = lattice constant]

So

$$a^3 \times 2.7 \times 10^3 = \frac{4 \times 119}{6.02 \times 10^{26}}$$

$$\text{or } a = \left[\frac{4 \times 119}{2.7 \times 10^3 \times 6.02 \times 10^{26}}\right]^{1/3} = 6.57 \times 10^{-10}\,\text{m} = 6.57\,\text{Å}.$$

6.7 In a crystal, a lattice plane cuts intercepts of $2a, 3b$ and $6c$ along the three axes, where a, b and c are primitive vectors of the unit cell. Determine the Miller indices of the given plane.

Solution. Let pa, qb and rc be respectively the intercepts of the plane on the three crystal axes, then

$$pa : qb : rc = 2a : 3b : 6c$$

$$\text{or } p : q : r = 2 : 3 : 6$$

$$\text{or } \frac{1}{p} : \frac{1}{q} : \frac{1}{r} = \frac{1}{2} : \frac{1}{3} : \frac{1}{6} = 3 : 2 : 1.$$

Hence, the Miller indices of the given plane are (321).

6.8 In an orthorhombic crystal a lattice plane cuts intercepts of lengths $3a$, $-2b$ and $3c/2$ along three axes. Deduce the Miller indices of the plane, where $\vec{a}, \vec{b}, \vec{c}$ are vectors of the unit cell.

Solution. If the intercepts of the plane in the three crystal axes are pa, qb and rc respectively, then

$$pa:qb:rc = 3a:(-2b):\frac{3c}{2}$$

$$\text{or } p:q:r = 3:(-2):\frac{3}{2}$$

$$\text{or } \frac{1}{p}:\frac{1}{q}:\frac{1}{r} = \frac{1}{3}:\left(-\frac{1}{2}\right):\frac{2}{3} = 2:(-3):4.$$

∴ Miller indices of the plane are **$(2\bar{3}4)$**.

6.9 Find the Miller indices of a plane having intercepts of $8a$, $4b$ and $2c$ on the a-, b- and c-axes respectively.

[BUH 2001]

Solution. If the intercepts of the plane on a-, b- and c-axes respectively be pa, qb and rc, then

$$pa:qb:rc = 8a:4b:2c$$

$$\text{or } p:q:r = 8:4:2$$

$$\text{or } \frac{1}{p}:\frac{1}{q}:\frac{1}{r} = \frac{1}{8}:\frac{1}{4}:\frac{1}{2} = 1:2:4.$$

∴ the Miller indices of the plane are (124).

6.10 In an orthorhombic crystal a plane makes intercepts 2.93 mm, 4.47 mm and 2.35 mm along the three crystallographic axes, the corresponding primitives being 3.05 Å, 6.99 Å and 4.90 Å. Deduce the Miller indices of the cleavage plane.

Solution.

$$p\times 3.05 : q\times 6.99 : r\times 4.90 = 2.93:4.47:2.35$$

$$\text{or } p:q:r = \frac{2.93}{3.05}:\frac{4.47}{6.99}:\frac{2.35}{4.90}$$

$$= 0.96:0.64:0.48$$

$$= 6:4:3.$$

$$\therefore \frac{1}{p}:\frac{1}{q}:\frac{1}{r} = \frac{1}{6}:\frac{1}{4}:\frac{1}{3} = 2:3:4$$

Hence, Miller indices of the plane are **(234)**.

6.11 Calculate the number of atoms per unit area of (111)-plane in an SC structure.

Solution. Area of the (111)-plane is

$$\tfrac{1}{2}\times\text{base}\times\text{height} = \tfrac{1}{2}\times a\sqrt{2}\times EO.$$

Now from Fig. 6.32 (b),

$$\cos\angle DEO = \cos 30^\circ = \frac{EO}{a\sqrt{2}}$$

$$\text{or } EO = a\sqrt{2}\times\cos 30^\circ = a\sqrt{2}\times 0.866.$$

∴ area of the (111)-plane is

$$\tfrac{1}{2}\times\left(a\sqrt{2}\right)\times\left(a\sqrt{2}\times 0.866\right)=0.866a^2.$$

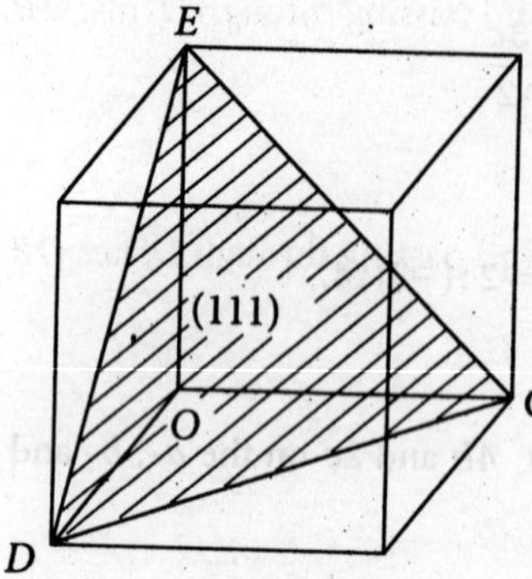

Fig. 6.32(a)

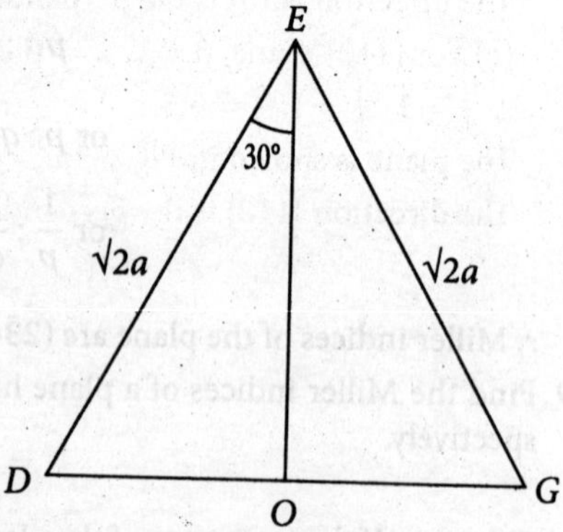

Fig. 6.32(b)

Now total number of atoms in this plane is

$$\left(\frac{1}{360}\times 60\right)\times 3=\frac{1}{2}=0.5.$$

Hence, number of atoms per unit area

$$=\frac{0.5}{0.866a^2}=\frac{0.58}{a^2}\text{ atoms/unit area.}$$

6.12 Draw the following planes and directions in the case of an FCC structure (a) (100) (b) (110) (c) (112).

Solution. (a) For (100)-plane, $h=1,\ k=0,\ l=0$.

The reciprocals are $\frac{1}{1},\frac{1}{0},\frac{1}{0}$, i.e., $1,\infty,\infty$.

The plane is shown in Fig. 6.33.

The direction [100] is given by the normal to (100)-plane and passing through the origin, i.e., *SR*.

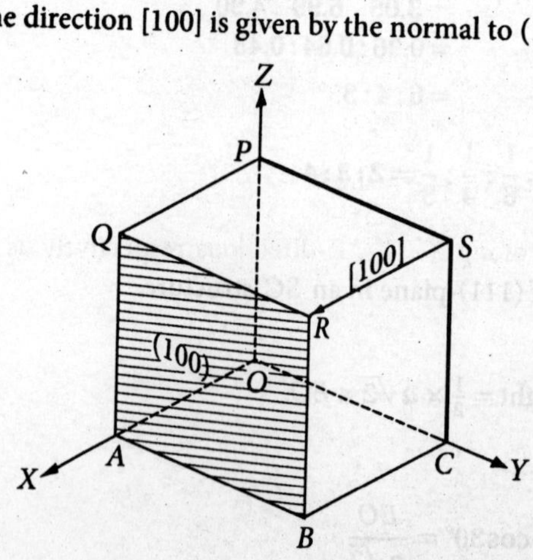

Fig. 6.33: (100)-plane.

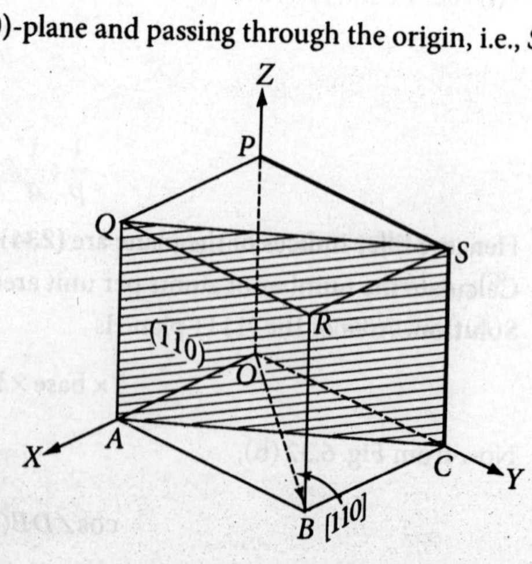

Fig. 6.34: (110)-plane.

(b) For (110)-plane, $h=1,\ k=1,\ l=0$.

$\therefore\ \frac{1}{h}=1,\ \frac{1}{k}=1,\ \frac{1}{l}=\infty$.

The plane is shown in Fig. 6.34.

The direction [110] is the perpendicular to (110)-plane and passing through O, i.e., OB.

(c) For (112)-plane, $h=1,\ k=1,\ l=2$.

$\therefore\ \frac{1}{h}=1,\ \frac{1}{k}=1,\ \frac{1}{l}=0.5$.

The plane is shown in Fig. 6.35.

The direction [112] is the perpendicular to (112)-plane and passing through O, i.e., OR.

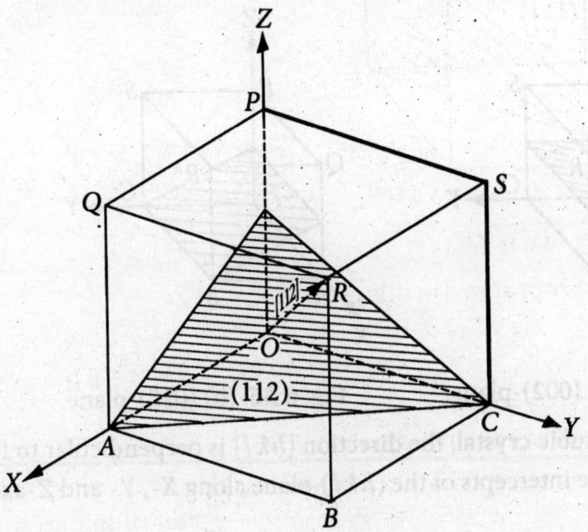

Fig. 6.35: (112)-plane.

6.13 Draw the crystal planes (002), (012) and (022) in an FCC lattice.

Solution.

(i) (022)-*Plane*: Here

$$h=0,\ \therefore\ \frac{1}{h}=\frac{1}{0}=\infty$$

$$k=0,\ \therefore\ \frac{1}{k}=\frac{1}{0}=\infty$$

$$\text{and } l=2,\ \therefore\ \frac{1}{l}=\frac{1}{2}=0.5.$$

So, the (002)-plane has intercepts $\infty, \infty, 0.5$ along X-, Y-, Z-directions respectively as shown in Fig. 6.36 (a).

(ii) (012)-*Plane*: Here

$$h=0,\ \therefore\ \frac{1}{h}=\frac{1}{0}=\infty$$

$$k=1,\ \therefore\ \frac{1}{k}=\frac{1}{1}=1$$

$$\text{and } l=2,\ \therefore\ \frac{1}{l}=\frac{1}{2}=0.5.$$

So, the (012)-plane has intercepts ∞, 1 and 0.5 along the X-, Y-, Z-directions respectively [Fig. 6.36 (b)].

(iii) (022)-*Plane*: Here

$$h=0, \therefore \frac{1}{h}=\frac{1}{0}=\infty$$

$$k=2, \therefore \frac{1}{k}=\frac{1}{2}=0.5$$

$$\text{and } l=2, \therefore \frac{1}{l}=\frac{1}{2}=0.5.$$

So, the (022)-plane has intercepts ∞, 0.5 and 0.5 along the X-, Y- and Z-directions respectively [Fig. 6.36 (c)].

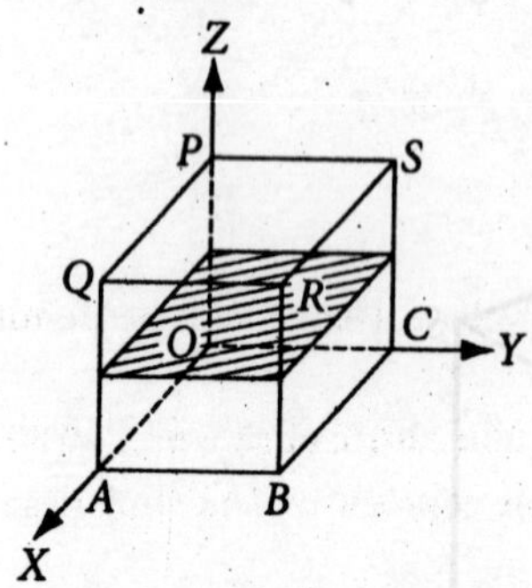

Fig. 6.36: (a) (002)-plane.

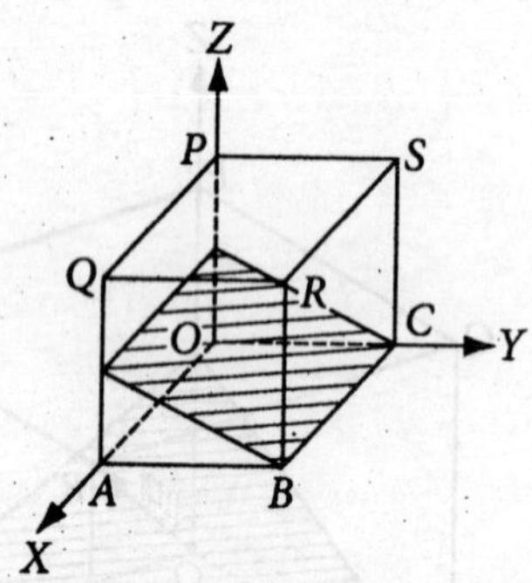

Fig. 6.36: (b) (012)-plane.

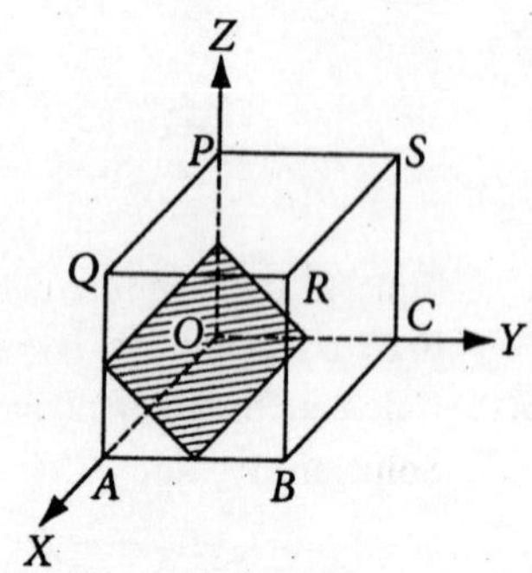

Fig. 6.36: (c) (022)-plane.

6.14 Prove that in a cubic crystal, the direction $[hkl]$ is perpendicular to the plane (hkl).

Solution. Let the intercepts of the (hkl)-plane along X-, Y- and Z-axes of the cubic unit cell be

$$\vec{R}_1=\frac{\vec{a}}{h},\ \vec{R}_2=\frac{\vec{b}}{k} \text{ and } \vec{R}_3=\frac{\vec{c}}{l},$$

where $\vec{a}$, $\vec{b}$ and $\vec{c}$ are basis vectors along X-, Y- and Z-directions respectively, Fig. 6.37.

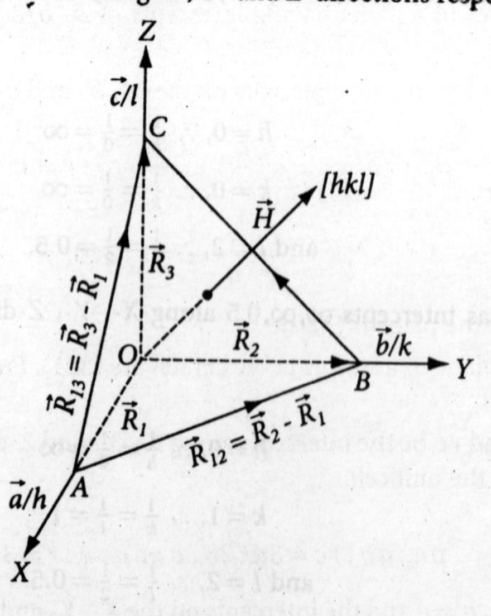

Fig. 6.37: (hkl)-plane in a cubic crystal.

The vector $\vec{H} = h\vec{a} + k\vec{b} + l\vec{c}$ represents the direction $[hkl]$ and $\vec{R}_{12}, \vec{R}_{13}$ lie in the plane (hkl). Now

$$\vec{H}\cdot\vec{R}_{12} = \left(h\vec{a}+k\vec{b}+l\vec{c}\right)\cdot\left(\vec{R}_2-\vec{R}_1\right) = \left(h\vec{a}+k\vec{b}+l\vec{c}\right)\cdot\left(\frac{\vec{b}}{k}-\frac{\vec{a}}{h}\right)$$
$$= \vec{b}\cdot\vec{b} - \vec{a}\cdot\vec{a} = \vec{a}\cdot\vec{a} - \vec{a}\cdot\vec{a}$$
$$= 0 \; [\because\; a = b = c \text{ for a cubic crystal}].$$

Also

$$\vec{H}\cdot\vec{R}_{13} = \left(h\vec{a}+k\vec{b}+l\vec{c}\right)\cdot\left(\frac{\vec{c}}{l}-\frac{\vec{a}}{h}\right) = \vec{c}\cdot\vec{c} - \vec{a}\cdot\vec{a}$$
$$= \vec{a}\cdot\vec{a} - \vec{a}\cdot\vec{a} = 0.$$

Thus, $\vec{H}$ is perpendicular to $\vec{R}_{12}$ as well as $\vec{R}_{13}$ which lie on (hkl)-plane. Hence, $\vec{H}$ is perpendicular to (hkl)-plane.

6.15 Calculate the inter-planar spacing d of planes (111) in a simple cubic lattice of side a. **[WBUT 2007]**

Solution. We know that the inter-planar spacing d_{hkl} between the consecutive (hkl)-planes is

$$d_{hkl} = \frac{1}{\sqrt{\dfrac{h^2}{a^2}+\dfrac{k^2}{b^2}+\dfrac{l^2}{c^2}}}.$$

In this case, $a = b = c$ and $h = k = l = 1$.

$$\therefore\; d_{hkl} = \frac{1}{\sqrt{\dfrac{1}{a^2}+\dfrac{1}{a^2}+\dfrac{1}{a^2}}} = \frac{a}{\sqrt{3}}.$$

6.16 Obtain the Miller indices of a plane having intercepts of a, $b/2$ and ∞ on the a-, b- and c-axes respectively. **[BUH 2004]**

Solution. Let pa, qb and rc be the intercepts on the a-, b- and c-axes respectively, then

$$pa : qb : rc = a : \frac{b}{2} : \infty \text{ or } p : q : r = 1 : \frac{1}{2} : \infty.$$

$$\therefore\; \frac{1}{p} : \frac{1}{q} : \frac{1}{r} = \mathbf{1 : 2 : 0}.$$

So, the Miller indices of the plane is **(120)**.

6.17 The Miller indices of a plane in a simple cubic crystal are (321). Find the intercepts of the plane and sketch it. **[BUH 2000]**

Solution. Let pa, qb and rc be the intercepts along X-, Y- and Z-axes respectively, where a, b, c are the primitive vectors of the unit cell.

Then

$$pa : qb : rc = 3a : 2b : c \text{ or } p : q : r = 3 : 2 : 1.$$

As the crystal is SC, $a = b = c$ and the intercepts on the X-, Y- and Z-axes are 3, 2 and 1 respectively as shown in Fig. 6.38.

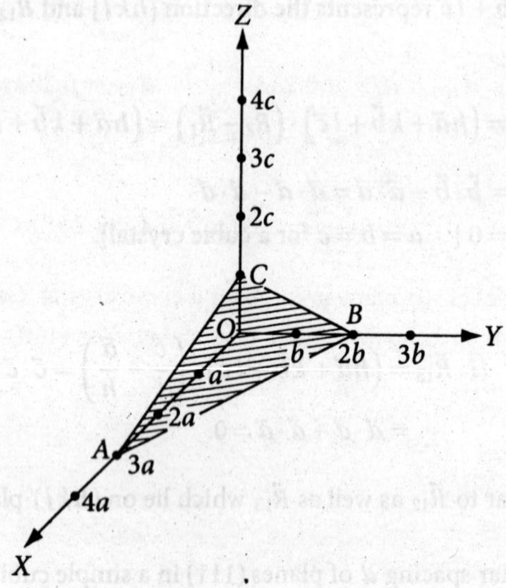

Fig. 6.38: The intercepts of (321)-plane.

6.18 Find the inter-planar spacing for a (321)-plane in a, simple cubic lattice whose lattice constant is 4.2×10^{-8} cm.

Solution. Here $a = b = c = 4.2 \times 10^{-8}\,\text{cm} = 4.8 \times 10^{-10}$ m and

$$d_{hkl} = \frac{a}{\sqrt{h^2+k^2+l^2}} = \frac{4.2 \times 10^{-10}}{\sqrt{3^2+2^2+1^2}} \quad [\because h=3,\ k=2 \text{ and } l=1]$$

$$= 1.1 \times 10^{-10}\,\text{m}.$$

6.19 In a tetrahedral lattice, $a = b = 2.5$ Å and $c = 1.8$ Å. Find the lattice spacing between (111)-planes.

Solution. For a given set of parallel planes (hkl) the lattice spacing is given by

$$d_{hkl} = \left[\frac{h^2}{a^2}+\frac{k^2}{b^2}+\frac{l^2}{c^2}\right]^{-1/2}.$$

In this case, $h = 1,\ k = 1,\ l = 1$ and $a = b = 2.5$ Å, $c = 1.8$ Å.

$$\therefore d_{hkl} = \left[\frac{1^2}{(2.5)^2}+\frac{1^2}{(2.5)^2}+\frac{1^2}{(1.8)^2}\right]^{-1/2} = \left[\frac{2}{6.25}+\frac{1}{3.24}\right]^{-1/2} = 1.26\,\text{Å}.$$

6.20 The distance d_{100} between (100)-planes in BCC structure is 0.232 nm. What is the size of the unit cell? What is the radius of the atom? [WBUT 2006]

Solution. For a BCC crystal the radius of the atom is given by

$$r = \frac{\sqrt{3}a}{4},$$

where a is the lattice constant.

Now for a BCC structure,

$$d_{100} = a/2 \text{ or } 0.232 = \frac{a}{2} \quad [\because d_{100} = 0.232\,\text{nm}]$$

$$\text{or } a = 0.464\,\text{nm}.$$

$$\therefore\ r = \frac{\sqrt{3}a}{4} = \frac{\sqrt{3}}{4} \times 0.464 = 0.20\,\text{nm}.$$

6.21 The inter-planar spacing of (110)-plane is 2.2 Å for a cubic crystal. Calculate the atomic radius.

Solution. The inter-planar spacing for a cubic crystal, between (110)-planes is

$$d_{100} = \frac{a}{\sqrt{h^2+k^2+l^2}}$$

$$\text{or } a = d_{100} \times \sqrt{h^2+k^2+l^2} = 2.2 \times \sqrt{1^2+1^2+0^2}.$$

$[\because d_{100} = 2.2$ Å and $h = 1,\ k = 1$ and $l = 0]$

$$\therefore\ a = 3.11\,\text{Å}.$$

Now for an SC crystal, atomic radius is

$$r = \frac{a}{2} = \frac{3.11}{2} = 1.555\,\text{Å}.$$

For a BCC crystal,

$$r = \frac{\sqrt{3}}{4}a = 3.11 \times \frac{\sqrt{3}}{4} = 1.347\,\text{Å}$$

and for an FCC crystal,

$$r = \frac{\sqrt{2}}{4}a = 3.11 \times \frac{\sqrt{2}}{4} = 1.099\,\text{Å}.$$

6.22 In a simple cubic crystal find the ratio of (a) spacing of the (110)- and (111)-planes; (b) nearest neighbour distance to the next nearest neighbour distance.

Solution. (a) For an SC lattice,

$$d_{hkl} = \frac{a}{\sqrt{h^2+k^2+l^2}}.$$

For (110)-planes,

$$d_{110} = \frac{a}{\sqrt{1^1+1^2+0^2}} = \frac{a}{\sqrt{2}}.$$

For (111)-planes,

$$d_{111} = \frac{a}{\sqrt{1^2+1^2+1^2}} = \frac{a}{\sqrt{3}}.$$

$$\therefore\ d_{110} : d_{111} = \frac{a}{\sqrt{2}} : \frac{a}{\sqrt{3}} = \frac{1}{\sqrt{2}} : \frac{1}{\sqrt{3}}.$$

(b) In an SC lattice the nearest neighbour distance $d_1 = a$, where a is the side of the unit cube.

The next nearest neighbour is situated at the end of the face diagonal. Hence, the next nearest neighbour distance is

$$d_2 = \sqrt{a^2 + a^2} = \sqrt{2}a.$$

$$\therefore d_1 : d_2 = a : \sqrt{2}a = 1 : \sqrt{2}.$$

6.23 If the radius of Ni atom is 1.245 Å, calculate the inter-planar distance of (100)- and (111)-planes of Ni crystal.

Solution. If a and r be respectively the lattice constant and radius of the Ni atom, then

$$a = 2\sqrt{2}r = \frac{4}{\sqrt{2}}r = \frac{(4 \times 1.245)}{\sqrt{2}} = 3.52\,\text{Å}.$$

Now the inter-planar distance between (100)-planes is

$$d_{100} = \frac{a}{\sqrt{1^2 + 0^2 + 0^2}} = a = 3.52\,\text{Å}$$

and the inter-planar distance between (111)-planes is

$$d_{111} = \frac{a}{\sqrt{1^2 + 1^2 + 1^2}} = \frac{3.52}{\sqrt{3}} = 2.03\,\text{Å}.$$

6.24 Copper has an FCC structure and the atomic radius is 0.1278 nm. Calculate the inter-planar spacing for (111)-planes. [BUH 1998]

Solution. If a be the lattice constant and r the atomic radius, then for FCC cubic crystal,

$$a = \frac{4}{\sqrt{2}}r = \frac{4}{\sqrt{2}} \times 0.1278 \times 10^{-9}\,\text{m}.$$

$$[\because r = 0.1278\,\text{nm} = 0.1278 \times 10^{-9}\,\text{m}]$$

Now inter-planar spacing for (hkl)-planes is

$$d_{hkl} = \frac{a}{\sqrt{h^2 + k^2 + l^2}}.$$

Here $h = k = l = 1$ and

$$a = \frac{4}{\sqrt{2}} \times 0.1278 \times 10^{-9}\,\text{m}.$$

$$\therefore d_{111} = \frac{\frac{4}{\sqrt{2}} \times 0.1278 \times 10^{-9}}{\sqrt{1^2 + 1^2 + 1^2}} = \frac{4 \times 0.1278 \times 10^{-9}}{\sqrt{6}} = \frac{0.5112 \times 10^{-9}}{2.4495}$$

$$= 0.209 \times 10^{-9}\,\text{m} = 0.209\,\text{nm}.$$

6.25 Find the Miller indices for a plane with the intercepts $6\vec{a}, 2\vec{b}, 3\vec{c}$. [BUH 2003]

Solution. If pa, qb and rc be the intercepts of the plane along X-, Y- and Z-axes respectively, then

$$pa : qb : rc = 6a : 2b : 3c$$
$$\text{or } p : q : r = 6 : 2 : 3$$
$$\text{or } \frac{1}{p} : \frac{1}{q} : \frac{1}{r} = \frac{1}{6} : \frac{1}{2} : \frac{1}{3} = 1 : 3 : 2$$
$$\text{or } h : k : l = 1 : 3 : 2.$$

∴ the required Miller indices of the plane are **(132)**.

6.26 The density of BCC iron is 7900 kg-m^3 and its atomic weight is 56. Calculate the side of the unit cell and the interatomic spacing. [BUH 1993]

Solution. The side of a unit cell of a BCC lattice is

$$a = \left(\frac{nM}{\rho N}\right)^{1/3},$$

where n = number of atoms per unit cell
M = atomic weight
ρ = the density of the material and
N = Avogadro number.

Here $n = 2$, $M = 56$, $\rho = 7900$ kg-m^3 and $N = 6.023 \times 10^{26}$/kg-mol.

$$\therefore a = \left[\frac{2 \times 56}{7900 \times 6.023 \times 10^{26}}\right]^{1/3} = \left[\frac{224}{47581.7 \times 10^{26}}\right]^{1/3}$$
$$\simeq 2.86 \times 10^{-10}\ \text{m} = \mathbf{2.86\ Å}.$$

Now interatomic spacing of a BCC lattice is

$$a_1 = \frac{\sqrt{3}}{2} a = \frac{\sqrt{3}}{2} \times 2.86 = \frac{1.732 \times 2.86}{2} = \mathbf{2.48\ Å}$$

6.27 Write down Miller indices of a plane having intercepts $2a$, ∞ and $c/2$ on the a-, b- and c-axes respectively. [BUH 2002]

Solution. Let pa, qb and rc be the intercepts of the plane along a-, b- and c-axes respectively. Then

$$pa : qb : rc = 2a : \infty : \frac{c}{2}$$
$$\text{or } p : q : r = 2 : \infty : \frac{1}{2}$$
$$\text{or } \frac{1}{p} : \frac{1}{q} : \frac{1}{r} = \frac{1}{2} : \frac{1}{\infty} : 2$$
$$\text{or } h : k : l = 1 : 0 : 4.$$

Hence, the Miller indices of the plane is **(104)**.

6.28 Find the Miller indices of a plane having intercepts of $8a$, $4b$ and $2c$ on the a-, b- and c-axes respectively. [BUH 2001]

Solution. If pa, qb and rc be the intercepts of the plane along a-, b- and c-axes respectively, then

$$pa:qb:rc = 8a:4b:2c$$

$$\text{or } p:q:r = 8:4:2$$

$$\text{or } \frac{1}{p}:\frac{1}{q}:\frac{1}{r} = \frac{1}{8}:\frac{1}{4}:\frac{1}{2} = 1:2:4$$

$$\text{or } h:k:l = 1:2:4.$$

So, the Miller indices of the plane is **(124)**.

6.29 Show that the maximum radius of the sphere that can just fit into the void at the body centre of the FCC structure coordinated by the facial atoms is $0.414r$, where r is the radius of the atom.

Solution. Let R be the radius of the sphere that can just fit into the void as shown in Fig. 6.39.

$$\therefore\ 2r + 2R = a \text{ or } r + R = \frac{a}{2} \text{ or } R = \frac{a}{2} - r.$$

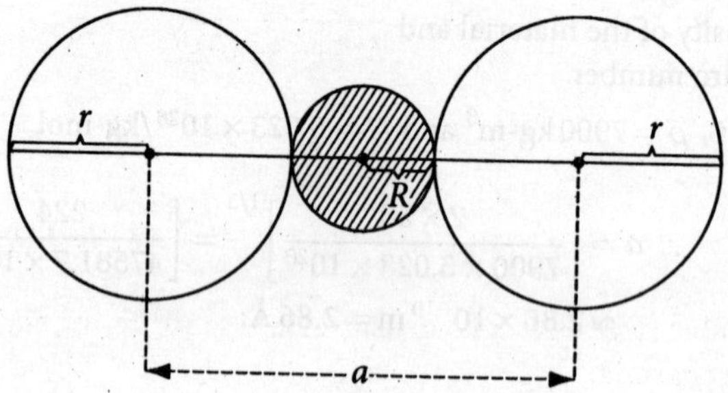

Fig. 6.39

For FCC structure,

$$a = \frac{4r}{\sqrt{2}}.$$

$$\therefore\ R = \frac{4r}{2\sqrt{2}} - r = \frac{2r}{\sqrt{2}} - r = r\left(\sqrt{2} - 1\right) = \mathbf{0.414r}.$$

6.30 Copper has FCC structure and the atomic radius is 0.1278 nm. Calculate the inter-planar spacing for (110)- and (212)-planes.

Solution. We know that inter-planar spacing d_{hkl} of a lattice having cubic structure of lattice constant a is

$$d = \frac{a}{(h^2 + k^2 + l^2)^{1/2}}.$$

For FCC structure $a = \frac{4r}{\sqrt{2}}$, where the atomic radius $r = 0.1278 \times 10^{-9}$ m.

$$\therefore\ a = \frac{4 \times 0.1278 \times 10^{-9}}{\sqrt{2}} = 0.3615 \text{ nm}.$$

Now inter-planar spacing for (110)-plane is

$$d_{110}=\frac{0.3615}{(1^2+1^2+0^2)^{1/2}}=\frac{0.3615}{\sqrt{2}}\text{ nm}=0.2556\text{ nm}.$$

Inter-planar spacing for (212)-plane is

$$d_{212}=\frac{0.3615}{(2^2+1^2+2^2)^{1/2}}=\frac{0.3615}{9^{1/2}}=0.1205\text{ nm}.$$

6.31 Show that in simple cubic lattice,

$$d_{100}:d_{110}:d_{111}=1:0.71:0.58.$$ [cf. CUH 2005]

Solution. Inter-planar spacing for (hkl)-planes in an SC structure of lattice constant a is

$$d=\frac{a}{(h^2+k^2+l^2)^{1/2}}.$$

$$\therefore d_{100}=\frac{a}{\sqrt{1^2+0^2+0^2}}=\frac{a}{1}=a,$$

$$d_{110}=\frac{a}{(1^2+1^2+0^2)^{1/2}}=\frac{a}{\sqrt{2}}$$

$$\text{and } d_{111}=\frac{a}{(1^2+1^2+1^2)}=\frac{a}{\sqrt{3}}.$$

$$\therefore d_{100}:d_{110}:d_{111}=1:\frac{1}{\sqrt{2}}:\frac{1}{\sqrt{3}}=\mathbf{1:0.71:0.58}.$$

6.32 KBr crystal has a cubic structure. If its density is 2.75×10^3 kg-m^3 and molecular weight is 119.01, calculate the lattice constant. [CUH 2006]

Solution. Given molecular weight of KBr is

$$=119.01\text{ gm-wt}=119.01\times10^{-3}\text{ kg-wt}.$$

Mass of KBr molecule

$$M=\frac{\text{Molecular weight}}{\text{Avogadro number}}=\frac{119.01\times10^{-3}}{6.023\times10^{23}}\text{ kg}=19.76\times10^{-26}\text{ kg}.$$

Number of KBr molecule/volume

$$=\frac{\text{Density}}{M}=\frac{2.75\times10^3}{19.76\times10^{-26}}=0.139\times10^{29}.$$

The number of atoms/volume of KBr

$$=2\times0.139\times10^{29}\ [\because \text{KBr is diatomic}]$$
$$=27.8\times10^{27}\text{ atoms/m}^3.$$

If a be the distance between two adjacent atoms, the volume a^3 will contain 1 atom due to the cubic structure.

So, the total number of atoms/volume $= \frac{1}{a^3}$.

$$\therefore \frac{1}{a^3} = 27.8 \times 10^{27} \text{ or } a = 3.30 \times 10^{-16} \text{ m.}$$

$\therefore$ lattice constant $2a = 6.6 \times 10^{-10}$ m $= 6.6$ Å.

EXERCISES

6(A) Objective Type Questions

Choose the correct alternatives.

6.1 Solids have definite shape and size because
(a) density of solids is larger compared to liquid or gas
(b) the atoms or molecules of solid are bound with strong forces of attraction
(c) solids are generally of high rigidity
(d) solids can be transformed into liquids and gases.

6.2 Most of the solids prefer to be in the crystalline state, because
(a) volume is generally minimum in crystalline state than liquid or gas state
(b) crystalline state is the low energy state
(c) crystal has definite shape and size
(d) none of these.

6.3 In amorphous solids the atoms and molecules are
(a) regularly arranged in periodic manner
(b) randomly arranged
(c) partly regularly and partly randomly arranged
(d) none of these.

6.4 The physical properties of solids depend on
(a) shape and size of the solids
(b) the atoms, ions or molecules forming the solid
(c) the presence of impurities
(d) the irregularities in the lattice.

6.5 Amorphous solids are isotropic because
(a) they do not have directional property
(b) they exhibit short range order
(c) their colour varies from solid to solid
(d) they are brittle.

6.6 Crystals are anisotropic because
(a) atoms, ions or molecules are closely packed in a small volume
(b) often they contain ions of different elements
(c) crystal has different periodic arrangements in different directions
(d) the interaction space is void.

6.7 The fundamental building block whose repetition in three dimensions generates a crystal is called

(a) unit cell
(b) lattice cell
(c) primitive cell
(d) none of these.

6.8 In a plane lattice the periodic arrangement of atoms occurs in
(a) three dimensions
(b) two dimensions
(c) a line
(d) both in 2 and 3 dimensions.

6.9 The number of basic crystal system is
(a) 2
(b) 3
(c) 4
(d) 7.

6.10 Basis is defined as
(a) an atom about a lattice point
(b) a group of atoms about a lattice point
(c) one or a group of atoms about a lattice point
(d) none of these.

6.11 Lattice + basis is
(a) unit cell
(b) primitive cell
(c) crystal structure
(d) molecular structure.

6.12 The number of Bravais lattices in three dimensions is
(a) 7
(b) 9
(c) 14
(d) 19.

6.13 The magnitudes of basis vectors $\vec{a}, \vec{b}, \vec{c}$ are called
(a) lattice parameters
(b) lattice vectors
(c) crystal vectors
(d) unit vectors.

6.14 In a primitive unit cell, the number of lattice points is
(a) 2
(b) 5
(c) 4
(d) 1.

6.15 For a cubic crystal systems,
(a) $a = b = c;\ \alpha = 90°, \beta \neq \gamma$
(b) $a \neq b = c;\ \alpha = \beta = \gamma = 90°$
(c) $a = b = c;\ \alpha = \beta = \gamma = 90°$
(d) $a = b \neq c;\ \alpha = \beta \neq \gamma$.

6.16 If $\vec{a}, \vec{b}$ and $\vec{c}$ represent three basis vectors, then the volume of Wigner-Seitz cell is
(a) $|\vec{a} \times \vec{b} \times \vec{c}|$
(b) $|(\vec{a} + \vec{b}) \times \vec{c}|$
(c) $|\vec{a} \cdot (\vec{b} + \vec{c})|$
(d) $|\vec{a} \times \vec{b} \cdot \vec{c}|$.

6.17 The effective number of lattice points in an SC lattice is
(a) 4
(b) 2
(c) 5
(d) 1.

6.18 In a BCC structure, the effective number of lattice points per unit cell is
(a) 2
(b) 4
(c) 1
(d) 3.

6.19 In an FCC lattice, the number of lattice points per unit cell is

(a) 2
(b) 4
(c) 5
(d) 1.

6.20 NaCl crystal is
(a) SC
(b) BCC
(c) FCC
(d) none of these.

6.21 If the number of Bravais lattice is three, then the crystal is
(a) monoclinic
(b) hexagonal
(c) cubic
(d) tetragonal.

6.22 The coordination number of an SC structure is
(a) 6
(b) 12
(c) 8
(d) 4.

6.23 The FCC structure is
(a) primitive
(b) non-primitive
(c) may be either primitive or non-primitive
(d) none of these.

6.24 The unit cell of a BCC structure is
(a) primitive
(b) non-primitive
(c) may be either primitive or non-primitive
(d) none of these.

6.25 Miller indices (hkl) represent
(a) a set of parallel planes
(b) a particular plane
(c) a set of arbitrarily oriented planes
(d) none of these.

6.26 The Miller indices of a plane having intercepts 2, ∞, ∞ units on the $\vec{a}, \vec{b}$ and $\vec{c}$ axes respectively are
(a) 101
(b) 001
(c) 200
(d) 020.

6.27 Miller indices of a plane having intercepts 2, 3, 4 units along X-, Y- and Z-axes are
(a) (643)
(b) (436)
(c) (231). [WBUT 2006]

6.28 The inter-planar distance d_{100} for (100)-plane of SC lattice is
(a) $\frac{a}{\sqrt{3}}$
(b) $\frac{a}{\sqrt{2}}$
(c) a
(d) $\frac{a}{\sqrt{5}}$.

6.29 The inter-planar distance d_{111} for a (111)-plane of an SC cubic lattice is
(a) a
(b) $\frac{a}{\sqrt{3}}$
(c) $\frac{a}{\sqrt{2}}$
(d) $\frac{a}{\sqrt{0.5}}$.

6.30 The inter-planar distance (d_{100}) for (100)-plane of a BCC lattice is
(a) $\frac{a}{\sqrt{2}}$
(b) $\frac{a}{\sqrt{3}}$
(c) a
(d) $\frac{a}{2}$.

6.31 The inter-planar spacing d_{111} for (111)-plane of an FCC lattice is

(a) $\frac{a}{\sqrt{2}}$ (c) $\frac{a}{\sqrt{3}}$

(b) a (d) $\frac{a}{2}$.

6.32 If a and r be respectively the lattice constant and radius of an atom in an SC structure, then

(a) $a=\frac{4}{\sqrt{3}}r$ (c) $a=\frac{2}{\sqrt{2}}r$

(b) $a=2r$ (d) $a=\frac{r}{2}$.

6.33 If a and r be respectively the lattice constant and radius of an atom in a BCC structure, then

(a) $r=\frac{\sqrt{3}}{4}a$ (c) a

(d) none of these. [WBUT 2007]

(b) $\frac{\sqrt{2}}{4}a$

6.34 For an FCC structure the relation between lattice constant (a) and radius (r) of the atom is

(a) $r=\frac{\sqrt{3}}{4}a$ (c) $r=a$

(b) $r=\frac{a}{2}$ (d) $r=\frac{\sqrt{2}}{4}a$.

6.35 The coordination number of SC structure is

(a) 6 (c) 2

(b) 3 (d) 4.

6.36 The coordination number of BC structure is

(a) 1 (c) 6

(b) 8 (d) 2.

6.37 For FCC cubic structure the coordination number is

(a) 6 (c) 12

(b) 8 (d) 10.

6.38 For SC structure atomic packing factor is

(a) 74% (c) 68%

(b) 52% (d) 47%.

6.39 If the atomic packing factor for a crystal is 68%, then the structure of the crystal is

(a) SC (c) BCC

(b) FCC (d) none of these.

6.40 The atomic packing factor for FCC structure is

(a) 74% (c) 68%

(b) 52% (d) 49%.

6.41 For NaCl crystal having lattice constant 5.6 Å the interatomic distance is

(a) 2.8 Å (c) 5.6 Å

(b) 1.4 Å (d) 7.2 Å.

6.42 If for a crystal $d_{hkl}=1.414$ nm and $a=1$ nm, then the Miller indices of the planes are

(a) (111)
(b) (101)
(c) (110)
(d) (100).

6.43 (111)-plane of a cubic crystal is
(a) perpendicular to X-axis
(b) perpendicular to Y-axis
(c) perpendicular to Z-axis
(d) none of these. [WBUT 2005]

6.44 $d_{100}:d_{110}:d_{111}$ for an SC structure is
(a) $1:\sqrt{3}:\sqrt{6}$
(b) $\sqrt{2}:\sqrt{5}:\sqrt{6}$
(c) $\sqrt{6}:\sqrt{3}:\sqrt{2}$
(d) $\sqrt{2}:\sqrt{6}:\sqrt{5}$.

6.45 The (111)- and (222)-planes in a lattice are
(a) perpendicular to each other
(b) parallel to each other
(c) inclined at 60° to each other
(d) none of these.

6.46 The element which crystallizes in a BCC structure is
(a) Sodium
(b) Zinc
(c) Copper
(d) Iron.

6.47 If the interatomic distance of NaCl crystal is 0.28 nm, then the lattice parameter is
(a) 0.14 nm
(b) 0.28 nm
(c) 0.56 nm
(d) none of these.

6.48 The crystal structure of aluminium is
(a) SC
(b) BCC
(c) FCC
(d) HCP.

6.49 If the primitives $a \neq b \neq c$ and interfacial angles $\alpha \neq \beta \neq \gamma \neq 90°$, then it belongs to the system
(a) monoclinic
(b) orthorhombic
(c) tetragonal
(d) cubic.

6.50 In a cubic unit cell, the angles between the normals to the planes $(\bar{1}11)$ and (101) is
(a) 50°
(b) 64.5°
(c) 54.69°
(d) 70°.

6.51 Volume of a unit cell in an FCC crystal is
(a) $16\sqrt{2}r^3$
(b) $4\sqrt{2}r^3$
(c) $\dfrac{16\sqrt{2}}{r^3}$
(d) $2\sqrt{2}r^3$.

6.52 The density of an FCC structure is
(a) $\rho = \dfrac{M}{Na^3}$
(b) $\rho = \dfrac{MN}{a^3}$
(c) $\rho = \dfrac{4M}{Na^3}$
(d) $\rho = \dfrac{Ma^3}{4N}$.

the different letters have their usual meanings.

6.53 The nearest neighbour distance in case of BCC structure is

(a) $\frac{a\sqrt{3}}{2}$ (c) $\frac{2a}{\sqrt{3}}$

(b) $\frac{a\sqrt{2}}{\sqrt{2}}$ (d) $\frac{2a}{\sqrt{3}}$. [WBUT 2008]

6.54 The primitives of a crystal are 1.2 Å, 1.8 Å, 2 Å, 2 Å along the three axes, where a plane with Miller indices (231) cuts intercept 1.2 Å along x-axis. What will be the lengths of intercept along y- and z-axes?

(a) 1.5 Å and 5 Å (c) 1.2 Å and 4 Å

(b) 1.9 Å and 2.5 Å (d) 1.5 Å and 4 Å. [WBUT 2008]

ANSWERS

6.1 (b)	6.8 (b)	6.15 (c)	6.22 (d)	6.29 (b)	6.36 (b)	6.43 (d)	6.50 (c)
6.2 (b)	6.9 (d)	6.16 (d)	6.23 (b)	6.30 (d)	6.37 (c)	6.44 (c)	6.51 (a)
6.3 (b)	6.10 (c)	6.17 (d)	6.24 (b)	6.31 (c)	6.38 (b)	6.45 (b)	6.52 (c)
6.4 (b)	6.11 (c)	6.18 (a)	6.25 (a)	6.32 (b)	6.39 (c)	6.46 (a)	6.53 (a)
6.5 (a)	6.12 (c)	6.19 (b)	6.26 (c)	6.33 (a)	6.40 (a)	6.47 (b)	6.54 (c).
6.6 (c)	6.13 (b)	6.20 (c)	6.27 (a)	6.34 (d)	6.41 (a)	6.48 (c)	
6.7 (a)	6.14 (d)	6.21 (c)	6.28 (c)	6.35 (a)	6.42 (c)	6.49 (a)	

6(B) Short Answer Type Questions

6.1 Why most of the solids prefer to be in the crystalline state? Why the solids have definite shape and size?

6.2 Why crystalline solids are isotropic in nature but the amorphous are not? [CUH 1996]

6.3 Distinguish between crystalline and amorphous solids.

6.4 What do you mean by crystal lattice? What are plane lattice and space lattice?

6.5 Explain the terms *lattice translation operation* and *fundamental translation vectors.* When are the translation vectors are called *primitive?* [BUH 1999]

6.6 What is unit cell? [WBUT (Old) 2005]

6.7 If a unit cell has the following characteristics $a = b = 10.5$ Å, $c = 6$ Å, $\alpha = \beta = \gamma = 90°$, identify to which crystal system does the unit cell belong. [BUH 2003]

6.8 What is a Wigner-Seitz primitive cell? How can it be constructed?

6.9 What is the difference between a primitive cell and unit cell? Show that all primitive cells are unit cells but all unit cells may or may not be primitive cell.

6.10 What are Bravais lattice? What is the number of Bravais lattice in three dimensions?

6.11 What is the basis of dividing the crystal systems into different types?

6.12 Draw the diagrams of SC, FCC and BCC structures.

6.13 Find the number of atoms per unit cell of SC, FCC and BCC types of crystals. [WBUT 2007]

6.14 What is meant by Miller indices. Explain with examples. [CUH 1990]

6.15 What is the utility of Miller indices? [BUH 1997]

6.16 Define a crystal lattice. What is a primitive cell? Is it unique?

6.17 What are basis vectors? How many different *Bravais lattices* are shown in crystal structure? [BUH 2006]

6.18 For an FCC lattice what are

(a) the volume of the primitive cell (b) lattice points per unit volume
(c) nearest neighbour distance and (d) the number of nearest neighbours? [BUH 1999]

6.19 Deduce the inter-planar spacing of a simple cubic lattice of side a. [WBUT 2007]

6.20 Show that in a cubic crystal of side a the inter-planar spacing between consecutive parallel planes of Miller indices (hkl) is $d_{hkl} = \frac{a}{\sqrt{h^2+k^2+l^2}}$. [WBUT 2005, 2004]

6.21 Calculate atomic packing fraction and atoms per unit cell in crystals having
(a) simple cubic
(b) body-centred cubic and
(c) face-centred cubic structure considering the atoms as hard spheres. [BUH 2003; WBUT 2003]

6.22 Calculate atomic radius for SC, FCC and BCC structures.

6.23 What do you mean by coordination number of atoms. Calculate the coordination number of SC, FCC and BCC structure.

6.24 Prove that in a cubic crystal, the direction $[hkl]$ is perpendicular to the plane (hkl).

6.25 Write down short note on Bravais lattice and Miller indices. [WBUT 2003]

6(C) Numerical Problems

6.1 Lead is face-centred cubic with an atomic radius $r = 1.746$ Å. Find the inter-planar spacing for (220)-plane. [WBUT 2005]

6.2 Aluminium has FCC structure. If the unit cell dimension and atomic weight of Aluminium is 4.048 Å and 26.98 respectively, then calculate its density. (*Ans.* 2700 kg/m^3)

6.3 In a cubic unit cell find the angle between the normals to the planes (111) and (101).

6.4 Copper has an FCC structure and the atomic radius is 0.1278 nm. Calculate the inter-planar spacing for (111)-planes. [BUH 1998]
(*Ans.* 0.209 nm)

6.5 Silicon crystallizes in the diamond cubic structure. The radius of silicon atom is 0.1176 nm. The atomic weight and density of silicon are 28.09 and 2.3×10^3 kg/m^3. What is the number of atoms present in a unit cell. (*Ans.* 8)

6.6 Compare the densities of lattice points in (111)- and (110)-planes in an SC lattice.
(*Ans.* $\rho_{111} : \rho_{110} = \sqrt{2} : \sqrt{3}$)

6.7 Aluminium is an FCC crystal with lattice constant $a = 0.405$ nm.
(a) How many unit cells are there in an aluminium foil 0.005 cm thick and side 25 cm square?
(b) If its weight is 0.0085 kg, then how many atoms are present?
(c) What is the number of atoms per unit cell? (*Ans.* 4.7×10^{22}; 1.9×10^{23}; 4)

6.8 Copper has FCC structure having atomic radius 1.278 Å. Find its density. (*Ans.* 8.94 gm/cc)

6.9 Calculate the inter-planar spacing for a (321)-plane in an SC lattice whose lattice constant is 4.2×10^{-8} cm. (*Ans.* 1.1×10^{-8} cm)

6.10 In a general lattice, $a = b = 2.5$, $c = 1.8$ Å, what is the lattice spacing between (111)-planes.
(*Ans.* 3.96 Å)

6.11 A lattice plane in an SC structure cuts the $\vec{a}, \vec{b}$ and $\vec{c}$ axes at distances $6a, 4b$ and $3c$. Find the Miller indices of the plane. (*Ans.* (234))

6(D) Broad Answer Type Questions

6.1 (a) Define unit cell.

(b) Show that the spacing d of planes (hkl) in an SC lattice of side a is $d = \frac{a}{\sqrt{h^2+k^2+l^2}}$. [CUH 2008]

(c) Lead is face-centred cubic with an atomic radius of $r = 1.746$ Å. Find the inter-planar spacing for (220)-planes. [WBUT 2005]

6.2 (a) Show that in a cubic crystal of side a the spacing between consecutive parallel planes of Miller indices (hkl) is $d_{hkl} = \frac{a}{\sqrt{h^2+k^2+l^2}}$.

(b) Copper has FCC structure and atomic radius is 0.1278 nm. Calculate the density and the inter-planar spacing for (321) planes. Take atomic weight of copper as 63.5. [WBUT 2005]

6.3 (a) What are Miller indices?

(b) Find the packing factor for SC structure and BCC lattice. [WBUT 2004]

(c) Calculate the inter-planar spacing d of planes (111) in an SC lattice of side a. [WBUT 2003, 2004, 2007]

6.4 (a) What are the differences between crystalline and amorphous solids? [CUH 1996]

(b) Show that an FCC structure is more closely packed than a BCC structure.

(c) Illustrate the SC, FCC and BCC structures with diagrams.

6.5 (a) Find the atomic radii of SC, FCC and BCC structures of lattice constant a.

(b) What are Miller indices? How can they be determined?

(c) What are the characteristics of Miller indices?

6.6 (a) Explain the terms *lattice translation operation* and *fundamental translation vectors.* When are the translation vectors called *primitive*?

(b) For an FCC lattice, what are (i) the volume of the primitive cell (ii) the lattice points per unit volume (iii) the nearest neighbour distance, (iv) the number of nearest neighbours?

(c) What fraction of total volume is occupied by atoms in an FCC structure? [BUH 1999]

6.7 (a) Find out expression for the atomic radius and coordination number of SC, FCC and BCC structures of lattice constant a.

(b) Illustrate the SC, FCC and BCC structures with diagrams.

(c) Find out the packing fraction for FCC structure. Find out the expression for the separation between the planes with Miller indices h, k, l and lattice constant a for a simple cubic structure. [WBUT 2008]

Chapter 7

Crystal Diffraction

7.1 Introduction

Von Laue in 1912 discovered that X-rays are diffracted by single crystal samples. The crystal acts as a three-dimensional space grating for the diffraction of X-rays. This X-ray diffraction technique can be successfully applied to determine the structure of the single crystals. In this technique the incident parallel monochromatic beam of X-rays is diffracted by the atoms of the single crystal. The intensity and the direction of the diffracted beam is related to the lattice structure and chemical composition of the crystal.

The three principal methods of crystal structure analysis are:

1. Laue Method,
2. Rotating Crystal Method and
3. Powder Method or Debye-Scherrer Method.

Professional crystallographers nowadays prefer to use Complicated Precision Camera Method.

7.2 Why X-rays are used for the Analysis of Crystal Structure?

There are several reasons to use X-rays for the analysis of crystal structure as stated below:

1. For diffraction effect to be observed *the wavelength of the radiation should be of the same order of the diffracting elements. Incidentally, the wavelength of the X-rays is comparable to the interatomic distances in crystals.*
2. X-rays are scattered elastically without any change of wavelength by the charged particles.
3. The wavelength of X-rays can be determined very accurately.
4. X-rays of suitable wavelength for diffraction by a specific crystal can be obtained by changing the voltage of the X-ray tube.
5. Radiation of very longer wavelength (e.g., visible light) and shorter wavelength (shorter than X-rays) are unsuitable for producing diffraction effect in single crystals.

7.3 Bragg's Law for X-ray Diffraction

W.L. Bragg established a relationship between the wavelength of X-rays diffracted by the atomic planes of a crystal and the interatomic distance of the same.

His work is based on a simple model with the following assumptions:

1. *The monochromatic X-rays are reflected by successive parallel planes of the atoms in the crystal, so that angle of incidence is equal to the angle of reflection.*

2. *Constructively interfering beams which are reflected by a set of parallel atomic planes, give rise to a strong diffracted beam.*
3. *The scattering of X-rays by the atoms is perfectly elastic, so that the wavelength of X-rays remains unchanged on reflection.*

Derivation of the law

Let a crystal be made up of equidistant parallel planes of atoms with inter-planar distance d as shown in Fig. 7.1. Let a monochromatic ray of X-rays PA of a wavelength λ be incident at an angle θ with the atomic planes. The corresponding reflected ray AQ must also be inclined at an angle θ to the plane. RB is another ray incident on the next plane at the same angle θ and the corresponding reflected ray be BS inclined also at θ with atomic plane.

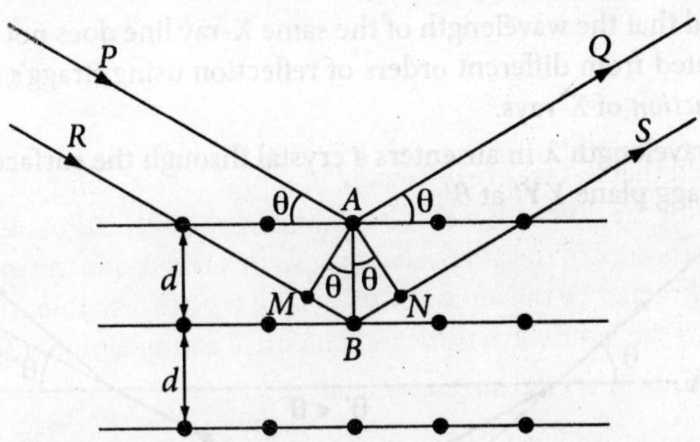

Fig. 7.1: Diffraction of X-rays by crystal planes.

The path difference between the two reflected rays AQ and BS is

$$\Delta = MB + BN, \tag{7.1}$$

where AM and AN are perpendiculars from A on RB and BS respectively.

In ΔMAB,

$$\sin\theta = \frac{MB}{AB} = \frac{MB}{d} \text{ or } MB = d\sin\theta.$$

Similarly, in ΔNAB

$$BN = d\sin\theta.$$

$\therefore$ from (7.1),

$$\Delta = d\sin\theta + d\sin\theta = 2d\sin\theta. \tag{7.2}$$

If the two consecutive crystal planes scatter X-rays in the same phase, then the path difference must be integral multiple of λ, i.e.,

$$\Delta = n\lambda, \tag{7.3}$$

where $n = 0, 1, 2, \ldots$ is the *order of reflection.*

Hence, the condition of constructive interference of the reflected X-rays by the planes in a crystal is

$$2d\sin\theta = n\lambda. \tag{7.4}$$

Since the maximum possible value of $|\sin\theta| = 1$,

$$\frac{n\lambda}{2d} \le 1. \tag{7.5}$$

Hence, the wavelength λ must not exceed twice the inter-planar spacing d for Bragg diffraction to occur.

Definition. *Bragg's law is defined as the X-rays reflected from different parallel planes of a crystal interfere constructively when the path difference is integral multiple of the wavelength of X-rays.*

Note. Bragg's law of X-ray reflection has the following utilities:

1. If λ of X-rays is known, then by measuring θ experimentally the value of inter-planar spacing d and hence lattice constant can be determined.
2. If d is supplied, then measuring θ we can find out the wavelength λ of the unknown X-rays.

7.3.1 Correction of Bragg's Law Due to Refraction

Stenstrom (1919) observed that the wavelength of the same X-ray line does not give *identical values* for the wavelength, when computed from different orders of reflection using Bragg's equation. This discrepancy was attributed to the *refraction* of X-rays.

Consider an X-ray of wavelength λ in air enters a crystal through the surface XX' at a glancing angle θ and be incident on the Bragg plane YY' at θ'.

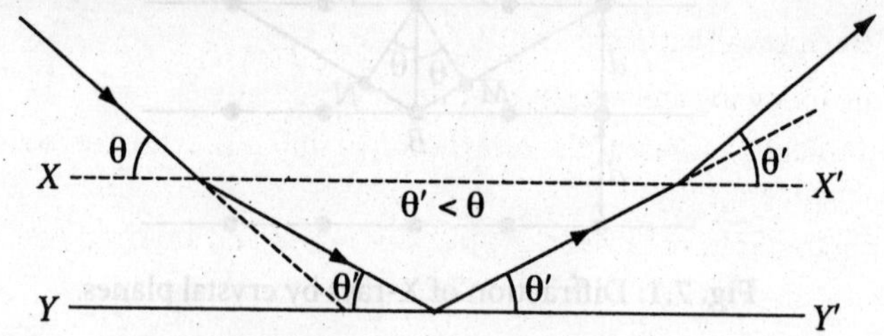

Fig. 7.2: Refraction of X-rays by a crystal.

From Bragg's equation

$$n\lambda' = 2d\sin\theta', \tag{7.6}$$

where λ' is the wavelength of X-rays in the crystal.

If μ be the refractive index of the crystal, then

$$\mu = \frac{\lambda}{\lambda'} \text{ also } \mu = \frac{\cos\theta}{\cos\theta'}. \tag{7.7}$$

Putting $\lambda' = \frac{\lambda}{\mu}$ in (7.6), we get

$$\frac{n\lambda}{\mu} = 2\sin\theta' = 2d\left(1-\cos^2\theta'\right)^{1/2}$$

$$\text{or } n\lambda = 2\mu d\left(1-\cos^2\theta'\right)^{1/2} = 2d\left(\mu^2-\mu^2\cos^2\theta'\right)^{1/2}$$

$$= 2d\left(\mu^2-\cos^2\theta\right)^{1/2} \quad \text{[from (7.7)]}$$

$$= 2d\left[\mu^2-(1-\sin^2\theta)\right]^{1/2} = 2d\sin\theta\left[1-\frac{1-\mu^2}{\sin^2\theta}\right]^{1/2}$$

$$\text{or } n\lambda \simeq 2d\sin\theta\left[1-\frac{1-\mu^2}{2\sin^2\theta}\right] \tag{7.8}$$

$$\text{or } n\lambda = 2d\sin\theta\left[1-\frac{1-\mu}{\sin^2\theta}\right] \quad \left[\because 1-\mu^2=(1+\mu)(1-\mu)\simeq 2(1-\mu), \because \mu\simeq 1\right]. \tag{7.9}$$

Again, from $n\lambda = 2d\sin\theta$,

$$\sin^2\theta = \frac{n^2\lambda^2}{4d^2}.$$

$\therefore$ from (7.9), we get

$$n\lambda = 2d\sin\theta\left[1 - \frac{4d^2(1-\mu)}{n^2\lambda^2}\right]. \quad (7.10)$$

(7.10) is corrected form of *Bragg's equation*. The correction term $\frac{4d^2(1-\mu)}{n^2\lambda^2}$ is usually very small and is progressively smaller for higher values of n. Thus uncorrected Bragg's law can be used without serious error.

7.4 Reciprocal Lattice

In the analysis of crystal structure by using X-rays, we have to deal with planes of several slopes in the same problem and it is difficult to visualize several slopes of such planes. In reciprocal lattice *the orientation of a set of parallel planes* and their *inter-planar spacing* are represented by a *reciprocal lattice point* and as such the problem-faced in direct lattice is reduced in reciprocal lattice.

7.4.1 Advantages of Reciprocal Lattice

Reciprocal lattice has the following advantages over direct space lattice.

1. We can very conveniently represent the orientation of different sets of crystal planes and their spacings in reciprocal lattice.
2. The diffraction photographs of crystals can be understood and analysed easily.
3. It provides a simple method of obtaining the diffraction condition.
4. The wave mechanical behaviour of electrons in periodic lattice can be conveniently understood in terms of reciprocal lattice.

These are the importance of reciprocal lattice in solid state physics.

7.4.2 Geometrical Construction of the Reciprocal Lattice

Let us consider a set of parallel planes (100) as shown in Fig. 7.3. The *reciprocal point* for these number of planes is formed by drawing a common normal to the planes from an origin taken on any lattice point

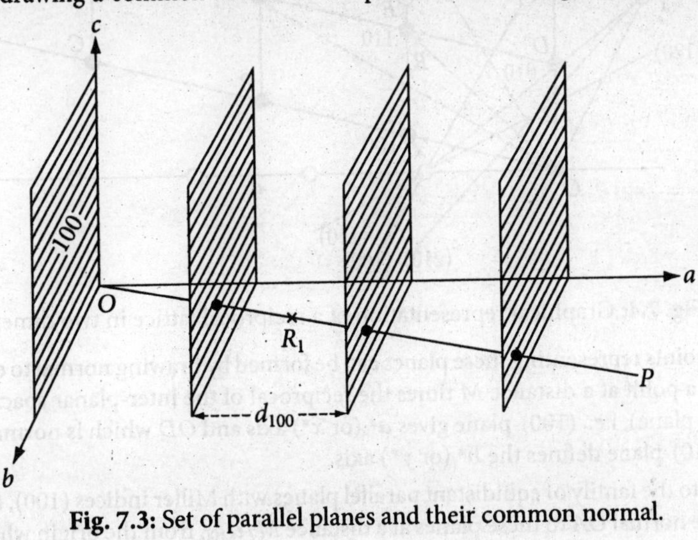

Fig. 7.3: Set of parallel planes and their common normal.

and locating a point on this line at a distance which is inversely proportional to the inter-planar spacing. In Fig. 7.3 the normal is represented by OP and the reciprocal lattice point by R_1, where $OR_1 = \frac{M}{d_{100}}$, M being a constant called the *scale (or magnification) factor* for OR_1 and d_{100} the *inter-planar spacing.* Thus the entire set of parallel plane is represented by a single point R_1 in Fig. 7.3.

7.4.3 Rules for Constructing the Reciprocal Lattice

1. Draw normal to every set of planes from the origin point in the direct lattice.
2. For a particular set of planes, set the length of each normal equal to the reciprocal of the inter-planar spacing.
3. Put a point at the end of each normal.

The collection of points so obtained gives us a new lattice array called the *reciprocal lattice,* because the distances in this lattice are *reciprocal* to those in the direct lattice.

7.4.4 Graphical Demonstration of Reciprocal Lattice in Two-Dimensional Space

Fig. 7.4 shows a monoclinic lattice ($a \neq b \neq c; \alpha = \beta = 90° \neq \gamma$) looking along the unique axis namely c-axis called the *first axis.* We suppose all the lattice planes parallel to the c-axis chosen perpendicular to the plane of the figure. OX and OY represent respectively the unit lattice parameters a and b along x- and y-axes of the direct lattice. The normals to all the planes in the zone described above will lie in the plane of the diagram. Fig. 7.4 shows several sets of planes of the said zone. Only that plane of each family of planes nearest to the origin has been shown in the figure for the sake of simplicity. The Fig. 7.4 shows the traces of (100), (010), (120), etc. planes which are perpendicular to the plane of the paper.

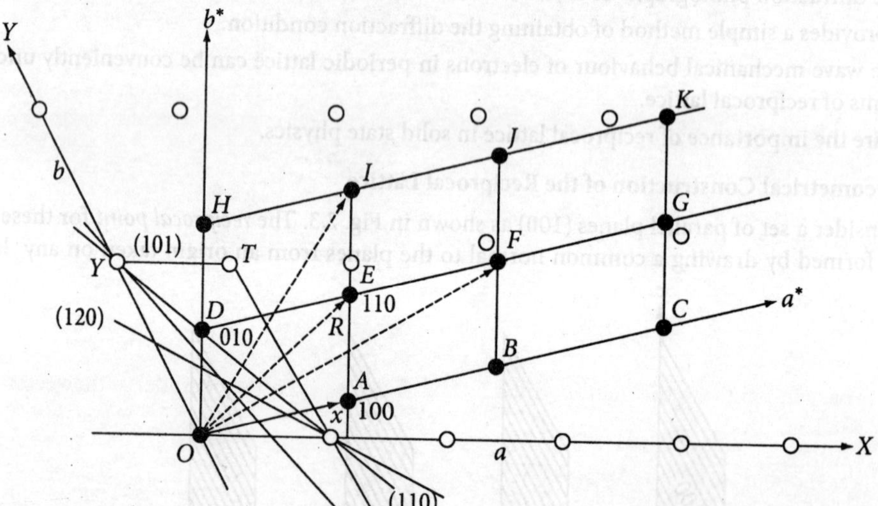

Fig. 7.4: Graphical representation of a reciprocal lattice in two dimensions.

The reciprocal points representing these planes can be formed by drawing normal to each family of planes and then locating a point at a distance M times the reciprocal of the inter-planar spacing. The normal OA to the a plane (bc-plane), i.e., (100)-plane gives a^* (or x^*) axis and OD which is normal to the p-plane (or ac-plane), i.e., (010)-plane defines the b^* (or y^*) axis.

Corresponding to the family of equidistant parallel planes with Miller indices (100), the reciprocal lattice points lie along the normal OA to these planes at a distance M/d_{100} from the origin where d_{100} is the inter-planar distance for (100)-planes. The reciprocal lattice point represented by the coordinates 100 (without parenthesis) is shown at A. Similarly, the planes (010) are represented by reciprocal point D which lies on the normal to these planes at a distance M/d_{010} and is assigned coordinates 010. Also the reciprocal lattice point E for (110)-planes is located on the normal at a distance M/d_{110} and is represented by the coordinates 110.

Applying the same geometrical constraction throughout, we can locate all the reciprocal lattice points relating to the crystal.

7.4.5 Definition of Reciprocal Lattice

The reciprocal lattice corresponding to a Bravais lattice point $\vec{R}$ is defined as the set of all wave vectors $\vec{k}$ that give plane waves of the form $e^{i\vec{k}\cdot\vec{r}}$ with the periodicity $(\vec{R})$ of the Bravais lattice, i.e.,

$$e^{i\vec{k}\cdot(\vec{r}+\vec{R})} = e^{i\vec{k}\cdot\vec{r}} \tag{7.11}$$

for any $\vec{r}$ and all $\vec{R}$ in the *Bravais lattice*

$$\text{or } e^{i\vec{k}\cdot\vec{r}} \times e^{i\vec{k}\cdot\vec{R}} = e^{i\vec{k}\cdot\vec{r}}, \text{ i.e., } e^{i\vec{k}\cdot\vec{R}} = 1. \tag{7.12}$$

So, the reciprocal lattice vector is represented by a set of lattice vectors $\vec{k}$ satisfying (7.12), for all $\vec{R}$ in the Bravais lattice.

Since $e^{i2\pi n} = 1$, where n is an integer, we can write,

$$e^{i\vec{k}\cdot\vec{R}} = e^{i2\pi n}, \text{ i.e., } \vec{k}\cdot\vec{R} = 2\pi n. \tag{7.13}$$

7.5 Reciprocal Lattice Vector

The reciprocal lattice vector $\vec{a}^*, \vec{b}^*$ and $\vec{c}^*$ are defined in terms of the primitive translation vectors $\vec{a}, \vec{b}$ and $\vec{c}$ of the direct lattice as follows:

$$\vec{a}^*\cdot\vec{a} = \vec{b}^*\cdot\vec{b} = \vec{c}^*\cdot\vec{c} = 1 \tag{7.14}$$

$$\text{and } \vec{a}^*\cdot\vec{b} = \vec{a}^*\cdot\vec{c} = \vec{b}^*\cdot\vec{c} = \vec{b}^*\cdot\vec{a} = \vec{c}^*\cdot\vec{a} = \vec{c}^*\cdot\vec{b} = 0. \tag{7.15}$$

From (7.15),

$$\vec{a}^*\cdot\vec{b} = \vec{a}^*\cdot\vec{c} = 0,$$

i.e., $\vec{a}^*$ is perpendicular to the plane defined by $\vec{b}$ and $\vec{c}$. It is, therefore, parallel to the vector $\vec{b}\times\vec{c}$ and can be expressed as

$$\vec{a}^* = A\left(\vec{b}\times\vec{c}\right), \tag{7.16}$$

where A is a scalar quantity.

Now from (7.14), we have

$$\vec{a}^*\cdot\vec{a} = A\left(\vec{b}\times\vec{c}\right)\cdot\vec{a} = 1. \tag{7.17}$$

$$\therefore A = \frac{1}{\vec{a}\cdot\left(\vec{b}\times\vec{c}\right)}. \tag{7.18}$$

$\therefore$ from (7.16) and (7.18),

$$\vec{a}^* = \frac{\vec{b}\times\vec{c}}{\vec{a}\cdot\left(\vec{b}\times\vec{c}\right)}. \tag{7.19}$$

In a similar way

$$\vec{b}^* = \frac{\vec{c}\times\vec{a}}{\vec{a}\cdot\left(\vec{b}\times\vec{c}\right)} \tag{7.20}$$

$$\text{and } \vec{c}^* = \frac{\vec{a}\times\vec{b}}{\vec{a}\cdot\left(\vec{b}\times\vec{c}\right)}. \tag{7.21}$$

Some authors prefer to define *primitive translation vectors* $\vec{a},\vec{b},\vec{c}$ of the direct lattice and *primitive translation vectors* of the reciprocal lattice in the following way:

$$\vec{a}^*\cdot\vec{a} = \vec{b}^*\cdot\vec{b} = \vec{c}^*\cdot\vec{c} = 2\pi \tag{7.22}$$

$$\text{and } \vec{a}^*\cdot\vec{b} = \vec{a}^*\cdot\vec{c} = \vec{b}^*\cdot\vec{c} = \vec{b}^*\cdot\vec{a} = \vec{c}^*\cdot\vec{a} = \vec{c}^*\cdot\vec{b} = 0.$$

Then we obtain in the above way

$$\left.\begin{aligned}\vec{a}^* &= 2\pi\frac{\vec{b}\times\vec{c}}{\vec{a}\cdot\left(\vec{b}\times\vec{c}\right)}\\ \vec{b}^* &= 2\pi\frac{\vec{c}\times\vec{a}}{\vec{a}\cdot\left(\vec{b}\times\vec{c}\right)}\\ \vec{c}^* &= 2\pi\frac{\vec{a}\times\vec{b}}{\vec{a}\cdot\left(\vec{b}\times\vec{c}\right)}\end{aligned}\right\}. \tag{7.23}$$

The common denominator in each case is

$$\vec{a}\cdot\left(\vec{b}\times\vec{c}\right) = V, \tag{7.24}$$

where V is the volume of the primitive unit cell in the direct lattice.

The vector $\vec{a},\vec{b}$ and $\vec{c}$ in the direct lattice have the dimension of [length], whereas the reciprocal lattice vectors $\vec{a}^*,\vec{b}^*$ and $\vec{c}^*$ have the dimension of $[\text{length}]^{-1}$. The crystal is a lattice in the *real space* but the reciprocal lattice is a lattice in the *Fourier space* [cf. (7.12)].

Choosing $\vec{a}^*,\vec{b}^*$ and $\vec{c}^*$ as the three reciprocal axes, any arbitrary set of primitive vectors $\vec{a},\vec{b},\vec{c}$ of a given crystal leads to the same set of reciprocal lattice points represented by the *reciprocal lattice vector*

$$\vec{G} = h\vec{a}^* + k\vec{b}^* + l\vec{c}^*, \tag{7.25}$$

where h,k,l are integers and $\vec{a}^*,\vec{b}^*,\vec{c}^*$ are the *primitive translation vectors* of the reciprocal lattice.

The reciprocal lattice vector $\vec{G}$ represents a set of points in the reciprocal lattice planes which are called the *reciprocal points.*

If $\vec{T}$ be the *crystal translation vector*, then

$$\vec{G}\cdot\vec{T} = 2\pi\times\text{integer} = 2\pi N \tag{7.26}$$

$$\text{or } e^{i\vec{G}\cdot\vec{T}} = 1. \tag{7.27}$$

But

$$e^{i\vec{G}\cdot\vec{T}} = \cos\left(\vec{G}\cdot\vec{T}\right) + i\sin\left(\vec{G}\cdot\vec{T}\right) = \cos(2\pi N) + i\sin(2\pi N) = 1 + 0 = 1.$$

$$\therefore \cos\left(\vec{G}\cdot\vec{T}\right) = 1 = \cos 0°.$$

$$\therefore \vec{G}\cdot\vec{T} = 0. \tag{7.28}$$

(7.28) shows that $\vec{G}$ *is perpendicular to* $\vec{T}$.

7.5.1 Properties of the Reciprocal Lattice

Reciprocal lattice is an invariant geometrical object and its properties are fundamental in the theory of solids. Some of the geometrical properties are as follows:

(i) *Every reciprocal lattice vector is normal to a lattice plane of the crystal lattice.*

Proof. Let $\vec{G}=h'\vec{a}^*+k'\vec{b}^*+l'\vec{c}^*$ be a reciprocal lattice vector and (hkl) be a set of parallel lattice planes in the direct lattice, then

$$\frac{h'}{h}=\frac{k'}{k}=\frac{l'}{l}=n. \tag{7.29}$$

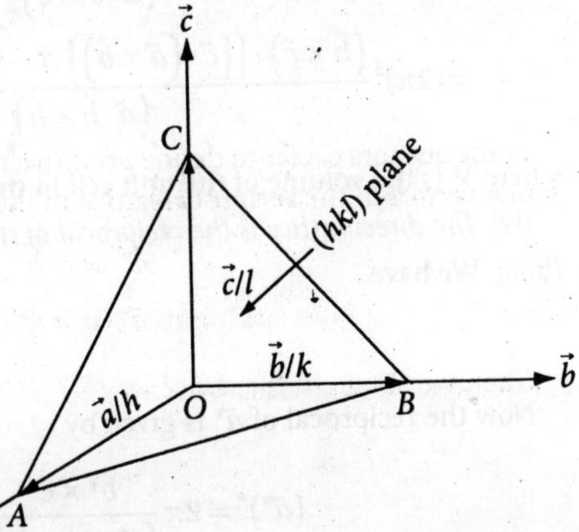

Fig. 7.5: An (*hkl*) plane in the direct lattice.

From Fig. 7.5,

$$\overrightarrow{AC}=-\frac{\vec{a}}{h}+\frac{\vec{c}}{l} \text{ and } \overrightarrow{AB}=-\frac{\vec{a}}{h}+\frac{\vec{b}}{k},$$

where the vectors $\overrightarrow{AB}$ and $\overrightarrow{BC}$ lie on the (hkl)-plane.

Now

$$\vec{G}\cdot\overrightarrow{AC}=\vec{G}\cdot\left[-\frac{\vec{a}}{h}+\frac{\vec{c}}{l}\right]=\left(h'\vec{a}^*+k'\vec{b}^*+l'\vec{c}^*\right)\cdot\left[-\frac{\vec{a}}{h}+\frac{\vec{c}}{l}\right]=2\pi\left(-\frac{h'}{h}+\frac{l'}{l}\right)=0. \tag{7.30}$$

Similarly

$$\vec{G}\cdot\left[-\frac{\vec{a}}{h}+\frac{\vec{b}}{k}\right]=0 \text{ [using (7.14) and (7.15)]}. \tag{7.31}$$

(ii) *If the components of $\vec{G}$ have no common factor, then $|\vec{G}|$ is inversely proportional to the spacing of the lattice planes normal to $\vec{G}$.*

The vectors $\overrightarrow{AC}$ and $\overrightarrow{AB}$ are in the (hkl)-plane. Hence, $\vec{G}$ is perpendicular to the (hkl) plane.

If $\hat{n}$ be a unit vector normal to (hkl)-plane, then $\hat{n}=\frac{\vec{G}}{G}$, where $\vec{G}$ is normal to the (hkl)-plane.

$$\begin{aligned}\therefore\ d&=\frac{\vec{a}\cdot\hat{n}}{h}=\frac{\vec{a}\cdot\vec{G}}{hG}=\frac{\vec{a}\cdot\left(h'\vec{a}^*+k'\vec{b}^*+l'\vec{c}^*\right)}{hG}\\&=\frac{2\pi h'}{hG}=\frac{2\pi n}{G}\left(\because\ \frac{h'}{h}=n\right),\end{aligned} \tag{7.32}$$

where d is the spacing between adjacent (hkl)-planes. (7.32) shows that $|\vec{G}|\propto 1/d$.

(iii) *The volume of a unit cell in the reciprocal lattice is inversely proportional to the volume of a unit cell of the direct lattice.*

Proof. If $\vec{a}^*, \vec{b}^*$ and $\vec{c}^*$ be the reciprocal lattice vectors, then the volume of its unit cell is

$$V^* = \left(\vec{a}^* \cdot \vec{b}^* \times \vec{c}^*\right) = (2\pi)^3 \frac{\vec{b} \times \vec{c}}{\left(\vec{a} \cdot \vec{b} \times \vec{c}\right)} \cdot \frac{\left[\left(\vec{c} \times \vec{a}\right) \times \left(\vec{a} \times \vec{b}\right)\right]}{\left(\vec{a} \cdot \vec{b} \times \vec{c}\right)\left(\vec{a} \cdot \vec{b} \times \vec{c}\right)} = (2\pi)^3 \frac{\left(\vec{b} \times \vec{c}\right) \cdot \left[\left(\vec{c} \times \vec{a}\right) \times \left(\vec{a} \times \vec{b}\right)\right]}{\left(\vec{a} \cdot \vec{b} \times \vec{c}\right)^3}$$

$$= (2\pi)^3 \frac{\left(\vec{b} \times \vec{c}\right) \cdot \left[\left\{\vec{c} \cdot \left(\vec{a} \times \vec{b}\right)\right\}\vec{a} - \left\{\vec{a} \cdot \left(\vec{a} \times \vec{b}\right)\right\}\vec{c}\right]}{\left(\vec{a} \cdot \vec{b} \times \vec{b}\right)} = \frac{(2\pi)^3}{\left(\vec{a} \cdot \vec{b} \times \vec{c}\right)^3} = \frac{(2\pi)^3}{V}, \tag{7.33}$$

where V is the volume of the unit cell in the direct lattice. $\therefore\ V^* \propto 1/V$.

(iv) *The direct lattice is the reciprocal of its own reciprocal lattice:*

Proof. We have

$$\vec{a}^* = 2\pi \frac{\vec{b} \times \vec{c}}{\vec{a} \cdot \vec{b} \times \vec{c}}.$$

Now the reciprocal of $\vec{a}^*$ is given by

$$(\vec{a}^*)^* = 2\pi \frac{\vec{b}^* \times \vec{c}^*}{\left(\vec{a}^* \cdot \vec{b}^* \times \vec{c}^*\right)} = 2\pi \left(\vec{a} \cdot \vec{a}^*\right) \cdot \frac{\vec{b}^* \times \vec{c}^*}{\vec{a}^* \cdot \vec{b}^* \times \vec{c}^*}, \quad [\because\ \vec{a} \cdot \vec{a}^* = 1]$$

$$= 2\pi \vec{a} \frac{\vec{a}^* \cdot \vec{b}^* \times \vec{c}^*}{\vec{a}^* \cdot \vec{b}^* \times \vec{c}^*} = 2\pi \vec{a}. \tag{7.34}$$

Hence, it is proved that *the direct lattice is the reciprocal of its own reciprocal lattice.*

7.5.2 Reciprocal Lattice to SC Lattice

If $\hat{i}, \hat{j}$ and $\hat{k}$ be taken as the unit vectors along primitive translation vectors $\vec{a}, \vec{b}$ and $\vec{c}$ of a simple cubic lattice, then $\vec{a} = \hat{i}a,\ \vec{b} = \hat{j}b,\ \vec{c} = \hat{k}c$.

Volume of the simple cubic cell is $\vec{a} \cdot \vec{b} \times \vec{c} = a^3\left(\hat{i} \cdot \hat{j} \times \hat{k}\right) = a^3$.

Now the reciprocal lattice vectors to the SC lattice are

$$\left.\begin{aligned} \vec{a}^* &= 2\pi \frac{\vec{b} \times \vec{c}}{\left(\vec{a} \cdot \vec{b} \times \vec{c}\right)} = 2\pi \frac{a\hat{j} \times a\hat{k}}{a^3} = \frac{2\pi}{a}\hat{i} \\ \text{Similarly, } \vec{b}^* &= 2\pi \frac{\vec{c} \times \vec{a}}{\vec{a} \cdot \vec{b} \times \vec{c}} = \frac{2\pi}{a}\hat{j} \\ \text{and } \vec{c}^* &= 2\pi \frac{\vec{a} \times \vec{b}}{\vec{a} \cdot \vec{b} \times \vec{c}} = \frac{2\pi}{a}\hat{k} \end{aligned}\right\} \tag{7.35}$$

It is evident from (7.35) that all the three reciprocal lattice vectors are equal in magnitude which means the *reciprocal lattice to SC lattice is also SC but with lattice constant equal to* $\frac{2\pi}{a}$.

7.5.3 Reciprocal Lattice to BCC Lattice

The primitive translation vectors of a body-centred cubic lattice [Fig. 7.6] are

$$\left.\begin{aligned} \vec{a}' &= \frac{a}{2}\left(\hat{i} + \hat{j} - \hat{k}\right) \\ \vec{b}' &= \frac{a}{2}\left(-\hat{i} + \hat{j} + \hat{k}\right) \\ \vec{c}' &= \frac{a}{2}\left(\hat{i} - \hat{j} + \hat{k}\right) \end{aligned}\right\}, \tag{7.36}$$

where $\hat{i}, \hat{j}$ and $\hat{k}$ are the *orthogonal unit vectors* along the edges of the cube and a is the length of the cube edge.

The volume of the primitive unit cell is

$$V = \vec{a}' \cdot \vec{b}' \times \vec{c}' = \frac{a}{2}(\hat{i}+\hat{j}-\hat{k}) \cdot \left[\frac{a^2}{4}(-\hat{i}+\hat{j}+\hat{k}) \times (\hat{i}-\hat{j}+\hat{k})\right]$$

$$= \frac{a}{2}(\hat{i}+\hat{j}-\hat{k}) \cdot \frac{a^2}{2}(\hat{i}+\hat{j}) = \frac{a^3}{2}.$$

$\therefore$ the reciprocal lattice vectors are

$$\left.\begin{aligned} \vec{a}^* &= 2\pi \frac{\vec{b}' \times \vec{c}'}{\vec{a}' \cdot \vec{b}' \times \vec{c}'} = \frac{2\pi\left(\frac{a^2}{2}\right)}{\frac{a^3}{2}}(\hat{i}+\hat{j}) = \frac{2\pi}{a}(\hat{i}+\hat{j}) \\ \text{Similarly, } \vec{b}^* &= 2\pi \frac{\vec{c}' \times \vec{a}'}{\vec{a}' \cdot \vec{b}' \times \vec{c}'} = \frac{2\pi}{a}(\hat{j}+\hat{k}) \\ \text{and } \vec{c}^* &= 2\pi \frac{\vec{a}' \times \vec{b}'}{\vec{a}' \cdot \vec{b}' \times \vec{c}'} = \frac{2\pi}{a}(\hat{k}+\hat{i}) \end{aligned}\right\} \tag{7.37}$$

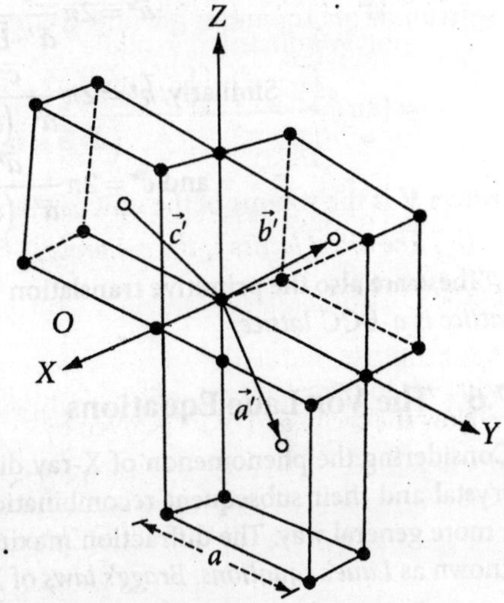

Fig. 7.6: Primitive translation vectors of a BCC lattice.

These are also the *primitive translation vectors of an FCC lattice*. Hence, *the reciprocal lattice to a BCC lattice is FCC lattice.*

7.5.4 Reciprocal Lattice to FCC Lattice

The primitive translation vectors of an FCC lattice are shown in Fig. 7.7 and given by

$$\left.\begin{aligned} \vec{a}' &= \frac{a}{2}(\hat{i}+\hat{j}) \\ \vec{b}' &= \frac{a}{2}(\hat{j}+\hat{k}) \\ \vec{c}' &= \frac{a}{2}(\hat{k}+\hat{i}) \end{aligned}\right\}. \tag{7.38}$$

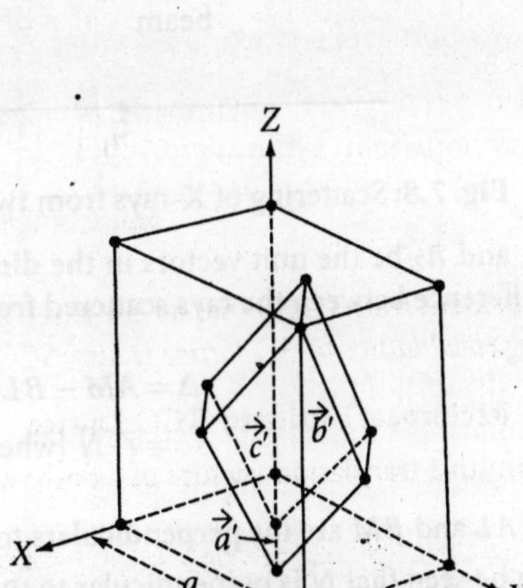

Fig. 7.7: Primitive translation vectors of an FCC lattice.

Volume of the unit cell is given by

$$V = \vec{a}' \cdot \vec{b}' \times \vec{c}' = \frac{a}{2}(\hat{i}+\hat{j}) \cdot \frac{a^2}{4}\left[(\hat{j}\times\hat{k}) \times (\hat{k}\times\hat{i})\right]$$

$$= \frac{a}{2}(\hat{i}+\hat{j}) \cdot \frac{a^2}{4}(\hat{i}+\hat{j}-\hat{k}) = \frac{a^3}{4}.$$

The primitive translation vectors of the reciprocal lattice are

$$\left.\begin{aligned} \vec{a}^* &= 2\pi\frac{\vec{b}'\times\vec{c}'}{\vec{a}'\cdot\vec{b}'\times\vec{c}'} = 2\pi\frac{\frac{a^2}{4}\left(\hat{i}+\hat{j}-\hat{k}\right)}{\frac{a^3}{4}} = \frac{2\pi}{a}\left(\hat{i}+\hat{j}-\hat{k}\right) \\ \text{Similarly, } \vec{b}^* &= 2\pi\frac{\vec{c}'\times\vec{a}'}{\vec{a}'\cdot\left(\vec{b}'\times\vec{c}'\right)} = \frac{2\pi}{a}\left(-\hat{i}+\hat{j}+\hat{k}\right) \\ \text{and } \vec{c}^* &= 2\pi\frac{\vec{a}'\times\vec{b}'}{\vec{a}'\cdot\left(\vec{b}'\times\vec{c}'\right)} = \frac{2\pi}{a}\left(\hat{i}-\hat{j}+\hat{k}\right) \end{aligned}\right\}. \quad (7.39)$$

These are also the primitive translation vectors of an FCC lattice. Thus, *the reciprocal lattice to an FCC lattice is a BCC lattice.*

7.6 The Von Laue Equations

Considering the phenomenon of X-ray diffraction as the scattering of X-rays by individual atoms in the crystal and their subsequent recombination, *Von Laue* obtained the directions of diffraction maxima in a more general way. The diffraction maxima occur in some specific direction obeying certain conditions known as *Laue's equations. Bragg's laws* of X-ray diffraction can be obtained from *Laue equations.*

Let a beam of X-rays be incident on two identical scattering centres A and B separated by a distance $\vec{r}$ in a crystal [Fig. 7.8].

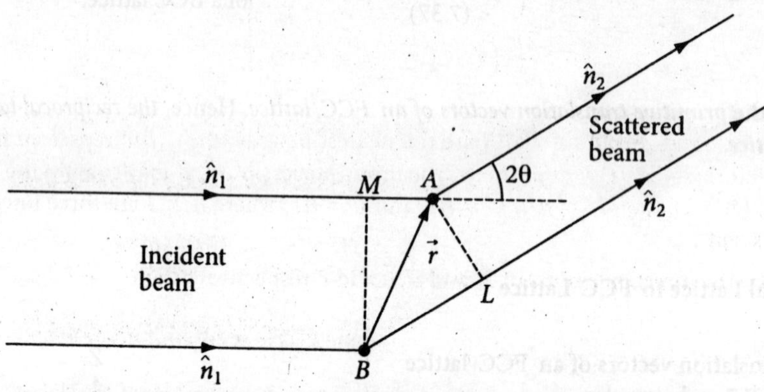

Fig. 7.8: Scattering of X-rays from two identical scattering centres separated by a distance $\vec{r}$.

If $\hat{n}_1$ and $\hat{n}_2$ be the unit vectors in the direction of incident and scattered beam respectively, then the path difference between the rays scattered from A and B is

$$\begin{aligned} \Delta &= AM - BL = \vec{r}\cdot\hat{n}_1 - \vec{r}\cdot\hat{n}_2 = \vec{r}\cdot(\hat{n}_1-\hat{n}_2) \\ &= \vec{r}\cdot\vec{N}\ [\text{where } \vec{N} = \hat{n}_1-\hat{n}_2]. \end{aligned} \quad (7.40)$$

Here AL and BM are the perpendiculars to the directions of incident and scattered beam respectively. It can be seen that $\vec{N}$ is perpendicular to the reflecting plane [Fig. 7.9].

The phase difference between the scattered and incident beams from B and A is

$$\phi = \frac{2\pi}{\lambda}\left(\vec{r}\cdot\vec{N}\right). \quad (7.41)$$

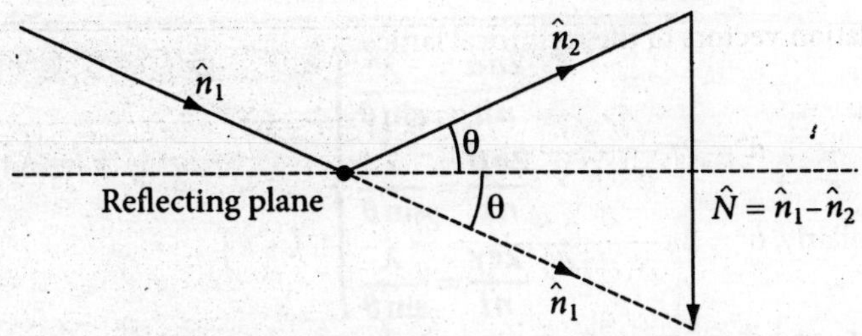

Fig. 7.9: Geometrical relationship between incident beam, scattered beam, reflecting plane and the normal.

The X-rays scattered from A and B will interfere constructively, if the phase difference ϕ be integral multiple of 2π. Due to periodicity of the atoms in the crystal radiation scattered from the other atoms may be in the same direction having exactly the same phase as those scattered from A and B. In a three-dimensional lattice $\vec{r}$ may coincide with any of the lattice vectors $\vec{a}, \vec{b}$ and $\vec{c}$.

So, for *constructive interference*,

$$\left.\begin{aligned}\frac{2\pi}{\lambda}\left(\vec{a}\cdot\vec{N}\right)&=2\pi h'=2\pi nh\\ \frac{2\pi}{\lambda}\left(\vec{b}\cdot\vec{N}\right)&=2\pi k'=2\pi nk\\ \frac{2\pi}{\lambda}\left(\vec{c}\cdot\vec{N}\right)&=2\pi l'=2\pi nl\end{aligned}\right\}, \tag{7.42}$$

where h', k' and l' are integers. As A and B are the nearest neighbours, a, b, c represent the interatomic distances along their *crystallographic directions.* The three integers h', k', l' may contain the largest integral common factor n $(n > 1)$, then $h' = nh$, $k' = nk$ and $l' = nl$, where h, k, l are three integers having no common factor except 1.

Let α, β and γ be the *direction cosines* of $\vec{N}$ and $\vec{a}, \vec{b}$ and $\vec{c}$ respectively, then

$$\left.\begin{aligned}\vec{a}\cdot\vec{N}&=aN\cos\alpha=2a\sin\theta\cos\alpha\\ \vec{b}\cdot\vec{N}&=bN\cos\beta=2a\sin\theta\cos\beta\\ \text{and } \vec{c}\cdot\vec{N}&=cN\cos\gamma=2a\sin\theta\cos\gamma\end{aligned}\right\}\left[\because\left|\vec{N}\right|=2\sin\theta \text{ and } \alpha=\cos\left(\vec{a},\vec{N}\right), \text{ etc.}\right]. \tag{7.43}$$

Then from the above set of equations,

$$\left.\begin{aligned}\vec{a}\cdot\vec{N}&=2a\alpha\sin\theta=h'\lambda=nh\lambda\\ \vec{b}\cdot\vec{N}&=2\beta b\sin\theta=k'\lambda=nk\lambda\\ \text{and } \vec{c}\cdot\vec{N}&=2\gamma c\sin\theta=k'\lambda=nk\lambda\end{aligned}\right\}. \tag{7.44}$$

The set of (7.44) is known as the *Laue's equations* representing *the conditions of X-ray diffraction.* These equations give the direction of *constructive interference* for a particular wavelength (λ). (7.44) can be written as

$$\left.\begin{aligned} \frac{2a\alpha}{nh} &= \frac{\lambda}{\sin\theta} \\ \frac{2b\beta}{nk} &= \frac{\lambda}{\sin\theta} \\ \frac{2c\gamma}{nl} &= \frac{\lambda}{\sin\theta} \end{aligned}\right\} \tag{7.45}$$

$$\left[\because \frac{2\pi}{\lambda}(\vec{a}\cdot\vec{N}) = 2\pi nh \text{ or } \frac{1}{\lambda}2a\alpha\sin\theta = nh \text{ or } \frac{2a\alpha}{n\lambda} = \frac{\lambda}{\sin\theta}\right]$$

$$\text{or } \frac{a\alpha}{h} = \frac{b\beta}{k} = \frac{c\gamma}{l} = \frac{n\lambda}{2\sin\theta}. \tag{7.46}$$

(7.45) can be expressed as

$$\left.\begin{aligned} \frac{2\alpha\sin\theta}{n\lambda} &= \frac{h}{a} \\ \frac{2\beta\sin\theta}{n\lambda} &= \frac{k}{b} \\ \frac{2\gamma\sin\theta}{n\lambda} &= \frac{l}{c} \end{aligned}\right\}. \tag{7.47}$$

This shows that in a direction of constructive interference, the direction cosines are proportional to $\frac{h}{a}, \frac{k}{b}$ and $\frac{l}{c}$. α, β and γ are also the direction cosines $\vec{N}$, the normal to the reflecting plane.

The set of (7.44) uniquely determine the values of θ and N and thus determine the scattering direction $(\hat{n}_2)$. Again, from (7.44) the direction cosines of $\vec{N}$ are proportional to $\frac{h}{a}, \frac{k}{b}$ and $\frac{l}{c}$. So (hkl)-planes are parallel to the *reflecting plane.*

7.6.1 Bragg's Law from Laue Equations

Since the inter-planar spacing (Art. 7.13) d is given by

$$d = \frac{a\alpha}{h} = \frac{b\beta}{k} = \frac{c\gamma}{l},$$

$\therefore$ from (7.44),

$$\begin{aligned} &2a\alpha\sin\theta = nh\lambda \\ \text{or } \; &2hd\sin\theta = nh\lambda \quad \left[\because d = \frac{a\alpha}{h}\right]. \\ \text{or } \; &2d\sin\theta = n\lambda. \end{aligned} \tag{7.48}$$

This is *Bragg's law of reflection*, where n is the *order of reflection*. Thus, one may consider different order reflection from (hkl)-plane or equivalently first order reflection from (nh, nk, nl)-planes. Putting $n = 1$, one gets rid of the factor n in structure determination of crystals.

7.6.2 Comparison of Bragg and Laue Approaches

1. Bragg considered the crystal as a set of equispaced parallel planes of ions or atoms.
   Laue did not consider a particular set of parallel lattice planes, rather he considered crystal ions or atoms placed at the lattice points.
2. Bragg considered that X-rays are specularly reflected by the atoms or ions in one lattice plane.
   According to Laue, incident X-rays are considered to be scattered elastically by the ions or atoms in all directions.

3. In Bragg reflection, the reflected ray from successive plane interferes constructively and produces maxima.

   In Laue's consideration constructive interference is equivalent to the Bragg assumption of specular reflection.

Thus, both the Bragg and Laue equations are based on the same physical principle.

7.7 Bragg's Law in Reciprocal Lattice: Ewald Construction

Bragg's law of X-ray diffraction can be expressed as a relation between vectors in the reciprocal lattice.

In Fig. 7.10, $\overrightarrow{AO}$ is a vector of length $\frac{1}{\lambda}$ in the direction of the incident X-ray beam and terminates at the origin O of the reciprocal lattice. With A as the centre we draw a sphere of radius $|\overrightarrow{AO}| = \frac{1}{\lambda}$, which intersects some point B in the reciprocal lattice.

The coordinates of B is (h', k', l') which may have a highest common factor n, i.e., $h' = nh, k' = nk$ and $l' = nl$, where h, k, l have common factor unity. $\overrightarrow{OB}$, then represents a vector joining the origin O of the reciprocal lattice and the point $B\,(h', k', l')$. $\overrightarrow{OB}$, therefore, must be normal to the plane $(h'k'l')$ and is of length

$$\left|\overrightarrow{OB}\right| = \frac{n}{d_{hkl}}. \tag{7.49}$$

Fig. 7.10 shows that AE is such a plane. Now from the ΔAOB,

$$OB = 2OE = 2OA\sin\theta = \frac{2\sin\theta}{\lambda}, \left[\because OA = \frac{1}{\lambda}\right], \tag{7.50}$$

where $\angle EAO = \theta$.

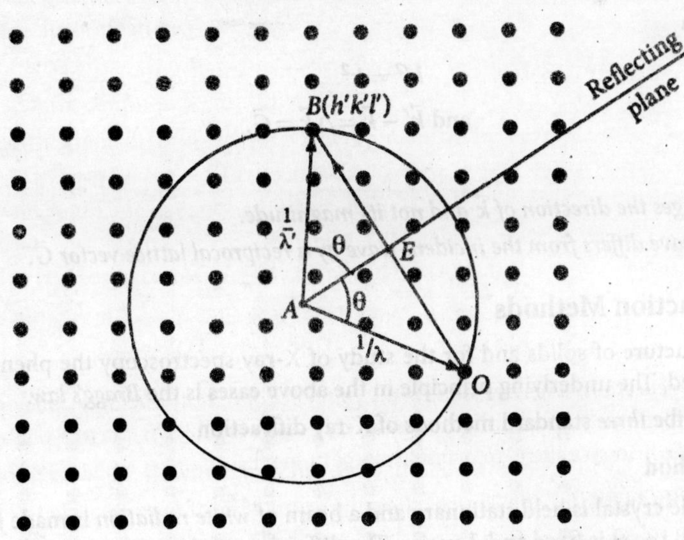

Fig. 7.10: Ewald Construction in the reciprocal lattice.

From (7.49) and (7.50),

$$\frac{2\sin\theta}{\lambda} = \frac{n}{d_{hkl}} \text{ or } 2d_{hkl}\sin\theta = n\lambda. \tag{7.51}$$

(7.51) is the *Bragg's law* and *n is the order of reflection.*

Note. (i) Since the spacing of the reciprocal lattice is $\frac{1}{a}$, if $\frac{1}{\lambda} < \frac{1}{2a}$ or $\lambda > 2a$, the sphere will not touch any other point and as such no diffraction will take place, if $\lambda > 2a$.

(ii) If $\frac{1}{\lambda} \geq 2a$ or, $\lambda \leq 2a$, the diffraction is possible.

The above construction in the reciprocal lattice giving Bragg's condition of X-ray reflection is called *Ewald construction.*

7.7.1 Vector Form of Bragg's Law

Let the vectors $\overrightarrow{A'O'}$, $\overrightarrow{O'B'}$ and $\overrightarrow{A'B'}$ be the magnified form by a constant factor of 2π such that $A'O' = 2\pi(OA)$, $O'B' = 2\pi(OB)$ and $A'B' = 2\pi(AB)$ [Fig. 7.11].

Since $OA = 1/\lambda$,

$\therefore$ $A'O' = \frac{2\pi}{\lambda}$ can represent the wave vector $\vec{k}$ by the vector $\overrightarrow{A'O'}$. Then $\overrightarrow{O'B'}$ is a reciprocal vector $\vec{G}$.

Now from the *triangle law of vectors,* $\overrightarrow{A'B'} = \overrightarrow{A'O'} + \overrightarrow{O'B'} = \vec{k} + \vec{G}$.

For the diffraction to take place

$$\left(\vec{k}+\vec{G}\right)^2 = k^2 \text{ or } k^2 + 2\vec{k}\cdot\vec{G} + G^2 = k^2$$

$$\text{or } 2\vec{k}\cdot\vec{G} + G^2 = 0. \qquad (7.52)$$

(7.52) represents the *vector form of Bragg's law* which is used in the construction of *Brillouin Zones.*

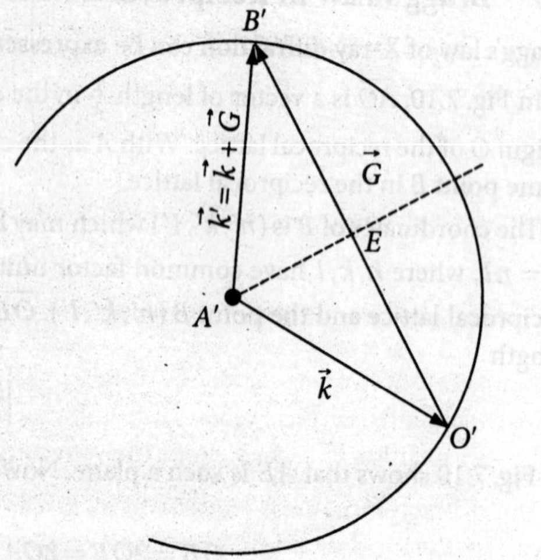

Fig. 7.11: Magnified Ewald Construction.

Let $\overrightarrow{A'B'} = \vec{k}'$, then

$$k'^2 = k^2 \qquad (7.53)$$

$$\text{and } \vec{k}' - \vec{k} = \Delta\vec{k} = \vec{G}. \qquad (7.54)$$

This shows that

1. *Scattering changes the direction of $\vec{k}$ and not its magnitude.*
2. *The scattered wave differs from the incident wave by a reciprocal lattice vector $\vec{G}$.*

7.8 X-ray Diffraction Methods

To determine the structure of solids and for the study of X-ray spectroscopy the phenomenon of X-ray diffraction is employed. The underlying principle in the above cases is the *Bragg's law.*

Here we shall describe *three* standard methods of X-ray diffraction.

7.8.1 The Laue Method

In this method a single crystal is held stationary and a beam of *white radiation* is made incident on it at a fixed glancing angle θ, i.e., θ is fixed and λ varies. The different wavelengths present in the incident beam select the appropriate reflecting planes out of the numerous ones present in the crystal so that the *Bragg's condition of diffraction* is satisfied.

The experiment

Fig. 7.12 shows the experimental arrangement of the Laue method. An X-ray beam of *continuous wavelength* (*white radiation*) is allowed to fall on a stationary single crystal (C). The crystal selects and diffracts the appropriate wavelength λ for which a family of planes of inter-planar spacing d and glancing angle θ, Bragg's condition, $2d \sin\theta = n\lambda$ is satisfied.

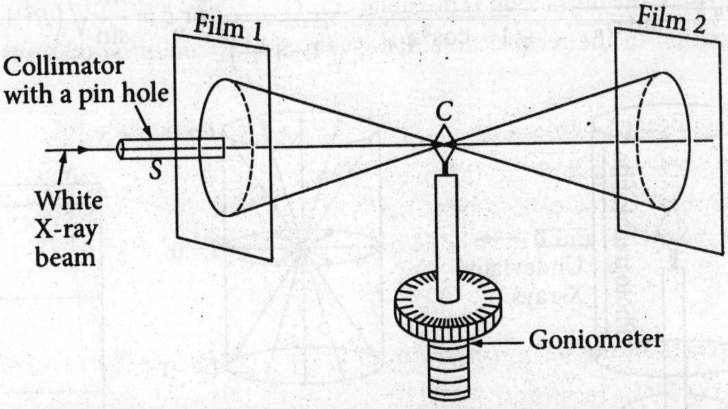

Fig. 7.12: A flat plate camera used in Laue diffraction method.

The source of X-ray is capable of producing X-rays of wide range of wavelengths (0.2 Å to 2 Å). The beam before incidence on the crystal is well collimated by a pinhole (S) arrangement. Film 1 and Film 2 are held in proper position to record either the reflected beam or transmitted diffracted beam.

The diffraction pattern consists of a series of spot whose position is related to the orientation of the crystal to the incident beam. The crystal can be rotated with the help of goniometer to change its orientation w.r.t. the incident beam when a new set of atomic planes satisfies the condition of diffraction. In this way different types of pattern is obtained on the film. In each pattern, different types of symmetry of the crystal is reflected.

Laue method is used to determine, the (i) *crystal symmetry* and (ii) *imperfection or strains in the crystal.*

7.8.2 Rotating Crystal Method

In this method a single crystal C is rotated about a fixed vertical axis in a beam of *mono-energetic X-ray* [Fig. 7.13 (a)]. The photographic film is mounted on the inner wall of a cylinder co-axial with the mount. One of the crystallographic axes is kept vertical and normal to the incident beam of X-rays. The X-rays are diffracted from a given crystal plane, when in the course of rotation, the value of θ satisfies the Bragg's condition $2d \sin\theta = n\lambda$.

The planes parallel to the axis of rotation diffract the X-rays in the horizontal plane in which the incident X-rays lie. So, these diffracted patterns cannot be observed. The planes which are inclined to the rotation axis produce diffraction pattern which lie above or below the horizontal plane containing incident X-rays, depending upon the angle of inclination.

If the c-axis of the crystal is made to coincide with the axis of rotation, then all planes of the type (hko) will produce the central *layer lines.* Similarly, the planes (hkl) and $(hk\bar{l})$ will produce the layer line above and below the central line. The vertical distance between the layer lines depends on the distance between the lattice points along c-axis of the crystal. Thus, the distance c can be measured from the film. In a similar way the lattice vectors $\vec{a}$ and $\vec{b}$ can be measured and the size of the unit cell can be determined.

If $\vec{c}$-axis of the crystal coincides with the axis of rotation of the crystal, then $S_n = R\cos\phi_n$, where $S_n =$ separation of layer lines and $R =$ the radius of the camera and $\phi_n = (90° - \theta_n)$, $\theta_n =$ the angle of diffraction for the nth order.

$$\therefore\ S_n = \frac{\cos\phi_n \cdot R}{\sin\phi_n} = \frac{\cos\phi_n \cdot R}{\sqrt{1-\cos^2\phi_n}} = \frac{\frac{n\lambda}{c}\cdot R}{\sqrt{1-\left(\frac{n\lambda}{c}\right)^2}} \text{ or } c = \frac{n\lambda}{\sin}\sqrt{\left(R^2+S_n^2\right)}.$$

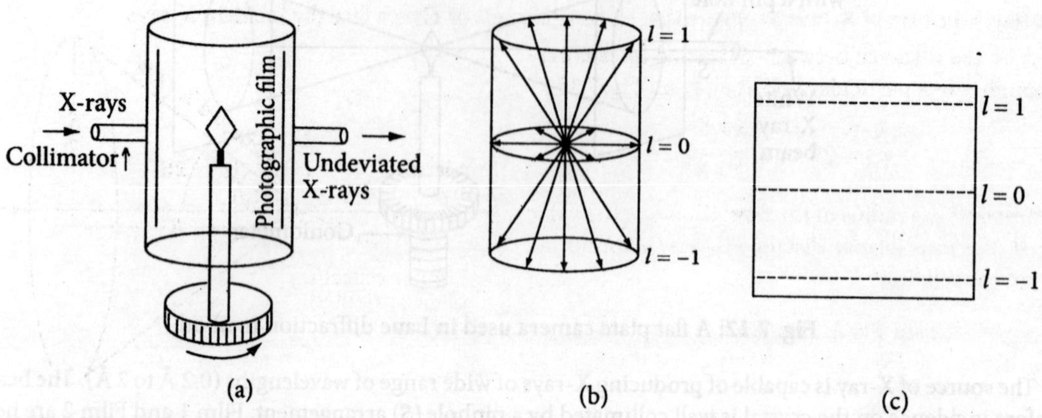

Fig. 7.13: (a) Apparatus for rotating crystal method (b) Cones of scattered X-rays from (hkl) planes (c) Layer lines on photoplate.

Knowing the values of n, λ, R and S_n the value of c can be found out. In a similar way, the values of b and a can be determined. Thus, the parameters a, b and c of the unit cell of the crystal can be determined.

7.8.3 Powder Method

In this method, a large number of small crystallites are exposed to X-rays. This lattice planes of the crystallites which satisfy the Bragg's condition, $2d\sin\theta = n\lambda$, diffract the X-rays. Here d and θ may have different values for a fixed value of λ.

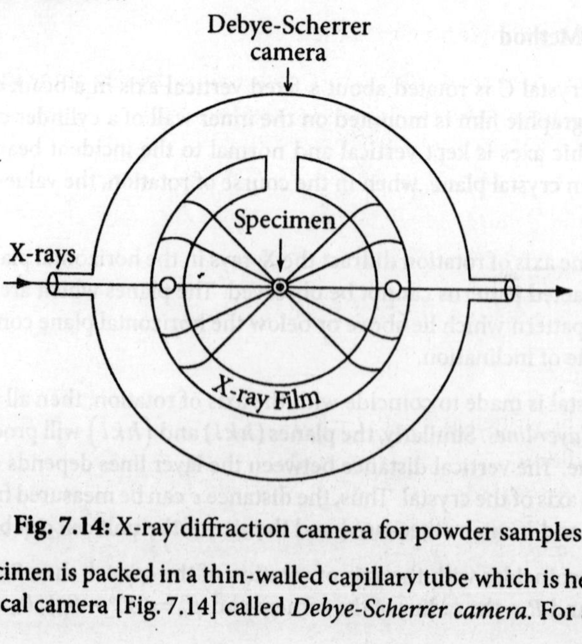

Fig. 7.14: X-ray diffraction camera for powder samples.

A finely powdered specimen is packed in a thin-walled capillary tube which is held on a movable mount at the centre of a cylindrical camera [Fig. 7.14] called *Debye-Scherrer camera*. For an arbitrary orientation of the sample w.r.t. the incident *monochromatic X-rays*, some sets of the crystal planes satisfy the condition of Bragg reflection, $2d \sin\theta = n\lambda$. For a particular value of λ the reflected rays for a particular set of lattice plane lie on the surface of a cone of semi-vertical angle 2θ. Additional cones of different semi-vertical angles arise due to other set of planes. These cones intersect the photographic film in concentric rings or arcs of circles with a sharp circular spot at the centre due to the direct beam. The crystal is slowly rotated during exposure of X-rays to ensure that all possible sets of planes face the incident X-rays.

If S be the distance between diffracted lines corresponding to a particular line, then from Fig. 7.15,

$$\frac{S}{2R} = 2\theta \text{ or } \theta = \frac{S}{4R},$$

where R = the radius of the film. Knowing the value of θ, the inter-planar distance d can be found out from the relation

$$2d \sin\theta = \lambda, \text{ (for } n = 1)$$

$$\text{or } 2d \times \frac{S}{4R} = \lambda, \ [\because \theta \to 0, \sin\theta \simeq \theta]$$

$$\text{or } d = \frac{2R\lambda}{S}. \tag{7.55}$$

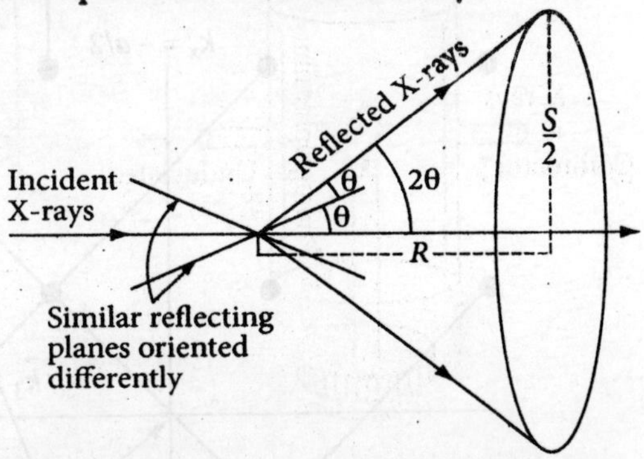

Fig. 7.15: X-ray diffracted cone of semi-vertical angle 2θ.

The d values are used to determine the space lattice of the crystal structure.

7.9 Brillouin Zones

We have seen in Art. 7.6 that all the k-values for which the reciprocal lattice points intersect the Ewald sphere are Bragg reflected.

A Brillouin Zone is, therefore, the locus of all those k-values in the reciprocal lattice which are Bragg reflected.

7.9.1 Brillouin Zones for a Simple Square Lattice

The primitive translation vectors are $\vec{a} = a\hat{i}$; $\vec{b} = a\hat{j}$, where a is the side of the square lattice.

The corresponding translation vectors of the reciprocal lattice are

$$\vec{a}^* = \left(\frac{1}{a}\right)\hat{i};\ \vec{b}^* = \left(\frac{1}{a}\right)\hat{j}.$$

$\therefore$ the reciprocal lattice vector is

$$\vec{G} = \left(\frac{2\pi}{a}\right)\left(h\hat{i} + k\hat{j}\right),$$

where h and k are integers.

The wave vector $\vec{k}$ can be written as $\vec{k} = k_x\hat{i} + k_y\hat{j}$.

Now from Bragg's condition

$$2\vec{k}\cdot\vec{G} + G^2 = 0 \text{ or } \frac{4\pi}{a}\left[\left(k_x\hat{i} + k_y\hat{j}\right)\cdot\left(h\hat{i} + k\hat{j}\right)\right] + \frac{4\pi^2}{a^2}\left(h^2 + k^2\right) = 0$$

$$\text{or } hk_x + kk_y = -\left(\frac{\pi}{a}\right)\left(h^2 + k^2\right). \tag{7.56}$$

Taking all possible combinations of h and k, we can get the values of k which will be Bragg reflected: If $h = \pm 1$, $k = 0$, then $k_x = \pm \frac{\pi}{a}$, k_y is arbitrary. Also, if $h = 0$ and $k = \pm 1$, then $k_y = \pm \frac{\pi}{a}$ and k_x is arbitrary.

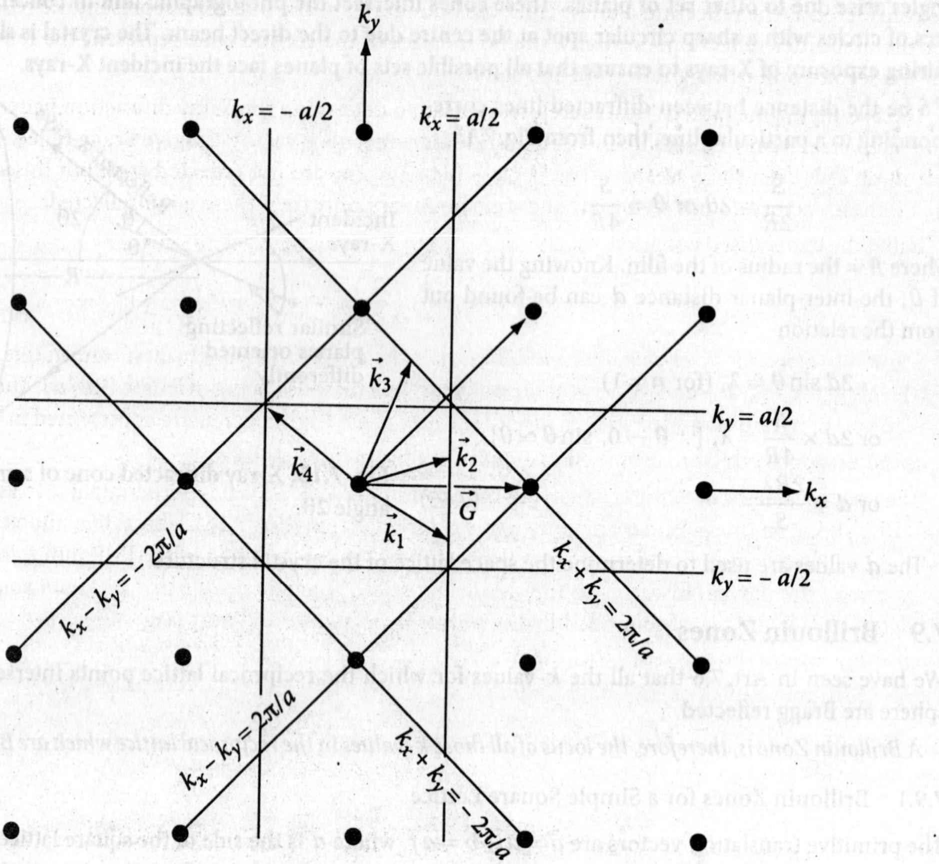

Fig. 7.16: Brillouin zones of a square lattice in its reciprocal lattice. The vectors $\vec{k}_x$, $\vec{k}_y$ and $\vec{k}_z$ are Bragg reflected whereas $\vec{k}_4$ is not. The vectors $\vec{k}_1$ and $\vec{k}_2$ have the same reciprocal lattice vector $\vec{G}_1$ whereas $\vec{G}_2$ is the reciprocal vector of $\vec{k}_3$.

These four lines, viz.,

$$\left.\begin{aligned} k_x &= +\frac{\pi}{a}; \quad & k_x &= -\frac{\pi}{a} \\ k_y &= +\frac{\pi}{a}; \quad & k_y &= -\frac{\pi}{a} \end{aligned}\right\} \tag{7.57}$$

are plotted in Fig. 7.16. All $\vec{k}$-vector originating at the origin and terminating on these lines will produce Bragg reflection.

Some other set of lines, viz.,

$$\pm k_x \pm k_y = \frac{2\pi}{a} \tag{7.58}$$

are possible. The regions included in such lines are called *Brillouin zones*. The square bounded by $k_x = \pm \frac{\pi}{a}$

and $k_y = \pm\left(\frac{\pi}{a}\right)$ is the *first Brillouin zone* which is a square of side $(2\pi/a)$. The additional area included by $\pm k_x \pm k_y = (2\pi/a)$ is the *second Brillouin zone* and so on.

The boundaries of the Brillouin zones represent the loci of $\vec{k}$-values which are Bragg reflected and may be considered as the reflecting planes. Thus, the boundaries of the first zone represent the reflecting planes for the first order reflection. Similarly, the boundaries of the second zone represent the reflecting planes for the second order Bragg reflection and so on.

The Brillouin zone pattern can, therefore, be used to determine the X-ray diffraction pattern of a crystal and vice versa. It may be said that within a Brillouin zone, no $\vec{k}$ vector can give Bragg reflection. This does not mean that even the $\vec{k}$-vectors with $|\vec{k}| > \frac{\pi}{a}$ (i.e., $\lambda < 2a$) are not reflected at all but this simply means that if these vectors do not terminate at a zone boundary only then these go unreflected.

Using the generalised equation

$$h k_x + k k_y + l k_z = -\left(\frac{\pi}{a}\right)\left(h^2 + k^2 + l^2\right), \tag{7.59}$$

the *Brillouin zones for three-dimensional cubic lattice can be constructed.* The first zone in this case is a cube intersecting the k_x-, k_y- and k_z-axes at the points (π/a), the sides being of length $(2\pi/a)$. The second zone is made up of pyramids added to each face of the first zone. Higher zones are constructed in a similar way.

Brillouin zones can be determined in another simple method as follows:

The reciprocal lattice vector $\vec{G}$ satisfying the relation $2\vec{k}\cdot\vec{G} + G^2 = 0$ is perpendicular to zone boundary and all $\vec{k}$'s lying on this boundary have the same $\vec{G}$ for reflection [Fig. 7.16]. Hence, it is sufficient to consider only the allowed $\vec{G}$-values and their normal bisector for the construction of the Brillouin zones. The region bounded by the normal bisectors to the shortest possible $\vec{G}$-vectors is the *first Brillouin zone.* The *second Brillouin zone* is the region bounded by the normal bisectors of the next larger $\vec{G}$-vectors and so on. We shall apply this method to construct the Brillouin zones of FCC and BCC lattices.

7.9.2 Brillouin Zones of the FCC Lattice

The primitive translation vectors of an FCC lattice are

$$\vec{a} = \frac{a}{2}\left(\hat{i} + \hat{j}\right)$$
$$\vec{b} = \frac{a}{2}\left(\hat{j} + \hat{k}\right)$$
$$\vec{c} = \frac{a}{2}\left(\hat{k} + \hat{i}\right).$$

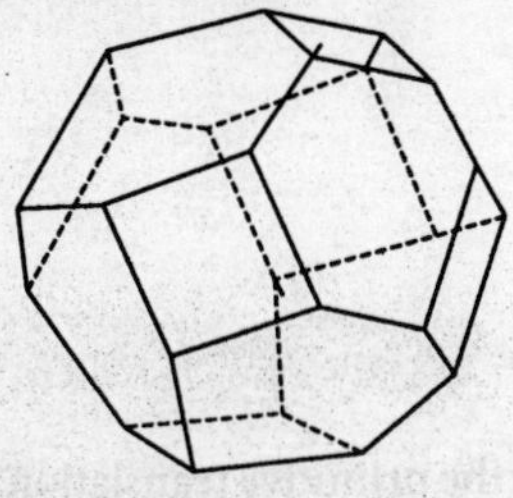

Fig. 7.17: Brillouin zones of an FCC lattice.

The primitive translation vectors of its reciprocal lattice are

$$\vec{a}^* = \left(\frac{2\pi}{a}\right)\left(\hat{i} + \hat{j} - \hat{k}\right)$$
$$\vec{b}^* = \left(\frac{2\pi}{a}\right)\left(-\hat{i} + \hat{j} + \hat{k}\right)$$
$$\vec{c}^* = \left(\frac{2\pi}{a}\right)\left(\hat{i} - \hat{j} + \hat{k}\right).$$

If h, k, l are integers, then the G-type vector is

$$\vec{G} = 2\pi\left(\vec{a}^* h + \vec{b}^* k + \vec{c}^* l\right) = \frac{2\pi}{a}\left[(h - k + l)\hat{i} + (h + k - l)\hat{j} + (-h + k + l)\hat{k}\right]. \tag{7.60}$$

The shortest non-zero $\vec{G}$'s are the following eight vectors:

$$(2\pi/a)\left(\pm\hat{i}\pm\hat{j}\pm\hat{k}\right).$$

The first zone boundaries are determined mostly by the eight planes normal to these vectors at their mid-points. But the corners of the octahedron so formed are truncated by the planes which are perpendicular bisectors of the six reciprocal lattice vectors:

$$\frac{2\pi}{a}(\pm 2\hat{i});\ \frac{2\pi}{a}(\pm 2\hat{j});\ \frac{2\pi}{a}(\pm 2\hat{k}).$$

Then the first Brillouin zone is the truncated octahedron fr the FCC lattice as shown in Fig. 7.17. This is also one of the primitive unit cell of a BCC lattice.

7.9.3 Brillouin Zones of BCC Lattice

The primitive translation vectors of a BCC lattice are

$$\vec{a}=\left(\frac{a}{2}\right)\left(\hat{i}+\hat{j}-\hat{k}\right)$$
$$\vec{b}=\left(\frac{a}{2}\right)\left(-\hat{i}+\hat{j}+\hat{k}\right)$$
$$\vec{c}=\left(\frac{a}{2}\right)\left(\hat{i}-\hat{j}+\hat{k}\right).$$

The primitive translation vectors of its reciprocal lattice are

$$\vec{a}^*=\left(\frac{2\pi}{a}\right)\left(\hat{i}+\hat{j}\right)$$
$$\vec{b}^*=\left(\frac{2\pi}{a}\right)\left(\hat{j}+\hat{k}\right)$$
$$\vec{c}^*=\left(\frac{2\pi}{a}\right)\left(\hat{k}+\hat{i}\right).$$

These are the primitive translation vectors of an FCC lattice. Thus, *the FCC lattice is the reciprocal lattice of the BCC lattice.*

The $\vec{G}$-type lattice vector is

$$\vec{G}=2\pi\left(h\vec{a}^*+k\vec{b}^*+l\vec{c}^*\right)=\frac{2\pi}{a}\left[(h+l)\hat{i}+(h+k)\hat{j}+(k+l)\hat{k}\right]. \tag{7.61}$$

The shortest non-zero $\vec{G}$'s are the following twelve vectors:

$$\left(\frac{2\pi}{a}\right)\left(\pm\hat{i}\pm\hat{j}\right);\ \left(\frac{2\pi}{a}\right)\left(\pm\hat{j}\pm\hat{k}\right);\ \left(\frac{2\pi}{a}\right)\left(\pm\hat{k}\pm\hat{i}\right).$$

The first Brillouin zone is the region enclosed by the normal bisector planes to these twelve vectors. This zone has the shape of a regular twelve-faced solid and is called *rhombic dodecahedron* [Fig. 7.18].

Fig. 7.18: First Brillouin zone of a BCC lattice.

7.10 Liquid Crystals

Liquid crystals show *intermediate mesomorphic (liquid crystalline) phase* of crystals and liquids. These are primarily *organic compounds resembling partly an ordered crystal and a disordered liquid.* They exhibit *anisotropic* properties like dielectric, optical, magnetic, etc. which are possessed by crystals. At the same time the property of fluidity, droplet forming capacity of liquid are also exhibited.

The transition from solid phase to meso-phase takes place at a specific temperature, called the *melting* or *(solidification) point* (T_s). The liquid crystal converts to the isotropic transparent liquid at T_t, called the *transparency temperature.* The values of T_s and T_t are different for different crystals and the interval $\Delta T = (T_t - T_s)$ is the range of temperature in which the liquid crystalline phase exists. The value of ΔT differs from crystal to crystal. Liquid crystals are organic materials made of rigid and moderately large rod like molecules having length 1 to 3 nm and breadth 0.3 to 1.0 nm. A high degree of polarizability is exhibited by a cluster of atoms situated at the ends of the molecules. Weak *van der Waals' force* makes the ordered molecular arrangement in the liquid crystal.

The electric displacement vector $\vec{D}$ and electric field $\vec{E}$ are not in the same direction in the anisotropic liquid crystal medium.

The following chart shows the classification of the liquid crystal:

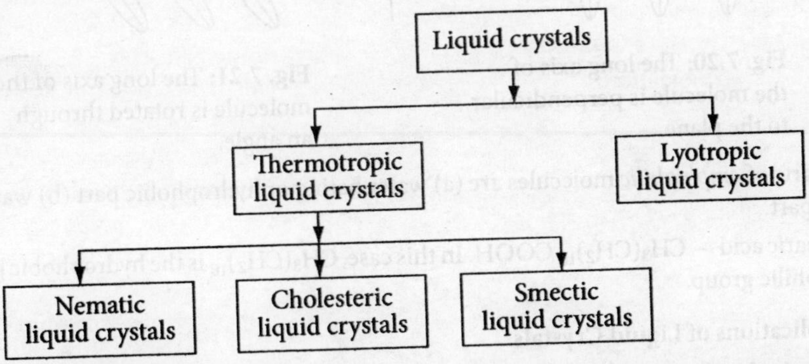

7.10.1 Nematic Liquid Crystals

The molecules of the *nematic liquid crystals* have a natural tendency to orient their long axes along a preferred *director axis* $(\vec{L})$ as shown in Fig. 7.19. The molecules align parallel to a unique axis in different small regions of this crystal. The direction of orientation generally varies in different regions of the liquid crystal. This property of the crystal is responsible for the *high optical inhomogeneity* of the crystal because of which it appears *opaque* both in reflected and transmitted light.

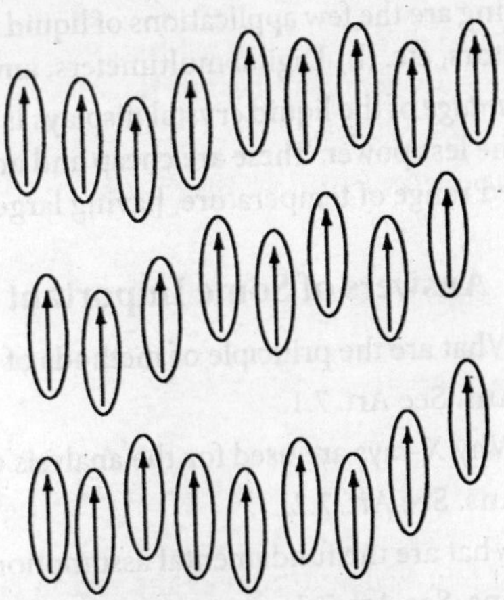

Fig. 7.19: Orientation of molecules along a preferred director axis (*L*).

Smectic liquid crystals

In *smectic liquid crystal,* the molecules orient themselves in parallel layers gliding one relative to other. This promotes fluidity of the system. The long axis of the molecules are either perpendicular to the planes [Fig. 7.20] or rotated through an angle ϕ with the normal to the planes [Fig. 7.21]. The rotation angle ϕ depends on the temperature which can be altered by applying an external electric field. The *smectic axis* represented by the vector $\vec{L}$ and the shear plane *AA* is shown in the Figs 7.20 and 7.21.

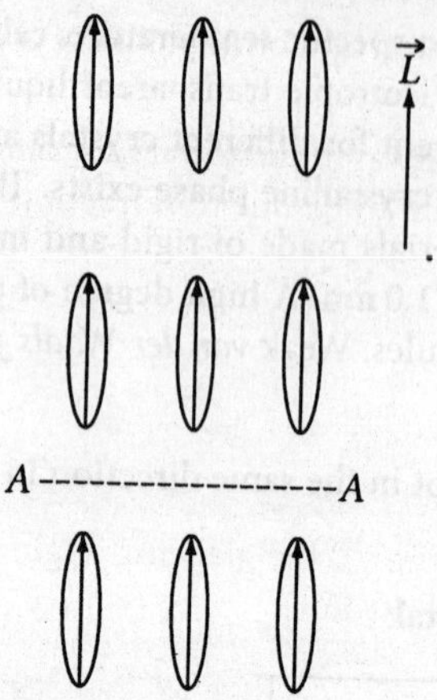

Fig. 7.20: The long axis of the molecule is perpendicular to the plane.

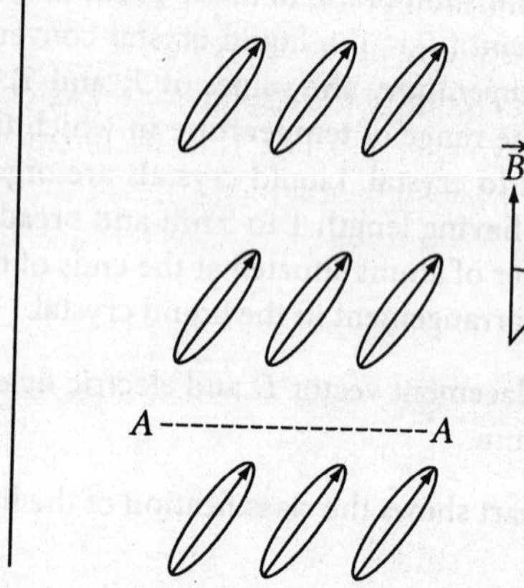

Fig. 7.21: The long axis of the molecule is rotated through an angle.

The two parts of *amphiphilic* molecules are (a) water hating or hydrophobic part (b) water loving part or hydrophilic part.

Example. Stearic acid — $CH_3(CH_2)_{16}$ COOH. In this case, $CH_3(CH_2)_{16}$ is the hydrophobic group and COOH is the hydrophilic group.

7.10.2 Applications of Liquid Crystals

Following are the few applications of liquid crystals: (a) Low power visual displays, viz., electronic watches, calculators, etc. (b) Digital multimeters, ammeters, voltmeters, etc.

Advantage of the liquid crystal displays is that they can be manufactured in any size and shape easily and consume less power. These are cheap and are of variety of colours. The *disadvantage* is that they operate in a limited range of temperature, having large response time, etc.

7.11 Answers of Some Important Questions

7.1 What are the principle of methods of crystal structure analysis?

Ans. See Art. 7.1.

7.2 Why X-rays are used for the analysis of crystal structure?

Ans. See Art. 7.2.

7.3 What are the fundamental assumptions in deriving Bragg's law of X-ray diffraction?

Ans. See Art. 7.3.

7.4 Show that for Bragg reflection to occur the wavelength of X-rays must not exceed twice the interplanar spacing.

Ans. We know that the Bragg condition for X-ray reflection by atomic planes of crystals is $2d \sin\theta = n\lambda$, where θ is the glancing angle of incidence, d is the inter-planar spacing, λ is the wavelength of X-rays and n is the order of reflection.

Since $|\sin\theta| \le 1$, $\frac{n\lambda}{2d} \le 1$.

For the first order reflection $n = 1$. $\therefore \lambda \le 2d$.

Thus, λ must not exceed twice the inter-planar spacing (d) for Bragg reflection to occur.

7.5 What led to introduce correction in Bragg's law of X-ray reflection by atomic planes of crystals. Write down the corrected form of the above law.

Ans. Experimental value of wavelength for the same X-ray line was found to vary for different orders when computed from Bragg's law. This discrepancy was found to be due to refraction of X-rays at different atomic layers in the crystal.

The corrected form of Bragg's law taking into account, refraction of X-rays is

$$n\lambda = 2d \sin\theta \left[1 - \frac{4d^2(1-\mu)}{n^2\lambda^2}\right],$$

where λ is the wavelength of X-rays, d the inter-planar distance, θ the glancing angle of incidence, n the order of reflection and μ refractive index of the crystal-medium.

7.6 What is reciprocal lattice?

Ans. See first para of Art. 7.4. Then add: The reciprocal lattice corresponding to a Bravais lattice point $\vec{R}$ is defined as the set of all wave vectors that give plane waves of the form $e^{i\vec{k}\cdot\vec{r}}$ with the periodicity $(\vec{R})$ of the Bravais lattice.

7.7 What is the importance of reciprocal lattice in solid state physics?

Ans. See Art. 7.4.1.

7.8 Write down Laue equation representing the conditions of X-ray diffraction by crystals.

Ans. See Art. 7.6 [(7.44)].

7.9 Obtain Bragg's Law from Laue equations.

Ans. See Art. 7.6.1.

7.10 Compare Bragg and Laue approaches as regards X-ray reflection is concerned.

Ans. See Art. 7.6.2.

7.11 Give the vector form of Bragg's Equation.

Ans. Art. 7.7.1.

7.12 What are Brillouin Zones?

Ans. All the values of wave vector $\vec{k}$ for which the reciprocal lattice points intersect the Ewald sphere are Bragg reflected. A Brillouin zone is, therefore, the locus of all those k-values in the reciprocal lattice which are Bragg reflected.

7.13 What are liquid crystals?

Ans. Liquid crystals are organic compounds exhibiting mesomorphic phase of crystals and liquids. They exhibit anisotropic properties like dielectric, optical, magnetic, etc.

7.14 How does anisotropy arise in liquid crystals?

Ans. In liquid crystal, the direction of the electric field $\vec{E}$ and the electric displacement vector $\vec{D}$ are not in the same direction which give rise to anisotropy.

7.15 Mention some applications of liquid crystals.

Ans. Various liquid crystal displays are used in electronic watch, calculators, digital electronic meters, etc.

7.12 Solved Problems

7.1 The spacing between principal planes of NaCl crystal is 2.82 Å. It is found that first order Bragg reflection occurs at an angle of 10°. What is the wavelength of λ-rays?

Solution. The Bragg's equation is $2d\sin\theta = n\lambda$.

Here the inter-planar distance $d = 2.82 \times 10^{-10}$ m, the order number $n = 1$ and the glancing angle of incidence is $\theta = 10°, \lambda = ?$

From Bragg's equation,

$$\lambda = \frac{2d\sin\theta}{n} = \frac{2\times(2.82\times10^{-10})\times\sin 10°}{1} = 0.98\times10^{-10}\,\text{m} = 0.98\,\text{Å}.$$

7.2 A beam of X-rays of wavelength 0.842 Å is incident on a crystal at a glancing angle of 8°35′ when the first order Bragg reflection occurs. Calculate the glancing angle of the third order reflection. [WBUT 2002]

Solution. The Bragg's equation is $2d\sin\theta_n = n\lambda$.

For the first order spectrum, $n = 1$, $\lambda = 0.842$ Å, $\theta_1 = 8°35' \simeq 8.6°$.

$$\therefore d = \frac{\lambda}{2\sin\theta_1} = \frac{0.842}{2\times0.1347} = 3.125\,\text{Å}.$$

If θ_3 be the glancing angle for the third order ($n = 3$) reflection, then

$$\sin\theta_3 = \frac{3\lambda}{2d} = \frac{3\times0.842}{2\times3.125} = 0.40416.$$

$$\therefore \theta_3 = \sin^{-1}(0.40416) = 26°29'.$$

7.3 X-rays of wavelength 0.71 Å are reflected from the (110)-plane of a rock salt crystal of lattice constant $a = 2.82$ Å. Calculate the corresponding glancing angle for second order reflection.

Solution. The inter-planar spacing between (111)-planes is

$$d = \frac{a}{\sqrt{h^2+k^2+l^2}}.$$

Here $a = 2.82$ Å, $h = 1, k = 1, l = 0$.

$$\therefore d = \frac{2.82}{\sqrt{1^2+1^2+0^2}} = \frac{2.82}{\sqrt{2}} = \frac{2.82}{1.41} = 2\,\text{Å}.$$

Now from Bragg's law, $2d\sin\theta = n\lambda$.

Here $d = 2$ Å, $n = 2$ and $\lambda = 0.71$ Å.

$$\therefore 2d\sin\theta = 2\times2\sin\theta = 2\times0.71 \text{ or } \sin\theta = \frac{2\times0.71}{2\times2} = 0.355 \text{ or } \theta = 20°42'.$$

7.4 The spacing between successive (100)-planes of NaCl is 2.82 Å. X-rays incident on the surface of the crystal is found to give rise to first order Bragg reflection at glancing angle 8.8°. Calculate the wavelength of X-rays. [BUH 2003]

Solution. The Bragg's equation is $2d\sin\theta = n\lambda$.

Here $d = 2.82$ Å, and $n = 1$, $\lambda = ?$

From Bragg's equation $1\times\lambda = 2\times 2.82\times\sin(8.8°)$ or, $\lambda = 2\times 2.82\times 0.1530 = 0.8629$ Å.

7.5 Using a crystal of KCl, Bragg found the following glancing angles for intense reflection from the lattice planes (100), (110) and (111); $\theta_1 = 5.22°$, $\theta_2 = 7.30°$ and $\theta_3 = 9.05°$ respectively. Show that KCl has SC structure. [cf. CUH 1996]

Solution. Bragg's law is $2d\sin\theta = n\lambda$.

$$\therefore \frac{1}{d} = \frac{2\sin\theta}{\lambda}, \ [\because n = 1 \text{ for the first order reflection}].$$

$$\therefore \frac{1}{d_{100}} : \frac{1}{d_{110}} : \frac{1}{d_{111}} = \sin 5.22° : \sin 7.30° : \sin 9.05°, \ \left[\because \tfrac{2}{\lambda} \text{ is constant}\right]$$

$$= 0.0910 : 0.1272 : 0.1570 = 1 : 1.40 : 1.73 \simeq 1 : \sqrt{2} : \sqrt{3}.$$

$\therefore$ KCl *has SC structure.*

7.6 The first order Bragg reflection in polonium appears at $\sin\theta = 0.225$, 0.317 and 0.389 for reflection from (100)-, (110)- and (111)-planes respectively. The wavelength of X-ray used is 1.54 Å. Show that the unit cell of polonium is cubic. What is the length of the side of the unit cell?

Solution. $\sin\theta = 0.225$, 0.317 and 0.389.

So, the values of $\sin\theta$ are in the ratio $0.225 : 0.317 : 0.389 = 1 : 1.41 : 1.73 = \mathbf{1 : \sqrt{2} : \sqrt{3}}$.

Thus, *the unit cell is cubic.*

If a be the length of the side of the cubic unit cell, then from Bragg's law for reflection at (100)-plane,

$$a = d_{100} = \frac{\lambda}{2\sin\theta} = \frac{1.54}{2\times 0.225} = 3.42 \text{ Å}.$$

7.7 If V is the volume of a primitive unit cell in the direct lattice, show that the volume of a primitive unit cell in the reciprocal lattice is $8\pi^3/V$. [BHH 1996]

Solution. If $\vec{a}^*, \vec{b}^*$ and $\vec{c}^*$ be the primitive vectors defining the unit cell in the reciprocal lattice, then its volume is

$$V^* = \left(\vec{a}^*\cdot\vec{b}^*\times\vec{c}^*\right) = (2\pi)^3 \frac{\vec{b}\times\vec{c}}{\left(\vec{a}\cdot\vec{b}\times\vec{c}\right)} \cdot \frac{\left\{(\vec{c}\times\vec{a})\times\left(\vec{a}\times\vec{b}\right)\right\}}{\left(\vec{a}\cdot\vec{b}\times\vec{c}\right)\left(\vec{a}\cdot\vec{b}\times\vec{c}\right)} = (2\pi)^3 \frac{\left(\vec{b}\times\vec{c}\right)\cdot\left\{(\vec{c}\times\vec{a})\times\left(\vec{a}\times\vec{b}\right)\right\}}{\left(\vec{a}\cdot\vec{b}\times\vec{c}\right)^3}$$

$$= (2\pi)^3 \frac{\left(\vec{b}\times\vec{c}\right)\cdot\left[\left\{\vec{c}\cdot\left(\vec{a}\times\vec{b}\right)\right\}\vec{a} - \left\{\vec{a}\cdot\left(\vec{a}\times\vec{b}\right)\right\}\vec{c}\right]}{\left(\vec{a}\cdot\vec{b}\times\vec{c}\right)^3} = \frac{(2\pi)^3}{\vec{a}\cdot\vec{b}\times\vec{c}} = \frac{(2\pi)^3}{V^3},$$

where $V = \vec{a}\cdot\vec{b}\times\vec{c}$ is the volume of the unit cell of the direct lattice. Thus, $V^* \propto 1/V$.

7.8 Prove that the direct lattice is the reciprocal of its own reciprocal lattice.

Solution. We consider the reciprocal of the reciprocal lattice vector

$$\vec{a}^* = 2\pi\frac{\vec{b}\times\vec{c}}{\vec{a}\cdot\vec{b}\times\vec{c}} \text{ as } (\vec{a}^*)^* = (2\pi)\frac{\vec{b}^*\times\vec{c}^*}{\vec{a}^*\cdot\vec{b}^*\times\vec{c}^*}.$$

Multiplying the RHS of the above equation by $(\vec{a}\cdot\vec{a}^*)$, we get

$$(\vec{a}^*)^* = (2\pi)(\vec{a}\cdot\vec{a})^*\cdot\frac{\vec{b}^*\times\vec{c}^*}{\vec{a}^*\cdot\vec{b}^*\times\vec{c}^*} = 2\pi\vec{a}\cdot\frac{\vec{a}^*\cdot\vec{b}^*\times\vec{c}^*}{\vec{a}^*\cdot\vec{b}^*\times\vec{c}^*} = 2\pi\vec{a}.$$

Thus, the statement is proved.

7.9 Prove that the reciprocal lattice of a BCC lattice is an FCC lattice. [BUH 2000]

Solution. The primitive basis vectors of a BCC lattice are

$$\vec{a}' = \frac{a}{2}(\hat{i}+\hat{j}-\hat{k}),\ \vec{b}' = \frac{a}{2}(-\hat{i}+\hat{j}+\hat{k}),\ \vec{c}' = \frac{a}{2}(\hat{i}-\hat{j}+\hat{k}),$$

where $\hat{i}$, $\hat{j}$ and $\hat{k}$ are the unit vectors along the edge of the cube whose a is the cube edge. The corresponding reciprocal lattice vectors are

$$\vec{a}^* = 2\pi\frac{\vec{b}'\times\vec{c}'}{\vec{a}'\cdot\vec{b}'\times\vec{c}'} = 2\pi\frac{a^2/2}{a^3/2}(\hat{i}+\hat{j}) = \frac{2\pi}{a}(\hat{i}+\hat{j})$$

$$\vec{b}^* = 2\pi\frac{\vec{c}'\times\vec{a}'}{\vec{a}'\cdot\vec{b}'\times\vec{c}'} = \frac{2\pi}{a}(\hat{j}+\hat{k})$$

$$\text{and } \vec{c}^* = 2\pi\frac{\vec{a}'\times\vec{b}'}{\vec{a}'\cdot\vec{b}'\times\vec{c}'} = \frac{2\pi}{a}(\hat{k}+\hat{i}).$$

For an FCC lattice, the primitive basis vectors are

$$\vec{a}' = \frac{a}{2}(\hat{i}+\hat{j}),\ \vec{b}' = \frac{a}{2}(\hat{j}+\hat{k}),\ \vec{c}' = \frac{a}{2}(\hat{k}+\hat{i}).$$

Comparing $\vec{a}^*, \vec{b}^*, \vec{c}^*$ and $\vec{a}', \vec{b}', \vec{c}'$, we find that $\vec{a}^*, \vec{b}^*, \vec{c}^*$, are the primitive basis vectors of an FCC lattice having cube edge $(4\pi/a)$. *This proves that the reciprocal lattice of a BCC lattice is an FCC lattice.*

7.10 Starting from $2\vec{K}\cdot\vec{G}+G^2=0$, obtain the diffraction condition, $2d\sin\theta = n\lambda$.

Solution. We know that spacing d_{hkl} between parallel lattice planes normal to the direction $\vec{G} = h\vec{A}+k\vec{B}+l\vec{C}$ is

$$d_{hkl} = \frac{2\pi}{|\vec{G}|}.$$

Now $2\vec{K}\cdot\vec{G} = G^2$ may be written as

$$2\left(\frac{2\pi}{\lambda}\right)\sin\theta = \frac{2\pi}{d_{hkl}} \text{ or } \mathbf{2d_{hkl}\sin\theta = \lambda}.$$

Here θ is the angle between the incident beam and lattice plane. The integers that define $\vec{G}$ may contain a common factor n. Then we obtain $2d\sin\theta = n\lambda$.

7.11 The wavelength of the k_α line of Ag is 0.563 Å. The radiation from a Ag target is analysed with a Bragg spectrometer using a calcite crystal (a simple cube of lattice constant 3.02945 Å). Determine the angle of reflection of the first order. What is the highest order for which this may be observed? [CUH 1990]

Solution. Bragg's law for the first order ($n=1$) is $2d\sin\theta=\lambda$.
Here $d=3.02945$ Å and $\lambda=0.563$ Å.

$$\therefore \sin\theta=\frac{\lambda}{2d}=\frac{0.563}{6.0589}=0.0929.$$

Hence, the angle of reflection for the first order is $\theta=\sin^{-1}(0.0929)=$ **5.33°**.
For the nth order reflection, $2d\sin\theta=n\lambda$.
For the highest volume of θ, i.e., 90°,

$$n=\frac{2d}{\lambda}=\frac{6.0589}{0.563}=10.8.$$

So, the highest order for which this reflection may be observed is **10.**

7.12 A simple cubic crystal illuminated with X-rays of wavelength 0.09 nm is rotated and the first order Bragg reflection occurs at a minimum glancing angle of 8.8°. Which set of crystal planes are responsible for this reflection? Find the spacing between these planes and the angle for the first order reflection from the (110) crystal planes.

Solution. Since the glancing angle is the minimum for the planes having largest spacing, the planes in question will be (100)-planes.
$\therefore\ 2d_{100}\sin\theta=\lambda$, (for $n=1$).

$$\therefore\ d_{100}=\frac{\lambda}{2\sin\theta}=0.294\,\text{nm}.$$

The spacing between the (110)-planes is

$$d_{110}=\frac{d_{100}}{\sqrt{1^2+1^2+0^2}}=\frac{d_{100}}{\sqrt{2}}=\frac{0.294}{\sqrt{2}}=\mathbf{0.208\ nm}.$$

So

$$2d_{110}\sin\theta_{110}=\lambda \text{ or } \sin\theta_{110}=\frac{0.09}{2\times 0.208}=0.2163.$$

$\therefore\ \boldsymbol{\theta_{110}=12.5°}$.

7.13 Show that greater the diffraction angle, greater will be the accuracy in determining the lattice parameters.

Solution. The Bragg's equation

$$2d\sin\theta=n\lambda \text{ or } d\sin\theta=\frac{n\lambda}{2}.$$

Differentiating

$$d\cos\theta\,d\theta+\sin\theta\,d(d)=0 \text{ or } \frac{d(d)}{d}=-\cot\theta\,d\theta.$$

The above equation shows that as $\theta\to\frac{\pi}{2}$, $\cot\theta\to 0$, i.e., the error in the measurement of $d\to 0$. *This shows that greater is the angle of diffraction, greater will be the accuracy in determining the lattice parameters.*

7.14 Given a square piece of X-ray film 10 cm × 10 cm. With Copper radiation of $\lambda = 1.52$ Å and powdered NaCl (cubic) with lattice parameter 4.57 Å, devise a diffraction experiment in such fashion that the rays from the (111)-planes will produce a circle of diameter 10 cm ($\sin 16.75° = 0.288$, $\tan 33.5° = 0.662$). **[CUH 1999]**

Solution. See Art. 7.8.3: (*Powder Method*).

Bragg's law for the first order diffraction is $2d \sin\theta = \lambda$.

Here

$$d = \frac{4.57}{\sqrt{3}} \text{ Å} \quad \text{and} \quad \lambda = 1.52 \text{ Å}.$$

$$\therefore \sin\theta = \frac{\lambda}{2d} = \frac{1.52 \times \sqrt{3}}{2 \times 4.57} = 0.288 = \sin 16.75°.$$

$\therefore \theta = 16.75°$.

The X-ray film is placed normally to the incident X-radiation at a distance D from the powdered sample such that the diffraction cone intersects the film in a circle of 5 cm.

Then

$$\tan 2\theta = \frac{5}{D} \text{ or } D = \frac{5}{\tan 2\theta} = \frac{5}{\tan 33.5°} = \frac{5}{0.662} = 7.55 \text{ cm}.$$

7.15 Monochromatic X-rays of wavelength 1.4 Å are incident on a crystal having 1.5 Å as interatomic spacing. Find the maximum order in which the diffraction takes place.

Solution. Bragg's equation

$$2d \sin\theta = n\lambda \text{ or } n = \frac{2d \sin\theta}{\lambda}.$$

Here $d = 1.5$ Å and $\lambda = 1.4$ Å

$$\therefore n = \frac{3}{1.4} \sin\theta \text{ or } n = 2.14 \sin\theta.$$

$\because |\sin\theta| \leq 1$, **the maximum order of the spectrum that can be observed is not more than 2.**

$\therefore n \leq 2$.

EXERCISES

7(A) Objective Type Questions

Choose the correct alternatives.

7.1 X-rays are

(a) mechanical waves
(b) electromagnetic waves of short wavelength
(c) matter waves
(d) none of these.

7.2 X-rays are used for diffraction studies in crystals because

(a) the wavelength of X-rays is of the same order as that of interatomic spacing
(b) X-rays penetrate the crystals
(c) Crystals have atoms/molecules capable of scattering X-rays
(d) none of these.

7.3 Monochromatic X-rays are used in

(a) Laue method
(b) powder method
(c) rotating crystal method
(d) none of these.

7.4 Every reciprocal lattice vector is
(a) parallel to a lattice plane
(b) normal to a lattice plane
(c) inclined at an angle of 45° to the lattice plane
(d) none of these.

7.5 The reciprocal lattice to an FCC lattice is
(a) an SC lattice
(b) a BCC lattice
(c) an FCC lattice
(d) none of these.

7.6 The reciprocal lattice to a BCC lattice is
(a) an SC lattice
(b) an FCC lattice
(c) a BCC lattice
(d) none of these.

7.7 The reciprocal lattice to an SC lattice is
(a) an SC lattice
(b) a BCC lattice
(c) an FCC lattice
(d) none of these.

7.8 Bragg's law in vector form is
(a) $2\vec{k}\cdot\vec{G}=0$
(b) $2\vec{k}\cdot\vec{G}+G^2=0$
(c) $2\vec{k}\cdot\vec{G}-G^2=0$
(d) $2\vec{k}\cdot\vec{G}+G^2=k^2$.

7.9. Bragg's law is modified due to
(a) reflection of X-rays
(b) refraction of X-rays
(c) polarisation of X-rays
(d) total internal reflection of X-rays.

7.10 The modified Bragg's law due to refraction of X-rays is
(a) $n\lambda=2d\sin\theta\left[\frac{4d^2(1-\mu)}{n^2\lambda^2}\right]$
(b) $n\lambda=2d\sin\theta\left[1-\frac{4d^2(1-\mu)}{\lambda^2}\right]$
(c) $n\lambda=2d\sin\theta\left[1-\frac{4d^2(1-\mu)}{n^2\lambda^2}\right]$
(d) $n\lambda=2d\sin\theta\left[1-\frac{4d^2(1-\mu)^2}{n^2\lambda^2}\right]$.

7.11 The refractive index (μ) of X-rays is given by
(a) $\mu>1$
(b) $\mu=0$
(c) $\mu<1$
(d) $\mu\geq 1$.

7.12 If V and V^* be the volume of a unit cell in direct and reciprocal lattice respectively, then
(a) $V=V^*$
(b) $V=\frac{V^*}{8\pi^3}$
(c) $V=8\pi^3 V^*$
(d) $V=\frac{8\pi^3}{V^*}$.

7.13 Liquid crystals are
(a) inorganic compounds
(b) organic compounds
(c) mixture of (a) and (b)
(d) none of these.

7.14 In anisotropic liquid crystals the directions of $\vec{D}$ and $\vec{E}$ are

(a) same
(b) different
(c) both (a) and (b)
(d) none of these.

7.15 An externally applied electric field converts the cholesteric liquid crystal into
(a) nematic type
(b) lyotropic type
(c) smectic type liquid crystal
(d) none of these.

ANSWERS

7.1 (b)	7.4 (b)	7.7 (a)	7.10 (c)	7.13 (b)
7.2 (a)	7.5 (b)	7.8 (b)	7.11 (c)	7.14 (b)
7.3 (b) and (c)	7.6 (b)	7.9 (b)	7.12 (d)	7.15 (a).

7(B) Short Answer Type Questions

7.1 What are the principal methods for crystal structure analysis?

7.2 Why X-rays are used for the analysis of crystal structure?

7.3 Point out the underlying assumptions in the derivation of Bragg's law.

7.4 What are the utilities of Bragg's law for X-rays diffraction?

7.5 Give the modification of Bragg's law due to refraction of X-rays. [CUH 2001]

7.6 What is a reciprocal lattice? How does it differ from direct lattice?

7.7 Why is the concept of reciprocal lattice in solid state physics so important?

7.8 Show that the direct lattice is the reciprocal of its own reciprocal lattice.

7.9 State Bragg's law of diffraction and discuss its importance in crystal structure analysis.

7.10 How does the Laue approach differ from the Bragg approach? [CUH 2004]

7.11 On the basis of Ewald's construction, find the geometrical significance of the Bragg's equation in the reciprocal lattice space.

7.12 Define (a) atomic scattering factor (b) geometrical structure factor.

7.13 Define Brillouin zone for a lattice. How would you construct the first Brillouin zone for a BCC lattice?

7.14 Prove that a reciprocal lattice vector is normal to a lattice plane of the crystal lattice.

7.15 What is liquid crystal? Discuss the principle of action of liquid crystal display.

7(C) Numerical Problems

7.1 Calculate the longest wavelength that can be analysed by a rock salt crystal of spacing $d = 2.81$ Å in the first order. *(Ans. 5.64 Å)*

7.2 Using a Bragg's X-ray spectrometer, the glancing angle for first order spectrum was observed to be equal to 6°. Find the wavelength of X-rays, if $d = 2.82 \times 10^{-8}$ cm. *(Ans. 0.59×10^{-8} cm)*

7.3 Using a crystal of KCl, Bragg found the following glancing angles for intense reflection from the lattice planes (100), (110) and (111) respectively: $\theta_1 = 5.22$ Å, $\theta_2 = 7.30$ Å and $\theta_3 = 9.05$ Å. Show that KCl has a simple cubic crystal structure. $\left(\textit{Ans. } \frac{1}{d_{100}} : \frac{1}{d_{110}} : \frac{1}{d_{111}} \simeq 1 : \sqrt{2} : \sqrt{3}\right)$

7.4 X-rays of wavelength 1 Å are incident on the cube face (111) of a NaCl crystal ($a = 1.702$ Å). Deduce the value of glancing angle corresponding to the first order diffraction maximum. *(Ans. 30°)*

7.5 The line *A* of an X-ray beam gives a first order reflection maximum when the glancing angle is 30° to the smooth face of a crystal. Line *B* with $\lambda = 0.97$ Å gives a third order reflection maximum at an angle of 60° from the same face of the same crystal. Determine the wavelength of the line *A*.[KU 2004] (*Ans.* 1.68 Å)

7.6 The primitive basis vectors of a lattice are $\vec{a} = \hat{i} + 2\hat{j}$, $\vec{b} = 4\hat{j}$ and $\vec{b} = \hat{k}$. What are the primitive translation vectors of its reciprocal lattice? (*Ans.* $\vec{a}^* = 2\pi\hat{i}$, $\vec{b}^* = \frac{\pi}{2}(-2\hat{i} + \hat{j})$, $\vec{c}^* = 2\pi\hat{k}$)

7.7 A beam of X-rays is incident on a NaCl crystal with lattice spacing 0.282 nm. Calculate the wavelength of X-rays, if the first order Bragg reflection takes place at a glancing angle of 8°35′. Also calculate the maximum order of diffraction possible. (*Ans.* $\lambda = 0.0842$ nm, $n = 6$)

7.8 A set of lattice planes reflects X-rays of 1.32 Å at glancing angle of 9.30°. Deduce the possible spottings of this set of planes. (*Ans.* 4 Å, 9 Å, 12 Å)

7.9 Calculate the wavelength of X-ray beam incident at 12° for the first order reflection from calcite crystal if grating constant for the crystal is 3.035 Å. (*Ans.* 126 Å)

7(D) Broad Answer Type Questions

7.1 (a) State Bragg's law of diffraction and discuss its importance in crystal structure analysis.[cf. CU 1998]
(b) Why X-rays are used for the analysis of crystal structure analysis?

7.2 (a) State the condition under which Bragg reflection can occur.
(b) Derive Bragg's law of X-ray diffraction clearly stating the underlying assumptions.
(c) Give the corrected form of Bragg's law taking into account the correction due to refraction.

7.3 (a) What is reciprocal lattice? Discuss its importance in solid state physics.
(b) Prove that reciprocal lattice vector is normal to a lattice plane of the crystal lattice.
(c) Prove that a direct lattice is the reciprocal of its own reciprocal lattice.

7.4 (a) Describe Ewald's construction and explain its importance.
(b) On the basis of Ewald's construction, find the geometrical significance of the Bragg's equation in the reciprocal lattice space. [BUH 1996]

7.5 (a) Derive Laue equations and hence deduce Bragg's law of X-ray diffraction.
(b) How does the Laue approach differ from the Bragg approach? [CUH 2004]

7.6 Discuss (a) Laue method and (b) the rotating crystal method of X-ray diffraction. Mention specific applications of each of them.

7.7 (a) What are Brillouin zones? Explain their significance.
(b) Discuss how Brillouin zones of FCC and BCC lattices can be constructed.

7.8 Explain what do you mean by Atomic Scattering Factor and Geometrical Scattering Factor. Find out their expressions.

7.9 (a) What are liquid crystals? Mention the name of different types of liquid crystals.
(b) Describe a cholesteric liquid crystal and mention some applications of LCD.

Chapter 8

Free Electron Theories of Metals

8.1 Introduction

The valence electrons of an atom in metal are loosely bound to the nucleus and at ordinary temperature become free of the atomic system. Such electrons in a metal move about freely through the volume of the metal and constitute what is called *free electron gas.* The loosely bound valence electrons are called *Conduction electrons.* Many important physical properties like electrical, thermal and magnetic properties of solids can be explained with the help of this *free electron gas model.* The *free electron* theory of metals has been developed in three different stages as follows:

1. *The Classical Free Electron Theory*: This theory was developed in 1900 by *Drude* and *Lorentz* according to which the free electrons move at *random chaotic motion* within the volume of the metal just like the ideal gas molecules in a chamber. These free electrons obey the *laws of classical mechanics.*

2. *The Quantum Free Electron Theory*: This theory was developed by *Sommerfeld* in 1928 which assumes that the free electrons obey the *laws of quantum mechanics.*

3. *The Zone Theory*: This is also known as the *Band Theory of Solids* and was formulated by *Bloch* in 1928.

8.2 Classical Free Electron Theory of Metals: Drude's Theory

The theory assumes that

1. In atoms the valence electrons are free to undergo *random chaotic motion* within the volume of the metal very much like the ideal gas molecules in a container. The collection of such electrons forms what is called *electron gas.*
2. The motion of these free electrons obeys *classical Maxwell-Boltzmann velocity distribution law* and *the laws of kinetic theory of gases.*
3. The free electrons move at random and collide *elastically* with each other or positive ions fixed to the lattice.
4. The electrons being negatively charged move in opposite direction to an externally applied electric field.
5. The free electrons are *non-interacting* and obey *Pauli's exclusion principle.*

8.2.1 Electrical Conductivity

Assuming the free electrons to obey *Kinetic theory of gases,* the *root mean square* (rms) *velocity* of an electron at temperature T is given by

$$C=\sqrt{\frac{3kT}{m}}, \text{ where } k \text{ is the } \textit{Boltzmann constant.} \text{ Thus, } C \propto \sqrt{T}. \tag{8.1}$$

The *electrical conductivity* in metals arise due to the motion of the conduction electrons under the influence of applied electric field. If E be the electric field and $-e$ the charge of an electron, then the force exerted on the electron is $F=-eE$.

The *accelerated electrons* collide with the rest of the medium and comes to rest. But due to the action of E, the electrons again are accelerated. Due to this collision we can attribute a *frictional force* acting on the electron.

If v be the velocity of the electron and τ, the time between two consecutive collisions, then the frictional force is $-mv/\tau$, where m is the *effective mass* of the electron. The equation of motion of the electron is

$$m\frac{dv}{dt}=-eE-m\frac{v}{t}. \tag{8.2}$$

At steady state condition,

$$\frac{dv}{dt}=0.$$

$$\therefore\ m\frac{v}{\tau}=-eE \text{ or } v=\frac{-e\tau}{m}E. \tag{8.3}$$

v is called the *steady state velocity* of the electron.

Drift Velocity and Charge Mobility: In absence of an electric field the electrons move randomly in all possible directions and the corresponding velocity is called *random velocity*.

But in presence of an electric field, in addition to random velocity there is an additional net velocity associated with the electrons and is called the *drift velocity* (v_d). The drift velocity of an electron of mass m in electric field is given by

$$v_d=-\frac{e\tau}{m}E \tag{8.4}$$

$$\text{or } v_d=-\mu E, \tag{8.5}$$

where $\mu=\frac{eE}{m}$ is called the *mobility* of the electron and is defined as *the average drift velocity in unit electric field.*

Now, if n be the number of conduction electrons per unit volume, then the charge in unit volume is $(-ne)$. The amount of charge crossing unit area per unit time, i.e., *current density* is

$$J=(-ne)v_d=-ne\left(-\frac{e\tau}{m}E\right)=\frac{ne^2\tau}{m}E. \tag{8.6}$$

But

$$J=\sigma\vec{E} \tag{8.7}$$

is the *vector form of Ohm's law*, where σ is the *electrical conductivity.*

Comparing (8.6) and (8.7), we get

$$\sigma=\frac{ne^2\tau}{m}. \tag{8.8}$$

Thus, the *electrical conductivity* of a metal increases with the increase of free electron density (n).

Mean Free Lifetime: The average lifetime between two consecutive collisions is called the *mean free lifetime.*

Mean Free Path: The average distance travelled by an electron between two consecutive collisions in presence of the applied electric field is called the *mean free path* (λ).

If $\lambda_1, \lambda_2, \lambda_3, \ldots$ be the successive free paths of the electron between two consecutive collisions, then its mean free path is given by

$$\lambda = \frac{\lambda_1 + \lambda_2 + \lambda_3 + \cdots + \lambda_n}{n} = \frac{1}{n}\sum_{i=1}^{n} \lambda_i. \quad (8.9)$$

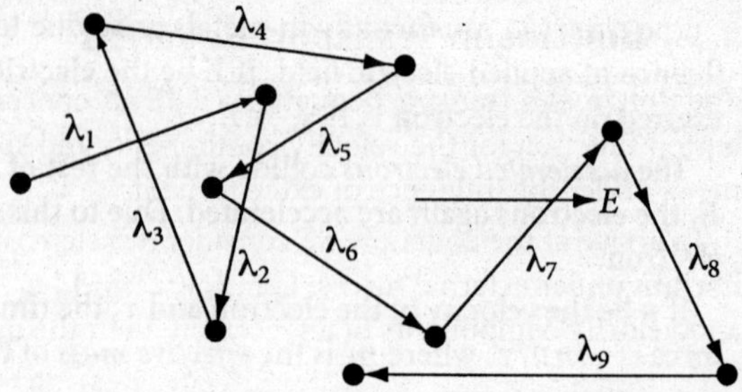

Fig. 8.1: Free paths of an electron in metal in presence of *E*.

Relaxation Time: Let the applied electric field *E* be cut off after the drift velocity of the electron has reached a steady value. Putting $E = 0$ in (8.2), we get

$$m\frac{dv_d}{dt} = -m\frac{v_d}{\tau} \text{ or } \frac{dv_d}{v_d} = -\frac{dt}{\tau}$$

$$\text{or } v_d(t) = v_d(0)\,e^{-t/\tau}, \quad (8.10)$$

where $v_d(0)$ is the drift velocity of the electron at the instant when the field is cut off. The variation of $v_d(t)$ with t is shown in Fig. 8.2.

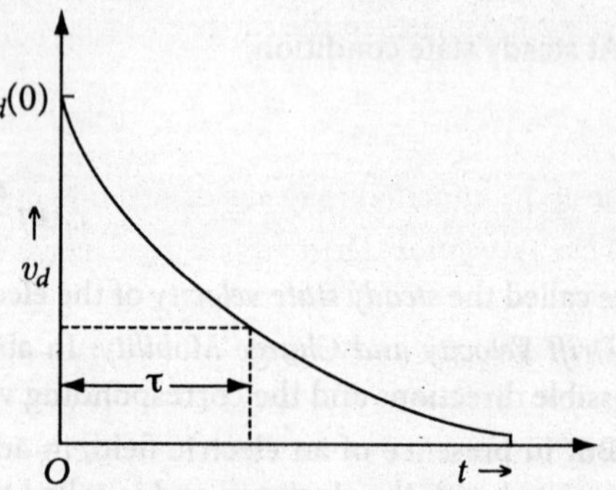

Fig. 8.2: Variation of v_d with *t*.

To define τ, let us put $t = \tau$ in (8.10),

$$v_d(t) = v_d(0)/e. \quad (8.11)$$

Hence, *relaxation time* τ is defined as *the time at which the drift velocity decays to* $\left(\frac{1}{e}\right)$ *times its initial value.*

Again, the electron mobility is the steady state drift velocity per unit electric field.

$\therefore$ electron mobility

$$\mu = \frac{\langle v_d \rangle}{E} = \frac{e\tau}{m}, \quad (8.12)$$

where $\langle v_d \rangle$ is the *average drift velocity* of the electron.

$\therefore$ from (8.3),

$$\sigma = \frac{ne^2\tau}{m} = ne\cdot(e\tau/m) \text{ or } \sigma = ne\mu. \quad (8.13)$$

The *resistivity of the material* of the conductor is

$$\rho = \frac{1}{\sigma} = \frac{1}{ne\mu} = \frac{m}{ne^2\tau}. \quad (8.14)$$

(8.13) shows that σ depends on n and μ, both of which depend on temperature of the conductor. In *metals* μ decreases slightly with temperature and n is practically constant. So, *the electrical conductivity of metals decreases with the rise of temperature.*

For *semiconductor,* n increases exponentially with temperature causing an increase of conductivity.

In insulators, n remains constant and above certain temperature μ increases exponentially leading to *dielectric breakdown.*

8.3 Boltzmann Transport Equation

The Boltzmann transport equation (BTE) is concerned with the determination of the steady state distribution function for the velocity components and spatial coordinates in the transport process occurring in metal under the influence of external fields.

To arrive at the equation we consider a system of particles (e.g., electrons) which are in dynamic equilibrium under external forces. Let (x, y, z) and $\left(v_x, v_y, v_z\right)$ represent respectively the position coordinates and velocity components of an electron, then the distribution function is given by

$$f\left(x, y, z, v_x, v_y, v_z\right). \tag{8.15}$$

The number of particles in the position coordinate range x to $x+dx$, y to $y+dy$, z to $z+dz$ and velocity coordinates in the range v_x to v_x+dv_x, v_y to v_y+dv_y, v_z to v_z+dv_z is represented as

$$f\left(x, y, z, v_x, v_y, v_z\right) dx\, dy\, dz. \tag{8.16}$$

The above function may vary with time due to:

(i) *Drift variation*: Drift variation is caused due to external field and takes place continuously. To evaluate such variation, we consider a system consisting of particles at an instant $t+dt$, which is drifted to a phase cell having coordinates $\left(x, y, z, v_x, v_y, v_z\right)$. The number of particles being the same as were at $(x-v_x\, dt)$, $\left(y-v_y\, dt\right)$, $(z-v_z\, dt)$, $(v_x-\alpha_x\, dt)$, $\left(v_y-\alpha_y\, dt\right)$, $(v_z-\alpha_z\, dt)$ at time t, α_x, α_y and α_z being the x-, y-, z-components of acceleration. The effect of collision has been neglected due to small interval of time so that the variations in velocity and accelerations are small, i.e., these are practically constant.

The change in the number of particles due to drift having coordinates $\left(x, y, z, v_x, v_y, v_z\right)$ in time *dt* is

$$(\Delta f)_d = f\left(x-v_x\, dt, y-v_y\, dt, z-v_z\, dt, v_x-\alpha_x\, dt, v_y-\alpha_y\, dt, v_z-\alpha_z\, dt\right). \tag{8.17}$$

Expanding in *Taylor's theorem* and retaining only the first order terms we get in the limit $\Delta t \to 0$,

$$(\Delta f)_d = \left(-v_x\frac{\partial f}{dx} - v_y\frac{\partial f}{dy} - v_z\frac{\partial f}{dz} - \alpha_x\frac{\partial f}{\partial v_x} - \alpha_y\frac{\partial f}{\partial v_y} - \alpha_z\frac{\partial f}{\partial v_z}\right) dt \tag{8.18}$$

$$\left(\frac{df}{dt}\right)_{\text{drift}} = -v_x\frac{\partial f}{dx} - v_y\frac{\partial f}{dy} - v_z\frac{\partial f}{dz} - \alpha_x\frac{\partial f}{\partial v_x} - \alpha_y\frac{\partial f}{\partial v_y} - \alpha_z\frac{\partial f}{\partial v_z}. \tag{8.19}$$

(ii) *Collision variation*: Let the probability that a particle changes its velocity from v_x, v_y, v_z to $v_x'+dv_x'$, $v_y'+dv_y'$, $v_z'+dv_z'$ in unit time be represented as

$$\theta\left(v_x, v_y, v_z;\ v_x', v_y', v_z'\right) dv_x'\, dv_y'\, dv_z'. \tag{8.20}$$

Then the total number of particles undergoing such change in velocity is

$$a = f\left(x, y, z, v_x, v_y, v_z\right)\int \theta\left(v_x, v_y, v_z, v_x', v_y', v_z'\right) dv_x'\, dv_y'\, dv_z'. \tag{8.21}$$

Similarly, the number of particles whose velocity changes from v_x, v_y, v_z to v_x'', v_y'', v_z'' is

$$b = \int f\left(v_x'', v_y'', v_z''\right)\theta\left(v_x'', v_y'', v_z'',\ v_x, v_y, v_z\right) dv_x''\, dv_y''\, dv_z''. \tag{8.22}$$

From (8.21) and (8.22), we get the rate of change of f due to collision

$$\left(\frac{df}{dt}\right)_{\text{coll}}=b-a. \tag{8.23}$$

Under condition of equilibrium

$$\left(\frac{df}{dt}\right)_{\text{drift}}+\left(\frac{df}{dt}\right)_{\text{coll}}=0 \tag{8.24}$$

$$\text{or } -v_x\frac{\partial f}{\partial x}-v_y\frac{\partial f}{\partial y}-v_z\frac{\partial f}{\partial z}-\alpha_x\frac{\partial f}{\partial x}-\alpha_y\frac{\partial f}{\partial y}-\alpha_z\frac{\partial f}{\partial z}+(b-a)=0$$

$$\text{or } v_x\frac{\partial f}{\partial x}+v_y\frac{\partial f}{\partial y}+v_z\frac{\partial f}{\partial z}+\alpha_x\frac{\partial f}{\partial x}+\alpha_y\frac{\partial f}{\partial y}+\alpha_z\frac{\partial f}{\partial z}=(b-a) \tag{8.25}$$

$$\text{or } \vec{v}\cdot\nabla f+\vec{\alpha}\cdot\nabla' f=(b-a). \tag{8.26}$$

(8.25) and (8.26) are *Boltzmann transport equations* (BTE).

8.3.1 Boltzmann Transport Equation for Electrons

If an electron of charge e and mass m moves in homogeneous electric field $\vec{E}$ and magnetic field $\vec{H}$ at right angles to each other, then the *Lorentz force* on it is given by

$$\vec{F}=m\vec{\alpha}=-e\left[\vec{E}+\frac{1}{c}\vec{v}\times\vec{H}\right], \tag{8.27}$$

here $\vec{\alpha}$ is the acceleration of the electron.

If $\vec{E}$ be in the xy-plane and $\vec{H}$ in the z-direction, then

$$m\alpha_x=-e\left[E_x+\frac{1}{c}v_yH_z\right] \tag{8.28}$$

$$m\alpha_y=-e\left[E_y+\frac{1}{c}v_zH_z\right] \tag{8.29}$$

$$\text{and } m\alpha_z=0. \tag{8.30}$$

The Boltzmann transport equation (8.25) then becomes

$$v_x\frac{\partial f}{\partial x}+v_y\frac{\partial f}{\partial y}+v_z\frac{\partial f}{\partial z}-\left[\frac{eE_x}{m}+\frac{e}{c}\frac{v_y}{m}H_z\right]\frac{\partial f}{\partial v_x}-\left[\frac{eE_y}{m}-\frac{e}{c}\frac{v_x}{m}H_z\right]\frac{\partial f}{\partial v_y}+0=\left(\frac{\partial f}{\partial t}\right)_{\text{coll}}. \tag{8.31}$$

(8.31) is the *Boltzmann transport equation for the electron.*

8.3.2 Lorentz Solution of Boltzmann Transport Equation for the Electron

Lorentz assumed that the electric field $\vec{E}$ is along x-direction and there is no magnetic field. Then the function f is independent of t, y and z. Under this condition (8.31) becomes

$$\left(\frac{\partial f}{\partial t}\right)_{\text{coll}}=v_x\frac{\partial f}{\partial x}-\frac{eE_x}{m}\frac{\partial f}{\partial v_x}. \tag{8.32}$$

Lorentz assumed that:

(a) *the collisions of electron is elastic,*

(b) *the scattering of electron is isotropic,*

(c) *the equilibrium distribution function f does not change due to electron scattering* and

(d) *the difference of equilibrium function f from the equilibrium forms f_0 so that*

$$\left(\frac{\partial f}{\partial t}\right)_{\text{coll}}=\frac{f-f_0}{\tau}, \tag{8.33}$$

where τ is called the *relaxation time*.

(8.33) represents the rate at which the distribution function approaches the condition of equilibrium when $\vec{E}$ or the temperature gradient is absent.

The solution of (8.33) is then

$$(f-f_0)=Axe^{-t/x}, \tag{8.34}$$

where A is a constant. (8.34) shows that the decay of $(f-f_0)$ is exponential.

From (8.33) and (8.34), we get

$$\left(\frac{\partial f}{\partial t}\right)_{\text{coll}}=v_x\frac{\partial f}{\partial x}-\frac{eE_x}{m}\frac{\partial f}{\partial v_x}=-\frac{f-f_0}{\tau}. \tag{8.35}$$

Replacing f by f_0 in the right-hand side of (8.35), we get

$$v_x\frac{\partial f_0}{\partial x}-\frac{eE_x}{m}\frac{\partial f_0}{\partial v_x}=-\frac{(f-f_0)}{\tau} \tag{8.36}$$

$$\text{or } f-f_0=-\tau\left[v_x\frac{\partial f_0}{\partial x}-\frac{eE_x}{m}\frac{\partial f_0}{\partial v_x}\right]$$

$$\text{or } f=f_0-\tau\left[v_x\frac{\partial f_0}{\partial x}-\frac{eE_x}{m}\frac{\partial f_0}{\partial v_x}\right]. \tag{8.37}$$

(8.37) is the solution of the *Boltzmann transport equation* for the electron. By the application of this solution we can obtain more refined results for electrical and thermal conductivities.

8.4 Sommerfeld's Theory of Electrical Conductivity

Sommerfeld in 1928 gave the theory of electrical conduction in metal using *Fermi-Dirac* (FD) *statistics* instead of *Maxwell-Boltzmann* (MB) statistics as the latter was unable to explain the results of *Wiedemann-Franz law.*

In presence of electric field $\vec{E}$,

$$v_x\frac{\partial f_0}{\partial x}-\frac{eE_x}{m}\frac{\partial f_0}{\partial v_x}=-\frac{f}{\tau}. \tag{8.38}$$

When steady state is reached,

$$-\frac{eE_x}{m}\frac{\partial f}{\partial v_x}=\frac{f}{\tau}. \tag{8.39}$$

Again,

$$\frac{1}{m}\frac{\partial f_0}{\partial v_x}=\frac{1}{m}\frac{\partial f_0}{\partial E}\frac{\partial E}{\partial v_x}=\frac{\partial f_0}{\partial E}\left(\frac{1}{m}\frac{\partial E}{\partial v_x}\right)=\frac{\partial f_0}{\partial E}v_x.$$

$$\therefore f=\tau e E_x v_x\left(\frac{1}{m}\frac{\partial f_0}{\partial v_x}\right)=\tau e E_x\frac{\partial f_0}{\partial E}v_x=\tau e E_x v_x\left(\frac{\partial f_0}{\partial E}\right). \tag{8.40}$$

(8.40) represents the *distribution function of electrons.*

To calculate the number of electronic states per unit volume associated with element $dp_x\, dp_y\, dp_z$ in momentum space is

$$g(p)=\frac{1}{h^3}\,dp_x\,dp_y\,dp_z. \tag{8.41}$$

Considering electron spin,

$$g(p)=\frac{2}{h^3}\,dp_x\,dp_y\,dp_z. \tag{8.42}$$

The average number of occupied electron states in *absence of field at thermal* equilibrium is

$$n_0=g(p)\,f_0(p)=\frac{2}{h^3}\,f_0(p)\,dp_x\,dp_y\,dp_z, \tag{8.43}$$

where $f_0(p)$ is the *Fermi-distribution function* in terms of total momentum p.

If now an electric field E_x be applied along x-direction, then *in the steady state* the average number of electrons per unit volume in the range p_x to p_x+dp_x, p_y to p_y+dp_y, p_z to p_z+dp_z is

$$n=\frac{2}{h^3}f(p)\,dp_x\,dp_y\,dp_z. \tag{8.44}$$

Therefore, the current density is

$$J_x=\sum -Nev_x=-\frac{2e}{h^3}\iiint v_x f(p)\,dp_x\,dp_y\,dp_z. \tag{8.45}$$

Substituting the value of $f(p)$ from (8.40) in (8.45), we get

$$J_x=-\frac{2e}{h^3}\iiint v_x\tau e E_x\, v_x\left(\frac{\partial f_0}{\partial E}\right)dp_x\,dp_y\,dp_z=-\frac{2e^2}{h^3}\iiint v_x^2\tau E_x\left(\frac{\partial f_0}{\partial E}\right)dp_x\,dp_y\,dp_z \tag{8.46}$$

$$=-\frac{2e^2E_x}{h^3}\int_0^\infty\frac{v^2}{3}\tau\left(\frac{\partial f_0}{\partial E}\right)4\pi p^2\,dp. \tag{8.47}$$

$$\because\ dp_x\,dp_y\,dp_z=4\pi p^2\,dp \text{ and } \langle v_x^2\rangle=v^{2/3}.$$

Again,

$$\because\ E=\frac{1}{2}mv^2,\ v^2=\frac{2E}{m}. \tag{8.48}$$

Also

$$E=\frac{p^2}{2m},\ p=\sqrt{2mE}.$$

$$\therefore\ dE=\frac{2p\,dp}{2m} \text{ or } p\,dp=m\,dE. \tag{8.49}$$

Substituting the value of v^2 and $p\,dp$ in (8.48), we get

$$J_x = -\frac{2e^2E_x}{h^3}\int_0^\infty \frac{1}{3}\left(\frac{2E}{m}\right)\tau\frac{\partial f_0}{\partial E}4\pi\sqrt{2mE}\;m\,dE = -\frac{16\pi e^2}{3h^3}(2m)^{1/2}E_x\int_0^\infty \tau E^{1/2}\frac{\partial f_0}{\partial E}\,dE$$

$$= \frac{16\pi e^2}{3h^3}(2m)^{1/2}E_x\tau E^{3/2}\left[\because \int_0^\infty \frac{\partial f_0}{\partial E}dE = -1\right]. \qquad (8.50)$$

The Fermi energy at 0 K is

$$E_{F(0)} = \frac{h^2}{2m}\left(3\pi^2N\right)^{2/3}.$$

$$\therefore\; J_x = \frac{16\pi e^2(2m)^{1/2}}{3h^3}E_x\tau\left(\frac{h^2}{2m}\right)^{3/2}\left(2\pi^2N\right). \qquad (8.51)$$

Now the *electrical conductivity*

$$\sigma = \frac{J_x}{E_x} = \frac{Ne^2\tau_F}{m}, \qquad (8.52)$$

where N is the *electron density.*

8.5 Thermal Conductivity

Thermal conductivity is defined as the ratio of thermal energy flux to the temperature gradient giving rise to energy flow.

If H be the thermal energy carried by the electron per unit area per second and $\frac{dT}{dx}$, the temperature gradient in the x-direction, then *thermal conductivity,*

$$K = \frac{H}{dT/dx}. \qquad (8.53)$$

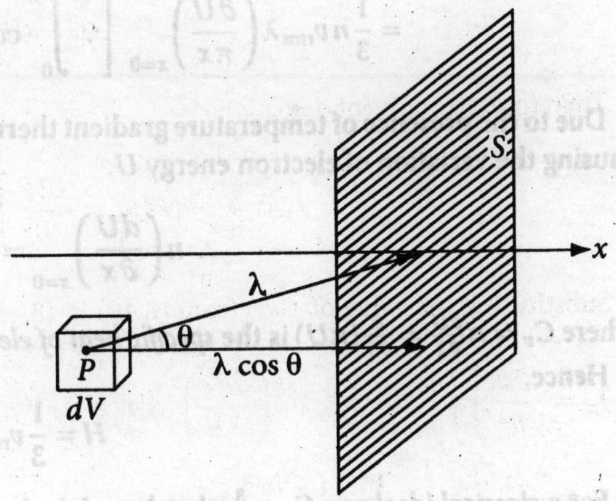

Fig. 8.3: Flux H crossing plane S.

To calculate the flux H across a plane (S) perpendicular to the x-axis, we assume that the average component of electron velocity in the x-direction is zero at the scattering centre but the velocity of the electron becomes v_{rms} after it has traversed a distance λ, the mean free path, at which it is scattered again by an ion of the crystal lattice.

We consider an element of volume dV containing the scattering centre at P. The amount of flux crossing the plane between solid angle Ω and $\Omega + d\Omega$ is given by

$$dH = Uf(\Omega)\,d\Omega, \qquad (8.54)$$

where U represents the energy carried by a single electron. $f(\Omega)\,d\Omega$ is the number of electron per unit area per unit time in the incremental solid angle $d\Omega$.

If the velocity distribution in dV be equally probable in all possible directions, then the tip of the velocity vector will lie on the surface of the sphere with centre at P. The number of electron crossing the plane S per unit area per unit time along the normal direction is $n v_{rms} \cos\theta$, where n is the number of electrons per unit volume. Then

$$f(\Omega)\, d\Omega = n v_{rms} \cos\theta \frac{d\Omega}{4\pi} = \frac{1}{2} n v_{rms} \cos\theta \sin\theta\, d\theta. \tag{8.55}$$

Now

$$U = U(0) + \lambda \cos\theta \left(\frac{\partial U}{\partial x}\right)_{x=0} + \dots \text{ [by Taylor's theorem]}. \tag{8.56}$$

If $\frac{\partial U}{\partial x}$ is small over a distance comparable with λ, then the series in (8.56) will be convergent and we may neglect the terms after the second. So

$$U = U(0) + \lambda \cos\theta \left(\frac{\partial U}{\partial x}\right)_{x=0}. \tag{8.57}$$

Now substituting the value of $f(\Omega)\, d\Omega$ and U from (8.56) and (8.57) in (8.54) and then integrating, we get

$$H = \int dH = \int_0^{\pi} \left[U(0) + \lambda \cos\theta \left(\frac{\partial U}{\partial x}\right)_{x=0}\right] \times \left[\frac{1}{2} n v_{rms} \cos\theta \sin\theta\right] d\theta$$

$$= \frac{1}{2} n v_{rms} \lambda \left(\frac{\partial U}{\partial x}\right)_{x=0} \int_0^{\pi} \cos^2\theta \sin\theta\, d\theta$$

$$= \frac{1}{3} n v_{rms} \lambda \left(\frac{\partial U}{\pi x}\right)_{x=0} \left[\because \int_0^{\pi} \cos^2\theta \sin\theta\, d\theta = \frac{2}{3}\right]. \tag{8.58}$$

Due to the presence of temperature gradient thermal diffusion of electrons along x-direction takes place causing the variation of electron energy U.

$$\therefore\ n\left(\frac{dU}{\partial x}\right)_{x=0} = n \frac{\partial U}{\partial T} \frac{\partial T}{\partial x} = C_v \frac{\partial T}{\partial x}, \tag{8.59}$$

where $C_v = n\frac{\partial U}{\partial T} = \frac{\partial}{\partial T}(nU)$ is the *specific heat of electron gas*.

Hence,

$$H = \frac{1}{3} v_{rms} \lambda C_v \frac{dT}{dx}. \tag{8.60}$$

For a *classical ideal gas*, $C_v = \frac{3}{2} nk$, where k is the *Boltzmann constant*.

Hence, from (8.53), *thermal conductivity*

$$K = \frac{H}{\left(\frac{dT}{dx}\right)} = \frac{1}{3} v_{rms} \lambda \times \frac{3}{2} nk \text{ or } K = \frac{1}{2} nk v_{rms} \lambda. \tag{8.61}$$

8.6 Wiedemann-Franz Law

Wiedemann and Franz in 1853 found that good conductors of electricity are also the good conductors of heat because both of these phenomena owe their origin to the common cause, viz., the electrons.

The states that, *at a particular temperature the ratio of thermal conductivity K to the electrical conductivity σ is the same for all metals, i.e.,*

$$\frac{K}{\sigma T} = \text{constant}. \tag{8.62}$$

This is known as *Wiedemann-Franz law.*

Now to obtain the relation, we know that the *electrical conductivity* of a metal at temperature T is given by

$$\sigma = \frac{ne^2\lambda}{\sqrt{3mkT}}, \tag{8.63}$$

where e and m are respectively the charge and mass of an electron, λ the mean free path, n the density of electrons and k the Boltzmann constant.

Again, the *thermal conductivity* K is given by

$$K = \tfrac{1}{3} C_v \lambda v_{\text{rms}}, \tag{8.64}$$

where $C_v = \frac{3}{2} Kn$, the *specific heat of electron gas.*

Hence, the ratio of K to σ is

$$\frac{K}{\sigma} = \frac{1}{3} C_v \lambda v_{\text{rms}} \times \frac{\sqrt{3mkT}}{ne^2\lambda} = \frac{1}{3}\left(\frac{3}{2}kn\right)\sqrt{\frac{3kT}{m}}\frac{\sqrt{3mkT}}{ne^2} = \left(\frac{3}{2}\frac{k^2}{n^2}\right)T \tag{8.65}$$

$$\text{or } \frac{K}{\sigma} = \left(\frac{3}{2}\frac{k^2}{e^2}\right)T = LT \quad \left[\because C_v = \frac{3}{2}kn;\ v_{\text{rms}} = \sqrt{\frac{3kT}{m}}\right] \tag{8.66}$$

where $L = \left(\frac{3}{2}\frac{k^2}{e^2}\right) = 1.12 \times 10^{-8}\ \text{W}\Omega/\text{K}^2$.

The above value of L largely deviates from the experimental value. Sommerfeld using *FD-statistics* to the free electron gas calculated the value of C_v as

$$C_v = \frac{n\pi^2 k^2 T}{2E_F}, \tag{8.67}$$

where

$$E_F = \tfrac{1}{2} V_F^2. \tag{8.68}$$

Then,

$$K = \frac{1}{6}\frac{n\pi^2 k^2 T \lambda V_F}{E_F}. \tag{8.69}$$

So

$$\frac{K}{\sigma} = \frac{1}{6}\left[\frac{n\pi^2 k^2 \lambda V_F}{\frac{1}{2} m V_F^2}\right] T \left[\frac{e^2}{3\pi^2}\frac{3\pi^2 n\lambda}{m v_F}\right] \text{ or } \frac{K}{\sigma T} = \frac{1}{3}\left(\frac{\pi^2 k^2}{e^2}\right) = L = 2.45 \times 10^{-8}\ \text{W}\Omega/\text{K}^2, \tag{8.70}$$

where $L = \frac{1}{3}\frac{\pi^2 k^2}{e^2}$ is known as *Lorentz number.*

(8.70) is in good agreement with the experimental results. Table 8.1 shows $\frac{K}{\sigma T}$ values for some of the metals.

Table 8.1

Metals	K(W/cm-K)	σ(℧/cm)	$\frac{K}{\sigma T}\times 10^8$ $(W\Omega/K^2)$
Li	0.72	1.17	2.22
Be	2.3	3.57	2.36
Fe	0.80	1.12	2.61
Ag	4.18	6.62	2.31
Cu	3.85	6.41	2.20

8.7 Merits and Demerits of Free Electron Theory

Merits

1. The theory proves the validity of *Ohm's law.*
2. It can explain the phenomena of *electrical and thermal conduction* in metals and verifies *Wiedemann-Franz law.*
3. It explains the high *lustre* and complete *opacity* of metals.

Demerits

1. Electrical conductivity of semiconductor or insulator could not be explained by this theory.
2. $\frac{k}{\sigma T}$ is constant in view of classical free electron theory but it is not so at low temperature.
3. The theory predicts that the resistivity ρ is proportional to $\sqrt{T}$ but actually ρ varies as T.
4. The theory fails to explain *heat capacity* and *paramagnetic susceptibility* of the conduction electrons.
5. The *Classical theory fails to account for the long mean free path* (more than one centimetre) at low temperature.

Temperature Dependence of Resistivity: If an electric field E is applied to the metal the electron will acquire an acceleration, $a = \frac{eE}{m}$, where e and m are the charge and mass of the electron. If C be the random velocity (rms velocity) and λ the mean free path, then the velocity acquired by the electron just before the next collision is

$$u = at = \frac{eE}{m}\left(\frac{\lambda}{C}\right),$$

where $t = \lambda/C$.

The *average drift velocity* of the electron is

$$v_d = \frac{u}{2} = \frac{eE\lambda}{2mC} = \frac{eE\lambda}{2m\times\sqrt{\frac{3kT}{m}}} \quad \left[\because C = \sqrt{\frac{3kT}{m}}\right]$$

$$= \frac{eE\lambda}{\sqrt{12mkT}}. \tag{8.71}$$

Using this value of v_d in (8.14), the resistivity

$$\rho = \frac{1}{ne\mu} = \frac{1}{ne\left(\frac{v_d}{E}\right)} = \frac{E}{ne\times\frac{eE\lambda}{\sqrt{12mkT}}} \text{ or } \rho = \frac{\sqrt{12mkT}}{ne^2\lambda}. \tag{8.72}$$

Since m, k, n, e, λ are constant, $\rho \propto \sqrt{T}$. Thus, the *resistivity of a conductor* is proportional to the square of the absolute temperature (T). But experimental result shows that $\rho \propto T$.

Molar Specific Heat: The KE of one gm-mol of monovalent metal is

$$U = \frac{1}{2}mC^2N_A = \frac{1}{2}mN_A\left(\frac{3kT}{m}\right) = \frac{3}{2}RT \; [\because N_A k = R],$$

where N_A is the *Avogadro number* and R, the *universal gas constant*.

$\therefore$ the molar specific heat of the electron is

$$C_v = \frac{dU}{dT} = 1.5R \text{ J/mol/K}. \tag{8.73}$$

The value of $C_v = 1.15R$ is about *ten times* the experimental value. *This disagreement is due to the wrong assumption that all the free electrons contribute to the specific heat.*

8.8 Quantum Theory of Free Electrons

We consider that the free electrons move freely in a solid in a constant potential V. Assuming $V = 0$, the *Schrödinger equation* for a free electron of mass m is

$$-\frac{\hbar^2}{2m}\nabla^2\psi = E\psi,$$

where E is the total energy of the electron.

The plane wave solution of the above equation is

$$\psi = \psi_0 e^{i\vec{k}\cdot\vec{r}}, \tag{8.74}$$

where $k = \frac{2\pi}{\lambda}$ is the *wave vector* and $r^2 = x^2 + y^2 + z^2$.

$$\therefore \quad \frac{\partial^2\psi}{\partial x^2} = -k_x^2\psi, \quad \frac{\partial^2\psi}{\partial y^2} = -k_y^2\psi, \quad \frac{\partial^2\psi}{\partial z^2} = -k_z^2\psi.$$

So

$$\nabla^2\psi = \frac{\partial^2\psi}{\partial x^2} + \frac{\partial^2\psi}{\partial y^2} + \frac{\partial^2\psi}{\partial z^2} = -\left(k_x^2 + k_y^2 + k_z^2\right)\psi = -k^2\psi. \; \left[\because k_x^2 + k_y^2 + k_z^2 = k^2\right] \tag{8.75}$$

$$\therefore -\frac{\hbar^2}{2m}\nabla^2\psi = -\frac{\hbar^2}{2m}(-k^2\psi) = E\psi, \quad \left[\because \frac{\hbar^2 k^2}{2m} = E\right]$$

$$\text{or } E = \frac{\hbar^2 k^2}{2m} = \left(\frac{h}{2\pi}\right)^2\left(\frac{2\pi}{\lambda}\right)^2 \cdot \frac{1}{2m} = \frac{h^2}{2m\lambda^2} = \frac{h^2 p^2}{2mh^2}, \quad \left[\because \lambda = \frac{h}{p}\right]$$

$$\therefore E = \frac{p^2}{2m} \tag{8.76}$$

(8.76) is the energy of a free particle. It is evident from Fig. (8.4) that the energy E is continuous.

If L be the length of the solid, then applying the boundary condition $k = \frac{2\pi n}{L}$, we get

$$E_n = \frac{\hbar^2 k^2}{2m} = \frac{n^2 h^2}{8mL^2} \tag{8.77}$$

For $L = 1$ cm, the permitted energy levels differ by about 10^{-19} eV which is very small, the energy levels are, therefore, said to form a *quasi-continuous band*.

8.8.1 Electrical Conductivity

If an external field E is applied to the metal, the force exerted on a free electron is $-eE$. If $\frac{dp}{dt}$ represents the rate of change of momentum, then

$$-eE = \frac{dp}{dt}.$$

But according to *de Broglie's hypothesis,*

$$p = \frac{h}{\lambda} = \frac{h}{2\pi}\frac{2\pi}{\lambda} = \hbar k.$$

$$\therefore\ -eE = \frac{d}{dt}(\hbar k) = \hbar\frac{dk}{dt} \text{ or } dk = -\frac{eE}{\hbar}dt. \tag{8.78}$$

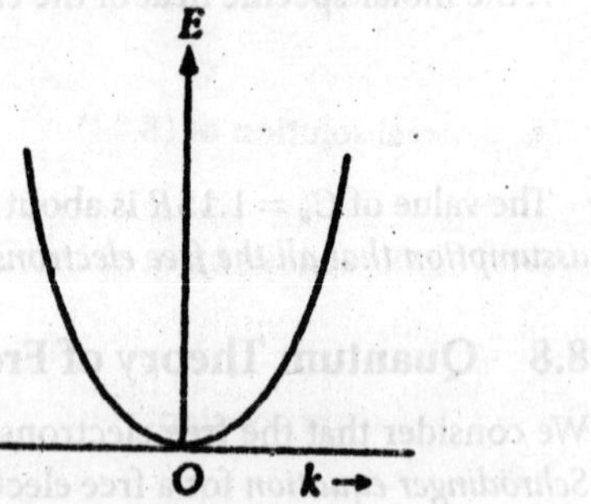

Fig. 8.4: E as a function of k.

(8.78) shows that, the origin of k-space moves through a distance dk in time dt in presence of the external field E. Due to collisions of the free electrons with the imperfections, the displacement of k-space becomes steady ($= \Delta k$) and then τ is the *average collision time.*

$$\therefore\ \Delta k = -\frac{eE\tau}{\hbar}.$$

Therefore, the change in velocity is

$$\Delta v = \frac{\hbar}{m}\Delta k = -\left(\frac{\hbar}{m}\right)\left(\frac{eE\tau}{\hbar}\right) = -\frac{eE\tau}{m}. \tag{8.79}$$

Therefore, the *current density,*

$$J = n(-e)\Delta v,\ (n = \text{electron density})$$

$$= \frac{ne^2E\tau}{m} = \sigma E. \tag{8.80}$$

(8.80) is the *Ohm's law,* where

$$\sigma = \frac{ne^2\tau}{m} \tag{8.81}$$

is the *electrical conductivity* of the metal.

8.9 Free Electron Gas in One-dimensional Box

We consider an electron of mass m confined in a one-dimensional line in a crystal of length L bounded by infinite potential barriers at $x = 0$ and $x = L$ as shown in Fig. 8.5. The problem is equivalent to an electron moving in one-dimensional potential box, the PE within the box being zero.

So

$$V(x) = \begin{cases} 0 & \text{for } 0 < x < L \\ \infty & \text{for } x \le 0 \text{ and } x \ge L \end{cases} \tag{8.82}$$

Wave Function: The *Schrödinger time-independent equation* for the electron is

$$\frac{d^2\psi_n}{dx^2} + \frac{2m}{\hbar^2}(E_n - V)\psi_n = 0, \tag{8.83}$$

where ψ_n is the *wave function* for the electron in the nth state of energy E_n. Since $V=0$,

$$\frac{d^2\psi_n}{dx^2}+\frac{2mE_n}{\hbar^2}\psi_n=0. \tag{8.84}$$

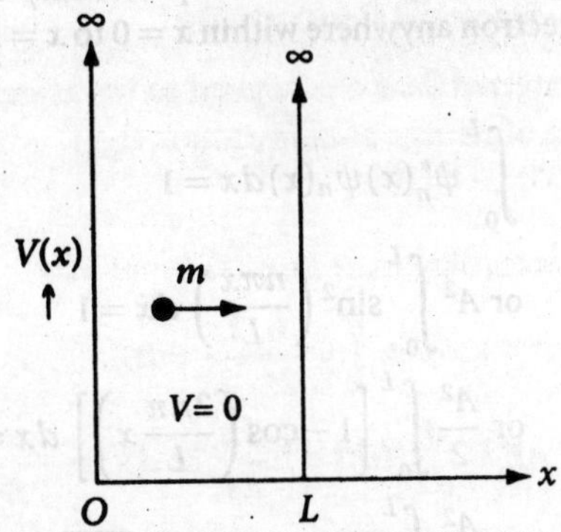

Fig. 8.5: An electron in one-dimensional potential box.

The general solution of (8.84) is

$$\psi_n(x)=A\sin kx+B\cos kx, \tag{8.85}$$

where A and B are constants and $k=\sqrt{\dfrac{2mE_n}{\hbar^2}}$.

Applying the *boundary condition*, $\psi_n(0)=0$ and $\psi_n(L)=0$, we get $B=0$.

$$\therefore \psi_n(x)=A\sin kx. \tag{8.86}$$

Again, since $\psi_n(L)=0$,

$$\therefore \sin kL=0,\ (\because A \text{ cannot be zero})$$

$$\text{or } k=\frac{n\pi}{L}, \tag{8.87}$$

where $n=1,2,3,\ldots$.

Therefore, the allowed wave function is

$$\psi_n(x)=A\sin\left(\frac{n\pi}{L}\right)x. \tag{8.88}$$

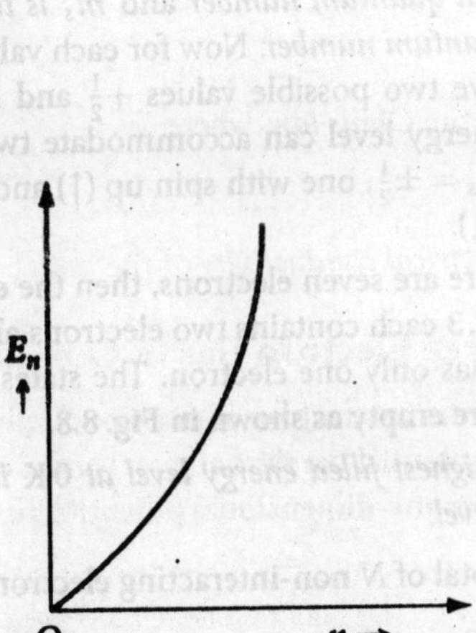

Fig. 8.6: Variation of E_n with n.

From (8.87),

$$\frac{2mE_n}{\hbar^2}=\frac{n^2\pi^2}{L^2}$$

$$\text{or } E_n=\frac{\hbar^2}{2m}\frac{n^2\pi^2}{L^2}=\frac{n^2h^2}{8mL^2}. \tag{8.89}$$

Thus, $E_n\propto n^2$. The plot of E_n against n is shown in Fig. 8.6.

The allowed wave function $\psi_n(x)$ and energy E_n exist for integral values of n, called the *quantum number*. Thus, the energy spectrum of the electron is *discrete* in nature.

Evaluation of A: Since the probability of finding the electron anywhere within $x = 0$ to $x = L$ is unity,

$$\therefore \int_0^L \psi_n^*(x)\psi_n(x)\,dx = 1$$

$$\text{or } A^2 \int_0^L \sin^2\left(\frac{n\pi x}{L}\right) dx = 1$$

$$\text{or } \frac{A^2}{2}\int_0^L \left[1-\cos\left(\frac{2n\pi}{L}x\right)\right] dx = 1$$

$$\text{or } \frac{A^2}{2}\int_0^L dx = 1 \text{ or } A = \sqrt{2/L}.$$

$$\therefore \psi_n(x) = \sqrt{2/L}\sin\left(\frac{n\pi x}{L}\right). \quad (8.90)$$

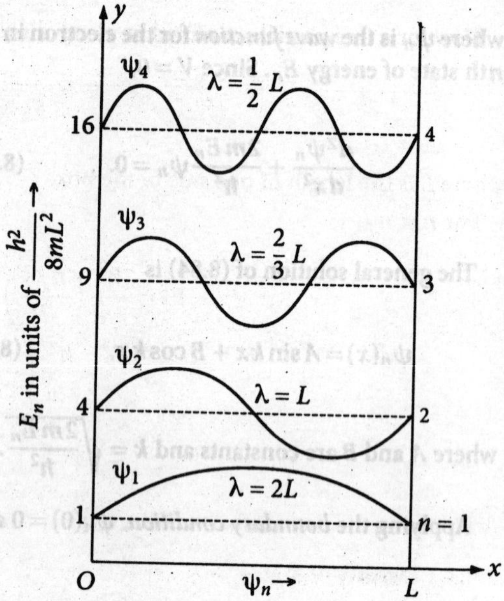

Fig. 8.7: ψ_n and E_n values of an electron in one-dimensional box.

The energy levels and the wave functions of the electron for $n = 1, 2, 3, 4, \ldots$ are shown in Fig. 8.7.

8.9.1 Fermi Level

Suppose we have N electrons to be accommodated on the line following *Pauli's exclusion principle.* According to this principle each quantum state can be occupied by at the most one electron.

In a solid a conduction electron is characterised by *two quantum numbers, n* and m_s, where n is the *principal quantum number* and m_s *is the magnetic spin quantum number.* Now for each value of n, m_s can have two possible values $+\frac{1}{2}$ and $-\frac{1}{2}$. Hence, each energy level can accommodate two electrons with $m_s = \pm\frac{1}{2}$, one with spin up (↑) and other spin down (↓).

If there are seven electrons, then the energy state $n = 1, 2, 3$ each contains two electrons and the state $n = 4$ has only one electron. The states $n = 5$ and above are empty as shown in Fig. 8.8.

The highest filled energy level at 0 K *is called the Fermi level.*

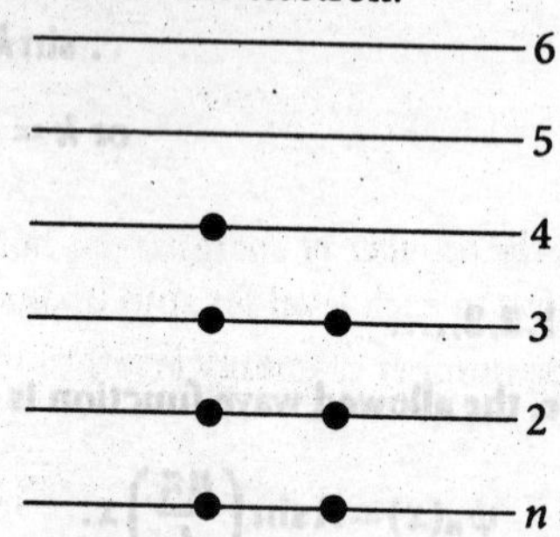

Fig. 8.8: Electron distribution in energy levels.

For total of N non-interacting electrons, the number of filled levels is $N/2$.

8.9.2 Fermi Energy

It is defined as the energy corresponding to the highest filled level. Fermi energy E_f depends on the temperature of the metal.

If N be the total number of electrons to be accommodated on a line, then for even N, $2n_f = N$, where n_f is the *principal quantum number* of the Fermi level.

Now for a free particle confined in a one-dimensional box, the energy in the nth quantum state is

$$E_n = \frac{\hbar^2}{2m}\left(\frac{n\pi}{L}\right)^2,$$

where L is the length of one side of the box.

For $n = n_f$,

$$E_f = \frac{\hbar^2}{2m}\left(\frac{n_f\pi}{L}\right)^2 = \frac{\hbar^2}{2m}\left(\frac{N\pi}{2L}\right)^2 \quad [\because n_f = N/2]$$

$$= \frac{1}{4\pi^2}\frac{h^2}{2m}\left(\frac{N}{2L}\right)^2\pi^2$$

$$\text{or } E_f = \frac{h^2}{2m}\left(\frac{N}{4L}\right)^2. \tag{8.91}$$

N/L represents the number of electrons per unit length.

8.9.3 Density of States

Density of states is the number of electronic states per unit energy range and is represented by $Z(E)$.

We have energy of the electron in the nth level

$$E = \frac{\hbar^2}{2m}\left(\frac{n\pi}{L}\right)^2 = \frac{\hbar^2}{2m}\left(\frac{\pi}{L}\right)^2 n^2.$$

$$\therefore \frac{dE}{dn} = \frac{\hbar^2}{2m}\left(\frac{\pi}{L}\right)^2 2n = \frac{h^2}{4\pi^2}\frac{1}{2m}\frac{\pi^2}{L^2}2n = \frac{h^2}{m}\frac{n}{4L^2}.$$

$$\therefore \frac{dn}{dE} = \frac{4m\,L^2}{h^2}\frac{1}{n}. \tag{8.92}$$

(8.92) gives *the number of energy levels* in the energy range E and $E + dE$. But there are two energy states corresponding to each level for spin up and spin down.

Hence, the number of energy levels in the energy range E and $E + dE$ is

$$Z(E) = 2\frac{dn}{dE} = \frac{8mL^2}{h^2}\cdot\frac{1}{n^2}. \tag{8.93}$$

Again,

$$\because E = \frac{h^2}{8m}\frac{n^2}{L^2},$$

$$\therefore n^2 = \frac{8mL^2E}{h^2} \text{ or } n = \sqrt{8mE}\frac{L}{h}$$

$$\text{or } \frac{1}{n} = \frac{h}{L\sqrt{8mE}}. \tag{8.94}$$

From (8.93) and (8.94),

$$Z(E) = \frac{8mL^2}{h^2} \times \frac{h}{L\sqrt{8mE}} = \frac{4L}{h}\left(\frac{m}{2E}\right)^{1/2}. \tag{8.95}$$

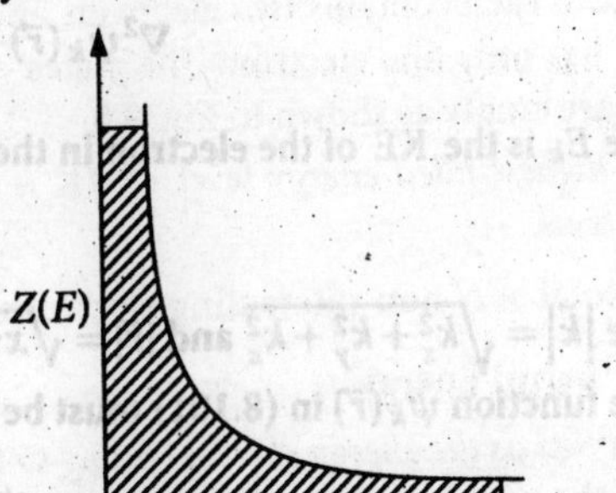

Fig. 8.9: Plot of $Z(E)$ versus E_f.

(8.95) represents the *density of states* in one-dimensional case. Fig. 8.9 shows that levels are filled from $E=0$ to $E=E_f$ and beyond E_f all the levels are vacant. Thus, *Fermi level* divides the filled and unfilled levels in the metals.

8.9.4 Average KE in the Ground State

The total energy E_0 of all the N electrons in ground state is given by

$$E_0 = 2\sum_{n=1}^{N/2} E_n.$$

The factor 2 is due to the fact that each level contains 2 electrons and for N electrons the number of levels is $N/2$.

$$\text{or } E_0 = 2\frac{\hbar^2}{2m}\left(\frac{\pi}{L}\right)^2 \sum_{n=1}^{N/2} n^2. \tag{8.96}$$

But

$$\sum_{n=1}^{s} n^2 = \frac{1}{6}s\left(2s^2+3s+1\right) \simeq \frac{1}{3}s^3, \text{ for } s \geq 1.$$

$$\therefore \sum_{n=1}^{N/2} n^2 \simeq \frac{1}{3}\left(\frac{N}{2}\right)^3. \tag{8.97}$$

$\therefore$ (8.96) becomes

$$E_0 = 2\frac{\hbar^2}{2m}\left(\frac{\pi}{L}\right)^2 \times \frac{1}{3}\left(\frac{N}{2}\right)^3 = \frac{1}{3}\frac{\hbar^2}{2m}\left(\frac{N\pi}{2L}\right)^2 N = \frac{1}{3}NE_f \left[\because E_f = \frac{\hbar^2}{2m}\left(\frac{N\pi}{2L}\right)^2\right]. \tag{8.98}$$

$\therefore$ the average KE in the ground state of the electron confined in a one-dimensional potential box is

$$\langle E_0 \rangle = \frac{E_0}{N} = \frac{1}{3}E_f. \tag{8.99}$$

8.10 Free Electron Gas in Three Dimensions

Considering the metal as a cubical electron gas container of side L, the ***three-dimensional Schrödinger equation*** for the electron of mass m is

$$\nabla^2\psi_k(\vec{r}) + \frac{2m}{\hbar^2}E_k\psi_k(\vec{r}) = 0, \ [\because \text{PE } V=0] \tag{8.100}$$

where E_k is the KE of the electron in the k state. The solution of (8.100) is

$$\psi_k(\vec{r}) = Ae^{i\vec{k}\cdot\vec{r}}, \tag{8.101}$$

where $\left|\vec{k}\right| = \sqrt{k_x^2+k_y^2+k_z^2}$ and $|\vec{r}| = \sqrt{x^2+y^2+z^2}$.

The function $\psi_k(\vec{r})$ in (8.101) must be periodic in x, y, z with period L such that

$$\left.\begin{aligned} \psi(x+L,y,z) &= \psi(x,y,z) \\ \psi(x,y+L,z) &= \psi(x,y,z) \\ \psi(x,y,z+L) &= \psi(x,y,z) \end{aligned}\right\}. \tag{8.102}$$

Applying the first boundary condition to the wave function $\psi_k(\vec{r})$, we get

$$e^{i[k_x(x+L)+k_y y+k_z z]} = e^{i[k_x x+k_y y+k_z z]}.$$

$$\therefore\ e^{iK_x L} = 1 \text{ or } k_x = 0, \pm\frac{2\pi}{L}, \pm\frac{4\pi}{L}, \ldots, \frac{2n\pi}{L}. \qquad (8.103)$$

Similar results follow for k_y and k_z. If m_s be the *spin magnetic quantum number* of the electron, then its state can be represented by a set of *four* quantum numbers k_x, k_y, k_z and m_s.

The value of A in (8.101) can be obtained from the condition of normalisation, i.e.,

$$\int_0^V \psi_k^*(\vec{r})\,\psi_k(\vec{r})\,dV = 1 \text{ or } \int_0^L\int_0^L\int_0^L A^2 e^{-i(\vec{k}\cdot\vec{r})}\cdot e^{i(\vec{k}\cdot\vec{r})}\,dx\,dy\,dz = 1 \text{ or } A^2L^3 = 1.$$

$$\therefore\ A = \left(\frac{1}{L^3}\right)^{1/2} = \left(\frac{1}{V}\right)^{1/2}.$$

$\therefore$ the normalised wave function for the electron in the kth state is

$$\psi_k(\vec{r}) = \left(\frac{1}{V}\right)^{1/2} e^{i(\vec{k}\cdot\vec{r})}. \qquad (8.104)$$

8.10.1 Fermi Energy and Fermi Sphere

We have

$$k^2 = \frac{2mE}{\hbar^2} \text{ or } E = E_f = \frac{\hbar^2 k^2}{2m}. \qquad (8.105)$$

E_f is the *Fermi energy* of the electron.

Fermi Sphere: A sphere of radius k_f in the k-space containing points representing the occupied states of a system of N free electrons in the ground state is called a *Fermi Sphere*.

The energy at the surface of the sphere is Fermi energy E_f given by

$$E_f = \frac{\hbar^2 k_f^2}{2m}.$$

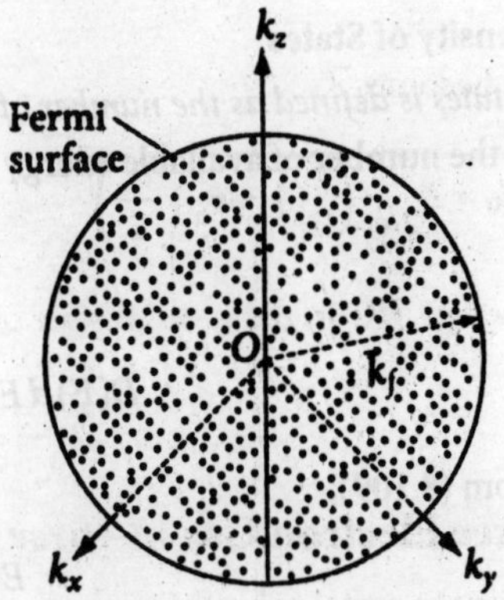

Fig. 8.10: The Fermi sphere.

The volume of the Fermi sphere is $\frac{4}{3}\pi k_f^3$ and the total number of electron state is

$$2\frac{\frac{4\pi}{3}k_f^3}{\left(\frac{2\pi}{L}\right)^3} = \frac{L^3k_f^3}{3\pi^2} = \frac{Vk_f^3}{3\pi^2} = N. \qquad (8.106)$$

The factor 2 comes because there are two allowed values of m_s for each value of k.

From (8.106),

$$k_f = \left(\frac{3\pi^2 N}{V}\right)^{1/3}. \qquad (8.107)$$

So, k_f depends on the concentration (N/V) of the electron.

Then the *Fermi energy* E_{f_0} at $T=0$ is expressed as

$$E_{f_0}=\frac{\hbar^2}{2m}k_f^2=\frac{\hbar^2}{2m}\left(\frac{3\pi^2N}{V}\right)^{2/3}=\frac{\hbar^2}{2m}(3\pi^2 n)^{2/3},\tag{8.108}$$

where $N/V=n$.

For $kT \ll E_f$, the variation of E_f with T is given by *Sommerfeld equation*

$$E_f(T)=E_{f_0}\left[1-\frac{\pi^2}{12}\frac{(kT)^2}{E_{f_0}^2}\right].\tag{8.109}$$

Thus, E_{f_0} depends on electron concentration $(N/V)=n$ as well as mass m of the electron, Therefore, E_{f_0} remains the same if two identical metals are joined together. Further, the electrons have some definite energy even at $T=0$ as $E_{f_0}\neq 0$.

8.10.2 Electron Velocity at Fermi Level

The velocity of an electron at the Fermi surface is given by

$$mv_f=\hbar k_f.$$

$$\therefore\ v_f=\frac{\hbar k_f}{m}=\frac{\hbar}{m}\left(\frac{3\pi^2N}{V}\right)^{1/3}\quad \text{[using (8.107)]}.\tag{8.110}$$

8.10.3 Density of States

Density of states is defined as the number of available energy states per unit energy range.

If dN be the number of available energy states between energy E to $E+dE$, then the density of states is

$$D(E)=\frac{dN}{dE}.\tag{8.111}$$

$$\therefore\ D(E)\,dE=dN \text{ and } \int_0^{E_f}D(E)\,dE=N.\tag{8.112}$$

Again, from (8.109),

$$E_f=\frac{\hbar^2}{2m}\left(3\pi^2\frac{N}{V}\right)^{2/3}.$$

$$\therefore\ N^{2/3}=\left(\frac{V}{3\pi^2}\right)^{2/3}\frac{2m}{\hbar^2}E_f \text{ or } N=\left(\frac{V}{3\pi^2}\right)\left(\frac{2m}{\hbar^2}\right)^{3/2}E_f^{3/2}.\tag{8.113}$$

From (8.112) and (8.113), we get

$$\int_0^{E_f}D(E)\,dE=\frac{V}{3\pi^2}\left(\frac{2mE_f}{\hbar^2}\right)^{3/2}.$$

Expressing the above integral in indefinite form,

$$\int D(E)\,dE=\frac{V}{3\pi^2}\left(\frac{2mE}{\hbar^2}\right)^{3/2}=\int\frac{V}{3\pi^2}\left(\frac{2m}{\hbar^2}\right)^{3/2}E^{1/2}\,dE$$

$$\text{or } D(E)=\frac{V}{3\pi^2}\left(\frac{2m}{\hbar^2}\right)^{3/2}E^{1/2}.\tag{8.114}$$

The plot of $D(E)$ against E is a parabola as shown in Fig. 8.11.

The density of states at the *Fermi surface* is obtained by putting $E = E_f$ in (8.114), i.e.,

$$D(E_f) = \frac{V}{3\pi^2}\left(\frac{2m}{\hbar^2}\right)^{3/2} E_f^{1/2}. \quad (8.115)$$

(8.115) is valid at 0 K.

If $f(E)$ be the *Fermi function*, then the density of filled energy states is given by $N(E) = D(E)f(E)$.

Now

$$f(E) = \frac{1}{e^{(E-E_f)/kT}+1},$$

where k is the *Boltzmann constant*.

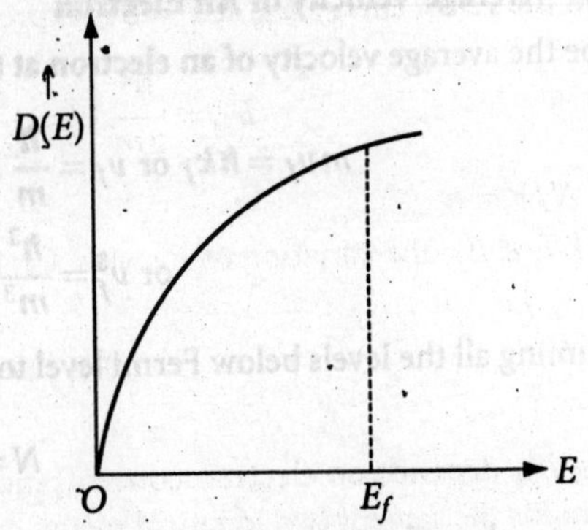

Fig. 8.11: Plot of $D(E)$ against E.

At $T = 0$, $f(E) = 1$ for $E < E_f$, this implies that all energy levels for $E < E_f$ are occupied.

For $E > E_f$, $f(E) = 0$, i.e., all energy levels for $E > E_f$ are vacant.

∴ the number of filled states between E and $E + dE$ at temperature T is

$$N(E)\,dE = D(E)f(E)\,dE = \left(\frac{V}{3\pi^2}\right)\left(\frac{2m}{\hbar^2}\right)^{3/2} \frac{E^{1/2}\,dE}{e^{(E-E_f)/kT}+1}.$$

For $E > E_f$, $(E - E_f) \gg kT$, i.e., $(E - E_f)/kT \gg 1$.

$$\therefore N(E)\,dE = \frac{V}{3\pi^2}\left(\frac{2m}{\hbar^2}\right)^{3/2} \frac{E^{1/2}\,dE}{e^{(E-E_f)/kT}} \quad (8.116)$$

$$= \frac{\pi V}{3}\left(\frac{8m}{h^2}\right) \frac{E^{1/2}\,dE}{e^{(E-E_f)/kT}} \quad \left[\because \hbar = \frac{h}{2\pi}\right]. \quad (8.117)$$

8.10.4 Average KE at $T = 0$

The average KE of an electron at absolute zero is given by

$$\langle E_0\rangle = \frac{1}{N}\int_0^\infty E D(E) f(E)\,dE = \frac{V}{3\pi^2}\left(\frac{2m}{\hbar^2}\right)^{3/2} \frac{1}{N}\left[\int_0^{E_{f0}} E E^{1/2} f(E)\,dE + \int_{E_{f0}}^\infty E E^{1/2} f(E)\,dE\right].$$

[Using the value of $D(E)$ from (8.114)]

$$\text{or } \langle E_0\rangle = \frac{V}{3\pi^2}\left(\frac{2m}{\hbar^2}\right)^{3/2} \frac{1}{N}\int_0^{E_{f0}} E^{3/2}\,dE = \frac{2}{5}\frac{V}{3\pi^2}\left(\frac{2m}{\hbar^2}\right)^{3/2} \frac{1}{N} E_{f0}^{5/2}. \quad (8.118)$$

Using the value of N from (8.113) and remembering that $E_f = E_{f0}$ at $T = 0$, we get

$$\langle E_0\rangle = \frac{3}{5} E_{f0}. \quad (8.119)$$

Hence, the average KE of an electron at $T = 0$ is $(3/5)E_{f0}$.

8.10.5 Average Velocity of An Electron

If v_f be the average velocity of an electron at the *Fermi surface*, then

$$mv_f = \hbar k_f \text{ or } v_f = \frac{\hbar}{m}k_f = \frac{\hbar}{m}\left(3\pi^2\frac{N}{V}\right)^{1/3} \quad \left[\because k_f = 3\pi^2\frac{N}{V}\right]$$

$$\text{or } v_f^3 = \frac{\hbar^3}{m^3}3\pi^2\frac{N}{V} \text{ or } N = \frac{V}{3\pi^2}\left(\frac{m}{\hbar}\right)^3 v_f^3.$$

Assuming all the levels below Fermi level to be occupied we can in general write,

$$N = \frac{V}{3\pi^2}\left(\frac{m}{\hbar}\right)^3 v^3. \tag{8.120}$$

$\therefore$ the number of energy states between v and $v + dv$ is

$$dN = \frac{V}{3\pi^2}\left(\frac{m}{\hbar}\right)^3 3v^2 dv.$$

So, the average speed of the electron at 0 K is given by

$$\langle v\rangle = \frac{1}{N}\int_0^{v_f} v\,dN = \frac{1}{N}\frac{V}{3\pi^2}\left(\frac{m}{\hbar}\right)^3\int_0^{v_f} 3v^2 v\,dv = \frac{1}{N}\frac{V}{3\pi^2}\left(\frac{m}{\hbar}\right)^3\left[\frac{3}{4}v^4\right]_0^{v_f} = \frac{1}{N}\frac{V}{3\pi^2}\left(\frac{m}{\hbar}\right)^3\frac{3}{4}v_f^4$$

$$\text{or } \langle v\rangle = \frac{3}{4}v_f, \quad \left[\because N = \frac{V}{3\pi^2}\left(\frac{m}{\hbar}\right)^3 v_f^3\right]. \tag{8.121}$$

8.10.6 Electronic Specific Heat

If E and E_0 be the internal energy of the solid at any temperature T and at absolute zero ($T = 0$), then it can be shown that

$$E = E_0 + \frac{\pi^2}{6}(kT)^2 g\left[E_{f_0}\right], \tag{8.122}$$

where

$$g\left[E_{f_0}\right] = \frac{3N}{2E_{f_0}} = \frac{3N}{2kT_f}, \tag{8.123}$$

where T_f is the *Fermi temperature* given by $kT_f = E_f(0)$.

Now substituting (8.122) in (8.123), we get

$$E = E_0 + \frac{\pi^2}{4}\frac{NkT^2}{T_f}. \tag{8.124}$$

$\therefore$ the electronic specific heat at constant volume is given by

$$C_v = \frac{dE}{dT} = \frac{\pi^2}{2}Nk\left(T/T_f\right). \tag{8.125}$$

If $E_{f_0} = 5$ eV and $T = 300$ K, then from (8.125), $C_v = Nk/40$, which is a small fraction of the classical value $1.5Nk$. The reason is that, *the thermal excitation of the free electron is possible if these find unoccupied states within a few times kT in energy. Electrons close to the Fermi level E_f can only be excited to such levels, whereas the electrons having much less energy than E_f, cannot be thermally excited. Hence, a small fraction of the total number of electrons can contribute to C_v. But in classical theory all the electrons are assumed to contribute to C_v.* This explains the small value of $C_v(= Nk/40)$ against $1.5Nk$ obtained from classical theory.

(8.125) shows that $C_v \propto T$. Also at low temperature *Debye theory* predicts that $C_v \propto T^3$. So, the *total specific heat* of metal at low temperature can be expressed as

$$C_v = AT + BT^3, \tag{8.126}$$

where A and B are constants

$$\text{or} \quad \frac{C_v}{T} = A + BT^2. \tag{8.127}$$

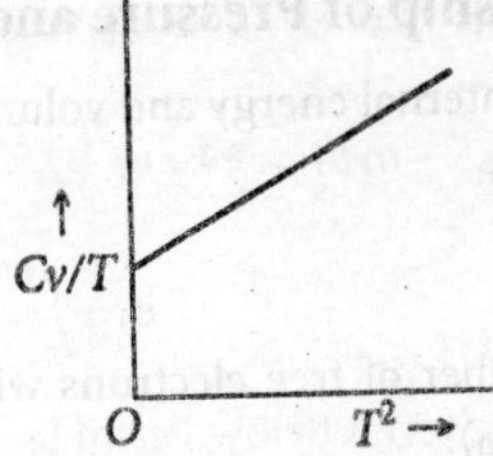

Fig. 8.12: Plot of Cv/T against T^2.

The variation of C_v/T against T^2 is shown in Fig. 8.12. This variation C_v with temperature is in agreement with experimental observation.

8.11 Fermi Distribution Function

To know how the electrons are distributed among different energy states, we consider the electrons in the metal as *electron gas* obeying Fermi Dirac (FD)-statics. According to FD-statistics the probability that an electron will occupy an energy level E is given by

$$f(E) = \frac{1}{e^{(E-E_f)/kT}+1}, \tag{8.128}$$

where E_f is the *Fermi level* which is assumed to be constant for a given metal and $F(E)$ is the *Fermi function*. From (8.128),

$$\left.\begin{array}{l} \text{at } T = 0\,\text{K for } E < E_f,\ f(E) = 1 \\ \qquad\qquad\quad\ E > E_f,\ f(E) = 0 \end{array}\right\} \tag{8.129}$$

This implies that at 0 K all the levels below E_f are filled and those above E_f are vacant as shown in Fig. 8.13. As T increases Fermi function plot shows a deviation as shown in Fig. 8.14. At $T > 0$, if $E = E_f$, then from (8.128),

$$f(E) = \tfrac{1}{2}. \tag{8.130}$$

Thus, *Fermi level is that energy state at which the probability of electron occupation is 1/2 at any temperature* $T > 0$ K. *It is also the level of maximum energy of the filled states at* 0 K.

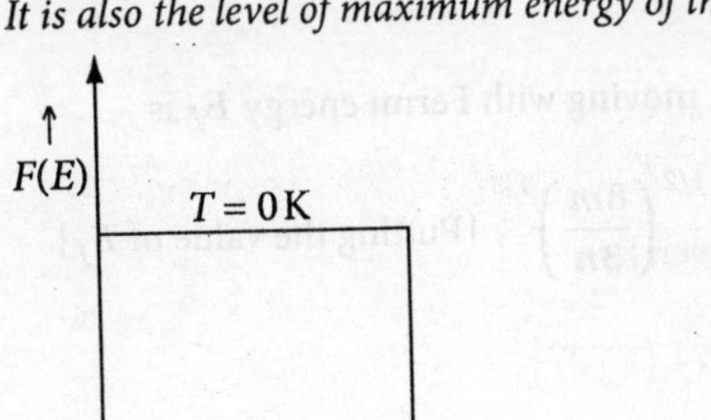

Fig. 8.13: Fermi distribution function at $T = 0$ K.

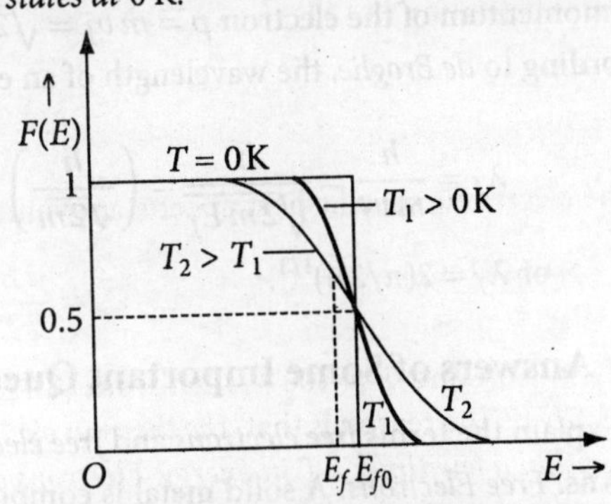

Fig. 8.14: Fermi distribution function at $T > 0$ K.

Fig. 8.14 shows that the position of Fermi level is not fixed but changes with temperature. An approximate relationship between $E_f(T)$ and $E_f(0)$ is given by (8.109).

8.12 Relationship of Pressure and Volume of the Electron Gas

If E and V be the internal energy and volume of a system of particles at pressure P, then from the *principle of thermodynamics,*

$$P = -\frac{\partial E}{\partial V}.$$

If N be the number of free electrons with average energy $\langle E_0 \rangle$, then the total internal energy of the electron gas is $N\langle E_0 \rangle$.

So

$$P = -N\frac{\partial}{\partial V}\langle E_0 \rangle.$$

But from (8.119),

$$P = -\frac{3}{5}N\frac{\partial E_{f_0}}{\partial V}.$$

Again,

$$E_{f_0} = \frac{\hbar^2}{2m}\left(\frac{3\pi^2 N}{V}\right)^{2/3}.$$

$$\therefore \quad \frac{\partial E_{f_0}}{\partial V} = -\frac{2}{3}\frac{\hbar^2}{2m}(3\pi^2 N)^{2/3}\left(\frac{1}{V}\right)^{5/3}.$$

$$\therefore \; P = \frac{2}{5}N\frac{\hbar^2}{2m}(3\pi^2 N)^{2/3}\left(\frac{1}{V}\right)^{5/3} \quad \text{or } P = \frac{2}{5}\frac{NE_{f_0}}{V}. \tag{8.131}$$

This gives the relation between pressure and volume of a free electron gas.

8.13 de Broglie Wavelength of an Electron having Energy E_f

If v_f and E_f be the velocity of an electron of mass m and Fermi energy respectively, then

$$\tfrac{1}{2}mv_f^2 = E_f \text{ or } m^2 v_f^2 = 2mE_f$$

or the momentum of the electron $p = mv_f = \sqrt{2mE_f}$.

According to *de Broglie,* the wavelength of an electron moving with Fermi energy E_f is

$$\lambda_f = \frac{h}{mv_f} = \frac{h}{\sqrt{2mE_f}} = \left(\frac{h}{\sqrt{2m}}\right)\left(\frac{2m}{\hbar^2}\right)^{1/2}\left(\frac{8m}{3n}\right)^{1/3} \quad \text{[Putting the value of } E_f]$$

or $\lambda_f = 2(\pi/3n)^{1/3}$. (8.132)

8.14 Answers of Some Important Questions

8.1 Explain the terms *free electrons* and *free electron gas.*

Ans. *Free Electrons*: A solid metal is composed of a number of atoms in which the outermost electrons (i.e., the valence electrons) are very loosely bound to the nucleus. At room temperature these electrons get free from the influence of the parent atom and move from atom to atom as well as in the volume of the metal at *random chaotic motion,* just like the ideal gas molecules in a container. Thus, the *valence electrons* of the metal become the conduction electrons. Since these electrons are free from mutual interaction, the PE is zero and whole energy is kinetic. With the application of an external electric field, these electrons move in a direction opposite to the field and give rise to electrical conduction.

Free Electron Gas: The free electron Fermi gas is a gas of free and non-interacting electrons obeying *Pauli's exclusion principle*. The free electron gas is negatively charged.

The free electrons move inside the metal very much like atoms or molecules of an ideal gas in a container. These electrons are referred to as a *free electron gas*. The free electron gas differs from an ordinary gas in the following respects:

(a) the free electron gas is negatively charged but the charge on ideal gas is zero

(b) the concentration of an electron gas is quite large compared to that of a perfect gas.

8.2 State the postulates of classical free electron theory of metals.

Ans. See Art. 8.2.

8.3 State some important phenomena which could be understood with the help of classical free electron theory of metal.

Ans. The classical free electron theory of metals successfully explains the following phenomena:

(a) The theory verifies Ohm's law in electrodynamics.

(b) It explains the electric and thermal conductivities of metals.

(c) It derives the relation between electric and thermal conductivities (i.e., Wiedemann-Franz law).

(d) It explains the optical properties like *complete opacity* and *high lustre* of metals.

8.4 Point out some of the *failures* of classical free electron theory of metals.

Ans. Some of the failures of classical free electron theory are as follows:

(a) Though classical theory correctly predicts the room temperature variation on electrical resistivity (ρ), it fails to account the variation of ρ with temperature at low temperature. The theory predicts that at low temperature $\rho \propto T^{-1}$ but actually $\rho \propto T$.

(b) The theory fails to explain *electronic heat capacity* of conduction electrons.

(c) It fails to explain the paramagnetic susceptibility of conduction electrons.

(d) $K/\sigma T$ is constant in view of classical electron theory but it is not so at low temperatures.

(e) The theory is unable to explain the existence of *long mean free path* (~ 10^8 to 10^9 interatomic spacings which is more than one centimetre) at low temperatures.

8.5 Indicate some of the successes of quantum theory of free electrons.

Ans. Following are some of the successes of *quantum theory of free electrons.*

(a) The theory can explain the electrical and thermal conductivities in metal, and semiconductors.

(b) It can explain the paramagnetic susceptibility and heat capacity due to free electrons in metals.

(c) It can explain the existence of long mean path at low temperatures.

(d) The theory can explain the various electron emission processes from metal like *Richardson's thermionic emission*, etc.

(e) It can explain the variation of $K/\sigma T$ at low temperature.

8.6 How can the difference between electronic specific heat of metal according to classical and quantum theory of free electron be explained? Which value is correct?

Ans. The molar electronic specific heat of metal according to classical theory is $C_v = \frac{3}{2} kT$, whereas that according to quantum theory is $C_v = \frac{3}{2} Nk\left(T/T_f\right) = (0.01)\frac{3}{2} kT$.

$\therefore$ for $T \to 0$, $C_v \to 0$.

The difference between the two values of C_v is due to the fact that *classical theory assumes, all the electrons contribute to specific heat but according to quantum theory only a small fraction contribute to it.*

Quantum theory suggests that only those electrons which are within an energy range kT below Fermi level contribute to C_v. This is because if the electron gas is heated at temperature T, these electrons gain energy $\sim kT$ and jump to the higher energy states. As the unfilled energy levels are not available to deep lying electrons, they cannot do so and cannot contribute to C_v.

The number of electrons which can contribute to C_v is of the order of

$$N\left(\frac{kT}{E_f}\right) = N\left(\frac{T}{T_f}\right)$$

$$\therefore\ C_v = \frac{3}{2}Nk = \frac{3}{2}\left(N\frac{T}{T_f}\right)k = \frac{3}{2}Nk\left(\frac{T}{T_f}\right).$$

For $T \to 0, C_v \to 0$.

Again, for $T = 100\,\text{K}$ and $T_f = 1000\,\text{K}, T/T_f = 0.1$.

$$\therefore\ C_v \simeq (0.01)\tfrac{3}{2}Nk.$$

This value of C_v obtained from quantum mechanical consideration agrees with experimental results and is a correct value of C_v.

8.7 What are Fermi sphere and Fermi surface?

Ans. The occupied states in the ground state of a system of N particles may be represented by points inside a sphere in the k-space [Fig. 8.10]. The energy corresponding to the surface of the sphere is the *Fermi energy* and the sphere is called the *Fermi sphere* and the surface of the sphere the *Fermi surface.* The volume of the Fermi sphere is $\left(\frac{4}{3}\pi\right)k_f^3$ and the total number of electrons in it is given by

$$2\times\frac{\frac{4}{3}\pi k_f^3}{\left(\frac{2\pi}{L}\right)^3} = \frac{L^3}{3\pi^2}k_f^3 = \frac{V}{3\pi^2}k_f^3 = N\ \text{(say)}.$$

The factor 2 is due to the fact that each state can accommodate two electrons with parallel and anti-parallel spin.

$$\therefore\ k_f = \left(\frac{3\pi^2 N}{V}\right)^{1/3}.$$

Thus, k_f depends on the concentration of electrons (N/V).

The Fermi energy is

$$E_f = \frac{\hbar^2}{2m}\left(3\pi^2\frac{N}{V}\right)^{2/3}$$

So, $E_f \propto (N/V)$. The Fermi energy does not change if identical metals are joined together.

8.8 Define density of states and write down its expression.

Ans. The number of energy levels (or orbitals) between unit energy interval is called the *density of states.* If dN be the number of energy levels in the energy interval E and $E + dE$, then density of states is given by

$$D(E) = \frac{dN}{dE}.$$

The expression for the density of states is given by

$$D(E) = \frac{V}{2\pi^2}\left(\frac{2m}{\hbar^2}\right)^{3/2} E^{1/2},$$

where V is the volume and m is the mass of an electron.

8.9 Find an expression for the bulk modulus of elasticity of an electron gas.

Ans. If an increase in pressure dP decreases the volume of the electron gas by an amount dV, then bulk modulus of elasticity of the gas is given by

$$B = -V\frac{dP}{dV}.$$

Since the pressure of the electron gas is given by

$$P = \frac{2}{5}N\frac{\hbar^2}{2m}\left(3\pi^2 N\right)^{2/3}\left(\frac{1}{V}\right)^{5/2}, \quad \text{[cf. (8.131)]}$$

$$\therefore \quad \frac{dP}{dV} = -\frac{5}{2}\left[\frac{2}{5}\frac{N\hbar^2}{2m}\left(3\pi^2 N\right)^{2/3}\right]\left(\frac{1}{V}\right)^{7/3} = -\frac{5}{3}\frac{P}{V}.$$

$$\therefore \; B = \frac{5}{3}P = \frac{2}{3}\frac{NE_{f_0}}{V} = \frac{10}{9}\frac{N\langle E_0\rangle}{V} \text{ or } B = \frac{10}{9}\frac{E}{V}\left[\because \langle E_0\rangle = \frac{3}{5}E_{f_0} \text{ and } P = \frac{2}{5}\frac{NE_{f_0}}{V}\right].$$

8.15 Solved Problems

8.1 Assuming that each silver atom provides one conduction electron, calculate (i) the density of electrons (ii) the mobility of electrons in silver. [Given: The density of silver = 10.5×10^3 kg/m³. The atomic weight of silver = 107.9. The conductivity of silver at 20°C is $6.8 \times 10^7\,\Omega^{-1}\,\text{m}^{-1}$].

Solution. (i) Volume of 1 gm-mole of silver $= \dfrac{M_A}{\rho_s}$.

$\therefore \left(\dfrac{M_A}{\rho_s}\right)$ m³ contains N_A electrons [N_A = Avogadro number].

Hence, 1m³ contains $\left[\dfrac{\rho_s N_A}{M_A}\right]$ electrons, i.e.,

$$n = \frac{\rho_s N_A}{M_A} = \frac{(10.5\times10^3)\times(6.02\times10^{26})}{107.9} = \mathbf{5.86\times10^{28}}.$$

(ii) We have electronic mobility $\mu = \frac{\sigma}{ne}$, where $\sigma = 6.8\times10^7\,\Omega^{-1}\,\text{m}^{-1}$; $n = 5.86\times10^{28}$; $e = 1.6\times 10^{-19}$ C.

$$\therefore \; \mu = \frac{6.8\times10^7}{(5.86\times10^{28})\times(1.6\times10^{-19})} = \mathbf{0.725\times10^{-2}\,m^2V^{-1}s^{-1}}.$$

8.2 Find the relaxation time and mean free path of conduction electron in copper. [Given: For copper, density of free electrons, $n = 8.5\times10^{28}$/m³. Resistivity, $\rho = 1.69\times10^{-8}\,\Omega$-m and average velocity of electrons $= 1.154\times10^5$ m/s.

Solution. If n and τ be the density of conduction electrons and relaxation time of an electron of mass m and charge e, then its resistivity is

$$\rho = \frac{m}{ne^2\tau} \text{ or } \tau = \frac{m}{ne^2\rho} \text{ or } \tau = \frac{9.1\times10^{-31}}{(8.5\times10^{28})\times(1.6\times10^{-19})^2\times(1.69\times10^{-8})} = \mathbf{2.47\times10^{-14}\ sec.}$$

Mean free path $\lambda = \tau\bar{c} = (2.47\times10^{-14})\times(1.154\times10^{5}) = \mathbf{2.85\times10^{-9}\ m.}$

8.3 Assuming that each aluminium atom contributes 3 free electrons for conduction, calculate the free electron concentration, mobility and drift velocity of electrons in aluminium wire of length 5 m and resistance 0.06Ω carrying current of 15 A. [Given: Resistivity of aluminium $\rho = 2.7\times10^{-8}\,\Omega$-cm. Atomic weight $M_A = 26.98$; Density $\rho_s = 2.7\times10^3\,\text{kg/m}^2$; Avogadro number $N_A = 6.025\times10^{23}$.]

Solution. Molar volume of aluminium is $\frac{M_A}{\rho_s}$. $\therefore$ $\frac{M_A}{\rho_s}\,\text{m}^3$ contains N_A atoms.

Since each atom contributes 3 free electrons,

$\therefore$ 1 m^3 contains $\left(\frac{\rho_s N_A}{M_A}\right)\times 3$ electrons

$$\text{i.e., } n = \frac{3\rho_s N_A}{M_A} = \frac{3\times(2.7\times10^3)\times(6.025\times10^{26})}{26.98} = \mathbf{1.8088\times10^{29}\ electrons/m^3}.$$

Again, $\frac{1}{\rho} = ne\mu$

$$\text{or } \mu = \frac{1}{ne\rho} = \frac{1}{(1.8088\times10^{29})\times(1.602\times10^{-19})\times(2.7\times10^{-8})} = \mathbf{1.278\times10^{-3}\ m/s/V.}$$

But $\mu = v_d/E$.

$\therefore$ drift velocity $v_d = \mu E$, where

$$E = \frac{V}{D} = \frac{IR}{D} = \frac{(15)\times(0.06)}{5} = 0.18\ \text{V/m}$$

$$v_d = \mu E = (1.278\times10^{-3})\times(0.18) = \mathbf{0.230\times10^{-3}\ m/s.}$$

8.4 Calculate the number of energy states for the free electrons in a cubical box of side 0.02 m lying between an energy of 0 to 3 eV. [Fermi energy of silver is 5 eV].

Solution. The number of energy states between E to $E+dE$ is

$$D(E)\,dE = \frac{\pi V}{2}\left[\frac{8m}{h^2}\right]^{3/2} E^{1/2}\,dE.$$

$\therefore$ the number of energy states between 0 and 3 eV is

$$n = \int D(E)\,dE = \frac{\pi V}{2}\left(\frac{8m}{h^2}\right)^{3/2}\int_0^{3/2} E^{1/2}\,dE = \frac{\pi V}{3}\left(\frac{8m}{h^2}\right)^{3/2}\left[E^{3/2}\right]_0^3$$

$$= \frac{\pi\times(0.02)^3}{3}\left[\frac{72.8\times10^{37}}{6.62\times6.62}\right]^{3/2}_{3/2}\times\left[4.8\times10^{-19}\right]^{3/2}$$

$$= 8.4\times10^{-6}\times6.677\times10^{55}\times3.3\times10^{-28} = \mathbf{1.89\times10^{21}}.$$

8.5 Consider silver in the metallic state with one free electron per atom. Calculate the Fermi energy. [Given: density of silver is 10.5 gm/cm^3 and atomic weight 108.] [CUH 1998, 2009; BU 1992; HPU 1996; VUH 2008]

Solution. Volume occupied by 1 gm-mole of silver $= \frac{108\,\text{gm}}{10.5\,\text{gm/cm}^3}$.

No. of atoms in 1gm-atom = Avogadro number $N_A = 6.02 \times 10^{23}$.

$$\therefore \text{ electron density } N/V = \frac{6.02 \times 10^{23} \times 10.5}{108} \times 1 = 5.85 \times 10^{22} \text{ free electrons/cm}^3$$

$$= \mathbf{5.85 \times 10^{28}\ electron/m^3}$$

Now Fermi energy

$$E_f = \frac{\hbar^2}{2m}\left(3\pi^2 \frac{N}{V}\right)^{2/3} = \frac{h^2}{8m}\left(\frac{3}{\pi}\frac{N}{V}\right)^{2/3} = \frac{(6.625)^2 \times 10^{-68}}{8 \times 9.1 \times 10^{-31}} \times \left[\frac{3 \times 5.85 \times 10^{28}}{3.142}\right]^{2/3}$$

$$= 0.6 \times 10^{-37} \times 14.6 \times 10^{18} = 8.77 \times 10^{-19}\,\text{J} = \frac{8.77 \times 10^{-19}}{1.6 \times 10^{-19}} = \mathbf{5.48\,eV}.$$

8.6 Calculate the Fermi energy at absolute zero of sodium metal if its atomic radius is 1.86 Å.

Solution. If the number of free electrons present in volume V be N, then the Fermi energy of sodium at $T = 0$ is $E_{f_0} = \frac{\hbar^2}{2m}\left(\frac{3\pi^2 N}{V}\right)^{2/3}$.

Now Sodium has BCC structure.

$\therefore$ volume of the unit cell of sodium is $V = a^3 = \left(\frac{4r}{\sqrt{3}}\right)^3$.

$\because$ interatomic distance of sodium is $a = \left(\frac{4r}{\sqrt{3}}\right)$ and r is the atomic radius of sodium.

$$\therefore\ V = \left(\frac{4 \times 1.86 \times 10^{-10}}{\sqrt{3}}\right)^3 = 7.93 \times 10^{-29}\,\text{m}^3.$$

Since there are two atoms/unit cell of sodium the number of free electrons in a unit cell is

$$N/V = \frac{2}{7.93 \times 10^{-29}} = 2.52 \times 10^{28}.$$

$$\therefore\ E_{f_0} = \frac{(1.05 \times 10^{-34})^2}{2 \times 9.1 \times 10^{-31}} \times \left(3\pi^2 \times 2.52 \times 10^{28}\right)^{2/3} = 4.98 \times 10^{-19}\,\text{J} = \frac{4.98 \times 10^{-19}}{1.6 \times 10^{-19}}\,\text{eV} = \mathbf{3.11\,eV}.$$

8.7 (a) Calculate the internal energy of electron gas per unit volume at 0 K. (b) Also calculate the molar specific heat due to free electrons at constant volume in metallic silver at 100 K. [Given: The density of silver = 10.5 gm/cm^3, Atomic weight = 108, Density of electron gas $N/V = 5.9 \times 10^{28}/\text{m}^3$ and $E_{f_0} = 5.5$ eV.]

Solution. (a) The internal energy of the electron gas per unit volume is

$$U = \frac{3}{5}\left(\frac{N}{V}\right) E_{f_0} = \frac{3}{5} \times \left(5.9 \times 10^{28}\right) \times \left(8.8 \times 10^{-19}\right)\,\text{J/m}^3 = \mathbf{3.11 \times 10^{10}\,J/m^3}.$$

(b) The electron molar specific heat is

$$C_v = \frac{\pi^2}{2} Nk\left(T/T_f\right) = \frac{\pi^2}{2} \times \left(6.02 \times 10^{23}\right) \times \left(1.38 \times 10^{-23}\right) \times \left(\frac{100}{63768}\right) = \mathbf{0.064\,J/K/mol.}$$

Using, $N/V = 6.02 \times 10^{23}$; $k = 1.38 \times 10^{-23}$ J/K;

$$T_f = \frac{E_f(0)}{k} = \frac{8.8 \times 10^{-19}\text{ J}}{1.38 \times 10^{-23}\text{ J/K}} = 63768\text{ K.}$$

8.8 Aluminium metal crystallizes in FCC structure. If each atom contributes single electron as free electron and the lattice constant a is 4 Å, calculate, treating conduction electrons as free electron Fermi gas (a) Fermi energy E_f and Fermi vector (k_f) (b) Total KE of free electron gas per unit volume at 0 K. [$\hbar = 1.054 \times 10^{-27}$ erg-sec; electron rest mass $= 9.11 \times 10^{-28}$ gm] [PU 1991]

Solution. The number of atoms per unit cell in FCC structure is 4.

Since each atom contributes one free electron, the number of electron per unit cell is 4.

Volume of the unit cell $V = a^3 = (4 \times 10^{-8})^3 = 64 \times 10^{-24}$ cc.

$\therefore$ free electron density $N/V = \frac{4}{64 \times 10^{-24}} = 6.25 \times 10^{12}$.

Now Fermi energy

$$E_f = \frac{\hbar^2}{2m}\left(3\pi^2 \frac{N}{V}\right)^{2/3} = \frac{(1.054 \times 10^{-27})^2}{2 \times 9.11 \times 10^{-28}}(3\pi^2 \times 6.25 \times 10^{22})^{2/3}$$

$$= 9.191 \times 10^{-12}\text{ erg-s} = \frac{9.191 \times 10^{-12}}{1.6 \times 10^{-12}} = 5.74\text{ eV.}$$

Fermi vector $k_f = \left(3\pi^2 \frac{N}{V}\right)^{1/3} = (3\pi^2 \times 6.25 \times 10^{22})^{1/3} = 1.227 \times 10^8\text{ cm}^{-1}$.

Now at 0 K, the total KE of free electron gas per unit volume

$$= \langle E_0 \rangle \times \frac{N}{V} = \frac{3}{5} E_f \times \frac{N}{V} = \frac{3}{5} \times 5.74 \times 6.25 \times 10^{22} = 21.52 \times 10^{22}\text{ eV.}$$

8.9 A copper wire of length 0.5 m and diameter 0.3 mm has a resistance of 0.12Ω at 20°C. If the thermal conductivity of copper at 20°C is 390 Wb/m^{-1} K^{-1}, calculate Lorentz number. Compare this value with the value predicted by free electron theory.

Solution. Resistance $R = \rho \frac{l}{a}$, where ρ is the resistivity of a wire of length l and diameter a.

$\therefore \rho = \frac{aR}{l}$ and conductivity

$$\sigma = \frac{l}{(\pi r^2)R} = \frac{0.5}{\pi \times (0.15 \times 10^{-3})^2 \times 0.12} = 5.89 \times 10^7\,\Omega^{-1}\text{ m}^{-1}.$$

Now

$$\frac{\sigma_T}{\sigma} = L = 1.5 \times \left(\frac{k}{e}\right)^2 = \frac{1.5 \times (1.38 \times 10^{-23})^2}{(1.6 \times 10^{-19})^2} = 1.116 \times 10^{-8}\text{ W}\Omega\text{K}^2.$$

The experimental value of L is obtained as

$$L = \frac{\sigma_T}{\sigma \times T} = \frac{390}{5.89 \times 10^7 \times 293} = 2.26 \times 10^{-8}\text{ W}\Omega\text{K}^{-2}.$$

EXERCISES

8(A) Objective Type Questions

Choose the correct alternatives.

8.1 Free electrons in metal are

(a) K-shell electrons
(b) valence electron
(c) L-shell electrons
(d) none of these.

8.2 Free electron gas is similar in many respect to ideal gas except that it is
(a) positively charged
(b) negatively charged
(c) both (a) and (b)
(d) none of these.

8.3 The ratio of the density of ideal gas to that of free electron gas is
(a) >1
(b) <1
(c) $=1$
(d) none of these.

8.4 Classical free electron theory was developed by
(a) Sommerfeld
(b) Bloch
(c) Dude and Lorentz
(d) Einstein.

8.5 Quantum theory of free electron was discovered by
(a) Drude and Lorentz
(b) Bloch
(c) Bohr
(d) Sommerfeld.

8.6 The zone theory of free electrons was developed by
(a) Sommerfeld
(b) Bloch
(c) Drude
(d) Lorentz.

8.7 The free electrons are non-interacting and obey
(a) Pauli's exclusion principle
(b) principle of superposition
(c) principle of least action
(d) none of these.

8.8 Free electrons in presence of an external electric field E, move in a direction
(a) parallel to E
(b) opposite to E
(c) perpendicular to E
(d) none of these.

8.9 If K and σ be the thermal and electrical conductivities of a metal at temperature T, then
(a) $\frac{KT}{\sigma} =$ constant
(b) $\frac{K\sigma}{T} =$ constant
(c) $\frac{\sigma}{kT} =$ constant
(d) $\frac{K}{\sigma T} =$ constant.

8.10 If n, v_d and e be the electron density, electron drift velocity and electronic charge respectively, then current density is given by
(a) $J = nv_d$
(b) $J = nev_d$
(c) $J = \frac{ne}{v_d}$
(d) $J = \frac{1}{nev_d}$.

8.11 Drift velocity for unit electric field is
(a) electrical conductivity
(b) thermal conductivity
(c) electric polarisation
(d) mobility of charge carriers.

8.12 Drift velocity of an electron in metal is given by

(a) $v_d = \frac{mE}{e\tau}$

(b) $v_d = \frac{e\tau}{m}E$

(c) $v_d = -\frac{e\tau}{m}E$

(d) $v_d = -\frac{e\tau}{mE}$.

8.13 Electrical conductivity of a metal is

(a) $\sigma = \frac{ne^2}{\tau m}$

(b) $\sigma = \frac{ne^2\tau}{m}$

(c) $\sigma = \frac{m}{ne^2\tau}$

(d) $\sigma = \frac{\tau m}{ne^2}$.

8.14 The electron mobility is given by

(a) $\mu = m/e\tau$

(b) $\mu = \frac{e}{\tau m}$

(c) $\mu = \frac{m\tau}{e}$

(d) $\mu = \frac{e\tau}{m}$.

8.15 Resistivity of a metal of the conductor is given by

(a) $\rho = \frac{m}{ne^2\tau}$

(b) $\rho = \frac{ne^2\tau}{m}$

(c) $\rho = \frac{n}{me^2\tau}$

(d) $\rho = \frac{me^2\tau}{n}$.

8.16 The relation between resistivity and absolute temperature of a metal is

(a) $\rho \propto T$

(b) $\rho \propto \sqrt{T}$

(c) $\rho \propto T^2$

(d) $\rho \propto \frac{1}{T}$.

8.17 According to quantum theory, the free electron gas obeys

(a) MB-statistics

(b) BE-statistics

(c) FD-statistics

(d) none of these.

8.18 At any temperature other than zero degree kelvin, the probability of finding an electron at Fermi level is

(a) 1

(b) 0

(c) 0.5

(d) any positive value.

8.19 The average KE in the ground state of the electron confined in one-dimensional potential is

(a) $3E_f$

(b) $\frac{1}{3}E_f$

(c) $\frac{3}{E_f}$

(d) $\frac{3}{5}E_{f_0}$.

8.20 If N be the number of free electrons in volume V, then the Fermi velocity

(a) $v_f \propto (N/V)$

(b) $v_f \propto (N/V)^{1/3}$

(c) $v_f \propto (N/V)^{1/2}$

(d) $v_f = N/V$.

8.21 In three-dimensional case, the average KE of an electron at 0 K is

(a) $\langle E_0 \rangle = 5/3E_{f_0}$

(b) $\langle E_0 \rangle = \frac{5E_{f_0}}{4}$

(c) $\langle E_0 \rangle = \frac{3}{5}E_{f_0}$

(d) $\langle E_0 \rangle = E_{f_0}$.

8.22 The variation of molar specific heat of free electrons at temperature T is given by

(a) $C_v/T = A + BT^3$ (c) $C_v/T = A + BT^2$
(b) $C_v/T = BT^3$ (d) $C_v/T = A$

where A and B are constants.

8.23 At Fermi level, the probability of electron occupation at $T > 0\,\text{K}$ is
(a) 0 (c) $\frac{1}{2}$
(b) 1 (d) $\frac{1}{4}$.

8.24 The pressure-volume relation of an electron gas is
(a) $P = \frac{1}{5}\frac{NE_{f_0}}{V}$ (c) $P = \frac{NE_{f_0}}{V}$
(b) $P = \frac{2}{5}\frac{NE_{f_0}}{V}$ (d) $P = \frac{3}{5}\frac{NE_{f_0}}{V}$

8.25 The de Broglie wavelength of an electron at the Fermi surface is
(a) $\lambda_f = 2\left(\frac{\pi}{3n}\right)^{1/3}$ (c) $\lambda_f = 2\left(\frac{\pi}{3n}\right)^{1/2}$
(b) $\lambda_f = \left(\frac{\pi}{3n}\right)^{2/3}$ (d) $\lambda_f = 2\left(\frac{\pi}{3n}\right)^{2/3}$.

8.26 According to quantum theory of free electrons the molar specific heat of free electrons is
(a) $C_v = \frac{3}{2}Nk$ (c) $C_v = (0.01)\frac{3}{2}Nk$
(b) $C_v = (0.01)Nk$ (d) $C_v = (0.001)\frac{3}{2}Nk$.

8.27 The density of states between energies E and $E + dE$ is
(a) $\frac{4\pi}{h^3}(2m)^{3/2}E^{1/2}\,dE$ (c) $\frac{4\pi}{h^3}(2m)^{2/3}E^{1/2}\,dE$
(b) $\frac{4\pi}{h^3}m^3 E\,dE$ (d) none of these.

8.28 The bulk modulus of elasticity for a free electron gas at 0 K is
(a) $B = \frac{9}{10}(E/V)$ (c) $B = \frac{10}{9}(V/E)$
(b) $B = \frac{10}{9}(E/V)$ (d) $B = \frac{10}{9}EV$.

ANSWERS

8.1 (b)	8.5 (d)	8.9 (d)	8.13 (b)	8.17 (c)	8.21 (c)	8.25 (a)
8.2 (b)	8.6 (b)	8.10 (b)	8.14 (d)	8.18 (c)	8.22 (c)	8.26 (c)
8.3 (b)	8.7 (a)	8.11 (d)	8.15 (a)	8.19 (b)	8.23 (c)	8.27 (a)
8.4 (c)	8.8 (b)	8.12 (c)	8.16 (b)	8.20 (b)	8.24 (b)	8.28 (b).

8(B) Short Answer Type Questions

8.1 Explain the terms *free electron* and *free electron* gas.

8.2 Explain the salient features of quantum theory of free electrons.

8.3 Define the terms: *Drift velocity* and *mean free path* of free electron in a metal. What is relaxation time?

8.4 What are the merits and demerits of free electron theory of metals after Drude?

8.5 Show that electrical resistivity of a metal is given by $\rho = \frac{m}{ne^2\tau}$, the letters have their usual significances.

8.6 Show that the energy levels of a free electron can be considered to from a quasi-continuous band.

8.7 Find out an expression for the total energy of an electron gas in the ground state.

8.8 Discuss the effect of temperature on the occupancy of energy levels.

8.9 What do you mean by *Fermi sphere* and *Fermi surface*? What is its volume?

8.10 Define Fermi energy, Fermi temperature and Fermi velocity of a free electron in a metal.

8.11 According to classical and quantum theories of free electrons the molar specific heats are $C_v = \frac{3}{2}Nk$ and $C_v = \frac{3}{2}Nk\left(T/T_f\right)$. How can the difference of the two values be accounted for?

8.12 Explain complete *opacity* and *high lustre* on the basis of the Drude-Lorentz free electron theory of metals.

8.13 Define density of states. Find out its expression for three-dimensional motion of free electrons.

8.14 Show that the total molar specific heat of free electrons at low temperature is given by $C_v = AT + BT^3$, where A and B are constants.

8.15 Show that the number of electrons in the Fermi surface is given by $\frac{V}{3\pi^2}\left(\frac{2m}{\hbar^2}\right)^{3/2} E_{fo}^{3/2}$, the different letters have their usual significances.

8(C) Numerical Problems

8.1 The resistivity of aluminium at room temperature is $2.62 \times 10^{-8}\,\Omega$-m. Calculate (a) the drift velocity (b) their mobility (c) their relaxation time (d) mean free path on the basis of classical free electron theory. *(Ans. (a) 0.396 m/s; (b) 3.96×10^{-3} m²/V/s; (c) 2.3×10^{-14} s (d) 2.65 nm)*

8.2 Show that if the mean free path is independent of the velocity, the electrical conductivity of Maxwell-Boltzmann free electron gas may be expressed as

$$\sigma = \frac{4}{3}\frac{nc^2\lambda}{(2\pi kTm)^{1/2}}.$$

8.3 The Fermi energy of silver is 5.51 eV. What is the average energy of free electron at 0 K? At what temperature will the average molecular energy in an ideal gas have this value? **[BUH 1998]**

8.4 Show that the average energy of an electron in a metal at $T = 0$ K is $\frac{3}{5}E_f$, where E_f is the Fermi energy. **[CUH 1997, 2004]**

8.5 There are 2.5×10^{28} free electrons per cubic metre of sodium. Calculate Fermi energy and Fermi velocity. *(Ans. $E_f = 3.1$ eV; $v_f = 1.048 \times 10^6$ m/s)*

8.6 If Fermi temperature is of the order of 5×10^4 K provide a rough estimate of electronic specific heat at 100 K. *(Ans. 0.025 J)*

8(D) Broad Answer Type Questions

8.1 (a) Describe the free electron gas model of metals.

(b) What is free electron approximation? In this approximation find the radius of the Fermi sphere for a crystal with Bravais lattice having Z electrons per atoms and n unit cells per unit volume.

(c) Find the wavelength of an electron at the Fermi level. **[BUH 1996]**

8.2 (a) Deduce an expression for the Fermi energy of a free electron gas in terms of the electron concentration *n*. What is the value of the Fermi energy in a typical metal (electron density $= 10^{28}\,m^3$)?

(b) Show that the conductivity of free electron gas can be expressed in the form $\sigma = \frac{ne^2\tau}{m}$, where the symbols have their customary connotations.

Explain: (a) Why only electrons near the Fermi surface contribute to the electrical conductivity of free electron metal?

(b) Why the resistance of a metal increases with temperature? **[BUH 1995]**

8.3 (a) What are the major difficulties in the classical theory of the free electron gas in metal ? Discuss how were these removed.

(b) Explain the significance of the Fermi energy of a metal.

(c) Prove that the average KE per electron in a metal is equal to $\frac{3}{5}E_f$. **[BUH 1994]**

8.4 (a) Obtain expressions for the electrical and thermal conductivities of metal in free electron approximation and hence establish the Wiedemann-Franz law. **[BUH 1993]**

(b) Show that the average KE of a conduction electron at 0 K is $(3/5)\,E_f$, where E_f is the Fermi energy and that the average speed is $\left(\frac{3}{4}\right) v_f$, where v_f is the velocity at the Fermi surface. **[BUH 2003]**

8.5 (a) Find the radius of the Fermi sphere and the wavelength of an electron at the Fermi level for a crystal having one electron per atom per unit cell and *N* unit cells per unit volume. **[BUH 1998]**

(b) What is meant by Fermi energy E_f? Obtain an expression for Fermi energy for free electron gas in one dimension at absolute zero. Hence, show that the average KE in the ground state is one-third of Fermi energy. **[BUH 2000]**

8.6 (a) What do you understand by Fermi gas? Can a metal be associated with two Fermi temperature? **[PU 1991]**

[*Hints.* $T_f = \frac{E_f}{k}$. Since there is only one volume of E_f at 0 K, there is only one value of T_f.]

(b) Does the Fermi energy of a metal depend upon temperature and total number of electrons in the sample? **[PU 1995, 1992]**

Chapter 9

Band Theory of Solids

9.1 Introduction

In a solid, the potential due to the fixed nuclei and the average potential due to all other electrons form a *periodic potential.* The motion of an electron in this periodic potential can be described by *Schrödinger equation,* whose solution gives a set of energy states. These states may be occupied by electrons according to *Pauli's exclusion principle.* All states are not permitted to electrons but bands of possible energies separated by *forbidden* energies are allowed.

9.2 Formation of Energy Bands

When two atoms are brought close to each other a single sharp energy level splits up into a pair of levels.

When two atoms are brought together their wave functions ψ_1 and ψ_2 overlap, the resultant wave function being either $(\psi_1 + \psi_2)$ or $(\psi_1 - \psi_2)$ as shown in Figs 9.1 (a) and (b).

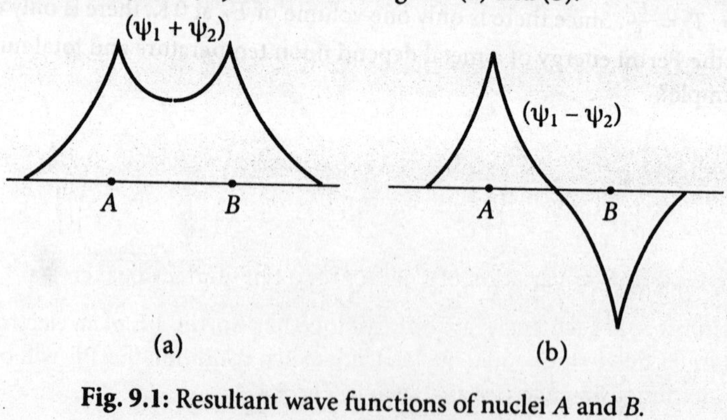

Fig. 9.1: Resultant wave functions of nuclei *A* and *B*.

The electron has a finite probability of existing midway between the nuclei for $(\psi_1 + \psi_2)$. Here the binding energy increases due to the binding force of both the electrons. But for $(\psi_1 - \psi_2)$ the probability of existence of the electron is zero, and the extra contribution of energy disappears. So, there is a difference in energy between the states having wave functions $(\psi_1 + \psi_2)$ and $(\psi_1 - \psi_2)$. Hence, when two atoms are brought close together each energy state splits up into two energy states. When N number of atoms are brought together each energy state splits into N energy states whose separation is very small if N is very large. These energy states, then form a *quasi-continuous* energy band. In other words, each energy level may split into a band of energy level when atoms are brought together.

The formation of band out of energy levels due to the decrease of separation between the two atoms is shown in Fig. 9.2. The *strength of interaction* and the *extent of overlap* determine the width of the band.

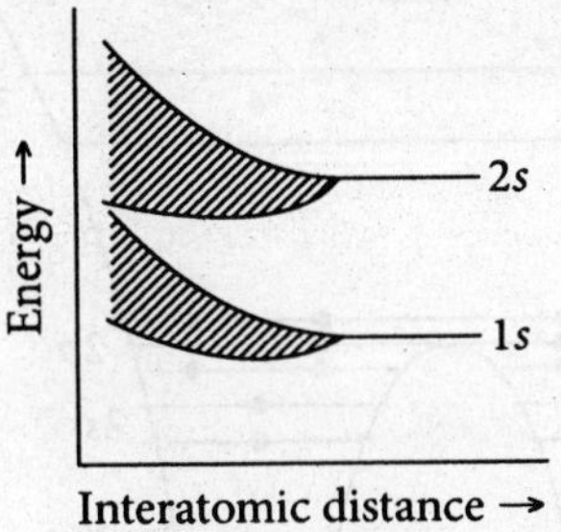

Fig. 9.2: Band formation of 1*s* and 2*s* states.

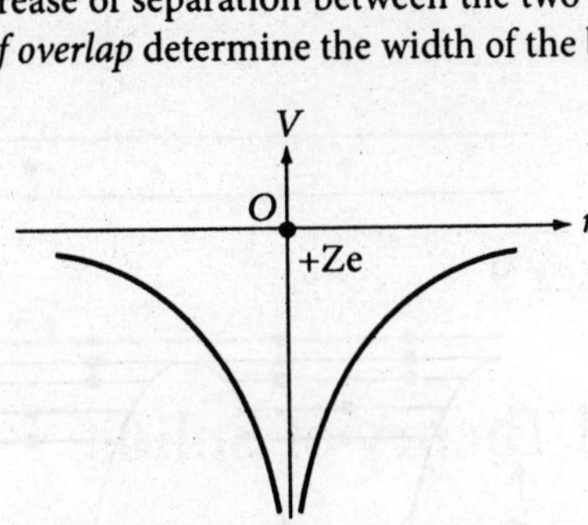

Fig. 9.3: Variation of *V* with *r* of an isolated atom.

9.3 Periodic Potential in a Crystalline Solid

The PE of an electron at a distance *r* from a nucleus of charge Ze is given by

$$V = \frac{Ze^2}{4\pi\varepsilon_0 r}, \tag{9.1}$$

where ε_0 is the *permittivity of the free space.* The graphical variation of *V* with *r* is shown in Fig. 9.3.

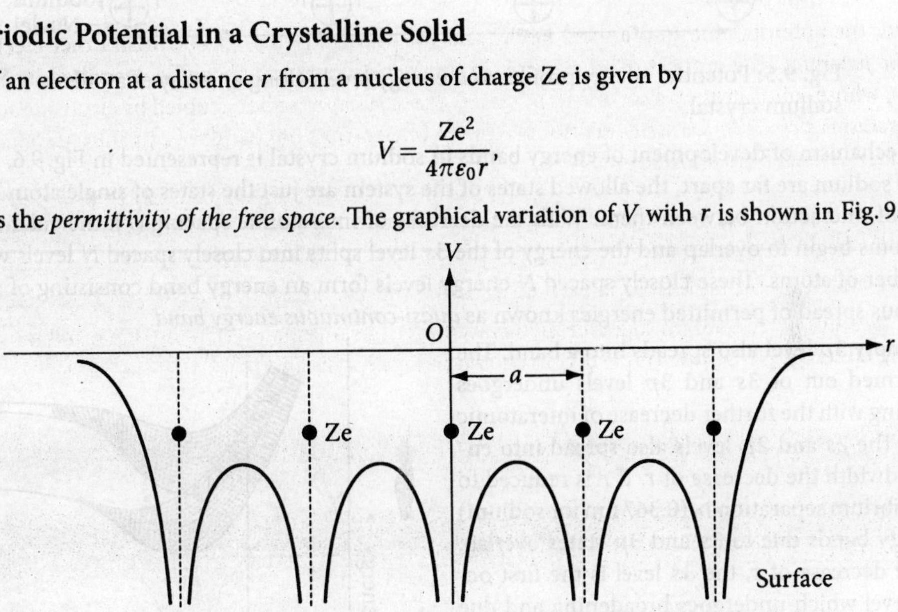

Fig. 9.4: Variation of *V* with *r* for a one-dimensional crystal.

In a solid a large number of such nuclei are brought together. So, the PE of an electron is the sum total of all the potential energies due to individual nuclei. Under such condition the PE will be a *periodic function of r* in an infinite one-dimensional lattice as shown in Fig. 9.4.

9.3.1 Development of Energy Bands in Sodium Crystals

Fig. 9.5 shows the PE profile for the linear array of sodium atoms in a crystal with r_0 as the equilibrium spacing between two neighbouring atoms.

Though outside the crystal the PE rises towards $E = 0$ asymptotically, the every *E* is considerable below $E = 0$ level between neighbouring atoms due to mutual interaction. The 2*s* and 2*p* levels of atoms are separated by a potential hump but the band formed by 3*s* electrons are only half-filled. The electrons in this band have energy higher than the potential hump and behave as free electrons and as such sodium metal is a good conductor of electricity.

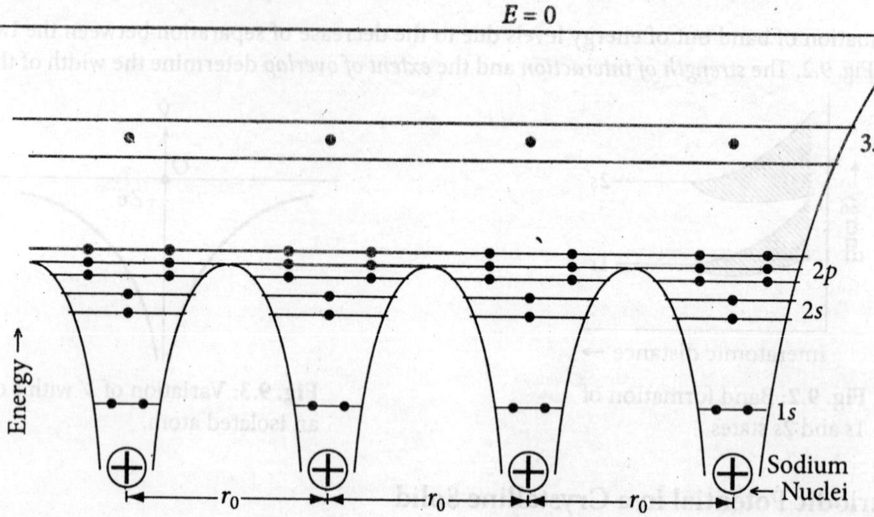

Fig. 9.5: Potential energy profiles and energy levels along an array of atoms in a sodium crystal.

The mechanism of development of energy bands in sodium crystal is represented in Fig. 9.6. When the atoms of sodium are far apart, the allowed states of the system are just the states of single atom due to the absence of interaction between them. With the decrease of interatomic spacing (r), the valence orbitals of the atoms begin to overlap and the energy of the $3s$ level splits into closely spaced N levels where N is the number of atoms. These closely spaced N energy levels form an energy band consisting of a virtually continuous spread of permitted energies known as *quasi-continuous energy band.*

The empty $3p$ level also spreads into a band. The band formed out of $3s$ and $3p$ levels undergoes broadening with the further decrease of interatomic spacing. The $2s$ and $2p$ levels also spread into energy bandwidth the decrease of r. If r is reduced to the equilibrium separation r_0 (0.367 nm for sodium) the energy bands due to $3s$ and $3p$ states overlap. With the decrease of r, the $3s$ level is the first occupied level which undergoes broadening and due to the effect of attractive forces between the atoms the average energy of the $3s$ level decreases. Hence, *we see that in a real solid, consisting of large number of atoms, splitting of energy levels into energy bands occurs with the decease of r due to the interaction between the atoms.*

In metals the energy bands so formed overlap but in other materials they may be separated by a gap called *band gap* (E_g). The energy of the highest filled level is called the *Fermi energy* (E_F). All energy levels below E_F are filled at 0 K but those above E_F are empty. According to *Pauli's principle* each s-band of a crystal having N atoms can accommodate $2N$ electrons.

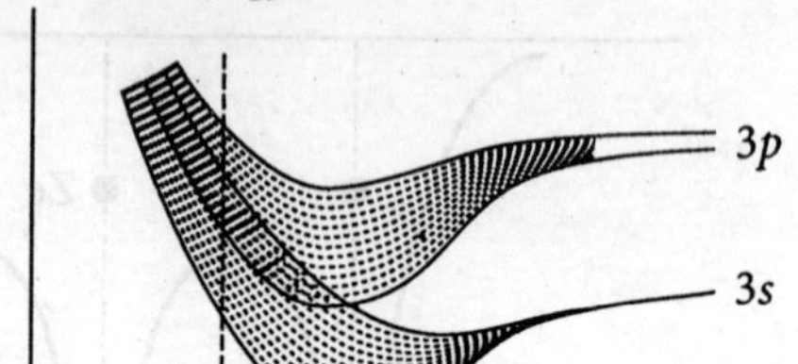

Fig. 9.6: Energy bands in sodium crystal.

If the highest s-band is completely filled, then the electron cannot drift within the crystal due to the application of an external electric field. Such solids are called *insulators*.

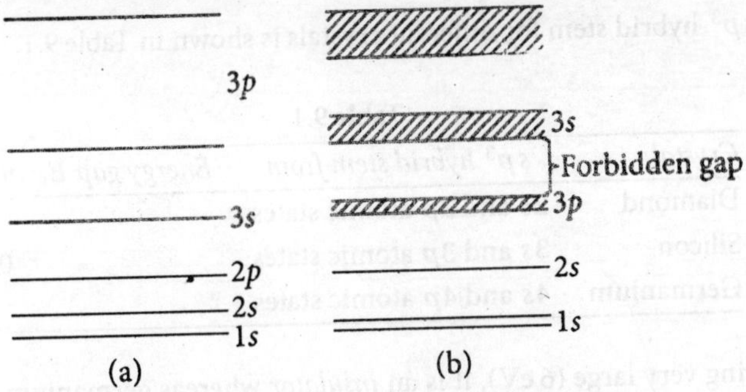

Fig. 9.7 (a) Energy levels of an atom (b) Energy bands of a solid.

9.3.2 Energy Bands in Diamond

The development of energy band in diamond is shown in Fig. 9.8. Diamond is the crystalline form of carbon.

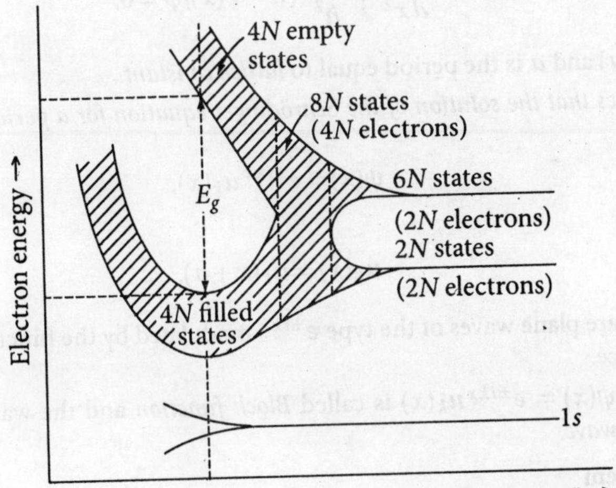

Fig. 9.8: Developments of energy bands in diamond.

The electronic configuration configuration of an isolated carbon atom is $1s^2\ 2s^2\ 2p^6$.

In the diamond crystal there is no modification of s and p-levels due to the formation of sp^3 hybrid, a mixture of $2s$ and $2p$-wave functions with covalent tetrahedral bonding.

From Fig. 9.8 it is clear that with the decrease of r the $6N$ states of the upper band and $2N$ states of the lower band merge together to form a wider energy band consisting of $8N$ states with $4N$ electrons. If the interatomic distance r be further decreased to r_0 (the equilibrium separation), the $2N$ states in $2s$ sub-shell and the filled $2N$ states in $2p$ sub-shell approaches each other and spread into a band of $4N$ states which are all filled with electrons. The remaining $4N$ states in $2p$ sub-shell which are all empty spread into another band being separated from the lower filled band by a forbidden energy gap $\left(E_g\right)$. The completely filled lower band is called the *valence band* and the upper empty band is called the *conduction band*. If an electron from the lower valence band transits to the upper conduction band by receiving energy E_g (= 6 eV for diamond) electrical conduction takes place.

The formation of sp^3-hybrid stem for different crystals is shown in Table 9.1.

Table 9.1

Crystal	*sp^3 hybrid stem from*	*Energy gap E_g in* eV
Diamond	$2s$ and $2p$ atomic states	6
Silicon	$3s$ and $3p$ atomic states	0.78
Germanium	$4s$ and $4p$ atomic states	1.2

E_g for diamond being very large (6 eV), it is an *insulator* whereas germanium and silicon are *semiconductor* as E_g values for them is 0.78 eV and 1.12 eV respectively which are quite low.

9.4 Bloch's Theorem

The *Schrödinger equation* for an electron moving in a one-dimensional periodic potential is given by

$$\frac{d^2\psi}{dx^2} + \frac{2m}{\hbar^2}[E - V(x)]\psi = 0, \tag{9.2}$$

where $V(x) = V(x+a)$ and a is the period equal to *lattice constant.*

Bloch's theorem states that the solution of the Schrödinger equation for a periodic potential must be of the form:

$$\psi(x) = e^{\pm ikx}u_k(x), \tag{9.3}$$

where

$$u_k(x) = u_k(x+a). \tag{9.4}$$

Thus, the solutions are plane waves of the type $e^{\pm ikx}$ modulated by the function $u_k(x)$ having the same periodicity as the lattice.

The wave function $\psi(x) = e^{\pm ikx}u_k(x)$ is called *Bloch function* and the wave vector $\vec{k}$ represents the direction of the *Bloch wave.*

Proof of Bloch's theorem

Let $f(x)$ and $g(x)$ be two *real* and *independent functions* satisfying (9.2). Then the general solution of (9.2) is

$$\psi(x) = Af(x) + Bg(x), \tag{9.5}$$

A and B being two arbitrary constants.

Since $V(x) = V(x+a)$, the functions $f(x+a)$ and $g(x+a)$ must also be solutions of (9.2). According to the theory of differential equation, a differential equation of second order can have only two independent solutions. So, $f(x+a)$ and $g(x+a)$ can be written as

$$f(x+a) = \alpha_1 f(x) + \alpha_2 g(x) \tag{9.6}$$

$$g(x+a) = \beta_1 f(x) + \beta_2 g(x), \tag{9.7}$$

where $\alpha_1, \alpha_2, \beta_1, \beta_2$ are functions of E.

Now the general solution of *Schrödinger equation* is $\psi(x) = Af(x) + Bg(x)$ and

$$\begin{aligned}\psi(x+a) &= Af(x+a) + Bg(x+a)\\ &= A\alpha_1 f(x) + A\alpha_2 g(x) + B\beta_1 f(x) + B\beta_2 g(x) \text{ [using (9.6) and (9.7)]}\\ &= (A\alpha_1 + B\beta_1)\, f(x) + (A\alpha_2 + B\beta_2)\, g(x). \end{aligned} \tag{9.8}$$

Now let

$$A\alpha_1 + B\beta_1 = \lambda A \tag{9.9}$$

$$\text{and } A\alpha_2 + B\beta_2 = \lambda B, \tag{9.10}$$

where λ is a constant.

$$\therefore\ \psi(x+a) = \lambda A f(x) + \lambda B g(x) = \lambda[Af(x) + Bg(x)] = \lambda\psi(x). \tag{9.11}$$

A and B will have non-zero values if the determinant of their coefficient vanishes, i.e.,

$$\begin{vmatrix} \alpha_1 - \lambda & \beta_1 \\ \alpha_2 & \beta_2 - \lambda \end{vmatrix} = 0 \tag{9.12}$$

$$\text{or } \lambda^2 - (\alpha_1 + \beta_2)\,\lambda + (\alpha_1\beta_2 - \alpha_2\beta_1) = 0. \tag{9.13}$$

To prove that $\alpha_1\beta_2 - \alpha_2\beta_1 = 1$, we have

$$\frac{d^2 f(x)}{dx^2} + \frac{2m}{\hbar^2}[E - V(x)]\, f(x) = 0 \tag{9.14}$$

$$\text{and } \frac{d^2 g(x)}{dx^2} + \frac{2m}{\hbar^2}[E - V(x)]\, g(x) = 0. \tag{9.15}$$

Multiplying, (9.14) by $g(x)$ and (9.15) by $f(x)$ and then subtracting, we get

$$f(x)\frac{d^2 g(x)}{dx^2} - g(x)\frac{d^2 f(x)}{dx^2} = 0$$

$$\text{or } f(x)\frac{dg(x)}{dx} - g(x)\frac{df(x)}{dx} = \text{Constant}$$

$$f(x)\frac{dg(x)}{dx} - g(x)\frac{df(x)}{dx} = W(x)$$

is the *Wronskian* of the solution and is constant in this case.

Now using (9.6) and (9.7) in the above equation, we have

$$\begin{aligned} W(x+a) &= f(x+a)\frac{dg(x+a)}{dx} - g(x+a)\frac{df(x+a)}{dx}\\ &= f(x+a)\left[\beta_1\frac{df(x)}{dx} + \beta_2\frac{dg(x)}{dx}\right] - g(x+a)\left[\alpha_1\frac{df(x)}{dx} + \alpha_2\frac{dg(x)}{dx}\right]\\ &= [\alpha_1 f(x) + \alpha_2 g(x)]\left[\beta_1\frac{df(x)}{dx} + \beta_2\frac{dg(x)}{dx}\right] - [\beta_1 f(x) + \beta_2 g(x)]\left[\alpha_1\frac{df(x)}{dx} + \alpha_2\frac{dg(x)}{dx}\right]\\ &= (\alpha_1\beta_2 - \alpha_2\beta_1)\left[f(x)\frac{dg(x)}{dx} - g(x)\frac{df(x)}{dx}\right]\\ &= (\alpha_1\beta_2 - \alpha_2\beta_1)\, W(x). \end{aligned} \tag{9.16}$$

$$\therefore\ W(x+a)=(\alpha_1\beta_2-\alpha_2\beta_1)\ W(x).$$

But $W(x+a)=W(x)=$ a constant.

Hence, $\alpha_1\beta_2-\alpha_2\beta_1=1$.

Now (9.13) reduces to

$$\lambda^2-(\alpha_1+\beta_2)\,\lambda+1=0. \tag{9.17}$$

The quantity $(\alpha_1+\beta_2)$ is a real function of energy E. Since (9.17) is quadratic it will have two roots and hence two wave functions $\psi_1(x)$ and $\psi_2(x)$ having the property $\psi_1(x+a)=\lambda\psi(x)$. The product of the roots, i.e., $\lambda_1\lambda_2=1$.

Special cases

Case I. *For energy ranges such that* $(\alpha_1+\beta_2)^2<4$.

In this case (9.17) will have *a pair of complex roots.* As $\lambda_1\lambda_2=1$, these complex roots will be complex conjugate of each other.

$$\left.\begin{aligned}\therefore\ \lambda_1&=e^{iKa}\\ \text{and } \lambda_2&=e^{-iKa}\end{aligned}\right\} \tag{9.18}$$

where k is real.

The corresponding functions $\psi_1(x)$ and $\psi_2(x)$ will satisfy the relation:

$$\left.\begin{aligned}\psi_1(x+a)&=e^{iKa}\psi_1(x)\\ \text{and } \psi_2(x+a)&=e^{-iKa}\psi_2(x)\end{aligned}\right\} \tag{9.19}$$

or, in general,

$$\psi(x+a)=e^{\pm ika}\,\psi(x). \tag{9.20}$$

The function of (9.20) can be shown to be *Bloch function* of the type $\psi(x)=e^{\pm ikx}\,u_k(x)$.

Replacing x by $(x+a)$ in (9.3), we get

$$\psi(x+a)=e^{\pm ik(x+a)}\,\acute{u}_k(x+a)=e^{\pm ika}\,e^{\pm ikx}\,u_k(x+a)=e^{\pm ika}\psi(x)=\lambda\psi(x), \tag{9.21}$$

where

$$\lambda=e^{\pm ika}. \tag{9.22}$$

Hence, (9.20) and (9.21) are the same and thus *Bloch's theorem is proved.*

Case II. *For the energy range such that* $(\alpha_1+\beta_2)^2>1$.

In this case (9.17) will have real roots λ_1 and λ_2 and may be written as

$$\lambda_1=e^{+\mu a} \text{ and } \lambda_2=e^{-\mu a}, \tag{9.23}$$

where μ is real. Then the corresponding solutions to the *Schrödinger equation* are

$$\psi_1(x)=e^{\mu x}\,u(x) \text{ and } \psi_2(x)=e^{-\mu x}\,u(x). \tag{9.24}$$

Since

$$\frac{\psi(x+na)}{\psi(x)}\to\infty,\ \text{as } n\to\infty$$

these solutions are not acceptable. This is contrary to the requirement because *the wave function should remain finite at all points in space due to their probabilistic interpretation.* For Case I, however,

$$|\psi(x+na)|=|\psi(x)|$$

and the ratio

$$\frac{\psi(x+na)}{\psi(x)} \text{ is finite.}$$

The roots $\lambda_1 = e^{iKa}$ and $\lambda_2 = e^{-iKa}$ are *allowed* roots whereas $\lambda_1 = e^{\mu a}$ and $\lambda_2 = e^{-\mu a}$ are *forbidden* roots though all are functions of $(\alpha_1 + \beta_2)$ in energy E. The allowed roots are related to allowed energy regions and the forbidden roots to forbidden energy regions.

Hence, *the energy spectrum of an electron moving in a periodic potential consists of allowed and forbidden energy regions (or bands).*

9.5 Kronig-Penney Model

The solution of the *Schrödinger equation* with periodic potential [Fig. 9.4] is not very easy. *Kronig* and *Penney,* therefore, introduced a simpler model for the shape of the potential. They assumed that the *potential inside the crystal is rectangular in shape as shown in* Fig. 9.9.

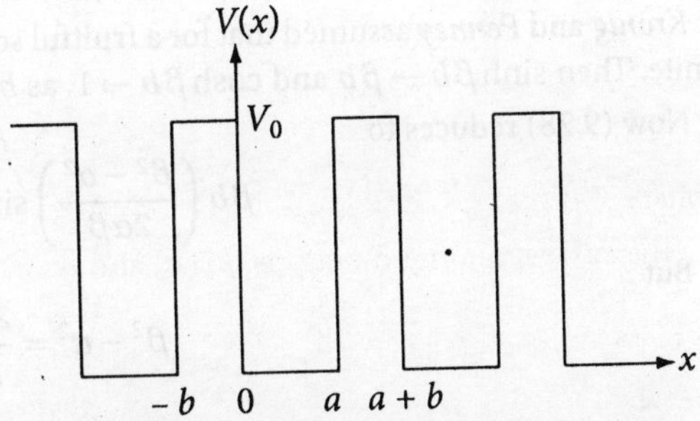

Fig. 9.9: Kronig-Penny model of potential.

The potential is defined as

$$V=0, \qquad 0<x<a$$

$$\text{and } V=V_0, \; -b<x<0.$$

The moving electrons in the lattice are not actually free but influenced to some extent by the periodic potential. The wave functions relating to this model may be obtained by solving *Schrödinger equation* for the following two regions:

$$\frac{d^2\psi}{dx^2}+\left(\frac{2m}{\hbar^2}\right)E\psi=0, \quad \text{for } 0<x<a \quad \text{with } V(x)=0$$

$$\text{and } \frac{d^2\psi}{dx^2}+\left(\frac{2m}{\hbar^2}\right)(E-V_0)\,\psi=0, \text{ for } -b<x<0 \text{ with } V(x)=V_0.$$

Putting $\alpha^2 = \frac{2mE}{\hbar^2}$ and $\beta^2 = \frac{2m}{\hbar^2}(V_0 - E)$, we get

$$\frac{d^2\psi}{dx^2}+\alpha^2\psi=0 \tag{9.25}$$

$$\text{and } \frac{d^2\psi}{dx^2}+\beta^2\psi=0. \tag{9.26}$$

According to *Bloch,* the solutions are of the form:

$$\psi(x)=u_k(x)\,e^{iKx}, \tag{9.27}$$

where $u_k(x)$ is a *periodic function* of x with periodicity of $(a+b)$, i.e.,

$$u_k(x)=u_k(x+a)=u_k(x+na),$$

where n is an integer and k the *wave number*, where

$$\frac{2\pi}{\lambda}=\frac{p}{\hbar}.$$

The solution applicable to (9.25) and (9.26) as shown by *Bloch* is

$$\psi(x)=u_k(x)\,e^{iKx},$$

where $u_k(x)=u_k(x+a)=u_k(x+na)$, n being an integer.

After a long mathematical calculations *Kronig* and *Penney* have shown that only those values of k are allowed which satisfy the following equation:

$$\left(\frac{\beta^2-\alpha^2}{2\alpha\beta}\right)\sinh\beta b\sin\alpha a+\cosh\beta b\cos\beta a=\cos k(a+b). \tag{9.28}$$

Kronig and *Penney* assumed that for a fruitful solution, $V_0\to\infty$ and $b\to 0$ in such a way that $V_0 b$ remains finite. Then $\sinh\beta b\to\beta b$ and $\cosh\beta b\to 1$, as $b\to 0$.

Now (9.28) reduces to

$$\beta b\left(\frac{\beta^2-\alpha^2}{2\alpha\beta}\right)\sin\alpha a+\cos\alpha a=\cos ka. \tag{9.29}$$

But

$$\beta^2-\alpha^2=\frac{2m}{\hbar^2}[(V_0-E)+E].$$

$$\therefore\ \frac{2mV_0}{2\hbar^2}(ab)\frac{\sin\alpha a}{\alpha a}+\cos\alpha a=\cos ka \tag{9.30}$$

$$\text{or}\quad \left(\frac{mV_0ab}{\hbar^2}\right)\left(\frac{\sin\alpha a}{\alpha a}\right)+\cos\alpha a=\cos ka \text{ or } P\frac{\sin\alpha a}{\alpha a}+\cos\alpha a=\cos ka, \tag{9.31}$$

where $P=\frac{mV_0ab}{\hbar^2}$.

The quantity $V_0\to\infty$, therefore, P is a measure of V_0b, the *area of the potential barrier*.

The physical significance of P is that, as P increases the area V_0b of the potential barrier increases. *This means, the electron is bound more strongly to a particular potential well. The potential barrier becomes very weak as $P\to 0$ and the electron becomes free.*

(9.31) will be satisfied for those values of αa for which $\left(P\frac{\sin\alpha a}{\alpha a}+\cos\alpha a\right)$ will have values between $+1$ and -1. Only such values of αa will give a solution of (9.31).

Fig. 9.10 shows the plot of

$$\left(P\frac{\sin\alpha a}{\alpha a}+\cos\alpha a\right)$$

against αa for $P=\frac{3\pi}{2}$. The part of the vertical axis lying between -1 and $+1$, as indicated by the horizontal lines, represents the values acceptable to

$$\left(P\frac{\sin\alpha a}{\alpha a}+\cos\alpha a\right).$$

The abscissa is a measure of energy because

$$\alpha^2=\frac{2mE}{\hbar^2}.$$

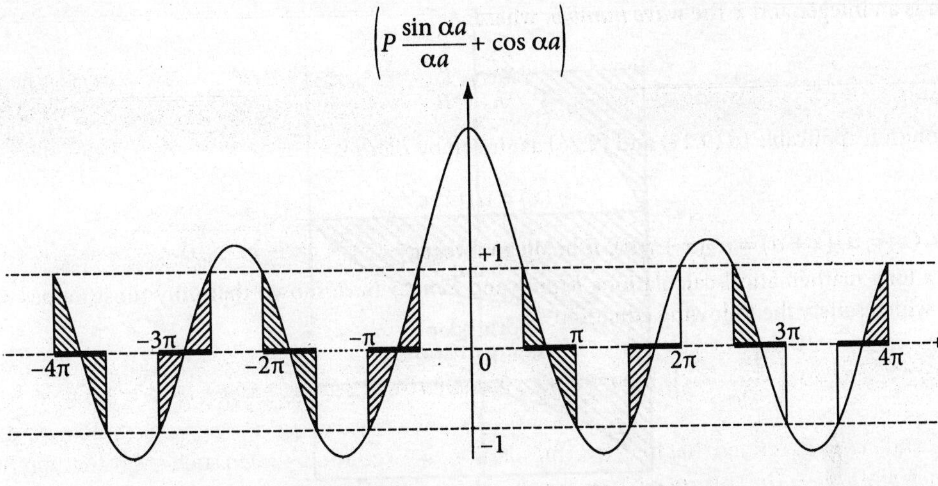

Fig. 9.10: Plot of $\left(P\dfrac{\sin \alpha a}{\alpha a} + \cos \alpha a\right)$ versus αa using $P = 3\pi/2$.

We can draw the following conclusions from Fig. 9.10:

1. The energy spectrum of the electron consists of alternate regions of *allowed energy bands* (solid lines on abscissa) and *forbidden energy bands* (broken lines). The boundaries of allowed ranges of αa corresponds to the values of $\cos ka = \pm 1$ or, $ka = n\pi$ or, $k = \frac{n\pi}{a}$.
2. The width of the allowed band increases with αa but the width of the forbidden band decreases.
3. P is a measure of the *barrier potential strength.* If P is large, the allowed energy bands are compressed into energy levels and a line spectrum is obtained. This is because for $P \to \infty$, the function $\left(P\frac{\sin\alpha a}{\alpha a} + \cos\alpha a\right)$ crosses $+1$ and -1 regions at a steeper angle. When $\alpha a = \pm n\pi$, $\cos\alpha a = \cos ka$ with $ka = \pm n\pi$ or, $k = \pm\frac{n\pi}{a}$. These values of k represent the *discontinuity* in the $(E\text{-}k)$ curve for electron.
4. The energy spectrum in two extreme cases using (9.31):

(a) When $P \to \infty$, (9.31) has a solution only for $\sin\alpha a = 0$ or, $\alpha a = \pm n\pi$, where n is an integer.

Now

$$\alpha = \pm\frac{n\pi}{a} \text{ or } \alpha^2 = \frac{n^2\pi^2}{a^2} = \frac{2mE}{\hbar^2} \text{ or } E = \frac{n^2\pi^2\hbar^2}{2ma^2} \text{ or } E = \frac{n^2h^2}{8ma^2}. \tag{9.32}$$

Thus, the energy is independent of k. *The electron is completely bound and the energy levels are discrete. The electron in this case will be confined to the potential well and moves only in the cell of width a.*

(b) When $P \to 0$,

$$\cos ka = \cos\alpha a \text{ or } \alpha^2 = k^2 = \frac{2mE}{\hbar^2} \text{ or } E = \frac{\hbar^2 k^2}{2m} = \frac{h^2}{8\pi^2 m}\left(\frac{2\pi}{\lambda}\right)^2. \tag{9.33}$$

$$\therefore E = \left(\frac{h^2}{2m}\right)\frac{1}{\lambda^2} = \frac{h^2}{2m}\left(\frac{p^2}{h^2}\right) = \frac{p^2}{2m} = \frac{1}{2}mv^2. \tag{9.34}$$

The energy E in this case corresponds to that of a *free particle.* All energies are allowed to the electrons as there is no energy level exists.

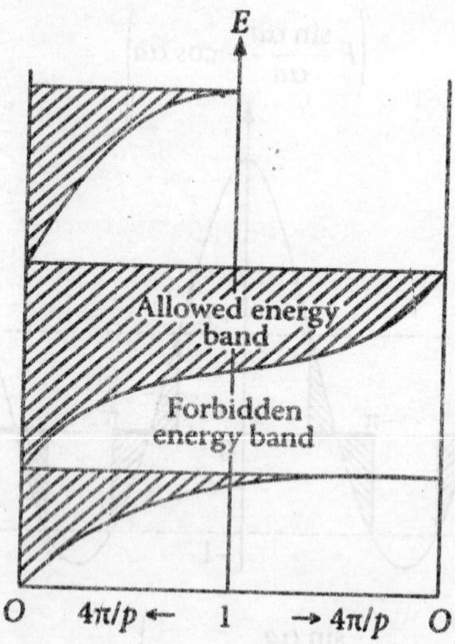

Fig. 9.11: Allowed and forbidden energy band as a function of *P*.

Fig. 9.11 shows the energy spectrum of *P*. At the extreme left for $P = 0$, the energy spectrum is *quasi-continuous* and on the extreme right for $P \to \infty$, a line spectrum is resulted. For $0 < P < \infty$, the position and width of allowed and forbidden energy bands can be shown by drawing a vertical line.

9.6 Energy (*E*) versus Wave Vector (*k*) Relationship

(9.33) gives a parabolic curve when *E* is plotted against *k* as shown by dotted lines in Fig. 9.12. An electron being acted by the lattice potential has a (*E-k*) curve as shown by solid lines. Near the energy gap at $ka = \pm n\pi$, the solid line deviates from the parabola considerably.

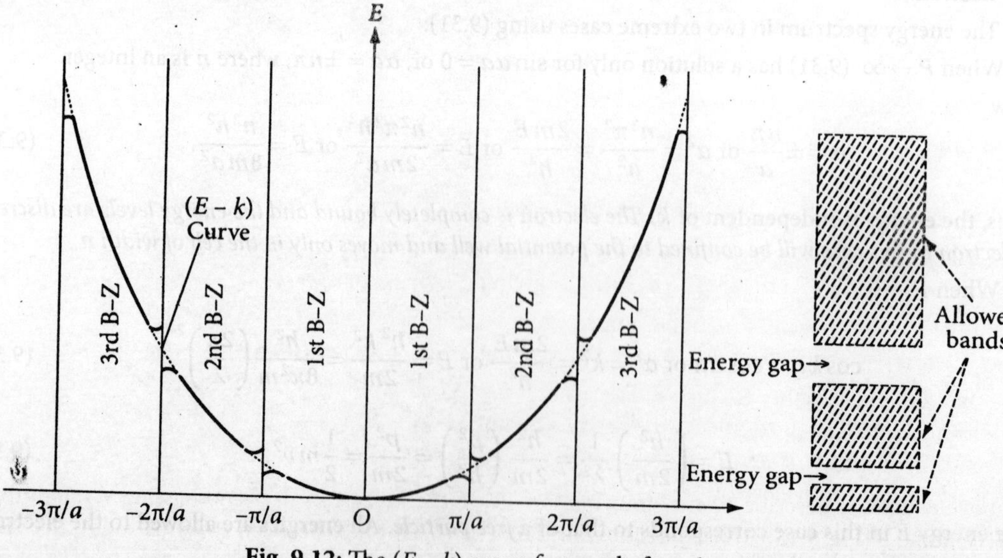

Fig. 9.12: The (*E* – *k*) curve for nearly free electron.

When $k=\pm\frac{n\pi}{a}$, the free electron energy is given by (9.34)

$$\text{i.e., } E=\frac{\hbar^2 n^2\pi^2}{2ma^2}=\frac{n^2h^2}{8ma^2}. \tag{9.35}$$

This is the same energy as that for a free particle in a potential well of width a.
If the size of the well is replaced by the distance between the atoms, say 0.3 nm, then from (9.35),

$$E=4n^2\,\text{eV}. \tag{9.36}$$

This shows that beyond a few discontinuities in the (E-k) curve, $E\to\infty$. The regions beyond the discontinuities are called *Brillouin zones.*

9.7 Brillouin Zones in Two Dimensions

The condition for any discontinuity is

$$k=\pm\frac{n\pi}{a}. \tag{9.37}$$

For *two-dimensional case,* the corresponding condition is

$$k_x n_1+k_y n_2=\frac{\pi}{2}\left(n_1^2+n_2^2\right),$$

where n_1 and n_2 are integers relating to x- and y-axes respectively. To sketch the first zone n_1 and n_2 are made in turn to ±1 or 0. The borders of the first zone are

$$n_1=+1,\quad n_2=0,\quad \text{so,}\quad k_x=\pm\frac{\pi}{a}$$

$$n_1=0,\quad n_2=\pm1,\quad \text{so,}\quad k_y=\pm\frac{\pi}{a}.$$

So, the *first Brillouin zone* is a square passing through the points A, B, C and D [Fig. 9.13].
For the *2nd Brillouin zone,* we have

$$n_1=+1,\quad n_2=+1,\quad \text{giving}\quad k_x+k_y=\frac{2\pi}{a}$$

$$n_1=-1\quad n_2=+1,\quad \text{giving}\quad -k_x+k_y=\frac{2\pi}{a}$$

$$n_1=+1,\quad n_2=-1,\quad \text{giving}\quad k_x-k_y=\frac{2\pi}{a}$$

$$n_1=-1,\quad n_2=-1,\quad \text{giving}\quad -k_x-k_y=\frac{2\pi}{a}$$

The above equations describe a set of four lines at 45° to the k_x and k_y axes and passing through the points P, Q, R and S [Fig. 9.13]. Thus, the *second Brillouin zone* is the area between P, Q, R, S and A, B, C, D. Putting n_1 and $n_2=0,\pm1$ and ±2, the *third Brillouin zone* may be obtained [Fig. 9.13].

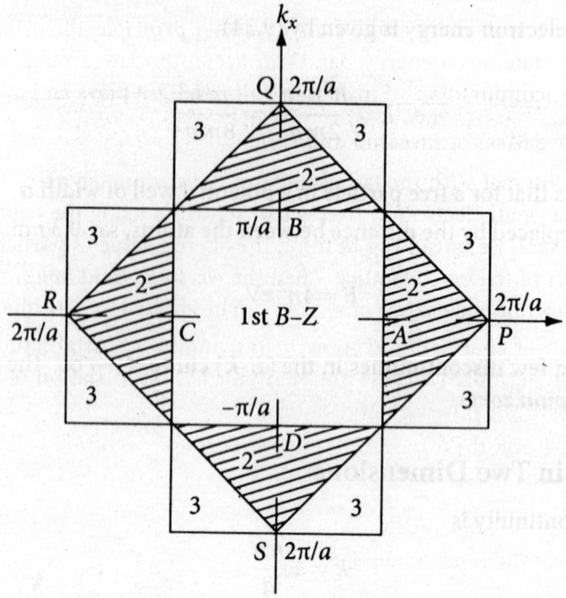

Fig. 9.13: Brillouin zones for a square lattice.

9.7.1 Number of Wave Functions in a Band

We have seen that in an infinitely long one-dimensional crystal there are certain allowed energy ranges which have a continuous distribution of energy.

For a linear crystals of length L, the periodic boundary condition is $\psi(x+L) = \psi(x)$.

As the wave function are *Bloch functions,* we have $e^{iK(x+L)}\, u_k(x+L) = e^{iKx}\, u_k(x)$, since $u_k(x+L) = u_k(x)$.

$\therefore\ e^{iK(x+L)} = e^{iKx}$ or $e^{iKL} = 1$.

$$\therefore\ k = \frac{2\pi n}{L} \tag{9.38}$$

with $n = \pm1, \pm2, \ldots$

$$\text{or } k = \pm\frac{2\pi}{L}, \pm\frac{4\pi}{L}, \ldots \tag{9.39}$$

So, the number of possible wave function in range dk is

$$dk = \frac{2\pi}{L}\, dn, \text{ [from (9.38)]}$$

$$\text{or } dn = \frac{L}{2\pi}\, dk. \tag{9.40}$$

In the first Brillouin zone, the maximum value of k is π/a, where a is the length of the *primitive cell.* Again, if in a crystal of length L, the number of primitive cells is N, then $a = L/N$. So, the maximum value of k in the *first Brillouin zone* is $N\pi/L$. This means that the series in (9.39) terminates at $N\pi/L$. Thus, the total number of allowed k-values in the *first Brillouin zone* is N. *Hence, the total number of possible states in any energy band is equal to the number of primitive unit cells N.*

Since an electron has *two spin states* and *Pauli's exclusion principle* allows at best two electrons with opposite spins in a single state, each energy band can accommodate a maximum of $2N$ electrons. For distinguishing metals, semiconductors and insulators this resultant plays an important role.

9.7.2 Concept of Effective Mass of Electron and Hole

In a crystal the electrons interact with the lattice and are not completely free. The wave particle motion, therefore, differs from that of an electron in free space. So, while using the equations of electrodynamics for charge carriers in a crystal, we must keep in mind the altered value of particle mass called the *effective mass* (m^*) due to the effect of the crystal lattice. Then the *electrons* (and *holes*) can be regarded as almost free charge carriers and *Newton's second law of motion* can be applied to describe their dynamic behaviour.

We consider an electron of charge e and mass m moving in a crystal lattice of electric field E. The acceleration $a=\frac{eE}{m}$ in the periodic lattice is not constant. Let the variation of acceleration is due to the variation of electron's mass while moving in the lattice.

$\therefore$ the acceleration $a=\frac{eE}{m^*}$ and the force on the electron is

$$F=m^*a. \tag{9.41}$$

Now the group velocity of the electron wave is

$$v_g=\frac{d\omega}{dk}=2\pi\frac{dv}{dk}=\frac{2\pi}{h}\frac{dE}{dk}=\frac{1}{\hbar}\frac{dE}{dk}, \tag{9.42}$$

where $E=h\nu$ and $\hbar=h/2\pi$.

Acceleration

$$a=\frac{dv_g}{dt}=\frac{1}{\hbar}\frac{d^2E}{dk\,dt}=\frac{1}{\hbar}\frac{d^2E}{dk^2}\frac{dk}{dt}.$$

But $\hbar k=p$ and $\frac{dp}{dt}=F$.

$$\therefore \quad \frac{dk}{dt}=\frac{F}{\hbar}.$$

So

$$a=\frac{1}{\hbar^2}\frac{d^2E}{dk^2}F \text{ or } F=\frac{\hbar^2}{\frac{d^2E}{dt^2}}a. \tag{9.43}$$

Comparing (9.41) and (9.43), we get

$$m^*=\frac{\hbar^2}{\frac{d^2E}{dk^2}}. \tag{9.44}$$

Thus, determining the value of $\frac{d^2E}{dk^2}$, we can calculate the value of the *effective mass* m^*.

(i) Variation of E with k: The variation of E with k for the first allowed band is shown in Fig. 9.14 (a). Using this type of variation of E with k, the group velocity $v_g=v$ can be calculated using (9.42).

(ii) Variation of v with k: Fig. 9.14 (b) shows the variation of v with k. For $k=0$, $v=0$ and with the increase of k (i.e., energy), v increases and reaches a maximum value at $k=k_0$, where k_0 is the value of k at the point of inflection of the (E-k) curve. Beyond k_0, v begins to decrease and finally assumes the zero value at $k=\pi/a$. The negative value of k shows a similar behaviour.

(iii) The variation of m^* with k: Fig. 9.14 (c) shows the variation of m^* with k. Around $k=0$, m^* increases, reaches its maximum value at the point of inflection of the (E-k) curve. Above the point of inflection, m^* is *negative* and as $k\rightarrow\pi/a$, m^* reduces to a small negative value.

The negative mass is due to the force on the electron (caused by Bragg reflection) in one direction leading to a gain in momentum in opposite direction. This results in negative effective mass.

The equation $E = \frac{\hbar^2 k^2}{2m}$ is satisfied near the bottom of the band and hence m^* has a constant value. But with the increase of k, m^* increases. Fig. 9.14 (b) shows that beyond the inflection point k_0, velocity decreases due to the decrease of acceleration. This implies that in this region of k the lattice exerts a large retarding force on the electron and it behaves as a positively charged particle called *hole*.

(iv) Variation of f_k with k: The degree of freedom of an electron is defined as

$$f_k = \frac{m}{m^*} = \frac{m}{\hbar^2}\left(\frac{d^2E}{dk^2}\right). \tag{9.45}$$

f_k gives a measure of extent to which the electron is free. When m^* is large, f_k is small, the particle, then behaves as a *heavy* particle. *If $f_k = 1$, the electron behaves as a free electron.*

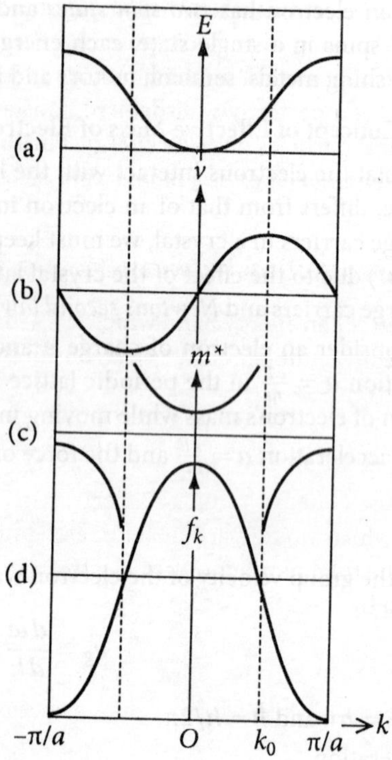

Fig. 9.14: E, m^* and f_k as a function of k.

9.8 Crystal Momentum

The momentum associated with the motion of a free electron is $\hbar\vec{k}$, where $\vec{k}$ is the wave vector. This momentum of free electron is called the *true momentum* which helps to understand the dynamical behaviour of it. For example, the energy of the free electron is

$$E = \frac{\hbar k}{2m} = \left(\frac{1}{2m}\right)\left(\frac{h^2}{4\pi^2}\right)\left(\frac{4\pi^2}{\lambda^2}\right) = \left(\frac{1}{2m}\right)\left(\frac{h^2}{\lambda^2}\right) = \frac{1}{2m}p^2. \tag{9.46}$$

This $\vec{k}$ was first introduced by *Sommerfeld* while formulating the theory of free electrons.

However, when an electron moves in a *periodic potential*, $\hbar\vec{k}$ does not represent the true momentum as the energy does not vary with $\vec{k}$ in the manner as it does in case of a free electron and as such known as the *crystal momentum.*

For an electron moving in a periodic potential, $\hbar\vec{k}$ is not the true electron momentum. This is also confirmed by the fact that, the Bloch wave functions which are eigenstate of the Hamiltonian, are not simultaneously eigenstates of the momentum operator

$$\vec{p}\psi_k(\vec{r}) = \frac{\hbar}{i}\nabla\psi_k(\vec{r}) = \frac{\hbar}{i}\nabla\left[e^{i\vec{k}\cdot\vec{r}}\,u_k(\vec{r})\right] = \hbar k\psi_k(\vec{r}) + e^{i\vec{k}\cdot\vec{r}}\frac{\hbar}{i}\nabla u_k(\vec{r}). \tag{9.47}$$

$\therefore$ $\vec{p}\psi_k(\vec{r})$ is not simply $\hbar k\psi_k(\vec{r})$, still $\hbar\vec{k}$ is an extension of momentum $\vec{p}$ to the periodic potential of the crystal. This is why $\hbar\vec{k}$ for a lattice is known as the *crystal momentum*, $\vec{k}$ being the wave vector introduced in *Bloch's theorem*.

It is to be remembered that, the free electron momentum $\vec{p} = \hbar\vec{k}$ is a constant of motion but for an electron moving in periodic potential $\hbar\vec{k}$ is not constant of motion and cannot be calculated directly by quantum mechanical method, only its average value can be calculated. On the other hand, the crystal momentum is a perfectly well-defined constant for a state of given energy just as the true momentum of a dynamical particles gives its energy. For this reason, *in dealing with interaction of electron with lattice, we use conservation of crystal momentum instead of true electron momentum.*

9.9 Distinction between Metals, Insulators and Semiconductors

We consider an energy band which is filled with electrons up to a value k_1 ($k_1 < \pi/a$) as shown in Fig. 9.15. The effective number of free electrons in the band is responsible to the conductivity of this band.

Now

$$f_k = \frac{m}{\hbar^2}\left(\frac{d^2E}{dk^2}\right)$$

gives the extent up to which an electron in the k state is free which would participate in electron conduction. The total number of free electrons in the band is given by

$$N_{\text{eff}} = \sum_k f_k, \tag{9.48}$$

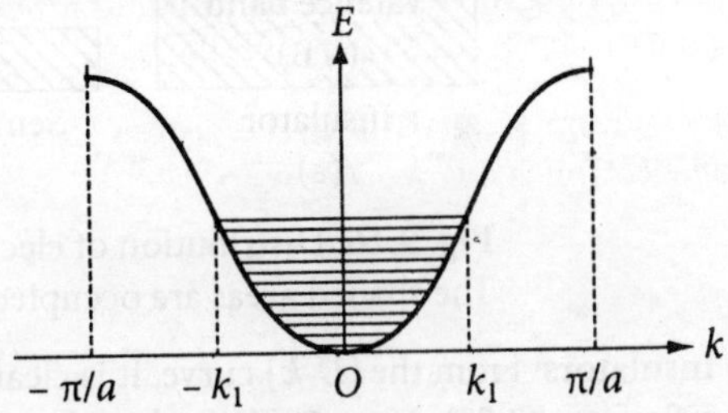

Fig. 9.15: Energy band filled up to the state k_1.

$\sum_k$ corresponds to various occupied energy states in the band.

Now the number of possible states in k to $k + dk$ for a one-dimensional lattice of length L is

$$dn = \frac{L}{2\pi}dk.$$

$\therefore$ the total number of free electrons present within the limit $-k_1$ to $+k_1$ is

$$N_{\text{eff}} = 2 \times \frac{1}{2\pi}\int_{-k_1}^{+k_1} f_k\, dk = \frac{L}{\pi}\int_{-k_1}^{k_1} f_k\, dk. \tag{9.49}$$

The factor 2 on the right side of (9.49) is due to the fact that each state can occupy a maximum of two electrons with opposite spin.

$$\therefore N_{\text{eff}} = \frac{L}{\pi}\int_{-k_1}^{k_1}\frac{m}{\hbar^2}\left(\frac{d^2E}{dk^2}\right)dk, \quad \left[\because f_k = \frac{m}{\hbar^2}\left(\frac{d^2E}{dk^2}\right)\right]$$

$$= \frac{2L}{\pi}\frac{m}{\hbar^2}\int_0^{k_1}\frac{d^2E}{dk^2}dk = \frac{2Lm}{\pi\hbar^2}\left(\frac{dE}{dk}\right)_{k=k_1}. \tag{9.50}$$

The behaviour of solids as *insulator, semiconductor* and *conductor* can be explained on the basis of *Band Theory* as follows:

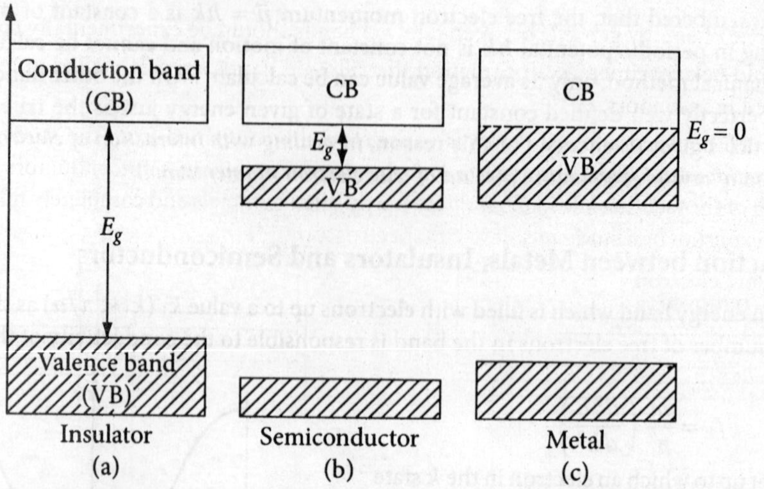

Fig. 9.16: Distribution of electrons in insulator, semiconductor and metal. The shaded areas are occupied by electrons.

(a) Insulators: From the $(E\text{-}k)$ curve, it is clear that, at the top (and bottom) of the band $\frac{dE}{dk} = 0$ as $k = 0$, π/a. So from (9.50), $N_{\text{eff}} = 0$. Thus when the valence band is completely filled (and the conduction band is completely empty) the solid behaves as a *perfect insulator* [Fig. 9.16(a)]. There is a *forbidden energy gap or band gap* (E_g) between the valence band and conduction band. For example, diamond is a good insulator having band gap $E_g \sim 6\,\text{eV}$. As the electrons in the valence band ordinarily do not possess this amount of energy they cannot transit to the conduction band and participate in the electrical conduction process. As a result, the solid behaves as an insulator. *All solids with large band gap* (E_g) *are good electrical insulators.*

(b) Semiconductors: There is a class of solids like Germanium (Ge) and Silicon (Si), etc. for which the band gap E_g is relatively small $\sim 1\,\text{eV}$. It is, therefore, possible for some electrons at room temperature to have thermal energy to jump across the forbidden gap into the empty conduction band. In presence of an electric field, these electrons constitute a limited current. The conductivity of such materials lies between good conductor and insulator [Fig. 9.16(b)].

(c) Conductors (Metals): Since the value of $\frac{dE}{dk}$ is the maximum at the point of inflection of $(E\text{-}k)$ curve [Fig. 9.15], the effective number of electrons N_{eff} is the maximum. So, in a material with completely filled valence band, there is a large number of the electrons available for conduction. Such materials are conductors like all metals. The valence band and conduction band in conductors partially overlap so that the band gap $E_g = 0$, Fig. 9.16 (c).

9.9.1 Effect of Temperature

In case of *good conductors* (metals) with the increase in temperature, more and more phonons are excited which scatter electrons and reduce their mobility. This, in turn, leads to the increase in resistivity and decrease the conductivity of metals. But in case of semiconductors, increase in temperature causes more and more electrons in the valence band to have sufficient energy to jump into the conduction band. Hence, *with the rise of temperature the conductivity of the semiconductors increases. This is an important distinction between metals and semiconductors.*

9.10 Direct Experimental Evidence for Band Structure in Solids

The essential requirement for the direct evidence of the band structure of the energy levels in solids are:

1. The existence of energy gap ΔE between allowed bands should be experimentally confirmed.
2. There should be experimental confirmation that allowed bands are partly filled in metals and completely filled in insulators.

The existence ΔE between allowed bands is experimentally confirmed from the study of variation of resistivity with temperature, light absorption and photo-conductivity in semiconductors and insulators. The confirmation of the fact that allowed bands are partly filled in metals and completely filled in insulators comes from the experimental study of soft X-ray spectroscopy.

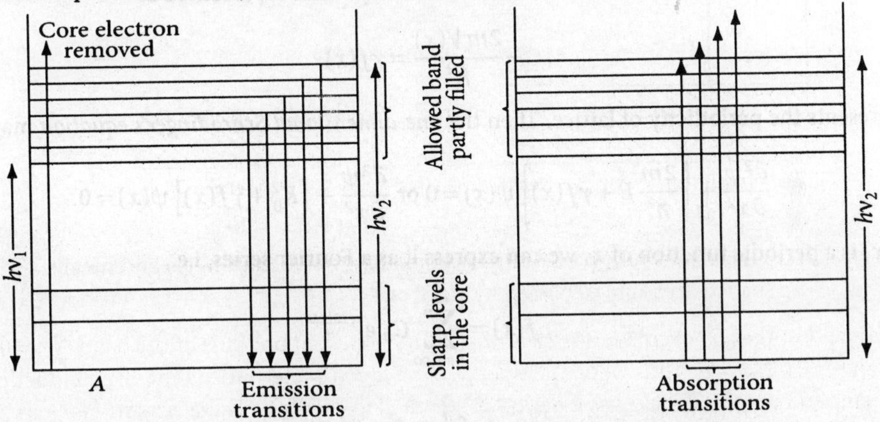

Fig. 9.17: Various transitions in soft X-ray spectroscopy.

In Fig. 9.17, A represents the energy level of a core electron, say, $1s$ (K-shell) state. This type of core levels are sharp. Let bombardment of a suitable cathode ray electron to the target atom, the core electron is removed and a vacancy is created there. This vacancy in the $1s$ state can be filled by the transition of an electron in the higher allowed band which is partly filled [Fig. 9.17 (a)].

Since there are no energy levels in higher band from where transition would take place to ν_2, there will be no emission, if $\nu > \nu_2$. Sudden emissions of all transitions will occur behaving as emission edge within $\nu_1 < \nu < \nu_2$. For $\nu < \nu_1$, no emission will occur due to the absence of transition levels. The observed emission bandwidth is a measure of the extent up to which the allowed band is filled.

Fig. 9.17 (a) shows the various absorption transitions of X-rays in the same specimen.

For $\nu < \nu_2$, there is no absorption of energy but for $\nu = \nu_2$, there is an absorption edge. For $\nu > \nu_2$, an absorption band arises. Hence, the existence of the unfilled levels in the allowed band is evidenced by the absorption of energy. In case of metals, the absorption and emission edges coincide in frequency.

The emission bands for non-metals are different from those of metals and have no sharp edges. In case of metals, the transitions from filled to unfilled levels occur over a region proportional to kT at the Fermi levels and hence, are not perfectly sharp. This is why the width of the emission edge is temperature dependent.

9.11 Limiting Cases of Periodic Potential

In the foregoing discussion of *Kronig-Penney model,* Sommerfeld's free electron theory is used which is poorly related to the actual form of potential. The effect of ion cores cannot be totally ignored. Therefore, two limiting cases arise where the periodic potential is either very weak or very strong. In case of very strong periodic potential each electron is almost bound to the minimum of the potential, the rest of the lattice is as being a perturbation to this minimum. This is *tight binding approximation.*

In case of solids where the above methods are inadequate, we use more complex methods like *orthogonalised plane wave* (OPW) or *augmented plane wave* (APW) *methods*.

9.11.1 Nearly Free Electron (NFE) Approximation

Here the crystal potential is assumed to be very weak in comparison to the electron's KE E, and the effect of the periodic potential as a *perturbation*. Under these conditions the allowed bands will be broad and the forbidden gaps quite narrow.

Let the periodic potential $V(x)$ in a one-dimensional crystal be represented as

$$-\frac{2m\,V(x)}{\hbar^2}=\gamma f(x). \tag{9.51}$$

$f(x)$ represents the periodicity of lattice. Then the *one-dimensional Schrödinger's equation* may be written as

$$\frac{\partial^2\psi}{\partial x^2}+\left[\frac{2m}{\hbar^2}E+\gamma f(x)\right]\psi(x)=0 \text{ or } \frac{\partial^2\psi}{\partial x^2}+\left[K_0^2+\gamma f(x)\right]\psi(x)=0. \tag{9.52}$$

As $f(x)$ is a periodic function of x, we can express it as a Fourier series, i.e.,

$$f(x)=\sum_{n=-\infty}^{\infty} C_n\, e^{-\frac{2\pi inx}{a}} \tag{9.53}$$

with

$$C_n=\frac{1}{a}\int_0^a f(x)\,e^{\frac{2\pi inx}{a}}\,dx, \tag{9.54}$$

where a is the lattice constant of the crystal.

If V_n be the Fourier expansion coefficient of $V(x)$, then according to (9.51)

$$V_n=-\frac{\hbar^2}{2m}\gamma C_n. \tag{9.55}$$

The wave function $\psi(x)$ must have the Bloch form:

$$\psi(x)=e^{iKx}\,U(x)=e^{iKx}\sum_{n=-\infty}^{\infty} b_n\, e^{-\frac{2\pi inx}{a}}, \tag{9.56}$$

where $U(x)$ is a periodic function of a, it can be expressed as a Fourier series as

$$U_k(x)=\sum_{-a}^{\infty} b_n\, e^{-\frac{2\pi lnx}{a}}. \tag{9.57}$$

In case the electron is completely free, $\gamma=0$ and $n=0$, when

$$\psi(x)=b_0\, e^{iKx}. \tag{9.58}$$

In the limit $U_k(x)\to b_0$ and $K\to K_0$. So, as $x\to 0$ all b_n (except b_0) approach zero. Then the approximate expression for the wave function is of the form:

$$\psi(x)=b_0\, e^{iKx}+\gamma\left[e^{iKx}\sum_{n\neq 0} b_n\, e^{-\frac{2\pi nx}{a}}\right] \tag{9.59}$$

(9.59) is valid for small values of γ

$$\text{or } \psi(x)=\left[b_0+\gamma\sum_{n\neq 0} b_n e^{-\frac{2\pi n x}{a}}\right] e^{iKx}=U(x)e^{iKx}, \tag{9.60}$$

where

$$U(x)=b_0+\gamma\sum_{n\neq 0} b_n e^{-\frac{2\pi n x}{a}}.$$

Substituting (9.59) in (9.52), we get

$$b_0\left(K_a^2-K^2\right)e^{iKx}+\gamma\sum_{n\neq 0}\left[\left(K_0^2-K_a^2\right)b_n+b_0 c_n\right]e^{iK_n x}+\gamma^2\sum_{n\neq 0}\sum_{n'\neq 0} b'_n C_n e^{i\left(K-\frac{2\pi n}{a}-\frac{2\pi n'}{a}\right)x}=0, \tag{9.61}$$

where

$$K_n=-K-\frac{2\pi n}{a}. \tag{9.62}$$

As a first approximation, we neglect γ^2 term in (9.61). Then multiplying the equation by $e^{-iK_m x}$ and integrating over unit cell from $x=0$ to $x=a$, we have

$$b_0\left(K_0^2-K^2\right)\int_0^a e^{\frac{2\pi i m x}{a}}dx+\gamma\sum_{n\neq 0}\left[\left\{\left(K_0^2-K_n^2\right)b_n+b_0 C_n\right\}\times\int_0^{a_0} e^{\frac{2\pi i(m-n)x}{a}}dx\right]=0. \tag{9.63}$$

For $m=0$,

$$\int_0^a e^{\frac{2\pi i(m-n)x}{a}}dx=0,$$

for all values of n in the summation. Then

$$b_0\left(K_0^2-K^2\right)=0 \text{ or } K=K_a. \tag{9.64}$$

Again, for $m\neq 0$,

$$\int_0^a e^{\frac{2\pi i m x}{a}}dx=0$$

and the second integral in (9.61) gives zero except $m=n$.

So

$$\gamma\left[\left(K_0^2-K_m^2\right)b_m+b_0 C_m\right]a=0 \text{ or } b_m=\frac{b_0 C_m}{K_m^2-K_0^2}=\frac{b_0 C_m}{K_m^2-K^2}. \tag{9.65}$$

$\therefore$ from (9.64),

$$K^2=K_0^2 \text{ or } \frac{2mE}{\hbar^2}=K_0^2 \text{ or } E=\frac{K_0^2\hbar^2}{2m}. \tag{9.66}$$

The relation is the same as that for a free particle.

Now substituting the value of b_m from (9.65) into (9.60), we get

$$\psi(x)=b_0 e^{iKx}\left[1+\gamma\sum_{n\neq 0}\frac{C_n}{K_n^2-K^2}e^{-\frac{2\pi i n x}{a}}\right]. \tag{9.67}$$

(9.64) shows that the contribution of the first order correction to the energy of a free particle due to the periodic potential is zero. To obtain second order energy correction, we multiply (9.61) by e^{-iKx} and then integrate from $x=0$ to $x=a$,

$$ab_0\left(K_0^2-K^2\right)+\gamma\sum_{n\neq 0}\left[\left(K_0^2-K_n^2\right)b_n+b_0C_n\right]\int_0^a e^{-\frac{2\pi inx}{a}}\,dx+\gamma^2\sum_{n\neq 0}\sum_{n'\neq 0}b'_{n'}C_n\int_0^a e^{-\frac{2\pi i(n+n')x}{a}}\,dx=0. \tag{9.68}$$

Now

$$\int_0^a e^{-\frac{2\pi inx}{a}}\,dx=0,$$

for all allowed values of n and

$$\int_0^a e^{-\frac{2\pi i(n+n')x}{a}}\,dx=0,\ \text{if } n\neq -n'.$$

Thus, (9.68) reduces to

$$b_0\left(K_0^2-K^2\right)a+\gamma^2\sum_{n'\neq 0}b'_nC'_na=0. \tag{9.69}$$

Again, replacing n by $-n$ in (9.53), we get

$$f(x)=\sum_{n=-\infty}^{\infty}C_{-n}\,e^{\frac{2\pi nix}{a}}. \tag{9.70}$$

The complex conjugate of (9.53) is

$$f^*(x)=\sum_{n=-\infty}^{\infty}C_n^*\,e^{\frac{2\pi inx}{a}}\ \text{or}\ f(x)=\sum_{n=-\infty}^{n=\infty}C_n^*\,e^{\frac{2\pi nix}{a}}, \tag{9.71}$$

since $f^*(x)=f(x)$. Comparing (9.70) and (9.71), we get

$$C_{-n}=C_n^*. \tag{9.72}$$

Now using (9.65) and (9.72) in (9.69), we get

$$K_0^2=K^2+\gamma^2\sum_{n\neq 0}\frac{C_n^*C_n}{K^2-K_n^2}=K^2+\gamma^2\sum_{n\neq 0}\frac{C^*C_n}{K^2-\left(K-\frac{2\pi n}{a}\right)^2}\left[\because K_n=K-\frac{2\pi n}{a}\right]. \tag{9.73}$$

The above relation can be expressed in terms of $E\left(=\frac{\hbar^2K^2}{2m}\right)$ as follows:

$$E=\frac{\hbar^2K^2}{2m}+\sum_{n\neq 0}\frac{|V_n|^2}{\left(\frac{\hbar^2K^2}{2m}\right)-\frac{\hbar^2}{2m}\left(K-\frac{2\pi n}{a}\right)^2}. \tag{9.74}$$

If $K^2\approx K_n^2$ for a particular value of n, then one of the quantities

$$\left(K^2-K_n^2\right)\ \text{in}\ \sum_{n\neq 0}\frac{C_n}{K_n^2-K^2}\,e^{-\frac{2\pi inx}{a}}$$

of (9.67) tends to zero and the corresponding term will tend to infinity despite γ is very small. Then the (9.59) is not a suitable form of the wave function $\psi(x)$. Further, if $K^2 \approx K_n^2$, then

$$K = \pm K_n = \pm K \mp \frac{2\pi n}{a}. \tag{9.75}$$

Since, for the upper sign, this equation gives $n=0$ and hence excluded from (9.67). For the lower sign, we get

$$K = \frac{n\pi}{a}. \tag{9.76}$$

(9.62) refers to all the band edge points as n may assume both positive as well as negative values.

Now let $K \approx n\pi/a$, whence $K_n = K - 2\pi n/a \simeq -\frac{n\pi}{a}$. Then $K_n^2 \simeq K^2$ and b_n will not be very large. Under this condition γb_n will not be very small. The wave function in this case can be written as

$$\psi(x) = b_0\, e^{iKx} + \gamma b_n\, e^{iKx}\, e^{-\frac{2\pi i n x}{a}} = b_0\, e^{iKx} + \gamma b_n\, e^{iK_n x}. \tag{9.77}$$

As $K_n = -K$ at the band edge, $\psi(x)$ in (9.77) in the result of superposition of a wave travelling along the positive x-direction and the other along the negative x-direction, i.e., $\psi(x)$ represents a standing wave.

Now putting

$$f(x) = \sum_{n=-\infty}^{+\infty} C_n\, e^{-\frac{2\pi i n x}{a}}$$

and $\psi(x)$ as given by (9.77) in (9.52), we get

$$b_0^2\left(K_0^2 - K^2\right) e^{iKx} + \gamma b_n \left(K_0^2 - K^2\right) e^{i\left(K - \frac{2\pi n}{a}\right)x} + \gamma b_0 \sum_{n'\neq 0} C_n^* e^{i(K - 2\pi n'/a)x}$$

$$+\gamma^2 b_n \sum_{n'\neq 0} C_0^* e^{i\left(K - \frac{2\pi n}{a} - \frac{2\pi n'}{a}\right)x} = 0. \tag{9.78}$$

Multiplying (9.78) by e^{-iKx} on both sides and then integrating within limits, $x=0$ to $x=a$, we find that the contributions of the second and third terms become zero and that from the fourth term is obtained only for $n'=n$. Finally,

$$\left(K_0^2 - K^2\right) b_0 + \gamma^2 C_n^* b_n = 0. \tag{9.79}$$

Similarly, multiplying both sides of (9.78) by

$$e^{-iK_n x} = e^{-i(k - 2\pi n/a)x}$$

and then integrating from $x=0$ to $x=a$, we get

$$C_n b_0 + \left(K_0^2 - K_n^2\right) b_n = 0. \tag{9.80}$$

(9.79) and (9.80) are a set of homogeneous equations whose solutions other than $b_0 = b_n = 0$ exist, if

$$\begin{vmatrix} \left(K_0^2 - K^2\right) & \gamma^2 C_n^* \\ C_n & K_0^2 - K_n^2 \end{vmatrix} = 0 \text{ or } \left(K_0^2 - K^2\right)\left(K_0^2 - K_n^2\right) - \gamma^2 C_n C_n^* = 0. \tag{9.81}$$

This is a quadratic equation having solution

$$K_0^2 = \frac{1}{2}\left[\left(K^2 + K_n^2\right) \pm \sqrt{\left(K^2 - K_n^2\right) + 4\gamma^2 C_n^* C_n}\,\right]. \tag{9.82}$$

Using the relations, $E = \frac{\hbar^2 K_0^2}{2m}$, (9.82) can be expressed as

$$E(k) = \frac{\hbar^2}{4m}\left[K + \left(K - \frac{2\pi n}{a}\right)^2 \pm \sqrt{\left\{K^2 - \left(K - \frac{2\pi n}{a}\right)^2\right\}^2 + \left\{\frac{4m|V_n|^2}{\hbar^2}\right\}}\right]. \quad (9.83)$$

At the band edge, $K - K_n = \frac{n\pi}{a}$, where (9.83) reduces to

$$E = E_n \pm |V_n| \quad (9.84)$$

with

$$E_n = \frac{\hbar^2}{2m}\left(\frac{n\pi}{a}\right)^2 \quad (9.85)$$

is the energy of a free particle at the band edge points.

The results of the foregoing discussion can be summarised as below:

1. *There exists an energy gap or discontinuity in the E versus K curve due to the internal Bragg reflection at the band edge points* $K = \pm\frac{n\pi}{a}$. *(9.81) shows that the energy gap for the forbidden band is of width* $2|V_n|$, V_n *being the nth Fourier coefficient in the Fourier expansion of the lattice potential.*

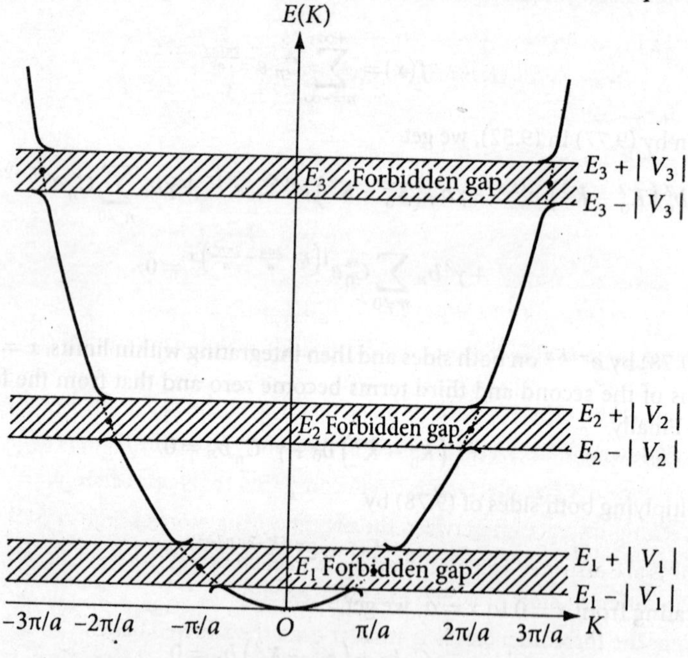

Fig. 9.18: E versus K curve for free electron approximation.

2. *For* $|K| > \frac{n\pi}{a}$, $E(K)$ *is greater than the value given by (9.64) where* $\left(K^2 - K_n^2\right)$ *is sufficient to reduce* γb_n *sufficiently small. Similarly, for* $|K| < \frac{n\pi}{a}$, $E(K)$ *must be less than the value given by (9.84). These conditions will be satisfied, if the positive sign and negative sign be chosen for* $|K| > \frac{n\pi}{a}$ *and* $|K| < \frac{n\pi}{a}$ *respectively in (9.84).*
3. *The schematic representation of variation of* $E(K)$ *with K for an electron is shown in* Fig. 9.18 *as obtained from (9.71) and (9.80).* Fig. 9.18 *shows that E versus K curves are nearly parabolic near the band edge points.*

Effective Mass

The effective mass of the electron (m_e^*) can be calculated by taking

$$K = \frac{n\pi}{a} + K',$$

where $K' \ll \pi/a$ in (9.83).

$$\therefore\ E(K) = \frac{\hbar^2}{2m}\left[\left(\frac{n\pi}{a}\right)^2 + K'^2 + m\frac{\Delta E}{\hbar^2}\sqrt{1 - 4K'^2\left(\frac{n\pi}{a}\right)^2\left(\frac{\hbar^2}{m\Delta E}\right)^2}\,\right], \tag{9.86}$$

where

$$\Delta E = 2|V_n|. \tag{9.87}$$

For points near the bottom of the upper band K' is very small. From (9.86) by expanding according to *binomial theorem* and neglecting higher order terms we get

$$E(K) = E_n + \frac{1}{2}\Delta E + \frac{\hbar^2 K'^2}{2m}\left(1 + \frac{4E_n}{\Delta E}\right). \tag{9.88}$$

Differentiating twice, we get

$$\frac{d^2E}{d^2k'} = \frac{\hbar^2}{m}\left(1 + \frac{\Delta E_n}{\Delta E_n}\right).$$

$$\therefore\ m_e^* = \frac{\hbar^2}{\frac{d^2E}{dK^2}} = \frac{\hbar^2}{\frac{d^2E}{dK'^2}} = \frac{m}{1 + \frac{4E_n}{\Delta E}}. \tag{9.89}$$

It can be shown that the effective mass of holes (m_h^*) near the top of the lower band is the same as m_e^*.

In conclusion, following facts were observed in the *nearly free electron approximation method*:

1. The PE of the nearly free electron is small compared to its total energy.
2. There is a strong interaction between the neighbouring atoms in the crystal.
3. As the atoms are assumed to be situated close to each other, the wave functions of the neighbouring electrons in different atoms overlap to a large extent.
4. The width of the forbidden bands is smaller than the allowed bands.
5. The wave functions and energy states of the crystal do not resemble with those of individual atoms.

9.11.2 The Tight Binding Approximation

Bloch in 1928 introduced an entirely new scheme for the construction of the crystal wave function known as the *tight binding approximation or the linear combinations of atomic orbitals* (LCAO) method.

This approximation deals with the situations in which the overlap of atomic wave function is enough to require corrections to the pictures of the isolated atoms. In this method, one starts from the wave functions for an electron in a free atom and then constructs a crystal orbital which just equals the LCAOs and describes the electron in the periodic field of crystal as a whole.

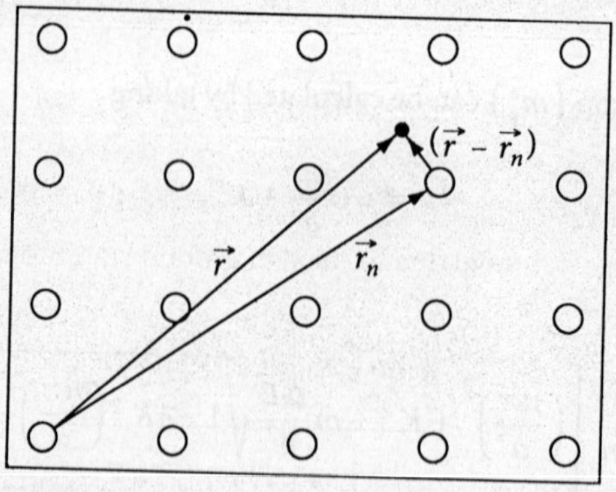

Fig. 9.19: Two-dimensional square lattice showing the positions of the nth atom and an electron.

We consider a two-dimensional square array of atoms in a crystal where $\vec{r}$ and $\vec{r}_n$ represent respectively the position vector of the electron and the nth atom respectively as shown in Fig. 9.19.

Fig. 9.20 shows the potentials of an electron in a free atom denoted by $V(\vec{r})$ and that in a crystal by $V_0(\vec{r}-\vec{r}_n)$.

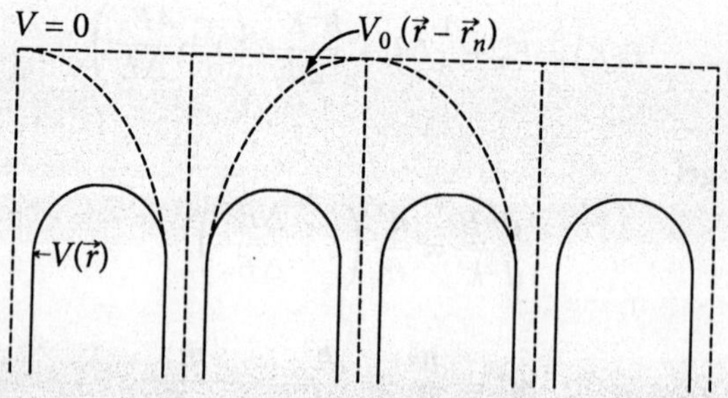

Fig. 9.20: Potential of an electron in an atom $[V(\vec{r})]$ and in a crystal $[V_0(\vec{r}-\vec{r}_n)]$.

In this method the electron wave function is assumed to be influenced by the nearest atoms and not by the other atoms in the crystal. If $\psi_k(\vec{r}-\vec{r}_n)$ be the wave function for an electron in a free atom and $\psi_k(\vec{r})$ the wave function of the electron in a crystal, then $\psi_k(\vec{r})$ may be expressed as a linear combination as

$$\psi_k(\vec{r})=\sum_n e^{(i\vec{k}\cdot\vec{r}_n)}\psi_0(\vec{r}-\vec{r}_n). \tag{9.90}$$

The summation extends over all the atoms in the crystal. It can be shown, the wave function $\psi_k(\vec{r})$ is a *Bloch function,* since

$$\begin{aligned}\psi_k(\vec{r}+\vec{r}_m)&=\sum_m e^{i(\vec{k}\cdot\vec{r}_n)}\psi_0(\vec{r}+\vec{r}_m-\vec{r}_n)=\sum_n e^{i(\vec{k}\cdot\vec{r}_n)}\psi_0(\vec{r}-\vec{r}_n+\vec{r}_m)\\&=e^{i(\vec{k}\cdot\vec{r}_m)}\sum_n e^{i[\vec{k}\cdot(\vec{r}_n-\vec{r}_m)]}\psi_0[\vec{r}-(\vec{r}_n-\vec{r}_m)]=e^{i(\vec{k}\cdot\vec{r}_m)}\psi_k(\vec{r}).\end{aligned} \tag{9.91}$$

The crystal *Hamiltonian H* can be written as

$$H=H_0+H' \tag{9.92}$$

with

$$H_0 = -\frac{\hbar^2}{2m}\nabla^2 + V_0(\vec{r}-\vec{r}_n) \tag{9.93}$$

and

$$H' = V(\vec{r}) - V_0(\vec{r}-\vec{r}_n), \tag{9.94}$$

where $V_0(\vec{r}-\vec{r}_n)$ represents the potential due to the nth ground state atom on the electron at $\vec{r}$.

The electron energy is

$$\vec{E}(k) = \frac{\int \psi_k^*(\vec{r})\,[H_0 + H']\,\psi_k(\vec{r})\,d\vec{r}}{\int \psi_k^*(\vec{r})\,\psi_k(\vec{r})\,d\vec{r}}. \tag{9.95}$$

Now

$$\int \psi_k^*(\vec{r})\,\psi_k(\vec{r})\,d\vec{r} = \sum_m \sum_n e^{i[\vec{k}\cdot(\vec{r}-\vec{r}_m)]} \times \int \psi_0^*(\vec{r}-\vec{r}_m)\,\psi_0(\vec{r}-\vec{r}_n)\,d\vec{r}. \tag{9.96}$$

If the overlap of the electron wave function between the neighbouring atoms be neglected, then

$$\int \psi_0^*(\vec{r}-\vec{r}_m)\,\psi_{0(\vec{r}-\vec{r}_n)}\,d\vec{r} = \delta_{mn} \tag{9.97}$$

and

$$\int \psi_k^*(\vec{r})\,\psi_k(\vec{r})\,d\vec{r} = N, \tag{9.98}$$

N is the total number of atoms in the crystal.

$$\therefore\ E\left(\vec{k}\right) = \frac{1}{N}\int \psi_k^*(\vec{r})\,(H_0 + H')\,\psi_k(\vec{r})\,d\vec{r} \tag{9.99}$$

$$= \frac{1}{N}\int \psi_k^*(\vec{r})\left[-\frac{\hbar^2}{2m}\nabla^2 + V_0(\vec{r}-\vec{r}_n)\right]\psi_k(\vec{r})\,d\vec{r} + \frac{1}{N}\int \psi_k^*(\vec{r})[V(\vec{r}) - V_0(\vec{r}-\vec{r}_n)]\,\psi_k(\vec{r})\,d\vec{r}$$

$$= \frac{E_0}{N}\int \psi_k^*(\vec{r})\,\psi_k(\vec{r})\,d\vec{r} + \frac{1}{N}\sum_m \sum_n e^{i[\vec{k}\cdot(\vec{r}_n-\vec{r}_m)]} \times \int \psi_0^*(\vec{r}-\vec{r}_m)[V(\vec{r}) - V_0(\vec{r}-\vec{r}_n)]\,\psi_0(\vec{r}-\vec{r}_n)\,d\vec{r}. \tag{9.100}$$

Every term in the summation from $n=0$ to $n=N-1$ contains N identical terms which may be found out by putting $r_n = 0$. Then

$$E\left(\vec{k}\right) = E_0 + \sum_m e^{-i(\vec{k}\cdot\vec{r}_m)} \int \psi_0^*(\vec{r}-\vec{r}_m)[V(\vec{r}) - V_0(\vec{r})]\,\psi_0(\vec{r})\,d\vec{r}. \tag{9.101}$$

If ψ_0 is spherically symmetric (like the ground state of alkali metal), then the contribution due to all the nearest neighbour may be assumed to be the same. The above integral for $\vec{r}_m = 0$ is

$$\int \psi_0^*(\vec{r})[V(\vec{r}) - V_0(\vec{r})]\,\psi(\vec{r})\,d\vec{r} = -\alpha \tag{9.102}$$

and for the nearest neighbour atoms

$$\int \psi_0^*(\vec{r}-\vec{r}_n)[V(\vec{r})-V_0(\vec{r})]\psi_0(\vec{r})\,d\vec{r}=-\beta, \tag{9.103}$$

where $\vec{r}_m$ is a vector connecting the atom at the origin with a nearest neighbour. α and β are constants called the *overlap integrals*. Thus, (9.101) reduces to the form:

$$E\left(\vec{k}\right)=E_0-\alpha-\beta\sum_m e^{i(\vec{k}\cdot\vec{r}_m)}. \tag{9.104}$$

(9.104) shows that the energy of the electron in the crystal differs from that of the free atom by a constant α plus a term which depends on the wave vector $\vec{k}$. *It is thus the last part which transforms the discrete atomic levels into an energy band in the solid.*

The summation in (9.104) extends only over the nearest neighbour. Since free ion potential $V_0(\vec{r})$ is more negative than the lattice potential $V(\vec{r})$, $[V(\vec{r})-V_0(\vec{r})]$ is a negative quantity and as such α and β are always positive. Thus, $E<E_0$.

9.11.3 Application of Tight Binding Approximation to Cubic Bravais Lattices

(a) **The Simple Cubic (SC) Lattice:** In this case there are six neighbours for any given ion at

$$\vec{r}_m=(\pm a,0,0),\ (0,\pm a,0),\ (0,0,\pm a), \tag{9.105}$$

where a is the lattice constant.

Therefore

$$E=(E_0-\alpha)-2\beta\left(\cos k_x a+\cos k_y a+\cos k_z a\right). \tag{9.106}$$

(b) **The Body-Centred Cubic (BCC) Lattice:** The eight nearest neighbours are at

$$r_m=(\pm a/2,\ \pm a/2,\ \pm a/2).$$

Therefore

$$E=(E_0-\alpha)-8\beta\cos\left(\frac{k_x a}{2}\right)\cos\left(\frac{k_y a}{2}\right)\cos\left(\frac{k_z a}{2}\right). \tag{9.107}$$

(c) **The Face-Centred Cubic (FCC) Lattice:** Here the twelve nearest neighbours are at

$$r_m=(0,\pm a,\pm a),\ (\pm a,0,\pm a),\ (\pm a,\pm a,0) \tag{9.108}$$

and

$$E=(E_0-\alpha)-4\beta\left[\cos\left(\frac{k_y a}{2}\right)\cos\left(\frac{k_z a}{2}\right)+\cos\left(\frac{k_z a}{2}\right)\cos\left(\frac{k_x a}{2}\right)+\cos\left(\frac{k_x a}{2}\right)\cos\left(\frac{k_y a}{2}\right)\right]. \tag{9.109}$$

Thus, the s-band is described by a constant energy term $(E_0-\alpha)$ and a term depending on the electron wave vector $\vec{k}$.

From (9.106), E is minimum for $k_x=k_y=k_z=0$.

Thus, the bottom of the energy band is given by

$$E_{\text{bottom}}=E_0-\alpha-6\beta. \tag{9.110}$$

The energy maximum corresponds to $k_x = k_y = k_z = \pm\pi/a$ and top of the energy band is given by

$$E_{\text{top}} = E_0 - \alpha + 6\beta \text{ (as } k_x = k_y = k_z = \pm\pi/a). \quad (9.111)$$

Thus, the width of the energy band is

$$E_{\text{top}} - E_{\text{bottom}} = (E_0 - \alpha + 6\beta) - (E_0 - \alpha - 6\beta) = 12\beta. \quad (9.112)$$

This corresponds to the s-state of the isolated atom. (9.112) shows that, the width of energy band increases with β which represents the overlap of the wave functions on the neighbouring atoms.

The inner electronic levels of free atoms give rise to *narrow bands* in the solid whereas the outer shell electrons give rise to *wider bands*.

Similarly, from (9.107) for the BCC lattice $E_{\text{bottom}} = E_0 - \alpha - 8\beta$ and $E_{\text{top}} = E_0 - \alpha + 8\beta$.

$\therefore$ energy bandwidth is

$$E_{\text{top}} - E_{\text{bottom}} = 16\beta \quad (9.113)$$

and for FCC lattice from (9.109), $E_{\text{bottom}} = E_0 - \alpha - 12\beta$ and $E_{\text{top}} = E_0 - \alpha + 12\beta$.

$\therefore$ bandwidth

$$E_{\text{top}} - E_{\text{bottom}} = 24\beta. \quad (9.114)$$

The variation of energy bandwidth in SC, BCC, FCC lattice is shown in Fig. 9.21.

Fig. 9.21: s-bands for (a) SC (b) BCC (c) FCC lattice in tight binding approximation.

9.11.4 Energy Surfaces

In tight binding approximation, the PE of the valence electrons in the field of their parent and neighbouring ions is larger than their KE. This indicates the separation between the ions is small compared with the electron in de Broglie wavelength, i.e., $|\vec{k}|a \ll 1$.

$$\therefore \cos ka \simeq 1 - \frac{k^2a^2}{2}. \quad (9.115)$$

Then

$$E(\vec{k}) \simeq (E_0 - \alpha) - 2\beta\left[\left(1 - \frac{k_x^2a^2}{2}\right) + \left(1 - \frac{k_y^2a^2}{2}\right) + \left(1 - \frac{k_z^2a^2}{2}\right)\right]$$

$$= (E_0 - \alpha) - 6\beta + \beta a^2\left(k_x^2 + k_y^2 + k_y^2\right)$$

$$\text{or } E(\vec{k}) = (E_0 - \alpha) - 6\beta + \beta a^2 k^2 + \ldots \text{ (for SC lattice)}. \quad (9.116)$$

Similarly, for BCC lattice,

$$E(\vec{k}) = (E_0 - \alpha) - 8\beta + \beta a^2 k^2 + \ldots \quad (9.117)$$

and for FCC lattice,

$$E(\vec{k}) = (E_0 - \alpha) - 12\beta + \beta a^2 k^2 + \ldots. \quad (9.118)$$

9.11.5 Effective Mass of the Electron

The effective mass of the electron is

$$m^* = \frac{\hbar^2}{\partial^2 E/\partial k^2} = \frac{\hbar^2}{2\beta a^2} \left[\because \frac{\partial^2 E}{\partial k^2} = 2\beta a^2 \text{ (for SC lattice)}\right]. \tag{9.119}$$

Thus, $m^* \propto 1/\beta$, this implies that *for a narrow energy band the electron behaves as a very heavy particle in SC lattice.*

From (9.113), the width of the energy band $\left(E_{\text{top}} - E_{\text{bottom}}\right) = 12\beta$ which refers to the bottom region of the band and corresponds to the constant spherical energy surfaces (i.e., $k = 0$) as shown in Fig. 9.22 (a). This means that the lowest energy state is at the centre of the zone and the higher energy states are at the corners [Fig. 9.22 (c)]. With the increase of energy ($k > 0$), the energy surfaces bulge more and more towards the centre of the zone faces, until they intersect the zone boundary. Since the normal derivative of $E\left(\vec{k}\right)$ is zero, the energy surfaces intersect the zone boundary orthogonally.

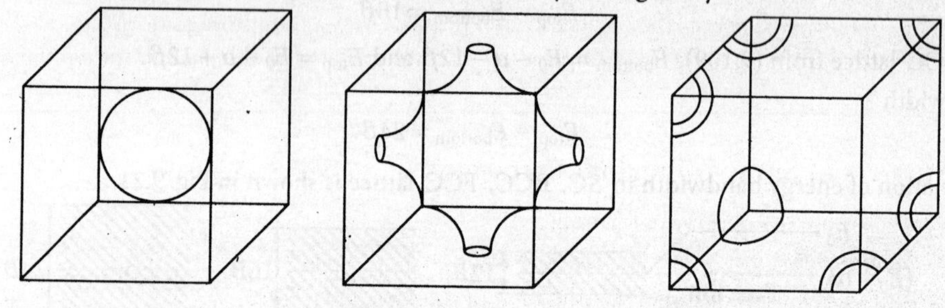

(a) (b) (c)

Fig. 9.22: Energy surfaces in tight binding approximation for SC lattice. (a) near the bottom of the band (b) half-way up the band and (c) near the top of the band.

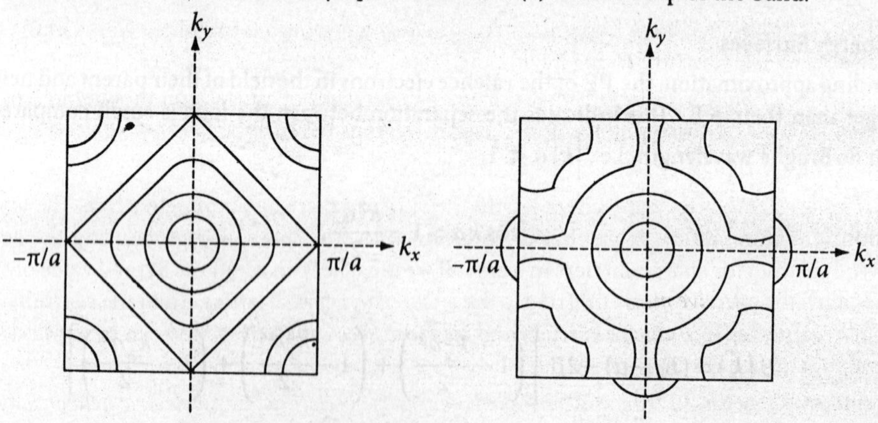

(a) (b)

Fig. 9.23: Electron energy contours for SC lattice. (a) for tight binding approximation (b) for nearly free electron approximation.

A cross section through the energy surfaces at equal intervals is depicted in Fig. 9.23. Fig. 9.23(a) shows that for small k the energy surfaces are spherical around the zone centre. The energy surfaces are also spherical with corners at the covers of the zone. Fig. 9.23 (b) shows that the energy surfaces are spherical up to large value of k (i.e., nearly free electron approximation).

Fundamental difference in the assumptions of nearly free electron approximation and tight binding approximation

Nearly free electron approximation (NFA)	*Tight binding approximation* (TBA)
1. The PE of the electron is small compared to its total energy.	1. The PE of the electron is nearly equal to its total energy.
2. The width of the forbidden bands is smaller compared to its allowed bands.	2. The width of the forbidden bands is larger compared to allowed energy bands.
3. As the atoms of the crystal are assumed to be closed spaced, the electron wave functions of neighbouring atoms overlap.	3. The atoms of the crystal are widely separated so that the electron wave functions of neighbouring atoms do not overlap to appreciable extent.
4. The interaction between the neighbouring atoms is strong.	4. The interaction between the neighbouring atoms is relatively weak.
5. The wave functions and the energy states in the crystal do not resemble with those of an individual atom.	5. The wave functions and energy as a whole are closely related to those of an individual atom.

9.12 Answers of Some Important Questions

9.1 Discuss how energy bands are formed?

Ans. See Art. 9.2.

9.2 State Bloch's Theorem. What is the outcome of the theory?

Ans. Bloch's theorem states that the solution of the Schrödinger equation for a periodic potential must be of the form

$$\psi(x) = e^{\pm ikx} u_k(x),$$

where $u_k(x) = u_k(x+a)$, a being the period. Thus, the solutions are plane waves of the type $e^{\pm ikx}$ modulated by the function $u_k(x)$ having the same periodicity as the lattice.

The *outcome of Bloch's theorem* is that the energy spectrum of an electron moving in a periodic potential consists of allowed and forbidden energy regions (or bands).

9.3 What is the concept of effective mass of the electron?

Ans. In a crystal the electrons interact with the lattice and are not completely free. The wave particle motion, therefore, differs from that of an electron in free space. So, while using the equations of electrodynamics for charge carriers in a crystal we must keep in mind the altered value of the particle mass called the *effective mass* (m^*) due to the effect of the crystal lattice. Then the *electrons* (and *holes*) can be regarded as free charge carriers and *Newton's second law of motion* can be applied to describe their dynamic behaviour.

Statement: *When an electron in a periodic potential is accelerated by an electric field or magnetic field, the mass of the electron varies and this mass is called the effective mass of the electron.*

If E be the externally applied electric field accelerating the electron in the lattice, then the effective mass of the electron is given by

$$m^* = \frac{\hbar^2}{(d^2E/dk^2)},$$

where $\hbar = h/2\pi$ and k is the wave vector.

9.4 What do you mean by valence band?

Ans. In a solid the highest filled energy band which includes electrons shared in *covalent bands* or electrons transferred in *ionic bonds* is known as *valence band.*

Valence band includes electrons which are transferred in case of *ionic bond* or shared in case of covalent bonds. These electrons are not free to gain energy from an external electric field and hence are unable to move through the solid. The sharp atomic levels are spread out in solids and all the energy states up to a forbidden energy region are occupied (at least at 0°K) and there is no more free electrons.

9.5 What is a conduction band?

Ans. The band of energy of valence electrons consisting of closely spaced energy levels is called conduction band. The bottom of the conduction band has energy E_c.

The electrons in the valence orbit of atoms are loosely bound to the nucleus and are free to move in the solid forming *free electron gas.* There is a change in energy of these electrons due to mutual interaction. The value of energy of these electrons varies from zero to a maximum value called the *Fermi energy* E_f. These electrons participate in the electric conduction process in presence of an external electric field. The energy band of these electrons is called the *conduction band.*

9.6 What is forbidden gap?

Ans. The energy gap between the valence band and conduction band is called *forbidden energy gap* or *band gap.* Electrons cannot have any energy in the forbidden energy gap. The energy of the forbidden energy is called *Band Gap* $\left(E_g\right)$. If E_v and E_c represent respectively the energy of the topmost level of the valence band and bottom level of conduction band, then $E_g = E_c - E_v$.

If E_g is very small, thermal excitation may excite the electrons in the valence band to the conduction band, so that the solid, although an insulator at 0 K, may become a conductor at temperature $T > 0$ K.

9.7 What are the sources of electrical resistance in solids?

Ans. The electrical conductivity depends mainly on two factors, viz., (a) temperature and (b) defects (e.g., impurities).

According to *Matthiessen's rule,*

$$\rho_{\text{total}} = \rho_{\text{pure}} + \rho_{\text{impure}},$$

where ρ_{pure} is the temperature dependent of resistivity and ρ_{impure} is the resistivity due to scattering of electrons by impurity atoms.

The *temperature dependence* of resistivity arises due to the scattering effect of thermal vibrations of the atoms on the electrons. *If A_0 be the amplitude of the thermal vibrations,* then the *mean free path* λ of the electron is

$$\lambda \propto 1/A_0^2.$$

Again, $A_0^2 \propto kT$. $\therefore \lambda_0^2 \propto 1/T$.

$$\therefore \rho \propto 1/\lambda \propto T,$$

i.e., *the resistivity is directly proportional to the temperature T of the solid.*

The *defect dependence* of resistivity is due to the additional scattering of electrons by the *defect or impurities.* Assigning two relaxation time, we can write,

$$\frac{1}{\tau} = \frac{1}{\tau_0} + \frac{1}{\tau'},$$

where τ_0 is independent of temperature while τ' depends on temperature.

$$\therefore \rho = \frac{m}{ne^2\tau} = \frac{m}{ne^2}\left(\frac{1}{\tau_0} + \frac{1}{\tau'}\right) = \frac{m}{ne^2\tau_0} + \frac{m}{ne^2\tau'} = \rho_0 + \rho_{\text{pure}}(T) \text{ [Matthiessen's rule]}$$

In Copper, Nickel, etc. both the contributions have been observed.

9.8 Give a comparative study of conductor, semiconductor and insulator.

Ans. The following table gives a comparative study of conductor, semiconductor and insulator.

Table 9.2

Conductor	*Semiconductor*	*Insulator*
1. The valence band (VB) and conduction band (CB) overlap.	The VB and CB are separated by a forbidden gap.	The VB and CB are separated by larger forbidden gap.
2. The band gap $E_g = 0$.	E_g ranges between 0.2 to 2.5 eV.	$E_g \sim 6\,\text{eV}$ (Diamond).
3. Electrical conductivity σ decreases with the increase of temperature (T).	σ increases with the increase of T.	$\sigma = 0$ and is independent of T.
4. Resistivity $\rho \propto T$.	Resistivity decreases with the increase of temperature.	Resistivity $\rho = \infty$ at all temperatures.

9.13 Solved Problems

9.1 An insulator has an optical absorption which occurs for all wavelengths shorter than 1800 Å. Find the width of the forbidden energy band for this insulator. [LU 1994]

Solution. Here the minimum energy required by an electron from conduction band to valence band across the forbidden energy gap is

$$E_g = h\nu = \frac{hc}{\lambda} = \frac{(6.626\times10^{-34})\times(3\times10^{8})}{1800\times10^{-10}}\,\text{J} = \frac{(6.625\times10^{-34})\times(3\times10^{8})}{(1800\times10^{-10})\times(1.6\times10^{-19})}\,\text{eV} = \mathbf{6.9\ eV.}$$

9.2 Find the lowest energy band using Kronig-Penney model for $P \ll 1$.

Solution. From the equation

$$P\frac{\sin\alpha a}{\alpha a} + \cos\alpha a = \cos ka,$$

the energy of the lowest band corresponds to $k = \pm\frac{\pi}{a}$.

$$\therefore\ P\frac{\sin\alpha a}{\alpha a} + \cos\alpha a = \pm1, \text{ (for lowest energy).}$$

Considering the magnitude of ±1, we get

$$\frac{P}{\alpha a}\sin\alpha a = 1 - \cos\alpha a \text{ or } \frac{P}{\alpha a}\left(2\sin\frac{\alpha a}{2}\cos\frac{\alpha a}{2}\right) = 2\sin^2\frac{\alpha a}{2} \text{ or } \frac{P}{\alpha a} = \frac{2\sin^2\frac{\alpha a}{2}}{2\sin\frac{\alpha a}{2}\cos\frac{\alpha a}{2}} = \tan\frac{\alpha a}{2}.$$

When $P \ll 1$,

$$\frac{P}{\alpha a} = \frac{\alpha a}{2} \text{ or } \alpha^2 = \frac{2P}{a^2} = \frac{2mE}{\hbar^2} \text{ or } E = \frac{\hbar^2}{ma}P = \frac{h^2}{4\pi^2 ma}P, \text{ i.e., } \boldsymbol{E = \frac{\hbar^2}{ma}P.}$$

9.3 The energy wave vector dispersion relation for a one-dimensional crystal of lattice constant a is given by $E(k) = E_0 - \alpha - 2\beta\cos ka$, where E_0, α, β are constants. Obtain the effective mass of the electron at the bottom and at the top of the band. [CUH 2003]

Solution. The effective mass of the electron m^* is given by

$$m^* = \frac{\hbar^2}{d^2E/dk^2} = \frac{\hbar^2}{2\beta a^2 \cos ka}.$$

At the bottom of the band $\cos ka = 1$.

$$\therefore\ \boldsymbol{m^* = \frac{\hbar^2}{2\beta a^2}}.$$

At the top of the band $\cos ka = -1$.

$$\therefore\ \boldsymbol{m^* = -\frac{\hbar^2}{2\beta a^2}}.$$

9.4 The energy near a valence band edge is given by $E(k) = -1 \times 10^{-26} k^2$ ergs. An electron is removed from the orbital $k = 1 \times 10^7 k_x\ \text{cm}^{-1}$. The band is otherwise full. Give the sign and magnitude of the effective mass of the hole.

Solution. The energy of the electron is given by

$$E = \frac{\hbar^2}{2m^*} k^2.$$

Here m^* is the effective mass of the electron.

But

$$E_k = -1 \times 10^{-26} k^2 \text{ ergs.} \ \therefore\ \frac{\hbar^2}{2m^*} = -1 \times 10^{-26}.$$

The effective mass of electron is

$$m^* = -\frac{\hbar^2}{2 \times 10^{-26}} \text{ gm.}$$

The effective mass of a hole is opposite to the effective mass of an electron at the same point on the energy band.

$\therefore\ m_h = -m_e.$

$$\therefore\ m_h^* = -m_e^* = \frac{\hbar^2}{2 \times 10^{-26}} = \frac{(1.05 \times 10^{-27}) \times (1.05 \times 10^{-27})}{2 \times 10^{-26}} = \mathbf{0.55 \times 10^{-28}\ gm.}$$

9.5 If the energy near the valence band of a crystal is given by $E = -Dk^2\ \left(D = 10^{-39}\ \text{Jm}^2\right)$. An electron with wave vector $\vec{k} = 10^{10} \hat{k}_x\ \text{m}^{-1}$ is removed from an orbital in the completely filled valence band. If the effective mass of the hole is $m_h^* = 5.5 \times 10^{30}$ kg, determine the velocity, momentum and energy of the hole.

Solution. The momentum of the hole is

$$p_h = \hbar k_h = -\left(1.05 \times 10^{-34}\right) \times 10^{10}\, \hat{k}_x = -1.05 \times 10^{-24} \hat{k}_x \text{ J-s/m.}$$

The velocity of the hole is

$$v_h = \frac{p_h}{m_h^*} = -\frac{1.05 \times 10^{-24} \hat{k}_x}{5.5 \times 10^{-30}} = \mathbf{-1.9 \times 10^5\, \hat{k}_x\ m/s.}$$

The energy of the electron with wave vector $\vec{k}_e$ is

$$E_e = -Dk_e^2 = -10^{-39} \times \left(10^{10}\,\hat{k}_x\right)^2 = -10^{-19}\,\text{J}.$$

Now the energy of the hole with zero at the top of the valence band is $E_h = -E_e = \mathbf{10^{-19}\,J}$.

9.6 An alloy of metal is found to have a resistivity of the order of 10^{-6} Ω-m at 27°C. When it is heated to a temperature of 1000 K, the resistivity increases by 6.5%. Using Matthiessen's rule, determine the resistivity of the alloy.

Solution. The total resistivity of the alloy $\rho = \rho_0 + \alpha T_1$, at temperature T_1 or, $\rho + 0.065\rho = \rho_0 + \alpha T_2$, at temperature $T_2 > T_1$ and

$$0.065\rho = \alpha(T_2 - T_1) \text{ or } \alpha = \frac{0.065\rho}{(T_2 - T_1)} = \frac{0.065\rho}{700} = 0.9 \times 10^{-10}\,\Omega\text{mK}^{-1}.$$

$$\therefore\ \rho_0 = \rho - \alpha T_1 = 10^{-6} - 0.9 \times 10^{-10} \times 300 = 10^{-6} - 2.7 \times 10^{-8} = 10^{-6}(1 - 0.0027)$$

or, $\rho_0 = \mathbf{0.9793 \times 10^{-6}}$ **Ω-m.**

EXERCISES

9(A) Objective Type Questions

Choose the correct alternatives.

9.1 According to the band theory of solids an electron can have
(a) all values of energy
(b) only the energy corresponding to energy bands
(c) the energy in the forbidden gap
(d) none of these.

9.2 The width of energy band depends on
(a) the extent of overlap of wave functions ψ_1 and ψ_2 corresponding to the two atoms
(b) the interaction between the two atoms
(c) both (a) and (b)
(d) none of these.

9.3 Bloch's theorem is applicable to
(a) periodic potential
(b) constant potential
(c) infinite potential
(d) none of these.

9.4 The energy spectrum of an electron moving in a periodic potential consists of
(a) continuous energy regions
(b) forbidden energy regions
(c) for allowed and forbidden energy regions
(d) zero energy regions.

9.5 The number of wave functions in an energy band in dk of an electron moving in a one-dimensional potential well is

(a) $dn = \frac{2\pi}{L} dk$

(b) $dn = \frac{L}{2\pi} dk$

(c) $dn = 2\pi L dk$

(d) $dn = \frac{2\pi L}{dk}$.

9.6 In an $(E\text{-}k)$ curve, the condition for energy discontinuity is

(a) $k = \pm \frac{n\pi}{a}$

(b) $k = \pm \frac{a}{n\pi}$

(c) $k = \frac{2a}{n\pi}$

(d) $k = \frac{n\pi}{a}$.

9.7 The K-electron of copper atom is so lightly bound to the nucleus that its effective mass is

(a) $m_e^* = 0$

(b) $m_e^* = \infty$

(c) $m_e^* = m_e$

(d) $0 < m_e^* < \infty$.

9.8 The effective mass of an electron moving in a periodic lattice is

(a) $m_e^* = \hbar^2 \Big/ \frac{d^2E}{dk^2}$

(b) $m_e^* = \hbar^2 \frac{d^2E}{dk^2}$

(c) $m_e^* = \frac{\hbar^2 d^2k}{dE^2}$

(d) $m^* = m_e$

where E and k are the field and wave vectors relating to the motion of the electron.

9.9 The extent to which an electron is free is given by f_k, where f_k is equal to

(a) $\frac{m}{\hbar^2}\left(\frac{d^2E}{dk^2}\right)$

(b) $m\hbar^2\left(\frac{d^2E}{dk^2}\right)$

(c) $\frac{\hbar^2}{m}\left(\frac{d^2E}{dk^2}\right)$

(d) $\frac{m}{\hbar^2}\left(\frac{d^2k}{dE^2}\right)$

9.10 If temperature is increased, then the electrical conductivity of a semiconductor

(a) decreases

(b) remains constant

(c) increases

(d) reduces to zero.

9.11 A semiconductor behaves at 0 K as

(a) an insulator

(b) a good conductor

(c) a semiconductor

(d) none of these.

9.12 The solution of Schrödinger equation for an electron moving in a periodic potential is of the form

(a) $\psi_k(\vec{r}) = e^{i\vec{k}\cdot\vec{r}}$

(b) $\psi_k(\vec{r}) = u_k(\vec{r})$

(c) $\psi_k(\vec{r}) = u_k(\vec{r})\, e^{i\vec{k}\cdot\vec{r}}$

(d) $\psi_k(\vec{r}) = u_k(\vec{r}) + Ae^{i\vec{k}\cdot\vec{r}}$.

9.13 In a crystal having N primitive cells, the maximum number of electrons per band is

(a) N

(b) $2N$

(c) ∞

(d) 0.

9.14 For a one-dimensional lattice of length L, the total effective number of free electrons in an energy band filled with electrons up to a certain volume k_1 $\left(k_1 < \frac{\pi}{a}\right)$ is

(a) $N_{\text{eff}} = \frac{\pi\hbar^2}{2Lm}\left(\frac{dE}{dk}\right)_{k=k_1}$

(b) $N_{\text{eff}} = \frac{2Lm}{\pi\hbar^2}\left(\frac{dE}{dk}\right)_{k=k_1}$

(c) $N_{\text{eff}} = \frac{Lm}{2\pi\hbar^2}\left(\frac{dE}{dk}\right)_{k=k_1}$

(d) $N_{\text{eff}} = \frac{2L}{\pi\hbar^2}\left(\frac{dE}{dk}\right)_{k=0}$.

9.15 The velocity of an electron from $(E\text{-}k)$ curve is

(a) $v = \frac{1}{\hbar}\frac{dE}{dk}$ (c) $v = \frac{1}{\hbar}\frac{d^2E}{dk^2}$

(b) $v = \hbar\frac{dE}{dk}$ (d) $v = \hbar\Big/\left(\frac{dE}{dk}\right)$.

ANSWERS

9.1 (b)	9.3 (a)	9.5 (b)	9.7 (b)	9.9 (a)	9.11 (a)	9.13 (b)	9.15 (a).
9.2 (c)	9.4 (c)	9.6 (a)	9.8 (a)	9.10 (c)	9.12 (c)	9.14 (b)	

9(B) Short Answer Type Questions

9.1 Discuss how energy bands are formed.

9.2 Compare the salient features of band theory and free electron theory. Discuss the merits and demerits of these theories.

9.3 State Bloch's theorem. What is Bloch function?

9.4 Explain the concept of conduction band, valence band and forbidden energy gap.

9.5 Distinguish metal, semiconductor and insulator in view of band theory of solids. [NBU 2001; CU 1983]

9.6 Explain the concept of effective mass of electron. [CU 1997]

9.7 What is negative effective mass? Explain with reference to the band theory of solids. [PU 2003]

9.8 Explain why the conductivity of metal decreases while that of the semiconductor increases with the rise of temperature.

9.9 Why does a completely filled band contribute nothing to an electric current in presence of an electric field?

9.10 Explain the concept of *holes*.

9.11 Show graphically the variation of energy with the wave number k for the first allowed band in a solid. Hence, explain the variation in sign of the effective mass of an electron in the band. What is a hole? [CU 2000]

9.12 Discuss how the effective mass (m^*) varies with wave vector $(\vec{k})$.

9.13 What are *Brillouin zones*? Discuss the formation of Brillouin zones for a linear lattice. [HPU 1991; PU 1997]

9.14 Prove that the total number of possible energy states in an energy band of a finite crystal is equal to the number of unit cells.

9.15 Discuss the origin of electrical resistance in a metal. What is *Matthiessen's rule* regarding temperature variation of resistivity?

9(C) Numerical Problems

9.1 Assuming a periodic potential of one-dimensional lattice of spacing a to be a box shaped waveform which has a value at each atom $-V$ and zero at $0.1a$ on either side of each lattice point, calculate the width of the first energy gap in the electron energy system. (*Ans.* 0.37 V)

9.2 If the origin be at the bottom of the conduction band, calculate the crystal momentum for a free electron of energy 0.02. [Effective mass $m_e^* = 0.2m_0$] (*Ans.* 3.416×10^{-26} kg-m/s)

9.3 An electric field of strength $E = E_0 \cos\omega t$ is applied along the x-direction of a crystal lattice. Find expressions for current density and electrical conductivity at frequency ω.

$\left(\textit{Ans. } J_x = \sigma_s(E_0)\left[\frac{\cos\omega t}{(1+\omega^2\tau^2)} + \frac{\omega\tau\sin\omega t}{(1+\omega^2\tau^2)}\right] \text{ and } \sigma_\omega = \frac{j_x}{E_0\cos\omega t}\right)$

9.4 Assuming the copper to be highly pure find the resistivity of copper at 700 K. [Given: the resistivity of copper at 300 K = $1.7 \times 10^{-8}\,\Omega$-m]. (*Ans.* $\rho = 552 \times 10^{-8}\,\Omega$-m)

9.5 Estimate the band gap of an insulator if it absorbs electromagnetic radiation of all wavelength below 3000Å. (*Ans.* 4.14 eV)

9(D) Broad Answer Type Questions

9.1 (a) Discuss briefly the origin of energy bands in solids.
(b) How would you distinguish between a semiconductor from an insulator on the basis of band theory? [BUH 1995]
(c) Prove that the motion of an electron through the periodic potential of solids gives rise to band structure. [PU 1992]

9.2 (a) Using the Kronig-Penney model, show that the energy spectrum of an electron consists of a number of allowed energy bands separated by forbidden regions.
(b) Discuss the nature of energy spectrum in the limiting cases of very weak and very strong potentials. [BUH 2004]
(c) How does the resistivity of a metal and a semiconductor vary with temperature? Explain qualitatively the temperature variation of resistivity in the two cases. [BUH 1999]

9.3 (a) Compare the salient features of the band theory and free electron theory. Discuss the merits and demerits of these theories. [BUH 2001]
(b) What are valence band and conduction band? What do you mean by forbidden band?
(c) Distinguish amongst metals, semiconductors and insulators. Why does a semiconductor acts as an insulator at 0 K? [BUH 2002]

9.4 (a) Discuss the periodic character of potential in a crystal.
(b) State and prove Bloch's theorem in periodic potential in a crystal. [HPU 1996; LU 1996]

9.5 (a) What do you mean by Brillouin zones?
(b) Discuss the formation of Brillouin zones for a (i) linear lattice (ii) two-dimensional lattice. [HPU 1991; LU 1994; PU 1997]

9.6 (a) What do you mean by effective mass of an electron? What is its significance?
(b) Show that the effective mass of an electron is inversely proportional to the second derivative of the $(E\text{-}k)$ curve.
(c) Discuss the condition when the effective mass of an electron becomes positive, negative and infinity.

9.7 (a) The band theory for a cubic crystal leads to discontinuities in the $E(k)$ surfaces whenever k satisfies the condition $nk = n^2\pi/a$. Show that this condition is equivalent with that for Bragg reflection of the electrons by the set of planes with miller indices (hkl).
(b) Explain the origin of the forbidden energy gap in crystalline solids and the concept of effective mass of an electron in a lattice.

9.8 (a) The solution of Schrödinger equation for a one-dimensional periodic lattice is given by

$$P\frac{\sin \alpha a}{\alpha a} + \cos \alpha a = \cos ka$$

with $\alpha^2 = (2mE/\hbar^2)$. The symbols have their usual significances. Discuss the formation of energy bands in a solid.

(b) Prove that the effective mass of an electron in any energy band is given by

$$\frac{1}{m^*} = \frac{1}{\hbar^2}\frac{\partial^2 E}{\partial k^2}.$$

9.9 (a) Describe band theory of solids on the basis of Kronig-Penney model. Distinguish among conductors, semiconductors and insulators.

(b) Discuss the temperature dependence of resistivity in metal and semiconductor.

(c) What is the concept of *hole*.

9.10 (a) Show that when an electron moves through a crystal lattice the effective mass of the electron can be expressed as

$$m^* = \hbar^2 \Big/ \frac{d^2 E}{d k^2},$$

where the symbols have their usual significances.

(b) According to simple model, the energy of a free electron in a metal is given by the following function of the wave number

$$E = A - B\cos ka,$$

where A and B are constants and a is the distance between adjacent atoms. Draw the E-k curve. What value of effective mass can you deduce from the above formula? With a spacing $a = 2.0$ Å what is the range of the electron wavelengths in the first Brillouin zone. [CUH 2007]

9.11 What is Bloch's theorem? Starting from the relation

$$P\frac{\sin \alpha a}{\alpha a} + \cos \alpha a = \cos ka,$$

discuss the motion of electrons in one dimension and the variation of energy, velocity and effective mass as a function of wave vector.

[Here $\alpha = \sqrt{\frac{2mE}{\hbar^2}}$, $P = \frac{mV_0 ba}{\hbar^2}$; $V_0 b =$ area of the potential barrier.] [VUH 2009]

9.12 Discuss the formation of allowed and forbidden energy bands on the basis of Kronig-Penney model. Set up the appropriate Schrödinger equation and boundary conditions. Use Bloch function as the wave function. [CUH 2006]

Chapter 10

Theory of Semiconductors

10.1 Introduction

In view of electrical conductivity, materials are classified as *conductors, semiconductors* **and** *insulators.* Semiconductors have electrical conductivity intermediate between conductors and insulators.

The resistivity of some typical materials at 300 K is shown in Table 10.1.

Table 10.1

Conductor Metal	*Resistivity* (Ω-m)	*Semiconductor*	*Resistivity* (Ω-m)	*Insulator*	*Resistivity* (Ω-m)
Cu	1.7×10^{-8}	Pure Ge	0.47	Glass	10^{10} to 10^{11}
Al	2.8×10^{-8}	Pure Si	3000	Diamond	10^{14}
Fe	10×10^{-8}	Sn (Grey)	2×10^{-6}	Mica	9×10^{14}

Some important characteristics of semiconductors are as follows:

1. Semiconductors have *negative* temperature coefficient unlike metals which have *positive* temperature coefficient of resistance.
2. Semiconductors have *small band gap* (Eg), which varies from = 0.2 to 2.5 eV, whereas for insulators, like Diamond, $E_g = 6\,\text{eV}$.

   This property determines which wavelength can be emitted or absorbed by the semiconductor and helps to construct *light emitting diodes* (LEDs).
3. Addition of a suitable chemical impurity may change the electrical properties of semiconductors manyfold. The process of addition of impurity is called *doping* which enhances the conductivity of the material.

Depending on the nature of the *dopant* semiconductors are classified as follows:

1. *Pure or intrinsic semiconductors.* In this type of semiconductors no impurity atoms are added.
2. *Impurity or extrinsic semiconductors.* Suitable impurity atoms are added to the semiconductors.

10.2 Intrinsic Semiconductors

Pure *germanium* (Ge) and pure *silicon* (Si) are important examples of intrinsic semiconductors, where electrons and *holes* are produced by *phonons* or *photons* of suitable frequency. Both Ge and Si are group IV elements of the periodic table and each has four valence electrons. Fig. 10.1 (a) shows a covalent Ge crystal in two dimensions.

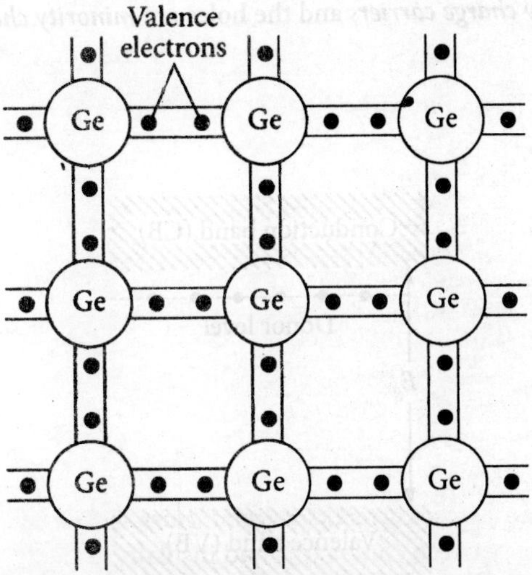

Fig. 10.1: (a) Two-dimensional Ge crystal.

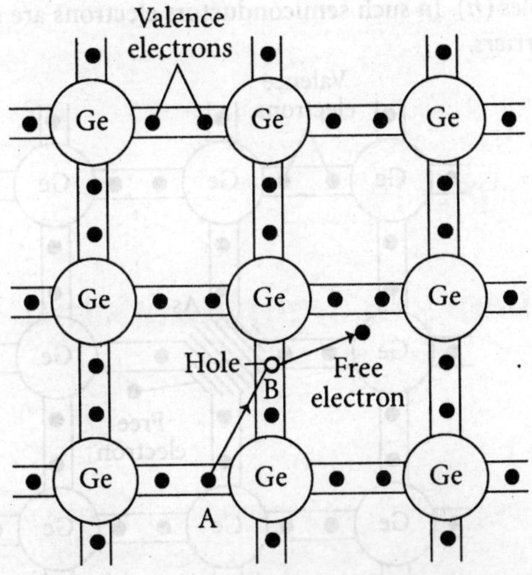

Fig. 10.1: (b) Production of electron and hole.

A valence electron breaks its *covalent bond* if it gets sufficient thermal energy and becomes free. The vacancy left behind at *B* serves as a *hole*. Thus, a free electron and a hole are created as shown in Fig. 10.1 (b). A valence electron in the neighbourhood at *A* after getting sufficient thermal energy jumps to the position of hole at *B* and the bond is reconstructed. In this process a *hole* is created at *A* and the *hole* moves from *B* to *A*. In this way the *electron* and *hole* move in the crystal in opposite direction to each other and conduction of electricity takes place. As the charge carriers are created due to the breaking of the *covalent bond*, the concentration of electrons (n_i) becomes equal to the concentration of holes (p_i), i.e., $n_i = p_i$.

Due to thermal agitation fresh electron-hole pair is generated continuously while electron-hole pair disappears by *recombination*. A free electron is converted into a bound electron creating a *covalent bond*. The Fermi level (E_f) in intrinsic semiconductor lies midway in the forbidden gap. The electron and hole so generated are called *intrinsic charge carriers*.

10.3 Extrinsic Semiconductor

If a small amount of selected *impurity* be added to the semiconductor, then free electrons and holes may be generated. Such materials are called *impurity* (or *extrinsic*) *semiconductor*. The impurity increases the carrier concentration and hence the electrical conductivity of the semiconductor. Impurity semiconductors are of the following two types:

(a) *n*-type semiconductors — The elements of group *V* such as *Phosphorous* (P), *Arsenic* (As) or *Antimony* (Sb) have five electrons in their outermost orbits. If any one of such element be added to *tetravalent Germanium* (Ge) or *Silicon* (Si) during crystal growing, four electrons of the impurity atom form covalent bonding with the four neighbouring semiconducting atom and the fifth electron is released with lesser thermal energy and the parent atom becomes *positively charged* [Fig. 10.2 (a)]. No hole is created during this process. As the pentavalent impurity atoms donate negative charges (i.e., electrons), so the semiconductors doped with pentavalent impurity is called *n-type semiconductor*.

The energy level of the fifth electron lies close to the conduction band [Fig. 10.2 (b)] and is called the *donor level*. That is why most of the electrons in the donor level are easily excited to the conduction band at room temperature and are called *majority charge carrier*. At higher thermal energy electrons and holes are also generated when the concentration of electrons (n) becomes much more than the concentration of holes (h). In such semiconductors electrons are *majority charge carriers* and the holes are *minority charge carriers.*

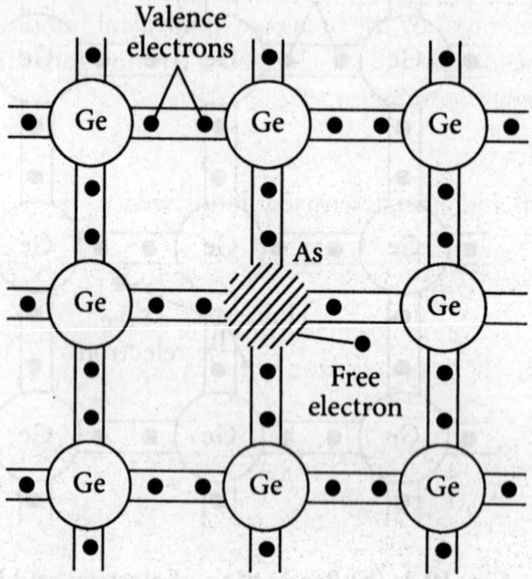

Fig. 10.2: (a) A pentavalent As in Ge crystal

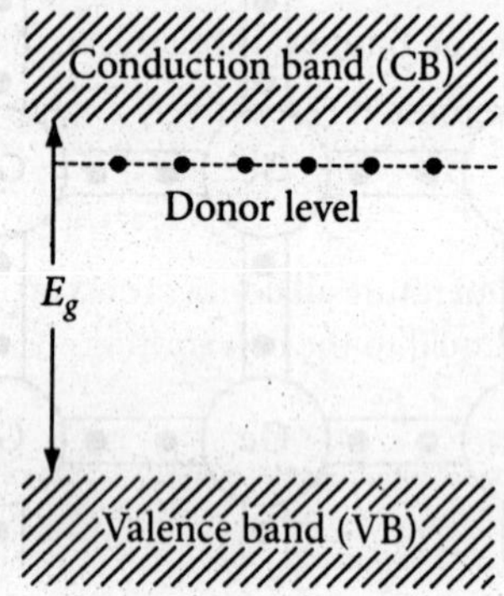

Fig. 10.2: (b) Donar level in energy diagram

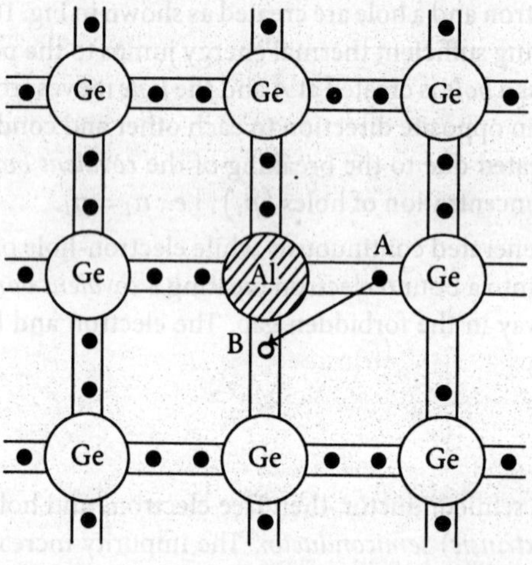

Fig. 10.3: (a) Trivalent impurity atom (Al) in a Ge crystal.

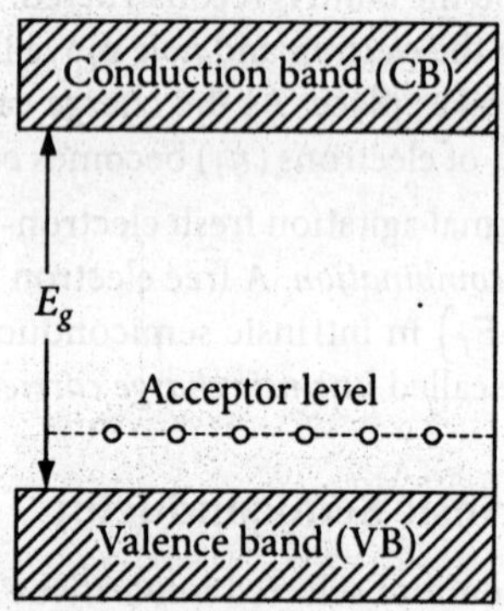

Fig. 10.3: (b) Acceptor level in energy diagram.

(b) *p*-type semiconductor — If a trivalent impurity atom of group III of the periodic table, such as *Aluminium* (Al), *Gallium* (Ga) or *Indium* (In) be added in small amounts to *Germanium* or *silicon*, a *p*-type semiconductor is formed [Fig. 10.3 (a)]. These trivalent atoms have three electrons in their outermost orbits. When a small amount of Al is added to a Ge crystal during its growing, all the three electrons of Al form covalent bond with three neighbouring Ge atoms leaving a vacancy at *B*. To complete the bond it needs one more electron from the neighbouring atoms. This electron may be supplied by Ge, thereby creating a *vacancy* or *hole* at *A* of the Ge atom. As Al accept one extra electron, the corresponding energy level is called the *acceptor level* which lies just above the *valence band* as shown in Fig. 10.3 (b). The trivalent impurity (Al) is called the *acceptor impurity* and the semiconductor is called the *p-type semiconductor.*

These acceptor atoms, even at low temperatures, get ionised taking electrons from the valence band and thereby leaving holes in the valence band. It is to be kept in mind that during this process of ionisation only holes and no electrons are generated. If, however, the temperature of the semiconductor is sufficiently high, then along with this process, electron-hole pairs are generated by the breakage of covalent bonds. That is why hole concentration (p) is much higher than the concentration of the electrons. The holes are the *major* and electrons, the *minor* carriers of electricity in *p-type* semiconductor.

10.3.1 Effect of Temperature on Extrinsic Semiconductor

If D and D^+ represent respectively a neutral donor atom and a singly ionised donor, then

$$e^- + D^+ \rightleftharpoons D + \Delta E_d. \tag{10.1}$$

At high temperature all donor atoms are thermally ionised and the concentration of conduction electrons (n) becomes equal to the concentration of donors (N_D). The concentration of holes is given by

$$np = n_i^2, \tag{10.2}$$

where

$$n_i = \sqrt{C T^{3/2}}\, e^{-E_g/2kT},$$

where E_g represents the binding energy of an electron to the semiconductor atom in the crystal, i.e., $e^- + e^+ \rightleftharpoons E_g$.

At higher temperature, $n_i > N_D$ and the semiconductor will behave as an intrinsic one.

The range, where $n = N_D$, independent of T is called *extrinsic.* The variation of electron concentration with T is shown in Fig. 10.4. At low temperature a point is reached at which thermal energy becomes too small to cause electron excitation. The electrons, then fall from CB to donor level and the conductivity of the material decreases considerably. This is known as *freeze-out,* i.e., the electrons are now *frozen* at the impurity sites. The temperature at which the freeze-out occurs can be evaluated from the relation $E_D \simeq kT$, which gives $T \simeq 100\,\text{K}$.

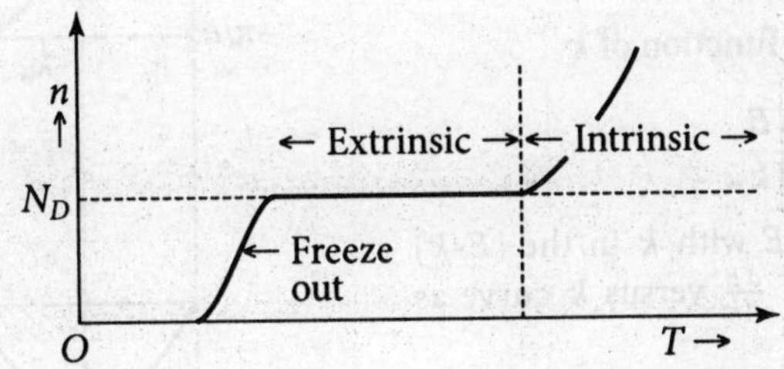

Fig. 10.4: Variation of n with T in an n-type semiconductor.

10.3.2 Effective Mass of the Electron

Due to the interaction of the electron with the *periodic potential* of the lattice in a crystal, the wave particle motion of an electron differs considerably from that of a free electron in space. So, while using the equations of electrodynamics for charge carriers in a crystal, the altered value of particle mass (the *effective mass*) is to be considered, so that the particle may be considered as a *free particle.* The dynamical behaviour of the particle (the charge carrier), then can be described by the *laws of classical physics.*

Let an electron at state k in a lattice be subjected to an external electric field X, the force acting on the electron being eX. If dx be the displacement of the electron in time dt, then the work done is equal to the energy gained by the electron, i.e., $dE = eX\,dx$.

The velocity of the electron is

$$v=\frac{dx}{dt} \text{ or } dx=v\,dt. \therefore dE=eXv\,dt.$$

But

$$v=\frac{1}{\hbar}\frac{dE}{dk}. \therefore dE=eX\frac{1}{\hbar}\frac{dE}{dk}dt \text{ or } \frac{dk}{dt}=\frac{eX}{\hbar}.$$

$$\because \hbar k=p \text{ and } \hbar\frac{dk}{dt}=\frac{dp}{dt}=F, \therefore \hbar\frac{dk}{dt}=eX=F.$$

Also

$$\frac{dv}{dt}=\frac{1}{\hbar}\frac{d^2E}{dk\,dt}=\frac{1}{\hbar}\frac{d^2E}{dk^2}\frac{dk}{dt}=\frac{1}{\hbar}\frac{d^2E}{dk^2}\frac{F}{\hbar}$$

or acceleration

$$a=\frac{1}{\hbar}\frac{d^2E}{dk^2}\frac{F}{\hbar} \text{ or } \frac{a}{F}=\frac{1}{\hbar^2}\frac{d^2E}{dk^2}.$$

Putting $\frac{a}{F}=m^*$, the *effective mass* of the electron

$$m^*=\frac{1}{\hbar^2}\frac{d^2E}{dk^2}. \tag{10.3}$$

10.3.3 Variation of m^* with k

Since the *effective mass* of an electron moving in the periodic potential of a crystal lattice is

$$m^*=\frac{1}{\hbar^2}\frac{d^2E}{dk^2}.$$

m^* depends on $\frac{d^2E}{dk^2}$ and is a function of k.

$$v=\frac{1}{\hbar}\frac{dE}{dk}$$

represents the variation of E with k in the (E-k) curve and $\frac{d^2E}{dk^2}$, the slope of $\frac{dE}{dk}$ versus k curve as shown in Fig. 10.5 (a).

Now the value of $\frac{d^2E}{dk^2}$ is positive for $k=-k_0$ to $+k_0$ representing the lower half of the energy band. So, the value of m^* is

$$m^*=\frac{1}{\hbar^2}\frac{d^2E}{dk^2},$$

which is *positive at the lower half of the energy band.*

Again, for $k=+k_0$ to $+\frac{\pi}{a}$ and $k=-k_0$ to $-\frac{\pi}{a}$, $\frac{d^2E}{dk^2}$ is *negative* representing that m^* is negative in the upper half of the energy band. At the *point of inflection* of (E-k) curve m^* becomes *infinite*. The variation of m^* with k is shown in Fig. 10.5 (c).

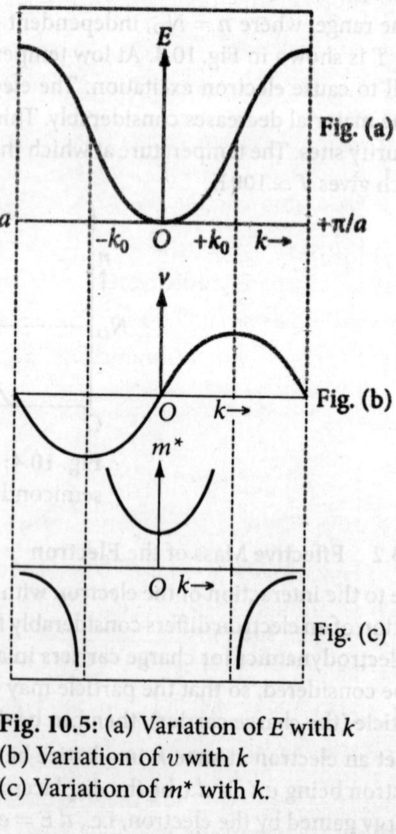

Fig. 10.5: (a) Variation of E with k (b) Variation of v with k (c) Variation of m^* with k.

The large effective mass can be due to *strong binding force between the electron and the lattice at specified region of k*. The negative m^* may be interpreted *due to Bragg reflection for $k \to \pm\pi/a$, $\pm 2\pi/a$ for which a force in one direction leads to a gain in momentum in opposite direction.*

10.4 Compound Semiconductors

Semiconductor elements are covalent having diamond-like structure and belong to Group IV of the periodic table. Zinc blend lattice and wurtzite lattice which occur in binary semiconducting compounds are closely related to them. The semiconducting compounds are strongly covalent and made up of elements placed symmetrically about Group IV in the periodic table.

The compound semiconductors like InSb, GaAs, InP, etc., the inter-metallic compounds of Groups III and V have energy gap between a few tenths to several electron volts and as such find important applications in electronic devices at wider ranges of temperature than Ge and Si.

The compounds such as PbS, PBTe and PbSe are formed between Groups IV and VI and CdS, ZnS and CdTe between Groups II and VI belong to other group of compound semiconductors. The crystalline defects or derivation from *stoichiometry* are responsible for conductivity in these materials. If there is an excess of cations or deficiency of anions, the *non-stoichiometric* compounds behave as an *n*-type semiconductors. But the presence of excess anions or deficiency of cations leads to a *p*-type semiconductor.

The energy gap E_g of some compound semiconductors at $T = 300$ K is shown in Table 10.2.

Table 10.2

Material	GaAs	GaP	InS	InAs	InP	PbS	GaSb
E_g (eV)	1.40	2.24	0.16	0.33	1.29	0.40	0.77

10.5 Direct and Indirect Semiconductors

An electron moving along the x-direction in a crystal lattice may be represented by the wave function

$$\psi_k(x) = V(\vec{k}, x)\, e^{iK_x x}.$$

This is a space dependence wave function and $V(\vec{k}, x)$ modulates $\psi_k(x)$ according to the periodicity of the lattice. As the periodicity of most lattices varies with direction we consider the $E(\vec{k})$ versus $\vec{k}$ curve for Si and GaAs shown in Figs 10.6 (a) and (b) respectively.

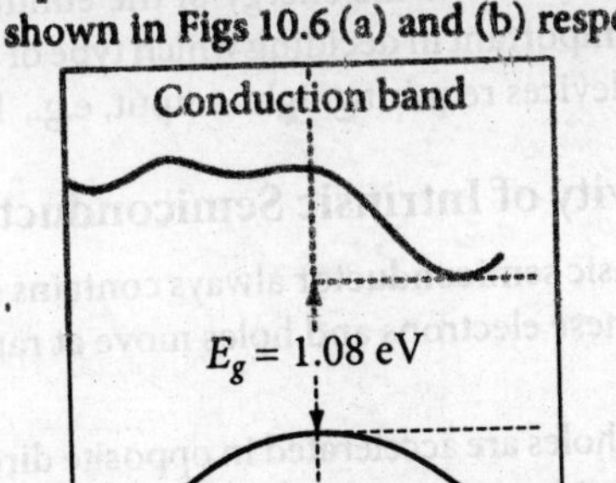

Fig. 10.6: (a) Si

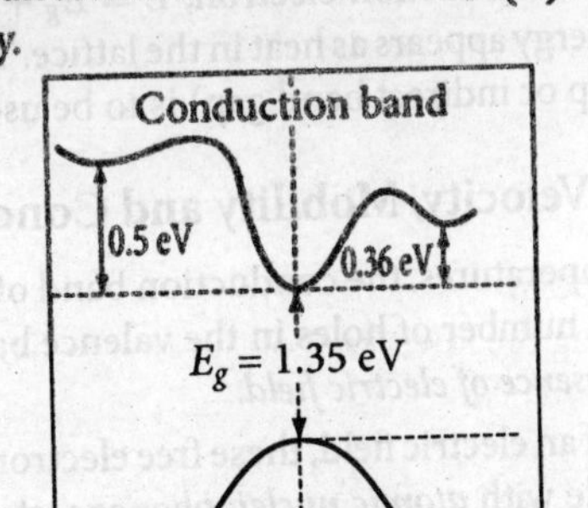

Fig. 10.6: (b) GaAs

Dispersion curves bounding the conduction and the valence bands in Si and GaAs.

Figs 10.6(a) and (b) show that the maxima of the dispersion curves for the valence bands of Si and GaAs lie in the middle of the Brillouin zone, resembling a parabola with the vertex facing upwards. In Si, the minima of the curve for the conduction band lie at a distance $k < \pi/a$ from the middle of the zone in the [100] direction. There are two maxima of this type of GaAs, one in the middle of the Brillouin zone and the other at a distance $0.75\pi/a$ from the middle in the direction [100]. The minima and the maxima of the dispersion curve are referred to as the *bottom* and *top* of the energy band. For $\vec{k} = 0$ there is a minimum in the conduction band and a maximum in the valence band of GaAs. But for Si, the valence band maximum is at a different value of $\vec{k}$ than its conduction band minimum. Hence, in GaAs, an electron can make a smallest energy transition from conduction band to valence band without a change in $\vec{k}$-value. But in Si, an electron can make a transition from conduction band minimum to valence band maximum with a change in $\vec{k}$-value. That is why GaAs and Si are referred to as *direct* and *indirect* band gap semiconductors [Fig. 10.7 (a)].

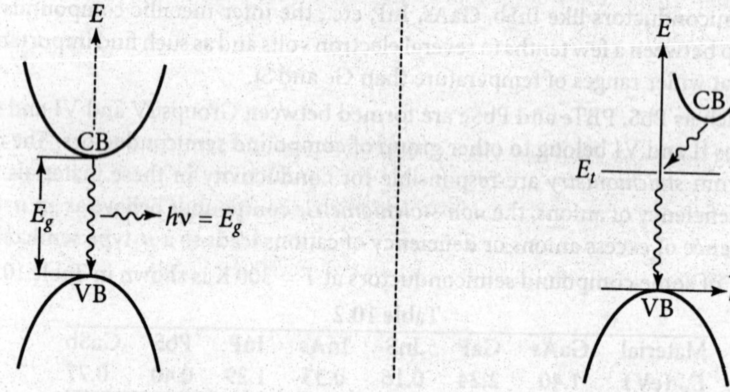

Fig. 10.7: (a) Direct BG semiconductor. Direct transition with a photon emission.

Fig. 10.7: (b) Indirect BG semiconductor. Indirect transition via a defect level.

In a direct band gap semiconductor, an electron in the conduction band can transit to an empty state in the valence, thereby giving rise to photon of energy $E_g = h\nu$. But in case of an indirect band semiconductor, an electron in the conduction band minimum cannot transit directly to the valence band instead undergoes a change in momentum as well as energy involving a defect state (or photon) [Fig. 10.7 (b)]. In this type of transition of a conduction electron, $E = E_g + \hbar\omega$, where $\hbar\omega$ is the energy of the emitted photon. The difference in energy appears as heat in the lattice. This is important in deciding which type of semiconductor (direct band gap or indirect band gap) is to be used in devices requiring light output, e.g., LED or Lasers.

10.6 Drift Velocity, Mobility and Conductivity of Intrinsic Semiconductors

At ordinary temperatures, the conduction band of intrinsic semiconductor always contains some free electrons and equal number of holes in the valence band. These electrons and holes move at random and give *no current in absence of electric field.*

In presence of an electric field, these free electrons and holes are accelerated in opposite directions to each other and collide with *atomic nuclei, phonons,* etc. present in the semiconductor. The electrons and holes acquire an *average velocity* which reduces to zero in the subsequent collision. This extra velocity gained by the charge carriers in presence of an electric field E is called the *drift velocity* (v_d). The drift velocity is proportional to the applied field strength E, i.e.,

$$v_d \propto E \text{ or } v_d = \mu E, \qquad (10.4)$$

where μ is a constant of proportionality and is called the *mobility. The mobility of the charge carrier is defined as the drift velocity gained by a carrier for unit electric field strength.*

In intrinsic semiconductor as the electrons move in *nearly empty conduction* band and the holes move in *nearly full valence band,* the *mobility, conductivity,* etc. of electrons and holes are, in general, different.

If v_{de}, μ_e and n denote the drift velocity, mobility and concentration of free electrons in the conduction band respectively, then the current density

$$J_e = nev_{de} \text{ or } J_e = ne\mu_e E \; [\because v_{de} = \mu_e E]. \tag{10.5}$$

But according to *Ohm's law,*

$$J_e = \sigma_e E, \tag{10.6}$$

where σ_e is the *electronic conductivity* of the material.

Now comparing (10.4) and (10.5), we get

$$\sigma_e = ne\mu_e. \tag{10.7}$$

Similarly, the *conductivity of holes* in the valence band is

$$\sigma_h = pe\mu_h, \tag{10.8}$$

where p and μ_h represent respectively the *concentration* and *mobility of holes.*

$\therefore$ the total conductivity is

$$\sigma = \sigma_e + \sigma_h = e\,(n\mu_e + p\mu_h). \tag{10.9}$$

For *intrinsic semiconductor,*

$$n = p = n_i \text{ (say)}.$$

$$\therefore \sigma = en_i\,(\mu_e + \mu_h). \tag{10.10}$$

The flow of current in semiconductors arises due to

1. the drift velocity of the carriers in presence of the applied field and the current is called the *drift current* and
2. diffusion of carriers due to *concentration gradient* of the dopant, the corresponding current is called the *diffusion current.*

In semiconductors with uniform distribution of impurities the diffusion current is nil. So, the drift current has been taken into account.

Temperature Dependence of Mobility

The electron and hole mobilities are greatly influenced by *lattice scattering* and *impurity scattering.* The effect of dislocation is, however, very small. With the rise of temperature the frequency of lattice scattering increases causing a decrease in mobility. On the other hand, at low temperature the effect is small as the thermal motions of the carriers is slower.

If μ_L and μ_I be the mobility of a carrier in presence of *lattice scattering* and *impurity scattering* respectively, then it has been found theoretically that

$$\mu_L = AT^{-3/2} \text{ and } \mu_I = BT^{3/2},$$

where A and B are constants for a given sample.

The resultant resistivity is given by

$$\rho = \frac{1}{\sigma} = \rho_L + \rho_I$$

$$\text{or } \frac{1}{\sigma} = \frac{1}{\sigma_L} + \frac{1}{\sigma_I}$$

$$\text{or } \frac{1}{ne\mu} = \frac{1}{ne\mu_L} + \frac{1}{ne\mu_I}$$

$$\text{or } \frac{1}{\mu} = \frac{1}{\mu_L} + \frac{1}{\mu_I}$$

$$\text{or } \frac{1}{\mu} = \frac{T^{3/2}}{A} + \frac{T^{-3/2}}{B}.$$

This gives the temperature dependence of carrier mobility.

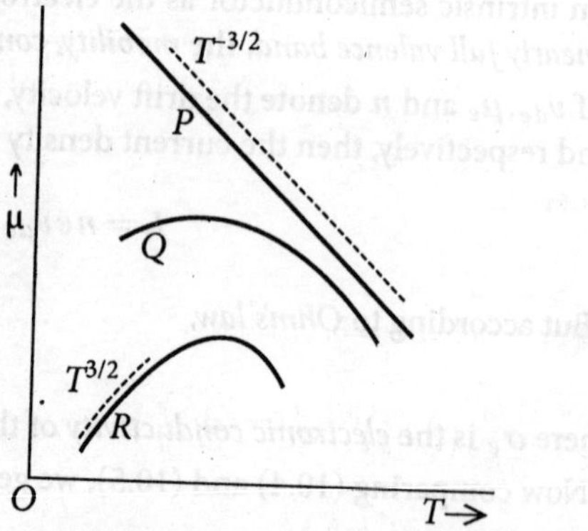

Fig. 10.8: Temperature dependence of μ for pure (*P*), less pure (*Q*) and least pure (*R*) materials.

10.6.1 Electron Concentration of Intrinsic Semiconductor in the Valence Band

If $D(E)$ be the *density of states* and $f(E)$, *the Fermi-distribution function*, then the free electron density in an energy interval E and $E+dE$ is

$$dn = D(E)f(E)dE. \qquad (10.11)$$

But

$$f(E) = \frac{1}{e^{(E-E_f)/kT}} \qquad (10.12)$$

and

$$D(E) = \frac{4\pi}{h^3}(2m)^{3/2}E^{1/2}. \qquad (10.13)$$

Using (10.13) in (10.11), we get

$$dn = \frac{4\pi}{h^3}(2m)^{3/2}E^{1/2}f(E)dE. \qquad (10.14)$$

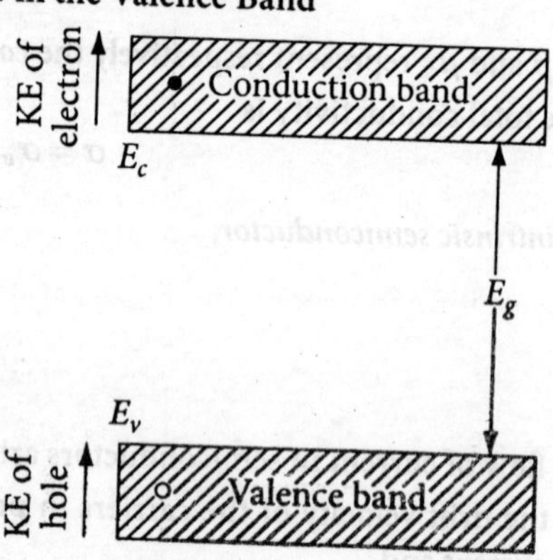

Fig. 10.9: Band model of an intrinsic semiconductor.

From Fig. 10.9 it is clear that an electron of energy E in the conduction band can have energy $(E - E_c)$.

∴ from (10.14),

$$dn = \frac{4\pi}{h^3}\left(2m_e^*\right)^{3/2}(E-E_c)^{1/2}\frac{1}{e^{(E-E_c)/kT}+1}dE, \qquad (10.15)$$

where m_e^* is the *effective mass of an electron* in the conduction band.

Therefore, the *electron concentration* in the conduction band is given by

$$n = \frac{4\pi}{h^3}\left(2m_e^*\right)^{3/2}\int_{E_c}^{\infty}\frac{(E-E_c)^{1/2}\,dE}{e^{(E-E_c)/kT}+1}. \qquad (10.16)$$

At room temperature, $kT \simeq 0.026\,\text{eV}$.

$$\therefore \text{ for } E > E_c,\ e^{(E-E_c)/kT}+1 \simeq e^{(E-E_c)/kT}.$$

So

$$n = \frac{4\pi}{h^3}\left(2m_e^*\right)^{3/2} e^{(E_f - E_c)/kT} \int_{E_c}^{\infty} (E - E_c)^{1/2} e^{-(E-E_c)/kT}\, dE.$$

Let

$$\frac{E - E_c}{kT} = x. \therefore dE = kT\,dx.$$

For $E = E_c$, $x = 0$.

$$\therefore n = \frac{4\pi}{h^3}\left(2m_e^*\right)^{3/2} e^{(E_f - E_c)/kT} \int_0^{\infty} x^{1/2}(kT)^{1/2} e^{-x}\, kT\, dx = \frac{4\pi}{h^3}\left(2m_e^* kT\right)^{3/2} e^{(E_f - E_c)/kT} \int_0^{\infty} x^{1/2} e^{-x}\, dx$$

$$= 2\left(\frac{2\pi m_e^* kT}{h^2}\right)^{3/2} e^{-(E_c - E_f)/kT} \left[\text{Using} \int_0^{\infty} x^{1/2} e^{-x}\, dx = (\pi/4)^{1/2}\right] \quad (10.17)$$

$$\text{or } n = 2\left(\frac{2\pi m_e^* kT}{h^2}\right)^{3/2} f(E_c), \quad (10.18)$$

where

$$f(E_c) = \frac{1}{e^{(E_c - E_f)/kT} + 1} \simeq e^{-(E_c - E_f)/kT}.$$

The factor

$$2\left(\frac{2\pi m_e^* kT}{h^2}\right)^{3/2}$$

represents the *effective density of states of electrons* at the conduction band edge.

$$\therefore n = N_c\, e^{-(E_c - E_f)/kT}, \quad (10.19)$$

where

$$N_c = 2\left(\frac{2\pi m_e^* kT}{h^2}\right)^{3/2}. \quad (10.20)$$

10.6.2 Concentration of Holes of Intrinsic Semiconductor in the Valence Band

The number of holes per unit volume in the energy range E to $E + dE$ is

$$dp = D(E)[1 - f(E)]\, dE, \quad (10.21)$$

where $[1 - f(E)]$ is the *probability* of finding a hole in the energy state E.

Now

$$1 - f(E) = 1 - \frac{1}{e^{(E-E_f)/kT}} = \frac{e^{(E-E_f)/kT}}{e^{(E-E_f)/kT} + 1}.$$

In the valence band, since $E < E_f$, $e^{(E-E_f)/kT}$ in the denominator may be neglected.

$$\therefore 1 - f(E) \simeq e^{(E-E_f)/kT}. \quad (10.22)$$

Therefore, the density of holes in the valence band is

$$D(E) = \frac{4\pi}{h^3}\left(2m_h^*\right)^{3/2} (E_v - E)^{1/2}, \quad (10.23)$$

where m_h^* is the *effective mass of the hole* in the valence band.

Substituting (10.22) and (10.23) in (10.21), we get

$$p = \frac{4\pi}{h^3}\left(2m_h^*\right)^{3/2}\int_{-\infty}^{E_v}(E_v - E)^{1/2}\,e^{(E-E_f)/kT}\,dE$$

$$= \frac{4\pi}{h^3}\left(2m_h^*\right)^{3/2}e^{(E_v-E_f)/kT}\int_{-\infty}^{E_v}(E_v - E)^{1/2}\,e^{(E-E_v)/kT}\,dE.$$

Let

$$\frac{(E_v - E)}{kT} = x. \therefore dE = -kT\,dx.$$

For $E = E_v, x = 0$.

$$\therefore p = \frac{4\pi}{h^3}\left(2m_h^*\right)^{3/2}e^{(E_v-E_f)/kT}\int_0^{\infty}x^{1/2}(kT)^{1/2}e^{-x}(-kT)dx$$

$$= \frac{4\pi}{h^3}\left(2m_h^*\right)^{3/2}e^{(E_v-E_f)/kT}(kT)^{3/2}\int_0^{\infty}x^{1/2}\,e^{-x}\,dx$$

$$= \frac{4\pi}{h^3}\left(2m_h^*kT\right)^{3/2}e^{(E_v-E_f)/kT}(\pi/4)^{1/2}\left[\because \int_0^{\infty}x^{1/2}\,e^{-x}dx = (\pi/4)^{1/2}\right]$$

$$= 2\left(\frac{2\pi m_h^* kT}{h^2}\right)^{3/2}e^{-(E_f-E_v)/kT} \qquad (10.24)$$

$$\text{or } p = N_v\,e^{-(E_f-E_v)/kT}, \qquad (10.25)$$

where

$$N_v = 2\left(\frac{2\pi m_h^* kT}{h^2}\right)^{3/2}, \qquad (10.26)$$

N_v is the *effective density of holes in the valence band edge.*

10.6.3 Fermi Level

Since $n = p = n_i$ for an *intrinsic semiconductor,* we have from (10.19) and (10.25)

$$N_c\,e^{-(E_c-E_f)/kT} = N_v\,e^{-(E_f-E_v)/kT}$$

$$\text{or } e^{(2E_f-E_c-E_v)/kT} = \frac{N_v}{N_c}$$

$$\text{or } \left(2E_f - E_c - E_v\right)/kT = \ln(N_v/N_c)$$

$$\text{or } E_f = \frac{E_c + E_v}{2} + \frac{1}{2}kT\ln(N_v/N_c). \qquad (10.27)$$

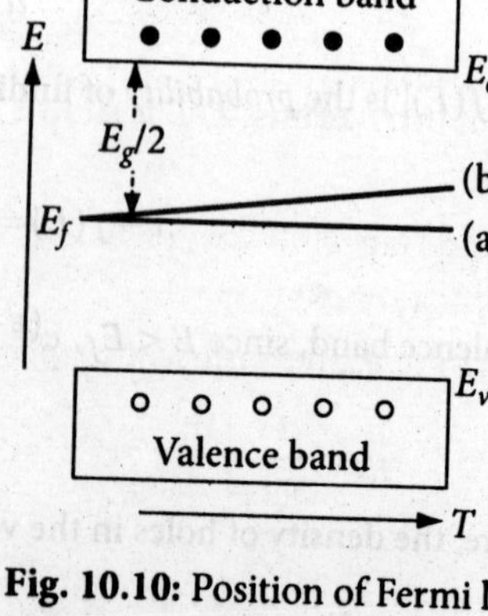

Fig. 10.10: Position of Fermi level in intrinsic semiconductor (a) at $T = 0$ K (b) at $T > 0$ K.

Using the values of N_c and N_v from (10.20) and (10.26) in (10.27), we get

$$E_f = \frac{E_c + E_v}{2} + \frac{3}{4}kT\ln\left(\mu_h^*/m_e^*\right). \qquad (10.28)$$

At $T = 0$ K,

$$E_f = \frac{E_c + E_v}{2}. \tag{10.29}$$

(10.29) shows that *Fermi level is located half-way between the valence and conduction bands.* If $\mu_h^* = m_e^*$, the position of the Fermi level is independent of temperature. In general, $\mu_h^* > m_e^*$ and the Fermi level is just above the middle and rises slightly with the increase of temperature.

10.6.4 Electrical Conductivity of Intrinsic Semiconductors

For an *intrinsic semiconductor,* the electrical conductivity σ is given by $\sigma = (ne\mu_e + pe\mu_h)$, where μ_e and μ_h denote the *mobilities of electrons and holes respectively.*

$$\because\ n = p = n_i.$$

$\therefore$ from (10.17) and (10.24),

$$n_i^2 = np = 4\left(\frac{2\pi kT}{h^2}\right)^3 \left(m_e^* \mu_h^*\right)^{3/2} e^{(E_v - E_c)/kT}$$

$$= 4\left(\frac{2\pi kT}{h^2}\right)^3 \left(m_e^* \mu_h^*\right)^{3/2} e^{-E_g/kT} \quad [\because (E_c - E_v) = E_g].$$

Therefore,

$$n_i = 2\left(\frac{2\pi kT}{h^2}\right)^{3/2} \left(m_e^* m_h^*\right)^{3/4} e^{-E_g/2kT}. \tag{10.30}$$

$$\therefore\ \sigma = (\mu_e + \mu_h)\, 2e\left(\frac{2\pi kT}{h^2}\right)^{3/2} \left(m_e^* m_h^*\right)^{3/4} e^{-E_g/2kT} \tag{10.31}$$

$$\text{or } \sigma_i = Ae^{-E_g/2kT}, \tag{10.32}$$

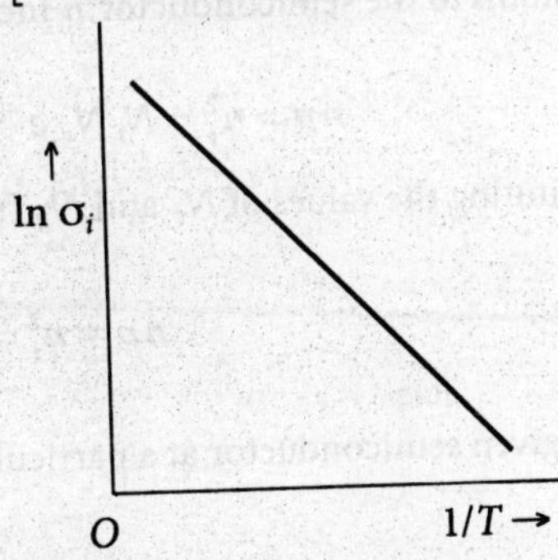

Fig. 10.11: Variation of σ_i with T.

σ_i is the *electrical conductivity of the intrinsic semiconductor,* where

$$A = (\mu_e + \mu_h)\, 2e\left(\frac{2\pi kT}{h^2}\right)^{3/2} \left(m_e^* \mu_h^*\right)^{3/4}.$$

Taking logarithm on both sides of (10.32), we get

$$\ln \sigma_i = \ln A - E_g/2kT. \tag{10.33}$$

The plot of $\ln \sigma_i$ against $1/T$ is shown in Fig. 10.11 which shows that σ_i increases with the decrease of temperature.

10.6.5 Band Gap

We have $\sigma_i = Ae^{-E_g/2kT}$.

$\therefore$ *Resistivity of the intrinsic semiconductor* is

$$\rho_i = \frac{1}{\sigma_i} = \frac{1}{A} e^{E_g/2kT} \text{ or } \frac{R_i a}{L} = (1/A)\, e^{E_g/2kT}, \left[\because \rho_i = \frac{R_i a}{L}\right]$$

$$\text{or } R_i = \frac{L}{aA} e^{E_g/2kt} = Ce^{E_g/2kT} \left[\text{where } C = \frac{L}{aA}\right],$$

where L is the length of the semiconductor of cross-sectional area a and resistance R_i.

Taking logarithm on both sides,

$$\ln R_i = \ln C + E_g/2kT. \qquad (10.34)$$

Plotting $\ln R_i$ against $\frac{1}{T}$ we get a straight line as shown in Fig. 10.12.

The slope of the curve is $E_g/2kT$. From Fig. 10.12,

$$\text{slope, } \frac{dy}{dx} = \frac{E_g}{2k}. \therefore E_g = 2k\left(\frac{dy}{dx}\right), \qquad (10.35)$$

where k is the *Boltzmann constant.*

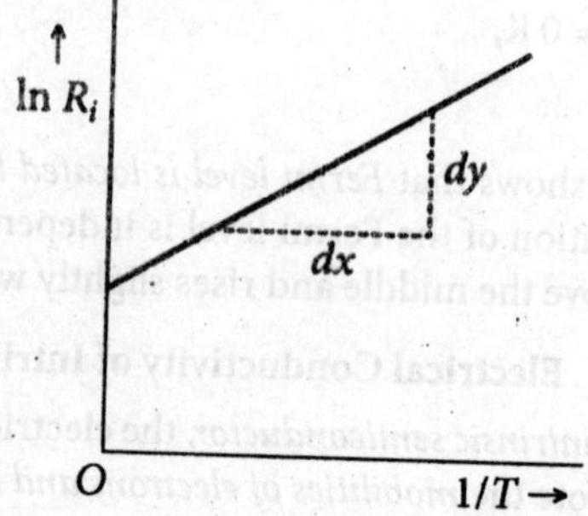

Fig. 10.12: Variation of R_i with $1/T$.

10.6.6 Law of Mass Action

Statement: *For a given semiconductor, the product of electron and hole concentrations is a constant at a given temperature and is equal to the square of the intrinsic carrier concentration.*

The *law of mass action* holds for both *intrinsic* and *extrinsic semiconductors.* Due to the addition of impurity atoms to the semiconductor n increases and corresponding p decreases so that $np = n_i^2$.

But

$$np = n_i^2 = N_c N_v \, e^{-(E_c - E_v)/kT} = N_c N_v \, e^{-E_g/kT} \quad [\because E_c - E_v = E_g]. \qquad (10.36)$$

Substituting the values of N_c and N_v from (10.20) and (10.26) in (10.36), we get

$$np = n_i^2 = 4\left(\frac{2\pi kT}{h^2}\right)^3 \left(m_e^* \mu_h^*\right)^{3/2} e^{-E_g/kT}.$$

For a given semiconductor at a particular temperature,

$$np = n_i^2 = \text{constant}. \qquad (10.37)$$

This proves the *laws of mass action.*

The *intrinsic carrier concentration* can be obtained from (10.17) and (10.24) as below:

$$n_i = (N_c N_v)^{1/2} e^{-E_g/2kT} = 2\left(\frac{2\pi kT}{h^2}\right)^{3/2} \left(m_e^* \mu_h^*\right)^{3/4} e^{-E_g/2kT}. \qquad (10.38)$$

The intrinsic electron concentration in pure Ge at $T = 300$ K is about $n_i = 2.4 \times 10^{19}/\text{m}^3$ when the concentration of Ge atom is $4.4 \times 10^{28}/\text{m}^3$. This indicates that at ordinary temperatures only about five covalent bonds per 10^{10} Ge atoms are broken contributing to intrinsic conduction. But for metals like Cu about 10^{28} electrons/m^3 are available for conduction.

10.7 Carrier Concentration in N-type Semiconductors

Fig. 10.13 shows the energy level diagram of an n-type semiconductor, where E_d is the donor energy level and N_d is the donor concentration of the material.

At very low temperature all *donor levels* are filled with electrons and with the increase in temperature more and more donor atoms get ionised. The released electrons from the donors increase the concentration of the electrons in the conduction band.

Neglecting the concentration of thermally generated holes in the valence band $n = N_d^+$, where N_d^+ is the *concentration of donors.*

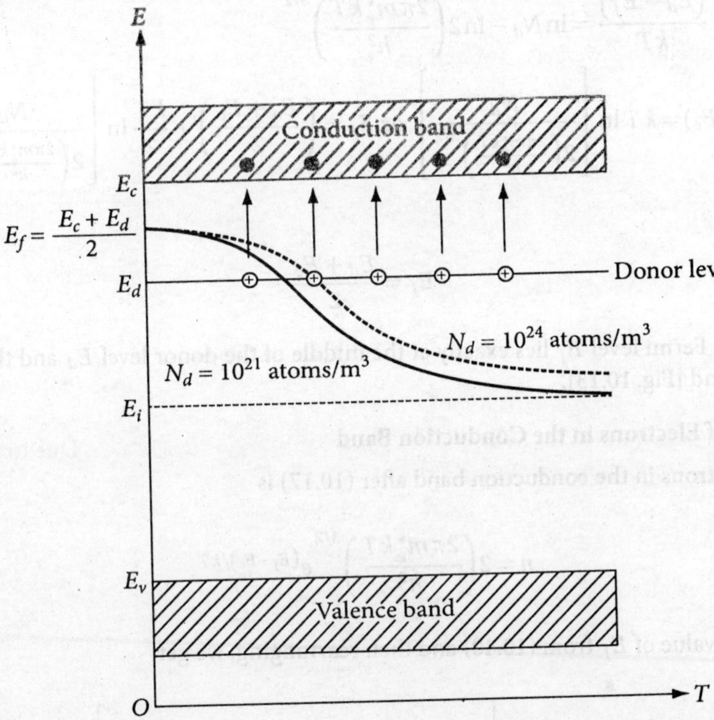

Fig. 10.13: Variation of Fermi level with temperature in n-type semiconductor.

The density of electrons in the conduction is given by (10.17),

$$n = 2\left(\frac{2\pi m_e^* kT}{h^2}\right)^{3/2} e^{(E_f - E_c)/kT}.$$

If the Fermi level E_f lies more than a few kT above the donor level E_d, then the density of ionised donors is given by

$$N_d\,[1 - F(E_d)] \simeq N_d\, e^{(E_d - E_f)/kT}. \tag{10.39}$$

At very low temperatures, the electron-hole pairs are not generated due to the breakage of covalent bonds. Then the number of electrons in the conduction band must be equal to the number of ionised donors in donor level E_d, i.e., $n = N_d^+$, N_d^+ is the *concentration of ionised donors.*

$$\therefore\ 2\left(\frac{2\pi m_e^* kT}{h^2}\right)^{3/2} e^{(E_f - E_c)/kT} = N_d\, e^{(E_d - E_f)/kT}$$

Taking logarithm on both sides of the above equation we get after rearranging

$$\frac{(E_f - E_c)}{kT} - \frac{(E_d - E_f)}{kT} = \ln N_d - \ln 2\left(\frac{2\pi m_e^* kT}{h^2}\right)^{3/2}$$

$$\text{or } 2E_f - (E_d + E_c) = kT\ln\left[\frac{N_d}{2\left(\frac{2\pi m_e^* kT}{h^2}\right)^{3/2}}\right] \text{ or } E_f = \left(\frac{E_d + E_c}{2}\right) + \frac{kT}{2}\ln\left[\frac{N_d}{2\left(\frac{2\pi m_e^* kT}{h^2}\right)^{3/2}}\right]. \quad (10.40)$$

At $T = 0\,\text{K}$,

$$E_f = \frac{E_d + E_c}{2}. \quad (10.41)$$

Thus, at 0 K, the Fermi level E_f lies exactly at the middle of the donor level E_d and the bottom level of the conduction band [Fig. 10.13].

10.7.1 Density of Electrons in the Conduction Band

The density of electrons in the conduction band after (10.17) is

$$n = 2\left(\frac{2\pi m_e^* kT}{h^2}\right)^{3/2} e^{(E_f - E_c)/kT}.$$

Substituting the value of E_f from (10.40) and then rearranging, we get

$$e^{(E_f - E_c)/kT} = \exp\left[(E_d - E_c)/2kT + \frac{1}{2}\ln\frac{Nd}{2\left(\frac{2\pi m_e^* kT}{h^2}\right)^{3/2}}\right]$$

$$= \exp\left[(E_d - E_c)/2kT + \ln\frac{N_d^{1/2}}{\left\{2\left(\frac{2\pi m_e^* kT}{h^2}\right)^{3/2}\right\}^{1/2}}\right]$$

$$= \exp\left(\frac{E_d - E_c}{2kT}\right)\frac{N_d^{1/2}}{\left\{2\left(\frac{2\pi m_e^* kT}{h^2}\right)^{3/2}\right\}^{1/2}}$$

$$\therefore\ n = 2\left(\frac{2\pi m_e^* kT}{h^2}\right)^{3/2}\frac{N_d^{1/2}}{\left\{2\left(\frac{2\pi m_e^* kT}{h^2}\right)^{3/2}\right\}^{1/2}} e^{(E_d - E_c)/kT}$$

$$\text{or } n = (2N_d)^{1/2}\left(\frac{2\pi m_e^* kT}{h^2}\right)^{3/4} e^{(E_d - E_c)/kT}. \quad (10.42)$$

Thus, density of electrons in the conduction band of an n-type semiconductor is proportional to $N_d^{1/2}$ at moderately low temperatures. At higher temperatures, the intrinsic behaviour dominates over donor concentration. Also, n increases exponentially as E_f moves away from E_i towards the conduction band.

10.7.2 Variation of Fermi Level with Temperature

(10.40) shows that E_f increases slightly with the increase of temperature (T), when T is not high. As temperature increases more and more, donor atoms become ionised more and more and at a particular temperature all the donor atoms get completely ionised. If temperature still goes on increasing, the breaking of covalent bond takes place and the material behaves as an intrinsic semiconductor, when $n_i \gg N_d$. At this condition the Fermi level gradually decreases and approaches the intrinsic Fermi level E_i as shown in Fig. 10.13.

The rate of lowering of Fermi level from $E_f = \frac{E_d+E_c}{2}$ to E_i with rise of temperature is more for lower donor concentration than the higher one. So, the highly doped semiconductor behaves intrinsically after all the donor atoms are ionised [Fig. 10.13].

10.7.3 Temperature Dependence of Electrical Conductivity of *n*-Type Semiconductor

The electrical conductivity of an n-type semiconductor is given by

$$\sigma = en\mu_e = eN_d^+\mu_e, \tag{10.43}$$

assuming the electrons are mainly responsible for conduction. The value of the electron concentration n and mobility μ_e can be determined at particular temperature and σ can be calculated using (10.43).

The variation of σ with T for an *n-type* semiconductor is represented in Fig. 10.14.

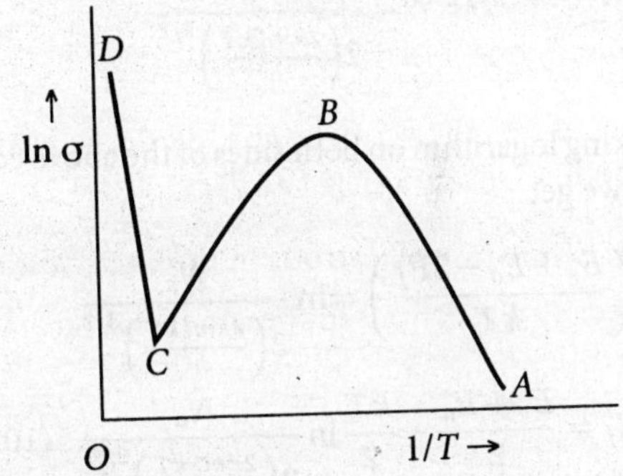

Fig. 10.14: Variation of σ with T for an n-type semiconductor.

The following points are apparent from Fig. 10.14:

1. From low temperature at A the conductivity increases and becomes maximum at B, where all the donor atoms are ionised.
2. With further increase of temperature, σ decreases and becomes minimum at C. This is due to *the decrease of mobility with the increase of temperature.*
3. From C to D, σ increases sharply. This is due to *the increase in electron concentration due to breaking of covalent bonds at high temperature dominating the decrease in mobility.*

10.8 Carrier Concentration in *p*-type Semiconductor

Fig. 10.15 shows the energy level diagram of a p-type semiconductor, where E_a and N_a represent respectively the acceptor energy level and concentration of acceptor atoms in the p-type semiconductor.

At very low temperatures all the acceptor levels are empty and at $T > 0\,\text{K}$ with the increase of temperature a part of acceptors is ionised by acquiring electrons from the valence band. Thus, holes are created in the valence band which causes p-type conduction. Neglecting the holes created due to thermal effect, the hole concentration (p) becomes equal to the concentration of the ionised acceptors $\left(N_a^-\right)$.

Thus

$$p = N_a^-. \tag{10.44}$$

Now the *density of holes* in the valence band is given by

$$p = 2\left(\frac{2\pi m_h^* kT}{h^2}\right) e^{\frac{(E_v - E_f)}{kT}}. \tag{10.45}$$

The Fermi level E_f lies below E_a, the density of ionised acceptors is given by

$$N_a F(E_a) = N_a\, e^{(E_f - E_a)/kT}.$$

Substituting the values of p and N_a^- is (10.44), we get

$$2\left(\frac{2\pi m_h^* kT}{h^2}\right)^{3/2} e^{(E_v - E_f)/kT} = N_a\, e^{(E_f - E_a)/kT},$$

i.e., $$e^{(E_v + E_a - 2E_f)/kT} = \frac{N_a}{2\left(\frac{2\pi m_h^* kT}{h^2}\right)^{3/2}}.$$

Taking logarithm on both sides of the above equation, we get

$$\left(\frac{E_v + E_a - 2E_f}{kT}\right) = \ln \frac{N_a}{2\left(\frac{2\pi m_h^* kT}{h^2}\right)^{3/2}}$$

$$\text{or } E_f = \frac{E_v + E_a}{2} - \frac{kT}{2} \ln \frac{N_a}{2\left(\frac{2\pi m_h^* kT}{h^2}\right)^{3/2}}. \quad (10.46)$$

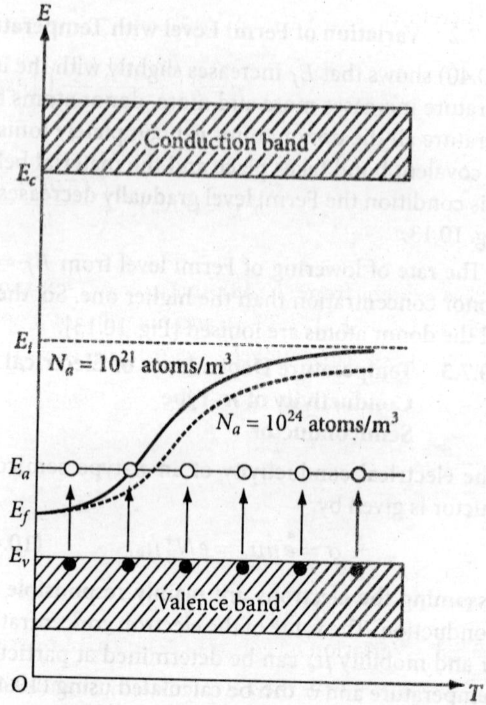

Fig. 10.15: Variation of Fermi level position with temperature.

At $T = 0$ K,

$$E_f = \frac{E_v + E_a}{2}. \quad (10.47)$$

Hence, *at* $T = 0$ K, *the Fermi level lies exactly at the middle of the acceptor level and top of the valence band as shown in* Fig. 10.15.

10.8.1 Density of Holes in the Valence Band of a *p*-Type Semiconductor

The density of holes in the valence band from (10.24) is

$$p = 2\left(\frac{2\pi m_h^* kT}{h^2}\right)^{3/2} e^{(E_v - E_f)/kT}.$$

Substituting the value of E_f from (10.46), we get

$$e^{(E_v - E_f)/kT} = \exp\left\{\frac{E_v}{kT} - \frac{E_v + E_a}{2kT} + \frac{1}{2}\ln \frac{N_a}{2\left(\frac{2\pi m_h^* kT}{h^2}\right)^{3/2}}\right\} = \exp\left\{\frac{E_v - E_a}{2kT} + \ln \frac{N_a^{1/2}}{\left[2\left(\frac{2\pi m_h^* kT}{h^2}\right)^{3/2}\right]^{1/2}}\right\}$$

$$= \left[\exp\frac{(E_v - E_a)}{2kT}\, \frac{N_a^{1/2}}{\left\{2\left(\frac{2\pi m_h^* kT}{h^2}\right)^{3/2}\right\}^{1/2}}\right].$$

Therefore,

$$p=2\left(\frac{2\pi m_h^* kT}{h^2}\right)\left[\exp\frac{(E_v-E_a)}{2kT}\right]\frac{N_a^{1/2}}{\left[2\left(\frac{2\pi m_h^* kT}{h^2}\right)^{3/2}\right]^{1/2}}$$

$$\text{or } p=(2N_a)^{1/2}\left(\frac{2\pi m_h^* kT}{h^2}\right)^{3/4}\exp(E_v-E_a)/kT. \tag{10.48}$$

It is evident from (10.48), that $p \propto N_a^{1/2}$ at moderately low temperature. At higher temperature, i.e., at $T \to \infty$, $\left(E_v - E_f\right)/kT \to 0$ and the value of p approaches the intrinsic concentration n_i. Also p increases exponentially as E_f moves away from E_i towards the valence band.

10.8.2 Temperature Dependence of Electrical Conductivity of p-Type Semiconductor

If μ_h be the mobility and p the concentration of holes, then the electrical conductivity is

$$\sigma_p = ep\mu_h = eN_a^- \mu_h. \tag{10.49}$$

The conductivity of the minority carriers has been neglected. The temperature variation of σ_p is similar to that of n-type semiconductor.

10.8.3 Variation of Fermi Level with Temperature of p-Type Semiconductor

(10.46) shows that, the Fermi level E_f slightly decreases with temperature at low temperature region. But with the increase of temperature more and more acceptor atoms are ionised. Again, electron-hole pair is created with the rise of temperature due to breaking of covalent bonds when the material behaves as an intrinsic semiconductor. The Fermi level E_f moves gradually towards the intrinsic Fermi level E_i as shown in Fig. 10.15.

The variation of E_f with acceptor concentration is also shown in Fig. 10.15.

10.9 Charge Neutrality Equation

We consider a mixed semiconductor containing both donor and acceptor impurities. We assume that at surrounding temperature the impurity atoms are ionised and no charge carriers are generated due to breaking of covalent bonds.

If n, p be the density of electrons and holes and N_d, N_a the density of donor and acceptor atoms respectively, then under the above condition $n = N_d$ and $p = N_a$.

Then for overall charge neutrality of the material

$$p + N_d = n + N_a. \tag{10.50}$$

According to the *law of mass action*, in any semiconductor *at thermal equilibrium* $np = n_i^2$, where n_i is the intrinsic carrier concentration.

$$\therefore\ n = \frac{n_i^2}{p} \text{ and } p = \frac{n_i^2}{n}.$$

From (10.50),

$$p = n + N_a - N_d = \frac{n_i^2}{p} + N_a - N_d \text{ or } p^2 - p(N_a - N_d) - n_i^2 = 0. \tag{10.51}$$

(10.51) is a quadratic equation having roots

$$p=\frac{N_a-N_d}{2}\pm\left[\frac{(N_a-N_d)^2}{4}+n_i^2\right]^{1/2}. \tag{10.52}$$

Similarly,

$$n=\frac{N_d-N_a}{2}\pm\left[\frac{(N_d-N_a)^2}{4}+n_i^2\right]^{1/2}. \tag{10.53}$$

Special Cases

Case I For intrinsic semiconductor

$$N_a=N_d=0, \therefore\ n=p=n_i. \tag{10.54}$$

Case II For n-type semiconductors, $N_a=0$,

$$\therefore\ p=-\frac{N_d}{2}\pm\left[\frac{(-N_d)^2}{4}+n_i^2\right]^{1/2}=\frac{-N_d\pm\left(N_d+4n_i^2\right)^{1/2}}{2}=\frac{-N_d\pm N_d\left[1+4\frac{n_i^2}{N_d^2}\right]^{1/2}}{2}$$

$$\text{or } p=\frac{-N_d+N_d\left(1+4\frac{n_i^2}{N_d^2}\right)^{1/2}}{2}.$$

$\therefore$ p cannot be negative.

Now

$$\left(1+4\frac{n_i^2}{N_d^2}\right)^{1/2}\simeq 1+\frac{2n_i^2}{N_d^2} \text{ (neglecting the higher order terms).}$$

$$\therefore\ p=-\frac{N_d}{2}+\frac{N_d}{2}\left(1+\frac{2n_i^2}{N_d^2}\right) \text{ or } p=\frac{n_i^2}{N_d}. \tag{10.55}$$

Similarly,

$$n=\frac{N_d}{2}+\frac{N_d}{2}\left(1+\frac{2n_i^2}{N_d^2}\right)=N_d+\frac{n_i^2}{N_d}.$$

At $T\to 0$, $n_i\simeq 0$.

$$\therefore\ n=N_d. \tag{10.56}$$

Case III For p-type semiconductor, $N_d=0$.

$\therefore$ we can show that

$$n\simeq\frac{n_i^2}{N_a} \tag{10.57}$$

$$\text{and } p=N_a. \tag{10.58}$$

Hence, at ambient temperatures, the density of majority carriers is approximately equal to the density of impurity atoms. This means, at low temperatures for large impurity density the charge carriers are totally due to ionisation of impurity atoms. On the other hand, if impurity density is low and temperature is high, then the contribution due to impurity atoms is less. Under such condition, charge carriers are generated due to breaking of covalent bonds, i.e., the material behaves as an intrinsic semiconductor.

10.10 Carrier Transport in Semiconductors

(a) Drift Current: In presence of an electric field E the free electrons and holes acquire certain drift velocity, $v_d = \mu E$, μ is the mobility of charge carrier. This movement of the charge carriers constitutes a current, called the *drift current.*

For the *holes,* the drift current density is

$$J_{hd} = pe v_{hd} = pe\mu_h E, \tag{10.59}$$

where v_{hd} and μ_h are the drift velocity and mobility of holes respectively.

Similarly, the drift current density for *electrons* is

$$J_{ed} = ne v_{ed} = ne\mu_e E, \tag{10.60}$$

where v_{ed} and μ_e represent respectively the drift velocity and mobility of the electron.

∴ the total drift current density is

$$J_d = J_{hd} + J_{ed} = ne\mu_e E + pe\mu_h E = (ne\mu_e + pe\mu_h)\, E \tag{10.61}$$

or electrical conductivity

$$\sigma = \frac{J_d}{E} = ne\mu_e + pe\mu_h. \tag{10.62}$$

For intrinsic semiconductor,

$$\sigma_i = n_i\,(\mu_e + \mu_h)\, e, \tag{10.63}$$

∵ $n = p = n_i$ for intrinsic semiconductor.

(b) Diffusion current: In a semiconductor, in addition to conduction current, transport of charge carriers also occurs by a process known as *diffusion* leading to *diffusion current.*

Diffusion is a process in which the particles (electrons or holes) move from a region of higher concentration to region of lower concentration. The resulting current due to concentration gradient of the charge carriers is called *diffusion current.*

The electron flux at any point due to diffusion is proportional to concentration gradient at that point, i.e., diffusion of electron flux

$$N \propto \left.\frac{dn}{dx}\right|_{x=x_1} \quad \text{or } N = -D_e\frac{dn}{dx},$$

where D_e is a constant of proportionality and is called *diffusion constant* for electrons.

∴ diffusion current density of electrons is

$$(J_e)_x = eD_e\frac{dn}{dx}. \tag{10.64}$$

Similarly, the diffusion current due to holes is

$$(J_h)_x = -eD_h\frac{dn}{dx}, \tag{10.65}$$

where D_h is the *diffusion coefficient for the holes.*

Therefore, the total current density is

$$J_x = (J_e)_x + (J_h)_x = e\left[\left(n\mu_e E_x + D_e\frac{dn}{dx}\right) + \left(p\mu_h E_x - D_h\frac{dp}{dx}\right)\right]. \tag{10.66}$$

10.10.1 Einstein's Equation

Mobility and diffusion constant for a particular type of charge carrier are intimately related. The higher the value of mobility, greater will be the tendency to diffusion. In the steady state the conduction current density (J_c) must be equal and opposite to the diffusion current, i.e.,

$$J_c = -J_{eD} \text{ or } \sigma E = -eD_e \frac{dn}{dx}. \tag{10.67}$$

The presence of electric field produces a potential difference $V(x)$ over a distance x. The variation of electron concentration (n) with x is given by *Boltzmann relation:*

$$n = Ce^{-\frac{eV}{kT}} = Ce^{-\frac{eEx}{kT}}.$$

$$\therefore \frac{dn}{dx} = Ce^{-\frac{eEx}{kT}}\left(-\frac{eE}{kT}\right) = -\frac{neE}{kT} \text{ or } \frac{dn}{dx} = -\frac{neE}{kT} \tag{10.68}$$

From (10.67) and (10.68), we get

$$ne\mu_e E = \left(\frac{eD_e}{kT}\right)(neE) \text{ or } D_e = \frac{\mu_e kT}{e} \quad [\because \sigma = ne\mu_e]. \tag{10.69}$$

This is known as *Einstein's equation.*

For hole diffusion,

$$D_h = \frac{\mu_h kT}{e}. \tag{10.70}$$

Hence,

$$\frac{D_e}{D_h} = \frac{\mu_e}{\mu_h} \tag{10.71}$$

The calculated values of D for Si are

$$D_e = 3.4 \times 10^{-3} \text{ m}^2/\text{s for electrons}$$

$$D_h = 10^{-3} \text{ m}^2/\text{s for holes}$$

$$\text{and for Ge, } D_e = 9.9 \times 10^{-3} \text{ m}^2/\text{s for electrons}$$

$$D_h = 10^{-3} \text{ m}^2/\text{s for holes.}$$

10.10.2 Lifetime and Diffusion Length

Though the electron and hole concentrations in a semiconductor under thermal equilibrium are steady, the equilibrium itself is dynamic. That is, at any instant of time, the rate of generation of electron-hole pair is exactly equal to the rate of recombination of such pair. Such equilibrium concentration changes by several processes.

The carriers do not exist for indefinite period of time but recombine to form neutral particles. It will be shown that the excess carrier density decreases exponentially with time, the carrier density at time τ being $\propto e^{-t/\tau}$, where τ is the *lifetime of the carriers.* So, *the lifetime τ can be defined as the time at which carrier density falls to $1/e$ of its initial value.*

The average distance L that a carrier diffuses before recombination is called the diffusion length.

The diffusion length L and lifetime τ of carriers are related as follows:

$$L_n = \sqrt{D_e \tau_e} \text{ (for electrons)} \tag{10.72}$$

$$L_h = \sqrt{D_h \tau_h}, \text{ (for holes).} \tag{10.73}$$

10.11 Theory of Generation and Recombination of Charge Carriers

If a semiconductor be supplied with sufficient energy, the covalent tetrahedral bonds are broken and free electrons are generated. These free electrons are raised to the conduction band subsequently and a conducting hole is left behind in the valence band. This process is called the *electron-hole pair production* or generation as shown in Fig. 10.16.

In *n*-type semiconductor the number of electrons is much more than the number of holes. During pair production the number *minority carriers* (holes) increases but the increase of *majority carriers* (electrons) is negligible above the equilibrium number already existing in the material.

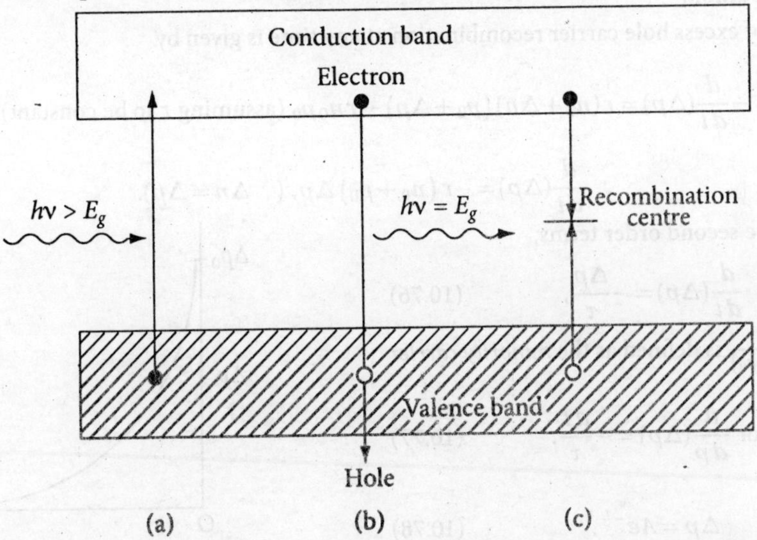

Fig. 10.16: (a) Generation of an electron hole pair (b) Direct recombination of an electron and a hole with the emission of a photon (c) An electron recombines with a hole via a recombination centre.

If the semiconductor be irradiated with light of frequency ν, such that $h\nu > E_g$, where E_g is the *forbidden energy gap*, electron-hole pair is created. When the light is cut off, the excess carriers disappear through electron-hole recombination and the material returns to thermal equilibrium. The recombination takes place when a conduction band electron is bound again with a hole in the valence band [Fig. 10.16 (b)]. The process is called *band to band recombination* which results with the emission of photon such that energy $h\nu = E_g$.

In some semiconductors (e.g., Ge and Si) the electron-hole recombination takes place via a recombination centre which is an energy level at about the centre of the gap. An electron from the conduction band and a hole from the valence band jump to the recombination centre and *annihilate* [Fig. 10.16 (c)].

Minority carrier lifetime: Under the thermal equilibrium condition, the densities of electrons (n_0) and holes (p_0) in a semiconductor remain constant in spite of the generation and recombination process taking place constantly. In other words, on an average, for every electron excited to conduction band, one electron recombines or drops to the valence band.

Let G and R denote respectively the rate of generation and recombination at equilibrium.

Then

$$R = G \text{ or } R - G = 0.$$

For band to band recombination, R is proportional to the product of n_0 and p_0, i.e.,

$$R \propto n_0 p_0 \text{ or } R = r n_0 p_0,$$

where r is the constant of proportionality.

$$\therefore\ r n_0 p_0 - G = 0. \tag{10.74}$$

Let Δp and Δn be the excess holes and electrons generated when the sample is disturbed from the equilibrium condition.

The net rate of excess hole carrier recombination at any time is given by

$$-\frac{d}{dt}(\Delta p) = r(n_0 + \Delta n)(p_0 + \Delta p) - r n_0 p_0 \text{ (assuming } r \text{ to be constant)} \tag{10.75}$$

$$\therefore\ \frac{d}{dt}(\Delta p) = -r(n_0 + p_0)\Delta p, \ (\because\ \Delta n = \Delta p).$$

Neglecting the second order terms,

$$\frac{d}{dt}(\Delta p) = -\frac{\Delta p}{\tau}, \tag{10.76}$$

where $\tau = \frac{1}{r(n_0+p_0)}$ is defined as the *minority carrier lifetime* τ.

$$\text{or } \frac{d}{dp}(\Delta p) = -\frac{dt}{\tau}. \tag{10.77}$$

Integrating,

$$\Delta p = A e^{-\frac{t}{\tau}}. \tag{10.78}$$

At $t = 0, \Delta p = (\Delta p)_0$,

$$\therefore\ (\Delta p)_0 = A.$$

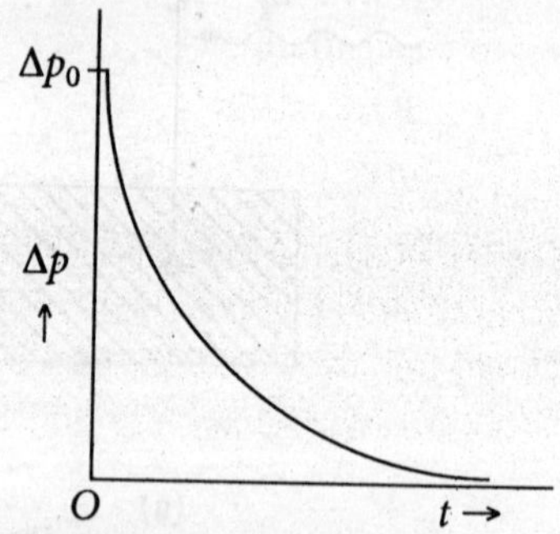

Fig. 10.17: Variation of Δp with t.

So

$$\Delta p = (\Delta p)_0 e^{-\frac{t}{\tau}}. \tag{10.79}$$

The equation indicates Δp *(or Δn) disappears exponentially with time as shown in* Fig. 10.17.
From (10.79), at $t = \tau$

$$\Delta p = (\Delta p)_0 e^{-1} \text{ or } \frac{\Delta p}{(\Delta p)_0} = \frac{1}{e}. \tag{10.80}$$

Thus, *the minority carrier lifetime is defined as the time taken for the excess charge carriers to reduce to $\frac{1}{e}$ times its initial value, once the source generating these excess charge is cut off.*

10.12 Hall Effect

When a piece of conductor (metal or semiconductor) carrying a current is placed in a transverse magnetic field, an electric field is produced inside the conductor in a direction normal to both the current and magnetic field. This phenomenon is known as Hall effect and the generated voltage is known as Hall voltage.

Let an n-type semiconductor carrying a current flow consisting entirely of electrons moving from right to left just opposite to the conventional flow of current. The semiconductor is placed in a magnetic field B in a direction perpendicular to the velocity (v) of the electrons. The *Lorentz force* acting on the electrons is $F =$

BeV which causes the electron to be deflected in the downward direction. As a result the negative charges will be accumulated on the bottom face of the semiconductor slab. The top surface will develop positive charge. A potential difference will, therefore, be established between the top and bottom surface with the bottom surface negative. The potential difference causes a field E_H, called the *Hall field*, in the negative y-direction as shown in Fig. 10.18 and exerts a force eE_H on the electron in the upward y-direction. In equilibrium condition the Lorentz force and the force due to E_H will balance each other when no further accumulation of electrons takes place and E_H attains a steady value.

In case of a p-type semiconductor, the charge carriers are holes and the bottom surface will be positively charged relative to the top surface due to the accumulation of holes.

At equilibrium, $eE_H = Bev$ or, $E_H = Bv$.

If J_x be the current density in the x-direction, then

$$J_x = nev,$$

where n is the concentration of the current carriers

$$\therefore\ E_H = \frac{BJ_x}{ne} \text{ or } \frac{E_H}{BJ_x} = \frac{1}{ne}. \tag{10.81}$$

The quantity $\frac{E_H}{BJ_x}$ giving the *Hall field per unit current density per unit magnetic flux density is called Hall coefficient* R_H, i.e.,

$$R_H = \frac{E_H}{BJ_x}. \tag{10.82}$$

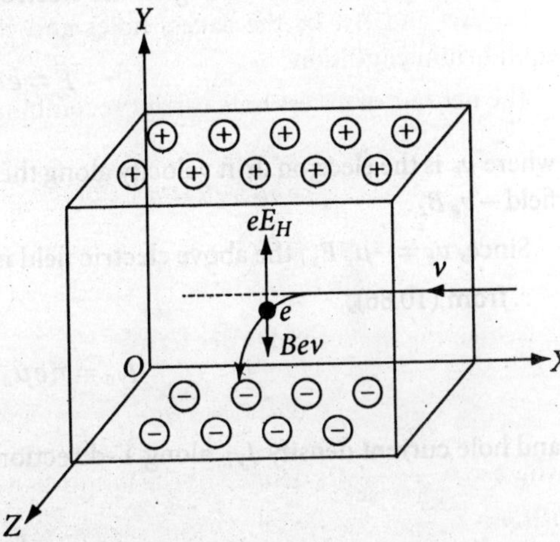

Fig. 10.18: Hall effect and motion of electrons in an n-type semiconductor.

Note. *A rigorous treatment shows that* $R_H = \frac{r}{ne}$, *where r is a numerical constant.*

Since for n-type semiconductor the carriers are negatively charged electrons,

$$\therefore\ R_H = -\frac{1}{ne}. \tag{10.83}$$

For p-type semiconductor the carriers are positively charged.

$$\therefore\ R_H = \frac{1}{pe}, \tag{10.84}$$

where p is the density of holes. (10.83) and (10.84) show that *the Hall coefficient R_H are of opposite sign for n-type and p-type semiconductors.*

When the contribution of both the electrons and holes to the current are appreciable, Hall coefficient may be proved to be

$$R_H = \frac{p\mu_h^2 - n\mu_n^2}{e\,(n\mu_e + p\mu_h)}, \tag{10.85}$$

since $\sigma = ne\mu_e + pe\mu_h$ and μ_e, μ_h are the mobilities of electrons and holes respectively.

To prove (10.85), the electron and hole current densities along x-direction can be expressed as

$$\left.\begin{aligned} J_{xe} &= ne\mu_e E_x \\ J_{xh} &= pe\mu_h E_x \end{aligned}\right\}, \tag{10.86}$$

since the magnitude of charge of both electron and hole is e. Then the total current density J_x is

$$J_x = J_{xe} + J_{xh} = e\left(n\mu_e + p\mu_h\right) E_x, \tag{10.87}$$

the change in μ_e and μ_h due to magnet resistance has been neglected.

Now the Lorentz force acting on the electron along the y-direction is

$$\vec{F} = e\left(\vec{v}_e \times \vec{B}\right) = e v_e B_z \hat{j}, \tag{10.88}$$

where v_e is the electron drift velocity along the y-direction. This Lorentz force $\vec{F}$ is equivalent to an electric field $-v_e B_z$.

Since, $v_e = -\mu_e E_x$, the above electric field is equal to $B_z \mu_e E_x$.

∴ from (10.86),

$$J_{ye} = ne\mu_e\left(B_z \mu_e E_x\right) = ne\mu_e^2 B_z E_x \tag{10.89}$$

and hole current density J_{yh} along Y-direction is

$$J_{yh} = pe\mu_p^2 B_z E_x. \tag{10.90}$$

So, the total current density along Y-direction is

$$J_y = J_{ye} + J_{yh} = e\left(n\mu_e^2 - p\mu_h^2\right) B_z E_x. \tag{10.91}$$

To make the net current density J_y zero a Hall field E_y is required.

$$\therefore \; \frac{E_y}{E_x} = \frac{J_y}{J_x} \text{ or } J_x E_y = J_y E_x \text{ or } e\left(n\mu_e + p\mu_h\right) E_y = -e\left(n\mu_e^2 - p\mu_h^2\right) B_z E_x$$

$$\text{or } E_y = \frac{p\mu_h^2 - n\mu_e^2}{n\mu_e + p\mu_h} B_z E_x. \tag{10.92}$$

The Hall coefficient R_H is given by

$$R_H = \frac{E_y}{B_z J_x} = \frac{p\mu_h^2 - n\mu_e^2}{e\left(n\mu_e + p\mu_h\right)^2}. \tag{10.93}$$

10.12.1 Measurement of the Hall Coefficient

A rectangular wafer of semiconductor having thickness t and width w is taken, through which a current I is passed along the x-direction. The wafer is placed between the pole pieces of an electromagnet so that the magnetic field B acts along the z-direction as shown in Fig. 10.19.

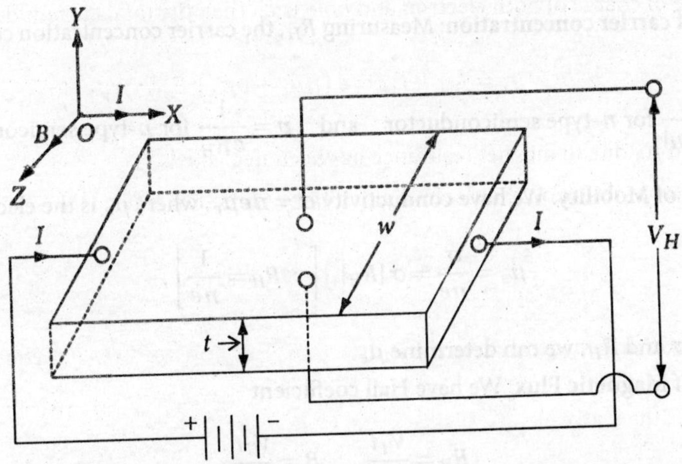

Fig. 10.19: Experimental arrangement for the measurement of R_H.

The Hall voltage V_H is measured by placing two probes at the top and bottom faces of the sample.
If E_H be the Hall field, then $E_H = \frac{V_H}{t}$.
Hence, $R_H = \frac{E_H}{B J_x} = \frac{V_H}{B J_x w}$. $\therefore V_H = R_H J_x B w$, [*w is the width of the wafer*].
Now $J_x = \frac{I}{a} = \frac{I}{wt}$, [$t$ = thickness of the wafer]

$$\therefore V_H = R_H B J_x w = R_H I B w / w t \text{ or } R_H = \frac{V_H t}{I B}. \tag{10.94}$$

If B is in Wb/m^2 and V_H is in volt, then R_H is in m^3/coulomb. The polarity of V_H will be *opposite for n-type and p-type materials.* It has been assumed in the above analysis that, *the charge carriers move in the energy bands with a steady drift velocity without any attractive forces.*

Hall angle (θ_H): The Hall angle θ_H is defined as

$$\tan\theta_H = \frac{E_H}{E_x}.$$

We have from (10.81),

$$E_H = \frac{J_x B}{ne} = \frac{(n e v_x) B}{ne} \quad [\because J_x = n e v_x]$$

$$\text{or } E_H = \frac{v_x B E_x}{E_x}.$$

$$\therefore \frac{E_H}{E_x} = \tan\theta = \frac{v_x B}{E_x} = \mu B, \; [\because v_x = \mu E_x] \tag{10.95}$$

μ = mobility of the charge carriers.

10.12.2 Applications of Hall Effect

(i) Determination of semiconductor type: Since *R_H is positive for p-type* and *negative for n-type semiconductor,* it can be used to determine the type of semiconductor.

(ii) Measurement of carrier concentration: Measuring R_H, the carrier concentration can be obtained from the relations:

$$n=\frac{1}{e|R_H|} \text{ for } n\text{-type semiconductor} \quad \text{and} \quad p=\frac{1}{eR_H} \text{ for } p\text{-type semiconductor.}$$

(iii) Determination of Mobility: We have conductivity $\sigma = ne\mu_e$, where μ_e is the electron mobility, then

$$\mu_e=\frac{\sigma}{ne}=\sigma|R_H|, \quad \left[\because R_H=\frac{1}{ne}\right].$$

Thus, measuring σ and R_H, we can determine μ_e.

(iv) Measurement of Magnetic Flux: We have Hall coefficient

$$R_H=\frac{V_H t}{IB}, \therefore B=\frac{V_H t}{IR_H}.$$

Thus, measuring V_H and R_H, B can be determined from the above relation.

(v) Apart from the above, Hall effect can be used to: (a) measure power in the electromagnetic wave and (b) determine Hall effect multiplier.

10.13 Cyclotron Resonance

If a constant magnetic field $\vec{B}$ be applied to a semiconductor, then the electrons in the conduction band and the holes in the valence band describe spiral (circular) orbits about $\vec{B}$. The angular frequency of rotation is given by

$$m^*\frac{v^2}{r}=\pm eVB \tag{10.96}$$

$$\text{or } \omega_c=\pm\frac{eB}{m^*}. \tag{10.97}$$

The $\pm$ sign is due to the fact that the electrons and holes rotate in opposite sense.

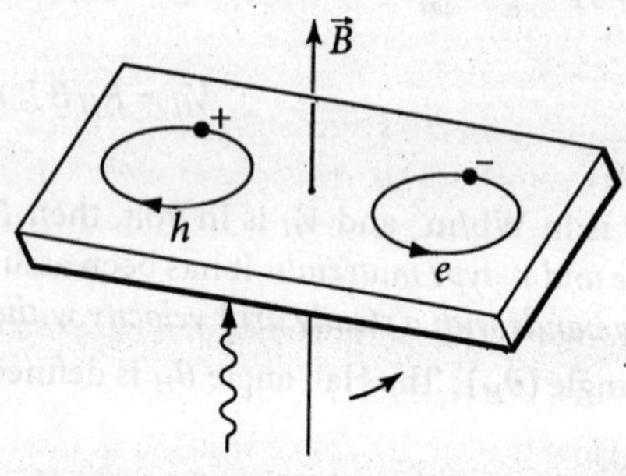

Fig. 10.20: Cyclotron motion of e and h in magnetic field B.

If now a radio frequency electric field is applied perpendicular to static magnetic field $\vec{B}$, then absorption of energy may take place under resonance condition,

$$\frac{d\vec{v}}{dt}=\frac{e}{m^*}\vec{v}\times\vec{B}. \tag{10.98}$$

The time period of rotation is.

$$T=\oint dt=\int\frac{dr}{v} \tag{10.99}$$

$$\vec{v}=\frac{\partial E}{\partial p}=\frac{1}{\hbar}\frac{\partial E}{\partial k}=\frac{1}{\hbar}=\frac{dE}{dk_\perp}, \tag{10.100}$$

where $dk_\perp$ is normal distance in $\vec{k}$-space projected on a plane perpendicular to $\vec{B}$.

But

$$\hbar\frac{dk}{dt}=e\vec{v}\times\vec{B} \text{ or } \vec{k}=\frac{e}{\hbar}\vec{r}\times\vec{B} \tag{10.101}$$

$$\therefore T=\oint\frac{dr}{v}=\hbar\oint\frac{d\vec{r}\cdot dk_{\perp}}{dE}=\frac{\hbar^2}{eB}\frac{\left|d\vec{k}\times d\vec{k}_{\perp}\right|}{dE} \tag{10.102}$$

$$\text{or } T=\frac{\hbar^2}{eB}\frac{dA}{dE} \quad [A=\text{area of the orbit in } k\text{-space}] \tag{10.103}$$

$$\text{or } \omega_c=\frac{eB}{m^*}=\frac{2\pi}{T}=\frac{2\pi eB}{\hbar^2}\frac{dE}{dA} \tag{10.104}$$

$$\text{or } m^*=\frac{\hbar^2}{2\pi}\frac{dA}{dE}. \tag{10.105}$$

Thus, measuring ω_c, in different direction of $\vec{B}$, we can measure the effective mass m^* in different directions. This technique is also used to distinguish between electron and hole. This is a standard technique. The cyclotron resonance experiments are performed at low temperatures ($\sim 4k$) using relatively pure samples by which the collision time τ is lengthened and $\omega_c\tau$ becomes larger.

10.14 Answers of Some Important Questions

10.1 What do you understand by the term semiconductor?

Ans. Semiconductors are materials having electrical conductivity intermediate between conductors and insulators. The resistivity of semiconductors varies between 10^{-5} to $10^{-4}\,\Omega$-m, whereas those for conductors and insulators vary from 10^{-8} to $10^{-6}\,\Omega$-m and 10^7 to $10^8\,\Omega$-m respectively. The band gap for semiconductors varies from 0.2 to 2.5 eV as compared to 6 eV for insulators (diamond). Temperature coefficient of semiconductors is negative, whereas for metals it is positive. Germanium, silicon, etc. are the examples of semiconductors which are widely used.

10.2 What are the important characteristics of semiconductors?

Ans. See Art. 10.1.

10.3 How the carrier concentration in a semiconductor can be increased?

Ans. The carrier concentration in a semiconductor can be increased by the following two processes:

(a) By increasing the temperature of the semiconductor and

(b) By the addition of a small amount of suitable impurity atoms to the semiconductor during growing.

10.4 Explain the concept of hole.

Ans. If an electron of an atom is removed, then a vacancy (or *hole*) is left behind which has a tendency to catch an electron due to its electropositive nature. By thermal excitation or by adding *dopants* to the semiconductor, an electron can be removed from the completely filled valence band of a semiconductor leaving behind a vacancy called *hole* in that band. The hole in the valence behave as a free positive electronic charge.

When a vacancy is created in the valence band, a lower energy electron in the valence band can be excited to the higher energy vacant level by applying an electric field when the hole moves down to the lower energy level. The movement of the hole in the electric field is just opposite to the direction of movement of the electron suggesting that the hole is a *positively charged particle.* The energy of a hole is increased when it moves from higher energy level to the lower energy level. Thus the movement of a hole in an applied electric field causes electrical conduction in the material. The conductivity in semiconductor arises due to the movement of electrons in conduction band and holes in valence band.

10.5 What is doping in semiconductors?

Ans. *Doping* in semiconductors is a process in which a small quantity of suitable impurity atom is added to the material during crystal growth. The dopants are added in the ratio 1 in 10^6 to 10^8 atoms which increases the electrical conductivity to a large extent. Doping does not change the structure of the semiconductor as the impurity atoms replace the regular atoms in the crystal. When a crystal is doped, the semiconductor is known as *extrinsic semiconductor.* Extrinsic semiconductors are of the following two types:

(a) *n-type* extrinsic semiconductor

Example: Ge or Si doped with pentavalent As or Sb.

(b) *p-type* extrinsic semiconductor

Example: Ge or Si doped with trivalent Al or Ga.

10.6 How the charge neutrality is maintained in an intrinsic semiconductor even on the addition of neutral pentavalent and trivalent atoms to it?

Ans. Intrinsic semiconductors like Ge or Si are tetravalent elements having four electrons in the outermost orbit. If a pentavalent element like As or Sb having five electrons in the valence band is added to it, the four electrons of As are used to form covalent bonds with four electrons of Ge and the fifth electron is free making the semiconductor *n*-type.

If a trivalent impurity like Al is added to Ge, three electrons of it form covalent bond with three valence orbit electrons of Al. But the fourth electron cannot find an electron to form a covalent bond. This gives rise to a hole which behaves like a positive electron and the material becomes a *p*-type semiconductor.

No charge is added or taken out of the material and hence the overall charge neutrality is maintained. Thus, the semiconductor even on the addition of pentavalent or trivalent impurity remains electrically neutral.

10.7 Discuss the effect of temperature on metal and semiconductor.

Ans. The effect of temperature creates an important distinction between metals (conductors) and semiconductors. In case of a metal the conductivity decreases with the rise of temperature as the temperature coefficient of resistance is *positive.* But the temperature coefficient of resistance for semiconductor is *negative* which causes an increase of conductivity with temperature. Further, as the temperature increases, more and more *phonons* are excited which scatter electrons and reduce their mobility. In semiconductor the decrease in mobility is more compensated by the increased number of charge carriers. The electrons in the valence orbit get sufficient energy with the increase of temperature and jump to the conduction band, thereby increasing the electrical conductivity of the semiconductor.

10.8 What do you mean by effective mass of an electron? How does it vary with the wave vector $\vec{k}$?

Ans. The wave particle motion of an electron is different from that of a free electron as the electron interacts with the *periodic potential* of the crystal lattice. For an electron with energy E and mass m_0 moving in free space,

$$m_0 = \frac{\hbar^2 k^2}{2E}$$

But the *effective mass* of the electron moving in a periodic potential is given by

$$m^* = \frac{1}{\hbar^2}\frac{d^2E}{dk^2}.$$

Hence, the dynamical behaviour of the electron is the same as that of a free electron of mass m^*. Hence, using the effective mass m^*, we can describe the motion of the electron in the crystal lattice by applying *principles of classical physics*.

Variation of m^* with $\vec{k}$: See last part of Art. 10.3.2.

10.9 Why does a semiconductor behave as an insulator at 0 K?

Ans. The electrons at 0 K in the valence band do not have sufficient energy to be excited from valence band to the conduction band across the forbidden gap of energy (E_g). The conduction band under such condition is empty and electrical conduction ceases to take place. That is why at 0 K, a semiconductor behaves as an insulator.

10.10 Why do all insulators behave as semiconductor at temperatures sufficiently higher than 0 K?

Ans. Insulators are characterised by large band gap (e.g., $E_g \sim 6$ V for Diamond) and at ~ 0 K the electrons in the filled valence band do not get sufficient energy (i.e., $kT \gg E_g$) to jump across the forbidden gap to the conduction band. But if temperature is sufficiently large so that $kT \gg E_g$, then the jump of the electrons from valence band to the conduction band is possible and electrical conduction can take place. Hence, all insulators may behave as semiconductor, if $T \gg 0$ K.

10.11 Discuss the variation of position of Fermi level in extrinsic semiconductor with temperature.

Ans. At low temperature the Fermi level E_f lies half-way between E_f and E_d in n-type semiconductor, whereas in p-type semiconductor E_f lies half-way between E_a and E_v. At high temperature E_f coincides with the *Fermi level* of intrinsic semiconductor (E_f).

10.12 State the law of mass action in semiconductor. Discuss the effect of addition of impurity atoms to the material.

Ans. The law of mass action states that, for a given semiconductor, the product of electron and hole concentrations is constant at a given temperature and is equal to the square of the intrinsic carrier concentration.

If n and p represent respectively the concentration of electrons and holes, then at a particular temperature, $np = n_i^2$, where n_i is the concentration of the intrinsic semiconductor.

If the addition of impurity atoms increases the value of n, then there will be a corresponding decrease in p, so that $np =$ constant.

10.13 Write down Einstein's equation and show that

$$\frac{D_e}{D_p} = \frac{\mu_e}{\mu_h},$$

where the symbols have their usual meanings.

Ans. If μ_e be the electron mobility of an electron of charge e at temperature T, then its *diffusion length* is given by $D_e = \mu_e kT/e$, where k is the *Boltzmann constant*.

For p-type semiconductor, $D_p = \mu_h kT/e$.

$$\therefore \quad \frac{D_e}{D_p} = \frac{\mu_e kT/e}{\mu_h kT/e} = \mu_e/\mu_h,$$

where μ_h is the mobility of holes.

10.14 Write down the expression for Hall coefficient. What is its unit in SI system?

Ans. If a wafer of semiconductor of thickness t be placed in a magnetic field B along Z-direction carrying a current I along y-direction, then the *Hall coefficient* is given by

$$R_H = \frac{V_H t}{IB},$$

where V_H is the *Hall voltage* developed between the faces perpendicular to y-axis.

The unit of R_H in SI system is m³/coulomb.

10.15 Mention some important applications of Hall effect.

Ans. See Art. 10.12.2.

10.15 Solved Problems

10.1 Calculate the resistivity of Ge at room temperature from the following data: Intrinsic carrier density of Ge at $RT = 2.37 \times 10^{19}/\text{m}^3$. The electron mobility $\mu_e = 0.38\,\text{m}^2/\text{V/s}$. The hole mobility $\mu_h = 0.18\,\text{m}^2/\text{V/s}$.

Solution. The intrinsic conductivity of Ge is given by $\sigma_i = n_i e(\mu_e + \mu_h)$ and resistivity $\rho = 1/\sigma$.

$$\therefore \sigma_i = \left(2.37 \times 10^{19}\right) \times \left(1.6 \times 10^{-19}\right) \times (0.38 + 0.18) = 2.1235\,\Omega\text{-m}.$$

$$\therefore \rho = \frac{1}{\sigma_i} = \frac{1}{2.1235} = \mathbf{0.471\,\Omega\text{-m}}.$$

10.2 Calculate the intrinsic carrier density and conductivity at 300 K in Germanium. Given: Atomic weight of Ge = 72.6, Density of Ge = 5400 kg/m³, Mobility of electrons $\mu_e = 0.4\,\text{m}^2/\text{V/s}$, Mobility of holes $\mu_h = 0.2\,\text{m}^2/\text{V/s}$ and Band Gap of Ge $E_g = 0.7\,\text{eV}$.

Solution. The intrinsic carrier density

$$n_i = 2\left[\frac{2\pi m k T}{h^2}\right]^{3/2}\left[e^{-E_g/2kT}\right] = 4.83 \times 10^{21} \times (300)^{3/2} \times e^{-\frac{0.7}{0.052}} = 3.5 \times 10^{19}/\text{m}^3.$$

Conductivity $\sigma = n_i e(\mu_e + \mu_h) = (3.5 \times 10^{19}) \times (1.6 \times 10^{-19}) \times 0.6 = \mathbf{3.36\,\Omega^{-1}m^{-1}}$.

10.3 Calculate the density of donor atoms that has to be added to an intrinsic germanium semiconductor of conductivity 4 mho/cm. Given: the electron mobility of the n-type semiconductor is 3900 cm²/V/s.

Solution. If N_d be the donor concentration, then assuming the concentration of donor atoms equal to the concentration of free electron n, we have

$$n = N_d = \frac{\sigma_n}{e\mu_e} = \frac{4}{(1.6 \times 10^{-19}) \times 3900} = \mathbf{0.64 \times 10^{16}/cm^3}.$$

10.4 The resistivity of an intrinsic semiconductor is 4.5 Ω-m at 20°C and 2.0 Ω-m at 32°C. Find the energy band gap.

Solution. The plot of $\ln \rho_i$ against $\frac{1}{T}$ is shown in the adjoining figure.

The band gap E_g is given by

$$E_g = 2k\left(\frac{dy}{dx}\right).$$

Now

$$dy = \ln\rho_2 - \ln\rho_1 = 0.6532 - 0.3010 = 0.3522$$

$$dx = \frac{1}{T_2} - \frac{1}{T_1} = \frac{1}{(273+20)} - \frac{1}{(273+32)}$$

$$= \frac{1}{293} - \frac{1}{305} = 0.134 \times 10^{-3}.$$

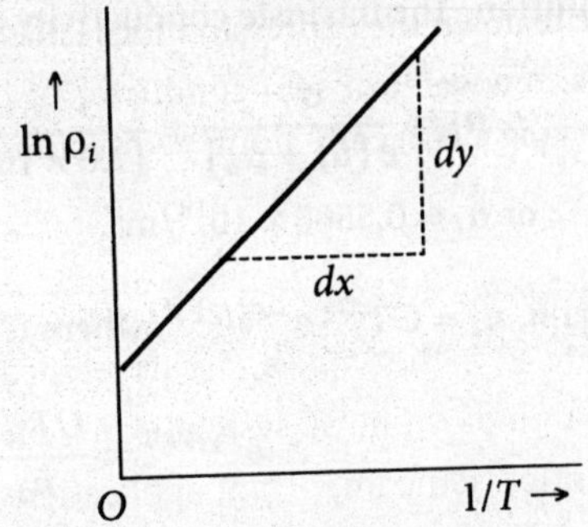

Fig. 10.21

$$\therefore \frac{dy}{dx} = \frac{0.3522}{0.134 \times 10^{-3}} = 2.628 \times 10^3.$$

Hence, $E_g = 2 \times (8.616 \times 10^{-5}) \times (2.628 \times 10^3) = \mathbf{0.453\ eV}$.

10.5 Show that the ratio of maximum resistivity to intrinsic resistivity is expressed as

$$\frac{\rho_m}{\rho_i} = \frac{(\mu_e + \mu_h)}{2\sqrt{\mu_e \mu_h}},$$

where μ_e and μ_h denote the mobilities of electron and hole respectively.

Solution. The intrinsic resistivity $\sigma_i = n_i e\,(\mu_e + \mu_h)$ and conductivity at any other dropping level is $\sigma = e\,(n_e\mu_e + n_h\mu_h)$, the different symbols having their usual meanings.

$$\text{or } \sigma = e\left[n_e\mu_e + \frac{n_i^2\mu_h}{n_e}\right] \quad [\because n_e n_h = n_i^2].$$

The condition for the resistivity to be maximum the conductivity is to be minimum, i.e.,

$$\frac{d\sigma}{dn_e} = 0 \text{ or } e\left[\mu_e - \frac{n_i^2\mu_h}{n_e^2}\right] = 0 \text{ or } \mu_e = \frac{n_i^2\mu_h}{n_e^2} \text{ or } n_e = n_i\sqrt{(\mu_h/\mu_e)} \text{ and } n_h = n_i\sqrt{\frac{\mu_e}{\mu_h}}.$$

$$\therefore \sigma_m = e\left[n_i(\mu_e\mu_h)^{1/2} + \frac{n_i^2\mu_h}{n_i}\sqrt{\frac{\mu_e}{\mu_h}}\right] = n_i e\left[(\mu_e\mu_h)^{1/2} + (\mu_e\mu_h)^{1/2}\right] = n_i e\left[2(\mu_e\mu_h)^{1/2}\right].$$

$$\therefore \frac{\sigma_i}{\sigma_m} = \frac{\rho_m}{\rho_i} = \frac{n_i e\,(\mu_e + \mu_h)}{2n_i e(\mu_e\mu_h)^{1/2}} \text{ or } \frac{\rho_m}{\rho_i} = \frac{(\mu_e + \mu_h)}{2\sqrt{\mu_e\mu_h}} \textbf{ (Proved).}$$

10.6 Holes were injected into a bar of germanium. 12 microsecond later the density decreased to $10^{18}/m^3$. The lifetime of hole in the bar is 4.5 μs. Compute the density of holes injected initially.

Solution. If $(\Delta p)_0$ and (Δp) be the change in density of holes at $t = 0$ and $t = t$, then

$$\Delta p = (\Delta p)_0\, e^{-t/\tau} \text{ or } 10^{18} = (\Delta p)_0\, e^{-\frac{12}{4.5}} = (\Delta p)_0\, e^{-2.7}.$$

$$\therefore (\Delta p)_0 = 10^{18} \times e^{+2.7} = \mathbf{1.4 \times 10^{19}}.$$

10.7 Compute the forbidden energy gap in germanium from the following data: Mobilities of electrons and holes in intrinsic germanium at 300 K = 0.36 m²/V/s and 0.17 m²/V/s respectively; Resistivity of Germanium at 300 K = 2.12 Ω-m.

Solution. The intrinsic conductivity of Ge, $\sigma_i = n_i e(\mu_e + \mu_h)$.

$$\therefore n_i = \frac{\sigma_i}{e(\mu_e+\mu_h)} = \frac{0.472}{(1.6\times10^{-19})\times(0.17+0.36)} \left[\because \sigma = \frac{1}{\rho} = \frac{1}{2.12} = 0.472\,\Omega^{-1}\text{m}^{-1}\right]$$

or $n_i = 0.5566 \times 10^{19}/\text{m}^3$.

Again, $n_i = CT^{3/2} e^{-E_g/2kT}$, where $C = 2\left[\frac{2\pi mk}{h^2}\right] = 4.83\times10^{21}$ and $2kT = 0.052$ at $T = 300$ K.

$$\therefore e^{E_g/2kT} = \frac{CT^{3/2}}{n_i} = \frac{(4.83\times10^{21})\times(300)^{3/2}}{0.5566\times10^{19}} = 4.509\times10^6.$$

$\therefore E_g/2kT = \ln(4.509\times10^6) = \ln(4.509) + 6\ln 10 = 15.322$ or, $E_g = 2kT\times15.322 = 0.052\times15.322$.

$\therefore E_g = 0.797$ eV.

10.8 The electron and hole mobilities of Si sample are 0.135 and 0.048 m²/V/s respectively. Determine the conductivity of intrinsic Si at 300 K if the intrinsic carrier concentration is 1.5×10^{16} atoms/m³. The sample is then doped with 10^{23} atoms/m³ of phosphorus. Determine the equilibrium hole concentration, conductivity and position of the Fermi level relative to the intrinsic level.

Solution. The conductivity of Si is given by $\sigma = n_i e(\mu_e + \mu_h)$.

Here $n_i = 1.5\times10^{16}$ atoms/m³, $\mu_e = 0.135\,\text{m}^2/\text{V/s}$, $\mu_h = 0.048\,\text{m}^2/\text{V/s}$.

$\therefore \sigma = (1.5\times10^{16})\times(1.6\times10^{-19})\times(0.135+0.048) = 1.5\times1.6\times0.183\times10^{-3} = \mathbf{0.4392\times10^{-3}\,\Omega\text{-m}}$.

Now the sample is doped with 10^{23} phosphorus atoms/m³. $\therefore N_d = 10^{23}/\text{m}^3$.

So, the hole concentration is

$$p = \frac{n_i^2}{N_d} = \frac{(1.5\times10^{16})^2}{10^{23}} = \mathbf{1.5\times10^9/m^3}.$$

Conductivity $\sigma_{ex} = N_d e\mu_e = 10^{23}\times(1.6\times10^{-19})\times0.135 = \mathbf{2.16\times10^3/\Omega\text{-m}}$.

$$E_f = \frac{E_g}{2} + \frac{3kT}{4}\left(\ln\frac{m_h^*}{m_e^*}\right) = \frac{E_g}{2} + \frac{3\times(1.38\times10^{-28})\times300}{4\times1.6\times10^{-19}}\left(\ln\frac{\mu_e}{\mu_h}\right) \left[\because \mu \propto \frac{1}{m}\right]$$

$$= \frac{E_g}{2} + 0.01935\times\ln\left(\frac{0.135}{0.048}\right) = \frac{E_g}{2} + 0.01935\times1.034 = \frac{E_g}{2} + 0.02.$$

$\therefore$ the Fermi level will be **0.02 eV** above the intrinsic level.

10.9 The intrinsic concentration of a semiconductor is $3.1\times10^{13}/\text{cm}^3$ at room temperature. It is doped with 10^{14} donor atoms/cm³. Assuming that the donors are all ionised, calculate the electron and hole concentrations and the percentage increase in conductivity of the doped material over the intrinsic values. Given: electron mobility = 3800 cm²/V/s hole mobility = 1800 cm²/V/s.

Solution. If n, p and N_d be the concentrations of electron, hole and donor, then from the condition of charge neutrality $n - p = N_d$.

Also, if n_i be the intrinsic concentration, then $np = n_i^2$.

$$\therefore n - \frac{n_i^2}{n} = N_d \text{ or } n^2 - nN_d - n_i^2 = 0.$$

$$\therefore n = \frac{N_d \pm \sqrt{N_d^2 + 4n_i^2}}{2} \text{ or } n = N_d + \sqrt{\left(\frac{N_d}{2}\right)^2 + n_i^2}.$$

Substituting

$$n_i = 3.1 \times 10^{13}/\text{cm}^3$$
$$N_d = 10^{14}/\text{cm}^3,$$

we get

$$n = \mathbf{1.09 \times 10^{14}/cm^3}.$$

$$\therefore\ p = \frac{n_i^2}{n} = \frac{(3.1\times10^{13})^2}{1.09\times10^{14}} = \mathbf{0.88 \times 10^{13}/cm^3}.$$

Now the total conductivity

$$\sigma = e\,(n\mu_e + p\mu_h) = \left(1.6\times10^{-19}\right)\left[\left(1.09\times10^{14}\right)\times 3800 + \left(0.88\times10^{13}\right)\times1800\right]$$
$$= 0.068\Omega^{-1}\,(\text{cm})^{-1}.$$

The intrinsic conductivity

$$\sigma_i = e n_i\,(\mu_e + \mu_h) = \left(1.6\times10^{-19}\right)\times\left(3.1\times10^{13}\right)[3800+1800] = 0.028\ \Omega^{-1}\,(\text{cm})^{-1}.$$

$\therefore$ % increase in σ over the intrinsic value is

$$\frac{\sigma-\sigma_i}{\sigma}\times100 = \frac{0.068-0.028}{0.028}\times100 = 143\%.$$

10.10 A semiconducting crystal 12 mm long, 5 mm wide and 1 mm thick has a magnetic flux density of $0.5\,\text{Wb/m}^2$ applied from front to back perpendicular to largest faces. When a current of 20 mA flows lengthwise through the specimen, the voltage measured across its width is found to be 37μV. Find the Hall coefficient of the conductor.

Solution. If J_x, E_y and B_z be respectively the current density along x-direction, electric field along y-direction and magnetic field along z-direction, then Hall coefficient

$$R_H = \frac{E_y}{J_x B_z} = \frac{1}{ne},$$

where n and e are the charge concentration and electronic charge respectively.

$$\text{or } R_H = \frac{V_y}{b\,J_x B_z}\left[\because\ E_y = \frac{V_y}{b}\right]$$

$$J_x = \frac{I_x}{bt}\ [b = \text{breadth and } t = \text{thickness of the sample}]$$

$$\therefore\ R_H = \frac{V_y}{bI_x}\frac{bt}{B_z} = \frac{V_y t}{I_x B_z}.$$

Here

$$V_y = 37\times10^{-16}\,\text{V}$$
$$t = 10^{-3}\,\text{mm}$$
$$B_z = 0.5\,\text{Wb/m}^2$$
$$I_x = 20\times10^{-3}\,\text{A}$$

$$\therefore\ R_H = \frac{(37\times10^{-6})\times(10^{-3})}{(20\times10^{-3})\times(0.5)} = \mathbf{3.7\times10^{-6}\,m^3/C}.$$

10.11 A current of 50 A is established in a slab of copper 0.5 cm thick and 2 cm wide. The slab is placed in a magnetic field $B = 1.5\,T$. The magnetic field is perpendicular to the plain of the slab and to the current. The free electron concentration in copper is 8.48×10^{28} electron/m^3. What is the magnitude of Hall voltage across the width of the slab?

Solution. The Hall coefficient is given by

$$R_H = \frac{1}{ne} \text{ and } V_H = \frac{R_H I B}{t}.$$

$$\therefore\ V_H = \frac{IB}{net} \text{ is the Hall voltage.}$$

Here

current $I = 50$ A

Magnetic field , $B = 1.5\,T$

Concentration of electrons $n = 8.4 \times 10^{28}/\text{m}^3$.

Thickness of the sample $t = 0.5\,\text{cm} = 0.5 \times 10^{-2}$.

$$\therefore\ V_H = \frac{50 \times 1.5}{(8.4 \times 10^{28}) \times (1.6 \times 10^{-19}) \times (0.5 \times 10^{-2})}$$

$$= 11.161 \times 10^{-7}\,V = 1.116 \times 10^{-6}\,V$$

$$= \mathbf{1.116\,\mu V.}$$

10.12 A current of 1 mA flows through an n-type Ge strip of 1 mm thick and 1 mm wide placed in a magnetic field B. If the Hall voltage produced inside the strip be 1 mV, what is the value of B? [Given: $R_H = 10^{-2}\,\text{m}^3/\text{coulomb}$.]

Solution. If w be the width and t be the thickness of the Ge strip, then Hall coefficient

$$R_H = \frac{V_H w}{IB} \text{ or } B = \frac{V_H w}{IR_H} = \frac{(1 \times 10^{-3}) \times (1 \times 10^{-3})}{(10^{-2}) \times (1 \times 10^{-3})} = \mathbf{0.1\,T.}$$

EXERCISES

10(A) Objective Type Questions

Choose the correct alternatives.

10.1 The temperature coefficient of semiconductor is

(a) positive
(b) negative
(c) 0 .
(d) none of these.

10.2 The band gap of semiconductor is of the order of

(a) 0.1 eV
(b) 10 eV
(c) 50 eV
(d) 100 eV.

10.3 When a semiconductor is doped with suitable impurity its conductivity

(a) remains constant
(b) decreases
(c) increases
(d) becomes zero.

10.4 In n-type Ge (or Si) semiconductor the impurity atom is
(a) trivalent
(b) tetravalent
(c) pentavalent
(d) hexavalent.

10.5 In p-type Ge (or Si) semiconductor the impurity atom is
(a) trivalent
(b) tetravalent
(c) pentavalent
(d) none of these.

10.6 In intrinsic semiconductor the carrier concentration varies as
(a) $1/T$
(b) $T^{3/2}$
(c) T^2
(d) T.

10.7 At 0 K pure Ge is an
(a) extrinsic semiconductor
(b) insulator
(c) intrinsic semiconductor
(d) a superconductor.

10.8 The electrical conductivity in a pure semiconductor, when an external electric field is applied, is due to
(a) holes
(b) electrons
(c) ions
(d) electrons and holes.

10.9 The majority carriers in p-type Ge are
(a) free electrons
(b) holes
(c) ions
(d) conduction electrons.

10.10 In a p-type semiconductor, the concentration of holes is proportional to the square root of the concentration of
(a) donor
(b) acceptor atom
(c) intrinsic impurity atom
(d) none of these.

10.11 The value of $\Delta T/T$ necessary to double the number of carriers (or conductivity) is
(a) $\dfrac{kT}{E_g}$
(b) $kT\ln 2$
(c) $\dfrac{2kT}{E_g}\ln T$
(d) $\dfrac{E_g}{2kT}$.

10.12 At 0 K the Fermi level in an intrinsic semiconductor lies at
(a) the centre of the top of the valence band and bottom of the conduction band
(b) the centre of the donor level and intrinsic Fermi level
(c) the centre of donor level and bottom of the conduction band
(d) the centre of the acceptor level and top of the valence band.

10.13 At 0 K, the resistivity of pure Si is 300 Ω-m. If $\mu_e = 0.14$ and $\mu_h = 0.05\ \text{m}^2\,\text{V}^{-1}\,\text{s}^{-1}$, then intrinsic carrier concentration per cubic metre is

(a) zero
(b) 10^{17}
(c) 2.19×10^{17}
(d) 1.095×10^{17}.

10.14 The density of a charge carrier in a pure semiconductor is proportional to
(a) $e^{-E_g/kT}$
(b) $e^{-2E_g/kT}$
(c) e^{-E_g/kT^2}
(d) $e^{-E_g/2kT}$.

10.15 At low temperatures and higher concentration of acceptor impurities, the density of holes (p) varies as
(a) N_a
(b) $\frac{1}{N_a}$
(c) $\sqrt{N_a}$
(d) N_a^2.

N_a being the concentration of donor atoms.

10.16 With increase of temperature the Fermi level E_f in n-type semiconductor moves towards the
(a) intrinsic Fermi level (E_i)
(b) bottom of conduction band (E_c)
(c) top of valence band E_v
(d) none of these.

10.17 With the increase of temperature the Fermi level E_f in a p-type semiconductor moves gradually towards the
(a) intrinsic Fermi level E_i
(b) bottom of the conduction band E_c
(c) top of the valence band (E_v)
(d) none of these.

10.18 The average distance an electron diffuse before recombination is related to the diffusion coefficient and lifetime by the equation
(a) $L_n = \sqrt{D_e \tau_n}$
(b) $L_n = \frac{1}{D_e \tau_n}$
(c) $L_n = D_e \tau_n$
(d) $l_n = \frac{D_e}{\tau_n}$.

10.19 The mean lifetime of the carrier injected into a semiconductor is the time in which the carrier value falls to
(a) 10% of its initial value
(b) 27% of its initial value
(c) 37% of its initial value
(d) 17% of its initial value.

10.20 Hall coefficient of a n-type semiconductor is
(a) $R_H = \frac{1}{ne}$
(b) $R_H = -\frac{1}{ne}$
(c) $R_H = \frac{r}{ne}$
(d) $R_H = rne$.

10.21 When both electrons and holes to the current is appreciable, Hall coefficient is given by
(a) $R_H = \frac{r}{\pm ne}$
(b) $R_H = \frac{p\mu_p^2 - n\mu_n^2}{e\left(n\mu_e + p\mu_p\right)}$
(c) $R_H = \frac{1}{ne}$
(d) $R_H = \frac{e\left(n\mu_e + p\mu_p\right)}{p\mu_p^2 - n\mu_n^2}$.

10.22 Which quantity cannot be calculated by measuring Hall coefficient?
(a) mobility of charge carriers
(b) Energy band gap of the semiconductor
(c) type of semiconductor
(d) concentration of charge carriers.

10.23 The effective mass of the electron is given by

(a) $m^* = \dfrac{\hbar^2 E}{2k^2}$

(b) $m^* = \dfrac{2E}{\hbar^2 k^2}$

(c) $m^* = \dfrac{\hbar^2 k^2}{2E}$

(d) $m^* = \dfrac{2\hbar^2 k^2}{E}$.

10.24 The law of mass action is

(a) $np = n_i^2$

(b) $nn_i^2 = p$

(c) $\dfrac{n}{p} = n_i^2$

(d) $n = n_i^2 p$.

10.25 If $m_p^* > m_e^*$, then

(a) $\mu_e < \mu_p$

(b) $\mu_e = \mu_p$

(c) $\mu_e > \mu_p$ (d) $\mu_e = \dfrac{1}{\mu_p}$.

ANSWERS

10.1 (b)	10.5 (a)	10.9 (b)	10.13 (d)	10.17 (a)	10.21 (b)	10.25 (c).
10.2 (a)	10.6 (b)	10.10 (b)	10.14 (d)	10.18 (a)	10.22 (b)	
10.3 (c)	10.7 (b)	10.11 (c)	10.15 (c)	10.19 (c)	10.23 (c)	
10.4 (c)	10.8 (d)	10.12 (a)	10.16 (a)	10.20 (b)	10.24 (a)	

10(B) Short Answer Type Questions

10.1 What is a semiconductor? How does it differ from metal and insulator?

10.2 What is intrinsic semiconductor? Show that in an intrinsic semiconductor both the electrons and holes contribute towards electrical conductivity. [LU 1996; PU 1997]

10.3 Distinguish between intrinsic and extrinsic semiconductors.

10.4 What is the concept of *hole*? How does it differ from a free electron?

10.5 What do you understand by *donor* and *acceptor*? Show their energy level in the energy band diagram. Give two examples of each.

10.6 What is effective mass of electron. Show that it depends on $\frac{d^2E}{dk^2}$ and is a function of the wave vector k.

10.7 Can a semiconductor containing both n-type and p-type impurities behave as an intrinsic semiconductor? Explain with reason.

10.8 Derive an expression for electron concentration in an intrinsic semiconductor.

10.9 Find out an expression for the hole concentration in an intrinsic semiconductor.

10.10 Discuss the effects of doping As and Al atoms separately in Ge crystal.

10.11 What is law of mass action? What is its significance?

10.12 Discuss the variation of Fermi level in n-type and p-type semiconductors with temperature.

10.13 Define (a) mean lifetime (b) diffusion length and write the relation between the two. [CUH 2004]

10.14 Describe the basic principle of Hall effect in semiconductors.

10.15 Show that the Hall coefficient for an intrinsic semiconductor is given by

$$R_H = -\frac{1}{ne}\left(\frac{\mu_e - \mu_h}{\mu_e + \mu_h}\right),$$

where μ_e, μ_h are the mobilities of electron and holes and n is the intrinsic carrier concentration.

10.16 Show that at a given temperature a semiconductor has a maximum conductivity σ_m given by

$$\sigma_m = 2en_i\sqrt{\mu_e - \mu_h},$$

where μ_e, μ_h are the mobilities of electrons and holes and n_i is the intrinsic carrier concentration.

10.17 What are mobility and conductivity? Obtain an expression for the conductivity of a doped semiconductor. [NBUH 2008]

10.18 What is Hall effect? Find an expression for Hall Voltage and Hall Coefficient. State the importance of this effect. [NBUH 2009]

10.19 Find an expression for (i) mobility of electrons and (ii) electrical conductivity on the basis of free electron theory of metals. [VUH 2009]

10.20 Qualitatively discuss the formation of energy bands for electrons in solids and distinguish between metals and semiconductors (Use necessary diagram). [VUH 2002]

10.21 What is Hall effect? Obtain an expression for Hall coefficient in a metal when the carriers are only electrons. Why is the Hall coefficient positive for some metals? [VUH 2007]

10.22 Show that the effective number of carriers is maximum when the band is half-filled and zero when filled or empty (Use effective mass). [CUH 2009]

10.23 Show that when an electron moves through a crystal, the effective mass of the electron can be expressed as

$$m^* = \frac{\hbar^2}{\left(\frac{d^2E}{dk^2}\right)},$$

where the symbols have their usual significances. [CUH 2007]

10.24 (a) Find the difference between the top and bottom of the energy band.

(b) Obtain an expression for the effective mass m^* of the electron as a function of k. Sketch m^* vs. k. [CUH 2003]

10.25 Show graphically the variation of energy with the wave vector k for the first allowed band in a solid. Hence, explain the variation in sign of the effective mass for an electron in the band. What is a hole? [CUH 2000]

10.26 Derive an expression for Hall coefficient. What information do you obtain from Hall effect experiment? [VUH 1995]

10(C) Numerical Problems

10.1 A wafer of Ge crystal of area 1 cm² and thickness 0.3 mm is placed in a potential difference of 2 volts. Calculate the current produced.

[Given: Concentration of free electrons in Ge $= 2 \times 10^{19}/\text{m}^3$, $\mu_e = 0.36\,\text{m}^2\text{V}^{-1}\text{s}^{-1}$ and $\mu_h = 0.17\,\text{m}^2\text{V}^{-1}\text{s}^{-1}$] *(Ans. 1.13 A)*

10.2 Estimate the temperature at which diamond has the same conductivity as Si at 27°C.

[Given: Band gap of diamond and silicon are 5.4 eV and 1.1 eV respectively). *(Ans. 1200°C)*

10.3 For intrinsic germanium at 300 K, calculate the resistivity of the sample from the given data: $n_i = 2.4 \times 10^{19}/\text{m}^3$; $\mu_e = 0.39\,\text{m}^2\text{V}^{-1}\text{s}^{-1}$, $\mu_h = 0.19\,\text{m}^2\text{V}^{-1}\text{s}^{-1}$. *(Ans. 0.449Ω-m)*

10.4 Find the resistance of an intrinsic Ge rod 1 cm long, 1 mm wide and 1 mm thick at 300 K.

[Given: $n_i = 2.5 \times 10^{19}/\text{m}^3$; $\mu_e = 0.39\,\text{m}^2\text{V}^{-1}\text{s}^{-1}$ and $\mu_h = 0.19\,\text{m}^2\text{V}^{-1}\text{s}^{-1}$ at 300 K] *(Ans. 4.31 × 10³Ω)*

10.5 If the effective mass of holes in a material is 4 times that of electrons, find the temperature at which the Fermi level will be shifted by 10% from the middle of the forbidden energy gap. [Given: $E_g = 1\,\text{eV}$.] (*Ans.* 1116 K)

10.6 In an n-type semiconductor, the Fermi level lies 0.5 eV below the conduction band. If the concentration of donor atoms is tripled, find the new position of the Fermi level, [$kT = 0.03\,\text{eV}$]

(*Ans.* 0.48 below the conduction band)

10.7 The Fermi energy of lithium at 0 K is 4.72 eV. Calculate the number of conduction electrons per unit volume in lithium. [CUH 1997]

(*Ans.* $4.66 \times 10^{22}/\text{cm}^3$)

10.8 An n-type Ge sample is 3.5 mm thick and is placed in a magnetic field $B = 0.5\,\text{T}$. If the donor density is $10^{21}/\text{m}^3$ and the current density $J = 500\,\text{A/m}^2$, find the Hall voltage. (*Ans.* 5.47 mV)

10.9 The R_H of a sample is $3.66 \times 10^{-4}\,\text{m}^3\text{C}^{-1}$ and its resistivity is $8.93 \times 10^{-3}\,\Omega\text{-m}$. Find μ and n.

(*Ans.* $\mu = 0.041\,\text{m}^2\text{V}^{-1}\text{s}^{-1}$, $n = 1.708 \times 10^{22}$ per m^3.)

10.10 Find the ratio of diffusion coefficient of electrons and holes of a silicon single crystal at 27°C. [Mobilities of electrons and holes at 27°C are $0.17\,\text{m}^2\text{V}^{-1}\text{s}^{-1}$ and $0.025\,\text{m}^2\text{V}^{-1}\text{s}^{-1}$ respectively]

(*Ans.* $D_e/D_p = 0.68$)

10.11 If the band gap in Ge is 0.47 eV, compare the intrinsic conductivities of Ge at 27°C and 47°C.

(*Ans.* 2.27)

10.12 Calculate the position of Fermi level E_f and the conductivity at 300 K for Ge crystal containing 5×10^{22} arsenic atoms/m³. Also calculate conductivity if the mobility of electron is $0.39\,\text{m}^2\text{V}^{-1}\text{s}^{-1}$.

(*Ans.* $E_f = 0.16\,\text{eV}$ below E_c, $\sigma = 3120\,\Omega^{-1}\,\text{m}^{-1}$)

10.13 Determine the average band gap of Si if the conductivity of intrinsic Si is 4.17×10^{-5} and 4×10^{-4} $(\Omega\text{-m})^{-1}$ at 0°C and 27°C respectively. (*Ans.* 1.11 eV)

10.14 Calculate the band gap of an intrinsic semiconductor having resistivity 4.5 Ω-m at 20°C and 2 Ω-m at 32°C. (*Ans.* 0.96 eV)

10.15 A copper strip 2.0 cm wide and 1.0 mm thick is placed in a magnetic field of $1.5\,\text{Wb/m}^2$ perpendicular to the strip. If a current of 200 A is set up along the length of the strip, what Hall voltage appears across the strip? [The free electron concentration of copper is 8.48×10^{28} electrons/m³.] [BUH 2003]

[*Hints.* $V_H = \frac{IB}{net}$, $I = 200\,\text{A}$, $B = 1.5\,\text{Wb/m}^2$, $n = 8.48 \times 10^{28}$ electrons/m³, $t = 1\,\text{mm} = 10^{-3}\,\text{m}$, $e = 1.6 \times 10^{-19}\,\text{C}$.

$$\therefore\ V_H = \frac{200 \times 1.5}{(4.48 \times 10^{28}) \times (1.6 \times 10^{-19}\,\text{c}) \times (0.5 \times 10^{-2})} = 7.81 \times 10^{-6}\,\text{V} = 7.81\,\mu\text{V}.]$$

10.16 In a metallic sodium there are 2.6×10^{22} free electrons per cm³. Compute the Fermi energy (in eV) in sodium. Find out E_f. [BUH 2006]

[*Hints.* Fermi energy $E_f = \frac{h^2}{8m}\left(\frac{3n}{\pi}\right)^{2/3}$.

Put $h = 6.625 \times 10^{-34}$ J-s; $m = 9.1 \times 10^{-31}$ kg; $n = 2.6 \times 10^{22}$ electron/cm³ $= 2.6 \times 10^{28}$ electron/m³.]

10(D) Broad Answer Type Questions

10.1 (a) What is intrinsic semiconductor? Explain why impure semiconductors exhibits intrinsic behaviour at high temperatures.

(b) Deduce expressions for the number density of electrons in the conduction band and holes in the valence band of an intrinsic semiconductor. Show that the product of these two quantities does not depend on E_f. Write down expression for conductivity of an intrinsic semiconductor.

(c) Deduce an expression for the Hall coefficient of a solid with one type of carrier only. [BUH 1994]

10.2 (a) What are p- and n-type semiconductors? Where is the Fermi level in the two cases compared to that in an intrinsic semiconductor?

(b) Write the expression for the electrical conductivity in an intrinsic semiconductor and discuss its temperature dependence.

(c) Define Hall constant. Derive its expression for a solid with only holes as carriers. [BUH 1996]

10.3 (a) What are intrinsic and extrinsic semiconductors? How can the band gap of an intrinsic semiconductors be determined from conductivity measurement?

(b) Define Hall constant. Find its expression for a solid with only one electron as carrier. [BUH 2000]

(c) Discuss briefly how energy bands are formed in solid. [BUH 1997]

10.4 (a) Deduce expression for the concentration of electrons in conduction band and holes in valence band of an intrinsic semiconductor at temperature T. Show that the product np is independent of the Fermi level. What is the implication of this result?

[You may assume the formula for the density of states of free electrons namely,

$$g(E)=\frac{1}{2\pi^2}\left(\frac{2m}{\hbar^2}\right)^{3/2}E^{1/2}].$$

(b) On the basis of deductions made in part (a) of the question, find the location of the Fermi level in an intrinsic semiconductor. Draw neat diagrams showing the location of the Fermi level in n- and p-type semiconductors.

(c) How will you experimentally determine whether a given semiconductor is n- or p-type? [BUH 1999]

10.5 (a) Show that the product of concentration of electrons in the conduction band and that of holes in the valence band of a semiconductor depends upon the energy gap.

(b) Deduce an expression for the density of holes in the valence band of a p-type semiconductor. Show the variation of position of the Fermi level. [BUH 2002]

10.6 (a) What is Hall effect?

(b) Find Hall coefficient in a metal, where the carriers are only electrons. Why is the Hall coefficient positive in some metals? [CUH 1992]

(c) Mention some important applications of Hall effect.

10.7 (a) What is the concept of effective mass? On what factors does it depend?

(b) State and establish the law of mass action in semiconductors.

(c) Discuss the effect of temperature and impurity towards electrical conductivity of semiconductors.

10.8 (a) Describe the drift and diffusion currents in a semiconductor. Derive their expressions.

(b) Deduce Einstein relation.

(c) Explain with necessary theory the concept of minority carrier lifetime in the extrinsic semiconductor.

10.9 (a) What do you understand by intrinsic and extrinsic semiconductors?

(b) Derive the expression:

$$R=\frac{1}{|e|}\frac{\left(n_h\mu_h^2-n_e\mu_e^2\right)}{\left(n_h\mu_h+n_e\mu_e\right)^2}$$

for Hall constant when n and μ are carrier concentration and mobility of charge carriers. Subscript e and h refer to electron and hole. [RUH 1986]

10.10 (a) Explain the Hall effect. Derive an expression for the Hall coefficient of semiconductor on two-band model of carriers in semiconductors.

(b) Describe with theory an experimental arrangement for the measurement of Hall coefficient.

(c) How the magnetic field can be determined by Hall effect experiment?

10.11 What is meant by Hall effect? Discuss the experiment with suitable diagram for finding Hall coefficient and Hall voltage. Mention two applications of Hall effect. [BUH 2008]

10.12 What do you mean by the effective mass of an electron? What can it mean physically for an electron to have zero or negative effective mass? Explain how it gives rise to the concept of hole conduction in solids. [VUH 1995]

Appendix A

Some Fundamental Constants

Constants	*Value*
Atomic mass unit	$1u = 1.66057 \times 10^{-27}$ kg = 931.502 MeV
Avogadro's number	$N_A = 6.02205 \times 10$ molecules/mole
Bohrmagneton	$\mu_B = 9.2741 \times 10^{-24}$ J/tesla
Bohr radius	$a_0 = 0.529177 \times 10^{-10}$ m
Boltzmann constant	$k = 1.3807 \times 10^{-23}$ JK
Stefan-Boltzmann constant	$\sigma = 5.67 \times 10^{-8}$ W/m^2/K^4
Compton wavelength for electron	$\lambda_c = 2.42631 \times 10^{-12}$ m
Electron volt	1 eV $= 1.60219 \times 10^{-19}$ J
Charge on an electron	$e = 1.60219 \times 10^{-19}$ C
Electron rest mass	$m_e = 9.10953 \times 10^{-31}$ kg
Electron rest mass energy	$m_e c^2 = 0.511003$ MeV
Permittivity of free space	$\epsilon_0 = 8.85 \times 10^{-12}$ F/m
Permeability of free space	$\mu_0 = 4\pi \times 10^{-7}$ H/m
Electron charge to mass ratio	$\left\lvert \frac{e}{m_e} \right\rvert = 1.76 \times 10^{11}$ C/kg
Proton electron mass ratio	$\frac{m_p}{m_e} = 1837$
Rydberg constant for infinite mass	$R_\infty = 1.0973732 \times 10^7$/m
Classical electron radius	$r_0 = 2.81794 \times 10^{-15}$ m
Proton rest mass	$m_p = 1.673 \times 10^{-27}$ kg $= 1.004\,u = 938.3$ MeV/C^2
Neutron rest mass	$m_n = 1.675 \times 10^{-27}$ kg $= 1.009\,u = 939.6$ MeV/C^2
Alpha rest mass	$m_a = 6.65 \times 10^{-27}$ kg $= 4.003\,u = 3728$ MeV/C^2
Velocity of light in vacuum	$c = 2.997925 \times 10^8$ m/s

Appendix B

Some Useful Mathematical Relations

1. $\int_0^\infty e^{-x}x^{n-1}dx = \Gamma(n)$
2. $\int_0^\infty e^{-x}dx = \sqrt{\pi}/2$
3. $\int_0^\infty e^{-x}x^2dx = \sqrt{\pi}/4$
4. $\int_0^\infty e^{-x}x^5dx = 3\sqrt{\pi}/8$
5. $\int_0^\infty e^{-ax^2}dx = 1/2\sqrt{\pi/a}$
6. $\Gamma(n+1) = n\Gamma(n)$; $\Gamma(1/2) = \sqrt{\pi}$
7. $\int_0^\infty e^{-x}x\,dx = 1/2$
8. $\int_0^\infty e^{-x}x^3dx = 1/2$
9. $\int_0^\infty e^{-x}x^5dx = 1$
10. $\ln N! \simeq N\ln N - N$
11. If $f(x) = \left(1/\sqrt{2\pi}\right)\int_{-\infty}^{+\infty} A(k)e^{ikx}dk$ then $A(k) = \left(1/\sqrt{2\pi}\right)\int_{-\infty}^{+\infty} f(x)e^{-ikx}dx$, where $A(k)$ is the Fourier transform of $f(x)$.
12. If $\rho = \psi^*\psi$ be the position probability density, then probability current density is $\vec{J} = i\hbar/2m\,[\psi\nabla\psi^* - \psi^*\nabla\psi]$.
13. Equation of continuity $\dfrac{\partial\rho}{\partial t} + \nabla\cdot\vec{J} = 0$.
14. Dirac delta function is defined as

$$\delta(x-a) = 0, \quad \text{if } x \neq a$$

$$\text{and} \quad \int_{a-\epsilon}^{a+\epsilon} \delta(x-a)\,dx = 1, \quad \text{if } \epsilon > 0.$$

Appendix C

Five-Fold Symmetry is Impossible

Let a two-dimensional lattice be invariant under the rotation operation.

The translation vector is given by

$$\vec{t} = u\vec{a} + v\vec{b}, \tag{C.1}$$

where $\vec{a}$ and $\vec{b}$ are the fundamental translation vectors, and u and v are integers.

If a point $P(x, y)$ with reference to a fixed origin be rotated through an angle θ to $P'(x', y')$, then

$$x' = x\cos\theta - y\sin\theta$$
$$y' = x\sin\theta + y\cos\theta.$$

If $\vec{a}$ be the shortest non-vanishing translation vector in the translation group and $\vec{a}$ is parallel to X-axis, then rotating $\vec{a}$ by an angle θ we get a new vector $\vec{a}'$ having components,

$$\vec{a}_x = a\cos\theta$$
$$\vec{a}_y = a\sin\theta.$$

If the lattice is invariant under rotation θ, then the vector $(\vec{a}' - \vec{a})$ must be of the form $\vec{t}$ in (C.1).

$$\therefore\ a'_x - a_x = a(\cos\theta - 1)$$
$$a'_y - a_y = a\sin\theta$$

and $|\vec{a}' - \vec{a}|^2 = \left(a'_x - a_x\right)^2 + \left(a'_y - a_y\right)^2 = a^2(\cos\theta - 1)^2 + a^2\sin^2\theta = a^2(2 - 2\cos\theta)$.

Since, $\vec{a}$ is the shortest vector in the translation group $|\vec{a}' - \vec{a}|^2 \geq a^2$,

$$\therefore (2 - 2\cos\theta)$$

$$\text{or } \cos\theta \leq \frac{1}{2}, \text{ so that } \pi/2 \leq \theta \leq \frac{5\pi}{2}. \tag{C.2}$$

This is the first condition of the acceptable angle θ. As $\theta = 0$ makes $\vec{a}' = \vec{a}$ so that $\vec{t} = \vec{0}$, the rotation $\theta = 0$ is allowed.

Again, if $(\vec{a}' + \vec{a})$ is of the form $\vec{t}$ and not smaller in magnitude than $\vec{a}$, then

$$(2 + 2\cos\theta) \geq 1 \text{ or } \cos\theta \geq -\frac{1}{2}$$

so that

$$-\frac{2\pi}{3} \le \theta \le \frac{2\pi}{3}. \qquad \text{(C.3)}$$

This represents the second condition of the acceptable angle θ. The rotation $\theta = \pi$ is allowed as $\vec{a}' = -\vec{a}$, so that $t = 0$.

The third restriction is when inverse rotation $-\theta$ is applied to $\vec{a}$ so that

$$a'_x + a''_x = 2a\cos\theta$$

$$a'_y + a_y = 0 \text{ and } |a' + a''|^2 = 4a^2\cos^2\theta.$$

Since, $\vec{a}$ is the shortest vector,

$$4a^2\cos^2\theta \ge 1 \text{ or } 2a\cos\theta \ge 1$$

$$\left.\begin{aligned} \text{or} \quad -\frac{\pi}{3} \le \theta \le \frac{\pi}{3} \\ \frac{2\pi}{3} \le \theta \le \frac{4\pi}{3} \end{aligned}\right\} \qquad \text{(C.4)}$$

$\theta = \frac{\pi}{2}$ and $\frac{3\pi}{2}$ are also allowed as $a' + a'' = 0$. From (2), (3) and (4) we conclude that,

$$\theta = 0, \frac{\pi}{3}, \frac{\pi}{2}, \frac{2\pi}{3}, \pi, \frac{3\pi}{2}, \frac{5\pi}{2}.$$

$\therefore$ the permissible θ values are multiple of $\frac{2\pi}{n}$, where $n = 1, 2, 3, 4$ or 6. Hence, five-fold symmetry rotation is impossible.

Appendix D

Excitons

An *exciton* is a neutral excited mobile state of a crystal that corresponds to the formation of a bound *electron-hole pair.* It may also be defined as a non-conducting electronic state in a perfect insulator, usually a *non-magnetic insulator.*

If a photon of energy $\hbar\omega_{\text{photon}} > E_g$ is absorbed by an insulating crystal a free hole and a free electron are created. The threshold condition for the above mechanism is $\hbar\omega_{\text{photon}} > E_g$ in a direct process. In the indirect phonon assisted process the threshold is $\hbar\omega_{\text{photon}} > \left(E_g - \hbar\omega_{\text{phonon}}\right)$. The exciton formation involves lowering of energy w.r.t. the finding energy of the exciton (i.e., 1 meV to 1 eV).

An exciton is electrically neutral and has no role in electrical conductivity. It can travel through the crystal transporting exciton energy but not transporting charge. It can have a polarisation vector when generated by light. Frenkel exciton is more tightly bound than *Mott-Wannier exciton.* Exitons can group themselves forming an *electron-hole droplet.*

Energy states: The energy of an exciton in the nth quantum state is

$$E_n = -\frac{\mu e^4}{2\hbar^2 \epsilon^2 n^2},$$

where n is the principal quantum number and $\mu = \frac{m_e m_h}{(m_e + m_h)}$ is the reduced mass of the exciton.

The allowed exciton states in the forbidden energy band is shown in Fig. D-1.

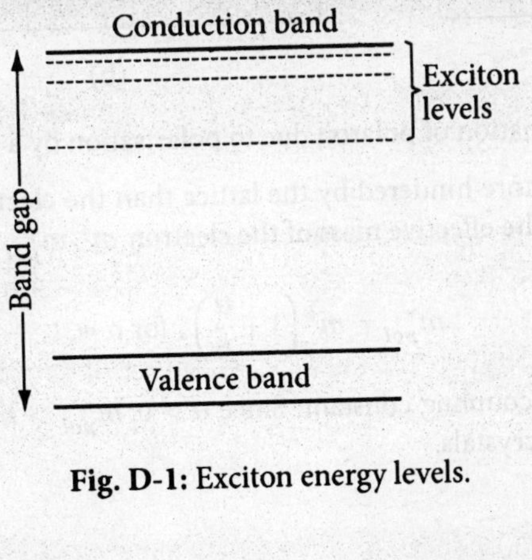

Fig. D-1: Exciton energy levels.

Appendix E

Polarons

A *polaron* is a combination of an electron and its associated strain field in a perfect ionic crystal like KCl. When an electron is introduced into the conduction band of an ionic crystal, it slightly shifts the ions from their equilibrium positions due to its own electric field. The lattice is then polarised [Fig. E-1]. Such polarisation would be unstable if the electron were absent. The displacement of the atoms in the lattice creates a potential well for the electron (due to its strain field). The energy of the electron is reduced and it gets trapped into the well. The total energy of the electron and that of the lattice are lowered and the complex entity is called the *polaron.*

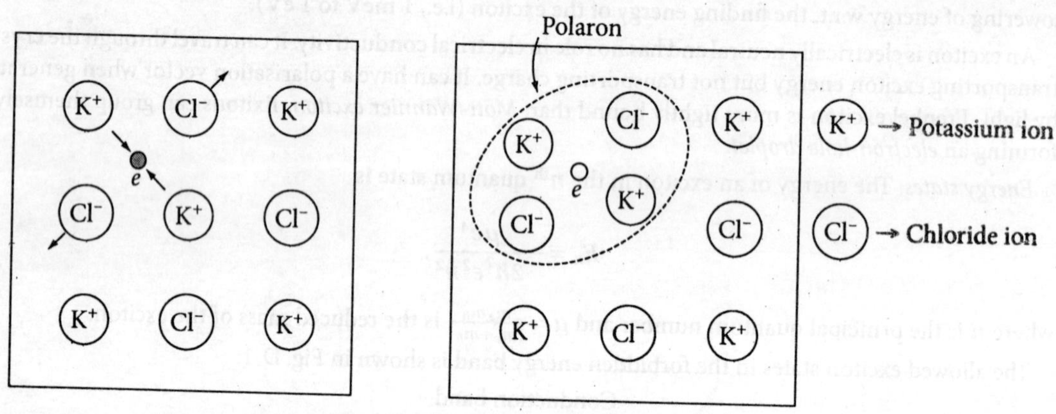

Fig. E-1: Formation of polaron due to polarisation by a conduction electron.

As the polaron motion is more hindered by the lattice than the electron motion, the effective mass of polaron m^*_{pol} is greater than the effective mass of the electron m^*. m^*_{pol} is related to m^* by the relation,

$$m^*_{pol} = m^*\left(1+\frac{\alpha}{6}\right), \text{ for } \alpha \ll 1, \tag{E.1}$$

where α is the dimensionless coupling constant. Since $\alpha > 0, m^*_{pol} > m^*$, Polaron effect is large in ionic crystals and weak in covalent crystals.

Appendix F

Magnons

Magnons are the quantised unit of spin in a ferromagnet. It is a quantum mechanical concept. Magnons are the elementary excitons of a spin system. When a magnon is created, total magnetisation decreases by one unit of spin.

The spin vectors precess around the saturation magnetisation vector $\vec{M}_0$ in such a manner that the orientations of the various spins along the line are correlated through exchange interaction as shown in Fig. F-1. As a result the exciton is transferred to the neighbouring spins and a spin wave is formed which propagates in the crystal. If a be the lattice constant, then wavelength (λ) of the spin wave is $\lambda = 16a$.

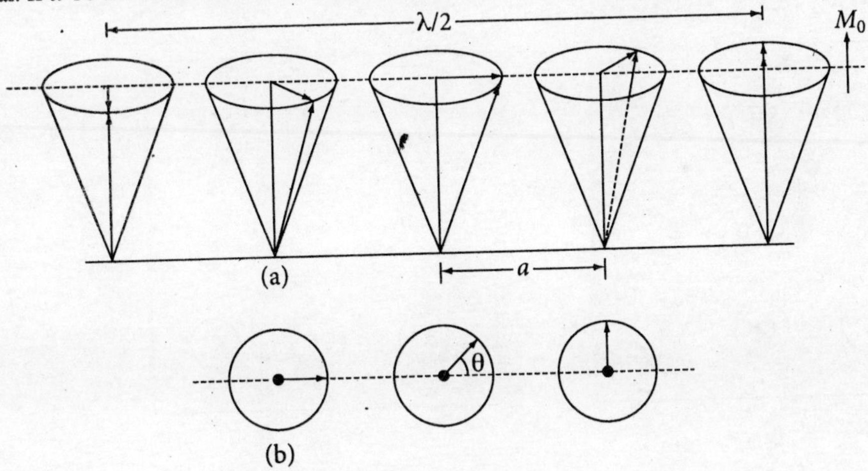

Fig. F-1(a): Spin wave motion
Fig. F-1(b): Angle θ between adjacent spins.

Spin waves are analogous to lattice waves. In lattice waves, atoms oscillate about their equilibrium position, the displacements being correlated through elastic forces. But in case of spin, the spins precess around $\vec{M}_0$ and the precessions are correlated through exchange forces.

A *magnon* of wave vector $\vec{q}$ carries an energy,

$$E = \hbar\omega(\vec{q}), \tag{F.1}$$

a momentum,

$$\vec{p} = \hbar\vec{q} \tag{F.2}$$

and a magnetic moment

$$m = g\mu_B. \tag{F.3}$$

The magnon exciton is equivalent to a reversal of one spin.

It can be shown that the magnon energy is

$$E \sim T^{5/2}$$

and the specific heat,

$$C_V = \left(\frac{dE}{dT}\right)_v \sim T^{3/2}. \tag{F.4}$$

This agrees with experimental result.

Due to thermal excitation the decrease in the magnetisation is given by

$$\Delta M = M(0) - M(T) \sim T^{3/2}. \tag{F.5}$$

This is known as *Bloch's law* which also agrees well with the experimental results.

Bibliography

1. Ali Omar, M., *Elementary Solid State Physics*, Addison-Wesley, Reading, 1975.
2. Baggott J., *The Meaning of Quantum Theory*, NY, Oxford University Press.
3. Beiser A., *Perspective of Modern Physics*, NY, McGraw-Hill, 1969.
4. Bohm D., *Quantum Theory*, N.J. Prentice Hall.
5. Bowley R. and Sanchez, M., *Introduction to Statistical Mechanics*, 1965.
6. Fano U. and Fano L., Physics of Atoms and Molecules, University of Chicago Press, Chicago, 1970.
7. Fermi E., *Notes on Quantum Mechanics*, Univ. of Chicago Press, Chicago, 1961.
8. Feynman R.P., Leighton, R.B. and Sands, M., *The Feynman Lecturers on Physics*, Addison-Wesley, Reading, Mass, 1965, Narosa Publishing House, India, 12th reprint, 1998.
9. Ghoshal S.N., *Atomic and Nuclear Physics*, Vol. I, S. Chand and Company Ltd, New Delhi, 1997.
10. H. Gold Stein, Classical Mechanics, Addison-Wiley Publishing Co., Inc. 1966.
11. Kakani S.L. and Hemarajan, C., *Solid State Physics*, S. Chand and Co., New Delhi, 1992.
12. Kittel C. and Kroemer, H., *Thermal Physics*, NY, Freeman, 1965.
13. Kittel C., *Introduction to Solid State Physics*, John Wiley, NY, 1976.
14. Kressel H., Ladany, H., Ettenberg, M. and Lockwood, H., *Light Wave Communications: Light Sources*, Phys. Today 29(5), 38, 1976.
15. Kuila S.P., *Perspective of Quantum Mechanics*, New Central Book Agency (P) Ltd, Kolkata, India, 2008.
16. Kuila, S.P., *Vector Analysis and its Applications*, Scitech Publication, Chennai, India, 2008.
17. Maitland A., Dunn, M.H., *Laser Physics*, North Holland Publ. Co. Amsterdam, 1969.
18. Mandl F., *Statistical Physics*, John Wiley, London, 1971.
19. Mani H.S. and Mehta, G.K., *Introduction to Modern Physics*, Affiliated East-West Preso Pvt. Ltd, New Delhi, 2006.
20. Merzbacher E., *Quantum Mechanics*, John Wiley, NY, 1961.
21. Morrison M., *Understanding Quantum Mechanics*, N.J. Prentice Hall, 1990.
22. Peaslee D.C., *Elements of Atomic Physics*, Prentice Hall, Englewood Cliffs, 1955.
23. Powell J.L. and Crasemann, B., *Quantum Mechanics*, Addition-Wesley Publishing Co., Inc., Reading, USA, 1971.
24. Schiff L.I., *Quantum Mechanics*, McGraw-Hill Book Company, NY, 1968.
25. Shurchliff W.A. and Ballard, S.S., *Polarised Light*, Van Nostrand, Princeton, N.J. USA, 1964.
26. Synge, John L. and Griffiths, Byron A. *Principle of Mechanics*, McGraw-Hill, NY, 1949.
27. Timoshenko, S and Young, D.H. *Advanced Dynamics*, McGraw-Hill, NY, 1949.

Index

QUESTION PAPERS

CS/B.TECH(EE-NEW/EEE-NEW/EIE-NEW/ICE-NEW)/SEM-4/PH(EE)-401/2012

2012

PHYSICS II

***Time Allotted*: 3 Hours** **Full Marks: 70**

The figures in the margin indicate full marks.

Candidates are required to give their answers in their own words as far as practicable.

Group A

Multiple Choice Type Questions

1. Choose the correct alternatives for any *ten* of the following: $10 \times 1 = 10$

(a) He^3 and muon are

(i) fermions
(ii) bosons
(iii) fermions and bosons respectively
(iv) bosons and fermions respectively.

(b) The degrees of freedom for a system of N particles with K constraint relations is given by

(i) $N - K$
(ii) $N - 3K$
(iii) $3N - K$
(iv) $3K - N$.

(c) The coordination number of FCC structure is

(i) 6
(ii) 8
(iii) 12
(iv) 5.

(d) The dielectric constant of a conductor is

(i) 0
(ii) 1
(iii) −1
(iv) infinity.

(e) Fermi-Dirac distribution approaches Maxwell-Boltzmann distribution at

(i) low temperature and high density
(ii) high temperature and low density
(iii) low temperature and low density
(iv) high temperature and high density.

(f) If E_1 is the energy of the ground state of a one-dimensional potential box of length l, and E_2 be the energy of the ground state when the length of the box is halved, then

(i) $E_2 = 2E_1$
(ii) $E_2 = E_1$
(iii) $E_2 = 4E_1$
(iv) $E_2 = 3E_1$.

(g) The reciprocal lattice of a body centred cubic (bcc) lattice is

(i) bcc
(ii) fcc
(iii) sc
(iv) hcp.

(h) The wave function of a particle is $\Psi = A\cos^2 x$ for $-\frac{\pi}{2} < x < \frac{\pi}{2}$. Then the value of A is

(i) $\sqrt{8/3\pi}$
(ii) $\sqrt{3/8\pi}$
(iii) $\sqrt{1/2\pi}$
(iv) $\sqrt{3/2\pi}$.

(i) The density of free electron states in a metal varies with energy E as

(i) $\sqrt{E}$
(ii) E^2
(iii) E^0
(iv) $1/E$.

(j) Curie-Weiss law is obeyed by

(i) paramagnetic materials
(ii) antiferromagnetic materials
(iii) ferromagnetic materials above the Curie temperature
(iv) ferromagnetic materials below the Curie temperature.

(k) The Miller indices of a plane parallel to XY-plane is

(i) (100)
(ii) (010)
(iii) (001)
(iv) (110).

(l) If σ and k be the electrical and thermal conductivities in a solid, then according to Wiedemann-Franz law.

(i) $\sigma/kT = \text{constant}$
(ii) $k\sigma/T = \text{constant}$
(iii) $k/\sigma T = \text{constant}$
(iv) $\sigma kT = \text{constant}$

(where T is the temperature).

(m) The product of generalised force (Q_i) and generalised displacement (δq_j) must have the dimension of

(i) force
(ii) work
(iii) power
(iv) length.

(n) The spacing between the nth energy state and next energy state in a one-dimensional potential box increases by

(i) $2n-1$
(ii) $2n+1$
(iii) $n-1$
(iv) $n+1$.

(o) In an n-type semiconductor, donor level is

(i) nearer to conduction band
(ii) at the middle between valence and conduction bands
(iii) nearer to valence band
(iv) not formed at all.

Group B

Short Answer Type Questions

Answer any *three* of the following: $3 \times 5 = 15$

2. (a) Describe briefly micro-state and macro-state with suitable examples.

(b) Show that the average energy of electrons at $T = 0\,\text{K}$ is given by $\frac{3}{5}E_F$ (where E_F is the Fermi energy).